Attosecond and Strong-Field Physics

Principles and Applications

Probing and controlling electrons and nuclei in matter at the attosecond timescale became possible with the generation of attosecond pulses by few-cycle intense lasers, revolutionizing our understanding of atomic structure and molecular processes. This book provides an intuitive approach to this emerging field, utilizing simplified models to develop a clear understanding of how matter interacts with attosecond pulses of light. An introductory chapter outlines the structures of atoms and molecules and the properties of a focused laser beam. Detailed discussion of the fundamental theory of attosecond and strong-field physics follows, including the molecular tunneling ionization model, the quantitative rescattering model, and the laser-induced electronic diffraction theory for probing the change of atomic configurations in a molecule. Highlighting cutting-edge developments in attosecond and strong-field physics, and identifying future opportunities and challenges, this self-contained text is invaluable for students and researchers in the field.

C. D. Lin is a University Distinguished Professor at Kansas State University. His research group has made important contributions to the field of attosecond science, including the development of the molecular tunneling ionization model and the quantitative rescattering model.

Anh-Thu Le is a research professor at Kansas State University. For more than twenty years, he has studied atomic, molecular, and optical physics, and, with C. D. Lin, developed the quantitative rescattering theory for high-order harmonic generation.

Cheng Jin is a professor at Nanjing University of Science and Technology who studies optimization of the generation of isolated attosecond pulses by synthesis of multicolor laser waveforms in the gas medium.

Hui Wei is a postdoctoral fellow at Kansas State University. His research interests include characterization and applications of attosecond pulses to molecules and solids.

"This is the book we were waiting for. The text covers strong field light-matter interaction and its effects for atoms and molecules in a concise but easy to read manner and serves as an excellent reference. Topics range from the fundamentals of ionization, the effects of molecular symmetry, high harmonic and attosecond pulse generation, ATI and LIED to attosecond dynamics in atoms and molecules. This book is equally useful for the experienced research as it is for a senior undergraduate student in physics of chemistry. Definitely a must-have!"

Professor Jens Biegert, The Institute of Photonic Sciences (ICFO)

"The book by Lin et al. provides a very accessible introduction to the emerging fields of attosecond science and strong field laser physics, combining a textbook-style presentation of the underlying physics with a presentation of selected recent research results. This book will be a valuable resource for both new students and experienced researchers in the field."

Professor Mark Vrakking, Max Born Institute, Berlin

"Attosecond science based on strong field physics are written comprehensively, including generation, measurement and application of attosecond pulse. This book can be not only used as a textbook for graduate students but also useful for researchers as the first reference in attosecond and strong field community."

Dr Katsumi Midorikawa, Director, RIKEN Center for Advanced Photonics

Attosecond and Strong-Field Physics

Principles and Applications

C. D. LIN
Kansas State University

ANH-THU LE
Kansas State University

CHENG JIN
Nanjing University of Science and Technology

HUI WEI
Kansas State University

CAMBRIDGE
UNIVERSITY PRESS

University Printing House, Cambridge CB2 8BS, United Kingdom

One Liberty Plaza, 20th Floor, New York, NY 10006, USA

477 Williamstown Road, Port Melbourne, VIC 3207, Australia

314–321, 3rd Floor, Plot 3, Splendor Forum, Jasola District Centre, New Delhi – 110025, India

79 Anson Road, #06–04/06, Singapore 079906

Cambridge University Press is part of the University of Cambridge.

It furthers the University's mission by disseminating knowledge in the pursuit of education, learning, and research at the highest international levels of excellence.

www.cambridge.org
Information on this title: www.cambridge.org/9781107197763
DOI: 10.1017/9781108181839

First published 2018

Printed in the United Kingdom by Clays, St Ives plc

A catalogue record for this publication is available from the British Library.

ISBN 978-1-107-19776-3 Hardback

Contents

Preface

The observation of second-harmonic generation by Franken and coworkers back in 1961 preceded 40 years of fast growth in the field of nonlinear optics as well as the first demonstration of the generation of attosecond pulses in 2001. Following the first report of high-order harmonic generation in rare gas atoms with intense-focused laser beams in 1987 and the subsequent emergence of titanium-sapphire lasers, it became clear that the generated high-order harmonics are phase locked. Much like how earlier mode-locked lasers led to femtosecond laser pulses, these phase-locked high-order harmonics were expected to lead to attosecond pulses. Still, fine-tuning the technique utilized to successfully measure the phases of harmonics took more than ten years of effort by pioneers in strong-field physics. During this time, laser-atom/molecule interactions and the properties of harmonics were widely studied both theoretically and in laboratories. Thus, the birth of attosecond science at the dawn of the twenty-first century owes much to the advance of strong-field physics in the decade before.

Today, attosecond pulses ranging from vacuum ultraviolet (VUV) to soft X-rays are available for applications, especially on atoms and molecules. Much of what has been understood in this field resulted from new experimental detection of electrons and ions together with high-resolution spectra borrowed from conventional atomic and molecular physics. The last fifteen years have witnessed the rapid growth of the field and that pace is accelerating.

Researchers now entering this field come with diverse backgrounds and expertise in physics, optics, chemistry, or engineering. This diversity speaks to the richness and potential of attosecond science that has found applications in systems ranging from simple atoms to complex molecules, nanostructures, and condensed materials. Attosecond pulses have been promised to probe electron dynamics in the attosecond timescale, but what parameters should be measured to characterize the dynamics? Can the theoretical tools and concepts developed for attosecond science and strong-field physics be scaled up to complex systems where the knowledge of the target is very limited?

The rapid developments in attosecond science in recent years have been widely reviewed in a number of journal articles and monographs. These publications tend to focus on specific knowledge obtained from experimental findings. In this book, we try to treat attosecond and strong-field physics as a cohesive subject. We examine the underlying physics behind the simple models used in this field. In addition, concepts used for qualitative understanding are distinguished from theoretical models that can make approximate predictions on experimental outcomes. Clear distinction of the roles of the concept and the theoretical model is essential in order to avoid the confusion that can lead to nonconstructive debates.

Attosecond physics forces practitioners away from the comfort zone of conventional energy-domain physics. Asking the right questions is an important part of this endeavor and this book does not shy away from addressing controversial issues such as the tomographic imaging of molecular orbitals, the photoionization time delay, and the description of charge migration or hole-hopping dynamics.

This book has been written with beginning graduate students in mind and can be used as a textbook for a two-semester course on strong-field and attosecond physics. Instructors may also select individual chapters for a one-semester course and serious students are advised to complete the exercises and fill out the missing steps in the derivations. Furthermore, we hope that this book will be used as a reference in the attosecond and strong-laser-field community.

We begin Chapter 1 with a brief review of atoms, molecules, and wave propagation. For atomic physics, wavefunctions of bound states and continuum states within the one-electron model are considered. Simple scattering theory and single-photon ionization theory are then presented. We then review helium atoms (including Fano resonances), many-electron atoms at the shell model level, and the simple density functional theory. For molecules, we discuss the rotational, vibrational, and electronic wavefunctions and spectroscopy for diatomic and polyatomic molecules. Furthermore, elementary group theory for describing polyatomic molecules is introduced to help readers understand the classification of molecular states. Finally, the propagation of a focused laser beam in space and through optical elements such as lenses or mirrors is treated.

Chapter 2 is devoted to the formal theories used in strong-field physics. Topics include weak-field and strong-field expansion methods, tunnel ionization theory, the Keldysh-Faisal-Reiss (KFR) theory and its various extensions, and the classical trajectory Monte Carlo (CMTC) method and its extension. These approximate theories serve as a basis for understanding the strong-field experiments treated in the following chapters.

Chapter 3 is devoted to ionization of atoms by an intense laser field. We cover total ionization, excitation, energy, and momentum distributions of photoelectrons and the surprising features of very low-energy electrons generated by mid-infrared lasers. Chapter 3 also discusses the theory of orientation and alignment of molecules and of creating vibrational wave packets.

In Chapter 4, rescattering phenomena are described. Beginning with the simple three-step picture, this concept is generalized to the quantitative rescattering (QRS) theory. The QRS model was first properly calibrated against results from solving the time-dependent Schrödinger equation. It is shown that high-energy photoelectron angular distributions can be used to extract elastic, differential-scattering cross-sections. This theory is then applied to laser-induced electron diffraction (LIED), where the bond lengths of a molecule can be extracted. Realization of LIED for small molecules in the laboratory are also illustrated in this chapter.

Chapters 5 and 6 are devoted to high-order harmonic generation. Harmonics generated within a single atom picture are first discussed in Chapter 5, using strong-field approximation, quantum orbits theory, and QRS theory. Then, descriptions of macroscopic propagation of the driving laser and harmonics in the gas medium are given. Chapter 5 closes by comparing experimental high-harmonic generation (HHG) spectra for rare gas atoms to QRS-based HHG theory including propagation effect. Applications of the QRS theory to HHG spectra from diatomic and polyatomic molecules are treated in Chapter 6. Here it is shown that photoionization dipole-transition matrix elements (including amplitude and phase) can be extracted from the HHG spectra under favorable conditions. Chapter 6 ends with a summary of recent efforts around the enhancement of harmonic yields and the extension of harmonics to the water-window region.

Chapters 7 and 8 are dedicated to attosecond physics. After a short discussion on the attosecond pulse train, we summarize the various methods of the generation of isolated attosecond pulses (IAP). Then, the various methods of the metrology of IAP are discussed. Chapter 7 concludes with a detailed analysis of the photoionization time delay. Attosecond pulses as a tool for probing electron dynamics are treated in Chapter 8. Starting with a description of Fano resonance in the time domain, Chapter 8 shows that an "electronic movie" (as depicted on the cover of this book) was recently filmed using the method of attosecond transient absorption spectroscopy (ATAS). Because it offers extreme temporal and spectral resolution, ATAS is a powerful method for attosecond sciences. The potential power of attosecond pulses for probing electron dynamics is also addressed in Chapter 8.

The choice of materials in this book is biased toward the expertise of the authors. It cannot cover all aspects of the recent important advances in attosecond and strong-field physics. Thus, for instance, only interactions with linearly polarized laser pulses are included. In addition, we do not feel comfortable covering recent developments in applications of attosecond pulses and strong fields on solid materials, nanostructures, and surfaces. A separate book by experts in this area is certainly necessary.

Many graduate students and postdoctoral scholars contributed to the materials covered in this book. They are recognized here: Zhangjin Chen, Wei-Chun Chu, Sam Micheau, Toru Morishita, Xiao-Min Tong, Xu Wang, Junliang Xu, Xi Zhao, and Zengxiu Zhao. Visiting scholars have also contributed to the topics covered in this book. These include Zigen Chen, Van-Hoang Le, Qianguang Li, Van-Hung Hoang, Shicheng Jiang, Ty Nguyen, Guoli Wang, M. Wickenhauser, Yan Wu, Shan Xue, Chao Yu, Song-Feng Zhao, Xiao-Xin Zhou, and Zhaoyan Zhou. Moreover, the time-dependent Schrödinger equation (TDSE) code from Toru Morishita and the molecular photoionization code from Robert Lucchese were undeniably essential to the success of the QRS theory.

Fruitful collaborations with experimental groups at Kansas State University and elsewhere on topics covered in this book should also be mentioned, in particular the groups of Ravi Bhardwaj, Jens Biegert, Ming-Chang Chen, Lew Cocke, Paul Corkum, Lou DiMauro and Pierre Agostini, Kyung-Han Hong, Andy Kung, Thomas Pfeifer, Kiyoshi Ueda, and David Villeneuve.

Lastly, Karin Lin and Trang "Tracy" Le have greatly aided in proofreading and improving the manuscript.

1 Elements of Atoms, Molecules, and Wave Propagation

1.1 One-Electron Atoms

1.1.1 Hydrogenic Atoms and Wavefunctions

This book begins with the simplest hydrogenic atom. Within the spectral resolution of its interaction with ultrafast light pulses, this atom can be considered as a nonrelativistic electron in the field of a Coulomb potential $V(r) = -Z/r$ from the nucleus of infinite mass with positive charge $Z(= 1)$. In spherical coordinates, the eigenstate can be expressed as

$$\psi_{Elm}(\mathbf{r}) = R_{El}(r)Y_{lm}(\theta, \phi), \tag{1.1}$$

where Y_{lm} is the familiar spherical harmonic. In this book, atomic units, where $m = |e| = \hbar = 1$ will be used unless otherwise noted. Here, m and e are the mass and charge of the electron, respectively. The inside back cover of the book gives fundamental constants and conversion factors that are useful for attosecond and strong-field physics.

Introducing $u_{El}(r) = rR_{El}(r)$, the radial equation for u_{El} reads

$$\left[-\frac{1}{2}\frac{d^2}{dr^2} + \frac{l(l+1)}{2r^2} + V(r)\right]u_{El}(r) = Eu_{El}(r). \tag{1.2}$$

For the bound states, the eigenvalues are

$$E_n = -\frac{Z^2}{2n^2}, \tag{1.3}$$

with the principal quantum number n.

The radial wavefunctions are given by

$$R_{nl}(r) = \frac{u_{nl}(r)}{r} = N_{nl}e^{-\frac{Z}{n}r}\left(\frac{2Z}{n}r\right)^l {}_1F_1\left(l+1-n, 2l+2, \frac{2Z}{n}r\right), \tag{1.4}$$

where ${}_1F_1$ is the confluent hypergeometric function. The normalization factor N_{nl} is

$$N_{nl} = \frac{1}{(2l+1)!}\sqrt{\left(\frac{2Z}{n}\right)^3 \frac{(n+l)!}{2n(n-l-1)!}}. \tag{1.5}$$

For the wavefunction of a continuum state, let $E = k^2/2$, the radial equation of a hydrogenic atom, where

$$\left[\frac{d^2}{dr^2} - \frac{l(l+1)}{r^2} + \frac{2Z}{r} + k^2\right]u_{El}(r) = 0 \tag{1.6}$$

has two linearly independent solutions: the regular Coulomb function $F_l(\gamma, kr)$ and the irregular Coulomb function $G_l(\gamma, kr)$. The former vanishes at $r = 0$ while the latter diverges. Therefore the allowed solution takes the form

$$u_{El}(r) = CF_l(\gamma, kr), \tag{1.7}$$

where the Sommerfeld parameter $\gamma = -\frac{Z}{k}$.

The asymptotic behaviors of F_l and G_l are

$$F_l(\gamma, kr) \overset{r\to\infty}{\sim} \sin(kr - \gamma \ln(2kr) - l\pi/2 + \sigma_{El}), \tag{1.8}$$

$$G_l(\gamma, kr) \overset{r\to\infty}{\sim} \cos(kr - \gamma \ln(2kr) - l\pi/2 + \sigma_{El}), \tag{1.9}$$

where the real quantity $\sigma_{El} = \arg[\Gamma(l + 1 + i\gamma)]$ is the Coulomb phase shift.

The radial wavefunctions $u_{El}(r)$ are not square integrable. They are usually momentum or energy normalized, which means

$$\int_0^\infty u_{El}(r)u_{E'l}(r)dr = \delta(k - k') \quad \text{momentum normalized,} \tag{1.10}$$

$$\int_0^\infty u_{El}(r)u_{E'l}(r)dr = \delta(E - E') \quad \text{energy normalized.} \tag{1.11}$$

For momentum normalization, the normalization constant $C = \sqrt{\frac{2}{\pi}}$ and for energy normalization $C = \sqrt{\frac{2}{\pi k}}$.

The hydrogenic wavefunction is also separable in parabolic coordinates defined by

$$\xi = r + z = r(1 + \cos\theta) \tag{1.12}$$

$$\eta = r - z = r(1 - \cos\theta) \tag{1.13}$$

$$\phi = \arctan\frac{y}{x} \tag{1.14}$$

with $0 \leq \xi \leq \infty$, $0 \leq \eta \leq \infty$, $0 \leq \phi \leq 2\pi$. The surfaces $\xi = const$ and $\eta = const$ are paraboloids of revolution about the z-axis.

In parabolic coordinates, the eigensolutions can be written as

$$\psi(\xi, \eta, \phi) = f(\xi)g(\eta)\Phi(\phi), \tag{1.15}$$

where the ϕ dependence is $\Phi(\phi) = \frac{1}{\sqrt{2\pi}}e^{im\phi}$ with $m = 0, \pm 1, \pm 2, \ldots$ being the magnetic quantum number.

For bound states these are called Stark states. For unbound states that exhibit azimuthal symmetry ($m = 0$), the continuum wavefunction is

$$\psi(\mathbf{r}) = Ce^{ik\frac{\xi-\eta}{2}}{}_1F_1(-i\gamma, 1, ik\eta) = Ce^{ikz}{}_1F_1(-i\gamma, 1, ik(r - z)), \tag{1.16}$$

where C depends on the normalization convention.

1.1.2 Single Active Electron Model for Atoms

In principle, when an atom (or ion) consists of N electrons, one needs to solve the complicated N-electron Schrödinger equation. However, for measurements where the interaction involves only one electron, treating such an electron alone is desirable. In this single "active" electron model, all of the other electrons are assumed to provide only an effective potential to this active one. In this way, the whole atom may be analyzed in the framework of a one-electron system in a central model potential $V(r)$. This potential can be written as a pure Coulomb potential plus a short-range potential

$$V(r) = -\frac{Z_c}{r} + V_{sr}(r). \tag{1.17}$$

The short-range potential $V_{sr}(r)$ must satisfy

$$\lim_{r\to\infty} r^2 V_{sr}(r) = 0. \tag{1.18}$$

Note that the parameter Z_c in Equation 1.17 is the asymptotic charge felt by the active electron. In a neutral atom, $Z_c = 1$.

The radial equation for the bound state with energy $E = -\beta^2/2$ for such a model one-electron system is

$$\left[\frac{d^2}{dr^2} - \frac{l(l+1)}{r^2} + \frac{2Z_c}{r} - 2V_{sr}(r) - \beta^2\right] u_{El}(r) = 0. \tag{1.19}$$

This equation can be solved numerically. The energy level E_{nl} can be expressed by

$$E_{nl} = -\frac{Z_c^2}{2(n-\Delta_{nl})^2}, \tag{1.20}$$

where Δ_{nl} is called the quantum defect as a result of the short-range potential $V_{sr}(r)$. For large n, the quantum defect is independent of n.

For the unbound state with energy $E = k^2/2$, the radial equation for this system is

$$\left[\frac{d^2}{dr^2} - \frac{l(l+1)}{r^2} + \frac{2Z_c}{r} - 2V_{sr}(r) + k^2\right] u_{El}(r) = 0. \tag{1.21}$$

This equation can also be solved numerically. In the asymptotic region in which V_{sr} vanishes,

$$\begin{aligned} u_{El}(r) &\overset{r\to\infty}{\sim} AF_l(\gamma, kr) + BG_l(\gamma, kr) \\ &\overset{r\to\infty}{\sim} C\sin[kr - \gamma\ln(2kr) - l\pi/2 + \sigma_{El} + \delta_{El}]. \end{aligned} \tag{1.22}$$

Here, $\gamma = -Z_c/k$ and $\sigma_{El} = \arg[\Gamma(l+1+i\gamma)]$ is the Coulomb phase shift. $C = \sqrt{A^2+B^2}$ is the normalization constant mentioned previously. $\delta_{El} = \arctan\frac{B}{A}$ is the additional phase shift caused by the short-range potential.

1.1.3 Scattering of an Electron by a Central Field Potential

Scattering by a Short-Range Potential

Consider an electron scattered by a target where the interaction can be modeled by a short-range potential $V_{sr}(r)$. In the asymptotic region where $V_{sr}(r)$ vanishes, the wavefunction should take the form:

$$\psi \overset{r\to\infty}{\sim} e^{ikz} + f(\theta)\frac{e^{ikr}}{r}. \tag{1.23}$$

Here, the energy of the incident electron $E = k^2/2$. The first term is an incident plane wave traveling along the z direction; the second term is an outgoing scattered spherical wave. The scattering amplitude $f(\theta)$ is a complex quantity depending on scattering angle θ and energy E. Due to cylindrical symmetry, f is independent of ϕ. The differential cross-section is

$$\frac{d\sigma}{d\Omega} = |f(\theta)|^2, \tag{1.24}$$

and the total cross-section is

$$\sigma = \int |f(\theta)|^2 d\Omega = 2\pi \int_0^\pi |f(\theta)|^2 \sin\theta d\theta. \tag{1.25}$$

The total wavefunction ψ can be expanded in a series of partial waves as

$$\psi = \sum_{l=0}^{\infty} A_l \frac{u_{El}(r)}{r} P_l(\cos\theta), \tag{1.26}$$

where $u_{El}(r)$ is the solution of the radial equation

$$\left[\frac{d^2}{dr^2} - \frac{l(l+1)}{r^2} - 2V_{sr}(r) + k^2\right] u_{El}(r) = 0. \tag{1.27}$$

In the asymptotic region as r goes to infinity

$$\begin{aligned} u_{El}(r) &\overset{r\to\infty}{\sim} Akrj_l(kr) + Bkrn_l(kr) \\ &\overset{r\to\infty}{\sim} A\sin(kr - l\pi/2) - B\cos(kr - l\pi/2) \\ &= C\sin(kr - l\pi/2 + \delta_{El}), \end{aligned} \tag{1.28}$$

where j_l and n_l are spherical Bessel and Neumann functions, $C = \sqrt{A^2 + B^2}$ and $\delta_{El} = -\arctan(B/A)$. The phase shift δ_{El} is an important quantity because it carries information about the short-range potential near the nucleus to physical effect in the asymptotic region. Note that δ_{El} depends on both the energy E and the angular momentum quantum number l.

With the help of

$$e^{ikz} = \sum_{l=0}^{\infty}(2l+1)i^l j_l(kr)P_l(\cos\theta) \overset{r\to\infty}{\sim} \sum_{l=0}^{\infty}(2l+1)e^{il\pi/2}\frac{\sin(kr - l\pi/2)}{kr}P_l(\cos\theta), \tag{1.29}$$

and comparison to Equation 1.26 in the asymptotic region (here, choose $C = 1$), one can obtain

$$A_l e^{-i\delta_{El}} = \frac{1}{k}(2l+1)e^{il\pi/2}. \tag{1.30}$$

By comparing Equations 1.23 and 1.26 in the asymptotic region, one can easily obtain

$$f(\theta) = \frac{1}{k}\sum_{l=0}^{\infty}(2l+1)e^{i\delta_{El}}\sin\delta_{El}P_l(\cos\theta). \tag{1.31}$$

The differential cross-section is $|f(\theta)|^2$. The total cross-section is

$$\sigma = 2\pi\int_0^{\pi}|f(\theta)|^2\sin\theta d\theta = \frac{4\pi}{k^2}\sum_{l=0}^{\infty}(2l+1)\sin^2\delta_{El}. \tag{1.32}$$

Each partial wave contributes

$$\sigma_l = \frac{4\pi}{k^2}(2l+1)\sin^2\delta_{El} \tag{1.33}$$

to the total cross-section. When evaluating the cross-section, it is not practical to include all partial waves. If the short-range potential has a range a, it is reasonable to cut off the partial-wave expansion at $l_{max} \approx ka$. Thus, in low-energy scattering, only a few partial waves are needed.

Scattering by a Pure Coulomb Potential

If the incident direction of the electron is chosen as the z direction, for a pure Coulomb potential $V(r) = -Z/r$, the total wavefunction can be expressed in parabolic coordinates as

$$\psi = Ce^{ikz}{}_1F_1(-i\gamma, 1, ik(r-z)), \tag{1.34}$$

where $\gamma = -Z/k$. By choosing $C = e^{-\pi\gamma/2}\Gamma(1+i\gamma)$, at large $|r-z|$ the wavefunction has the asymptotic form

$$\begin{aligned}\psi \overset{|r-z|\to\infty}{\sim} & e^{ikz+i\gamma\ln[k(r-z)]}\left(1+\frac{\gamma^2}{ik(r-z)}+\cdots\right)\\ & +f_c(\theta)\frac{e^{ikr-i\gamma\ln(2kr)}}{r}\left(1+\frac{(1+i\gamma)^2}{ik(r-z)}+\cdots\right).\end{aligned} \tag{1.35}$$

The Coulomb scattering amplitude is given by

$$f_c(\theta) = -\frac{\gamma}{2k\sin^2(\theta/2)}e^{-i\gamma\ln[\sin^2(\theta/2)]+2i\sigma_{E0}}, \tag{1.36}$$

in which $\sigma_{E0} = \arg[\Gamma(1+i\gamma)]$.

Clearly, the asymptotic expansion does not hold for $\theta = 0$ ($r = z$). For large $|r-z|$ the incident wave and the outgoing wave have each acquired a logarithmic phase. The differential cross-section is

$$\frac{d\sigma_c}{d\Omega} = |f_c(\theta)|^2 = \frac{\gamma^2}{4k^2\sin^4(\theta/2)} = \frac{Z^2}{16E^2\sin^4(\theta/2)}. \tag{1.37}$$

This result is identical to the Rutherford formula derived from classical mechanics. The Rutherford differential cross-section diverges strongly in the forward direction $\theta = 0$, and the total cross-section is not defined since the integral also diverges. This is the consequence of the long-range Coulomb potential. However, in actual situations, the Coulomb potential at large distance is screened by the environment, and thus the cross-section is finite.

Scattering by a Modified Coulomb Potential

For an electron scattered by a model potential, which consists of a Coulomb part and a short-range part $V(r) = -Z_c/r + V_{sr}(r)$, the scattering amplitude can be written as

$$f(\theta) = f_c(\theta) + f_s(\theta), \tag{1.38}$$

where f_c is the Coulomb scattering amplitude given by Equation 1.36 with $\gamma = -Z_c/k$, and f_s denotes the modification due to the short-range potential. Using partial-wave expansion, f_s is given by

$$f_s(\theta) = \frac{1}{k}\sum_{l=0}^{\infty}(2l+1)e^{2i\sigma_{El}}e^{i\delta_{El}}\sin\delta_{El}P_l(\cos\theta), \tag{1.39}$$

where $\sigma_{El} = \arg[\Gamma(l+1+i\gamma)]$ is the Coulomb phase shift. δ_{El} is the additional phase shift due to the short-range potential (see Equation 1.22). The differential cross-section is

$$\frac{d\sigma}{d\Omega} = |f_c(\theta) + f_s(\theta)|^2. \tag{1.40}$$

Scattering Cross-Section across a Resonance

In general, the phase shift δ_{El} varies smoothly with the incident energy E as does the partial cross-section σ_l. However, δ_{El} may vary rapidly near a resonance. As a result, the corresponding partial cross-section σ_l may also change dramatically in this energy interval. Resonance usually happens for a certain partial wave. Near the resonance, the total cross-section is dominated by this partial wave, that is, $\sigma \approx \sigma_l$. For example, the effective potential for the lth partial wave

$$V_{eff}(r) = V(r) + \frac{l(l+1)}{2r^2} \tag{1.41}$$

may have a potential barrier at large r that can support a metastable state with a positive energy E_r that is below the top of the barrier. An incoming particle with an energy close to E_r will be trapped for a long time before it tunnels out of the barrier. This metastable state has a finite lifetime or a resonance width Γ. In this example, the resonance is called a shape resonance because it is an effect of the shape of the effective potential.

Near the resonance energy, the phase shift δ_{El} can be expressed by

$$\delta_{El} = \xi + \arctan\frac{\Gamma/2}{E_r - E}, \tag{1.42}$$

where ξ is a background phase shift and the second term changes rapidly by π as the energy goes through the resonance from below to above.

If ξ is negligible, the cross-section has the Breit–Wigner form

$$\sigma_l = \frac{4\pi}{k^2}(2l+1)\sin^2\left(\arctan\frac{\Gamma/2}{E_r - E}\right)$$
$$= \frac{4\pi}{k^2}(2l+1)\frac{\Gamma^2/4}{(E-E_r)^2 + \Gamma^2/4}. \qquad (1.43)$$

Near the resonance energy, k can be taken to be constant and so σ_l has a Lorentzian shape characterized by the width Γ and the resonance energy E_r.

For a nonzero ξ, by introducing the reduced energy $\epsilon = \frac{E-E_r}{\Gamma/2}$ and the shape parameter $q = -\cot\xi$, the cross-section can be reduced to

$$\sigma_l = \frac{4\pi}{k^2}(2l+1)\sin^2\left(\xi + \arctan\frac{\Gamma/2}{E_r - E}\right)$$
$$= \frac{4\pi}{k^2}(2l+1)\frac{1}{1+q^2}\frac{(q+\epsilon)^2}{1+\epsilon^2}. \qquad (1.44)$$

This cross-section Equation 1.44 has the form of the Fano profile. When $q \to \pm\infty$ it reduces to the Lorentzian profile. For small q values, σ_l is smaller at the center of the resonance than at the wings. Such resonances are called window resonances.

The First Born Approximation

For collisions at high energies, partial-wave expansion is not practical. One can use the plane wave or first Born approximation. Let the incident wave be given by $\psi_i = e^{i\mathbf{k}_i\cdot\mathbf{r}}$ and the scattered wave by $\psi_f = e^{i\mathbf{k}_f\cdot\mathbf{r}}$ where the momentum $\mathbf{k}_i$ and $\mathbf{k}_f$ only differ in direction but not in magnitude. Then, a momentum transfer can be defined as

$$\mathbf{q} = \mathbf{k}_f - \mathbf{k}_i, \qquad (1.45)$$

with its magnitude given by

$$q = 2k\sin\frac{\theta}{2}. \qquad (1.46)$$

The scattering amplitude given by the first Born approximation reads

$$f^{B1} = -\frac{1}{2\pi}\langle\psi_f|V|\psi_i\rangle = -\frac{1}{2\pi}\int V(\mathbf{r})e^{-i\mathbf{q}\cdot\mathbf{r}}d\mathbf{r}. \qquad (1.47)$$

If the potential $V(\mathbf{r})$ is central symmetric, the scattering amplitude is

$$f^{B1} = -\frac{2}{q}\int_0^\infty rV(r)\sin(qr)dr. \qquad (1.48)$$

Consider the incident electron scattered by N-independent atoms. Each atom is modeled by a local potential $V_i(\mathbf{r}')$ and located at $\mathbf{r} = \mathbf{R}_i$. By making the first Born approximation, the total scattering amplitude has a simple form:

$$
\begin{aligned}
f^{B1} &= -\frac{1}{2\pi}\int\left[\sum_{i=1}^{N} V_i(\mathbf{r}-\mathbf{R}_i)\right] e^{-i\mathbf{q}\cdot\mathbf{r}} d\mathbf{r} \\
&= \sum_{i=1}^{N} e^{-i\mathbf{q}\cdot\mathbf{R}_i}\left[-\frac{1}{2\pi}\int V_i(\mathbf{r}-\mathbf{R}_i)e^{-i\mathbf{q}\cdot(\mathbf{r}-\mathbf{R}_i)} d\mathbf{r}\right] \\
&= \sum_{i=1}^{N} e^{-i\mathbf{q}\cdot\mathbf{R}_i}\left[-\frac{1}{2\pi}\int V_i(\mathbf{r}')e^{-i\mathbf{q}\cdot\mathbf{r}'} d\mathbf{r}'\right] \\
&= \sum_{i=1}^{N} f_i^{B1} e^{-i\mathbf{q}\cdot\mathbf{R}_i}
\end{aligned} \tag{1.49}
$$

in which f_i^{B1} is the individual scattering amplitude for the ith atom. The phase factor $e^{-i\mathbf{q}\cdot\mathbf{R}_i}$ accounts for the interference between different scattering centers. This interference depends on the momentum transfer $\mathbf{q}$ and the geometric configuration determined by $\mathbf{R}_i$.

1.1.4 One-Electron Atoms in Weak Electromagnetic Fields

Basic Formulation

With the one-electron potential $V(r)$, the Hamiltonian of a one-electron system in an electromagnetic field reads

$$H = \frac{1}{2}(\mathbf{p}+\mathbf{A})^2 - \phi + V(r), \tag{1.50}$$

where $\mathbf{A}$ and ϕ are the vector and scalar potential of the electromagnetic field, respectively. There is no external source for the cases considered in this book. Thus, it is convenient to choose the Coulomb gauge

$$\nabla\cdot\mathbf{A} = 0, \tag{1.51}$$

$$\phi = 0. \tag{1.52}$$

The vector potential satisfies the wave equation

$$\nabla^2\mathbf{A} - \frac{1}{c^2}\frac{\partial^2\mathbf{A}}{\partial t^2} = 0. \tag{1.53}$$

In general, $\mathbf{A}$ can be represented as a superposition of plane-wave components with propagation direction $\hat{k}$ and polarization direction $\hat{\epsilon}$. Each component takes the form

$$\mathbf{A} = \hat{\epsilon}A_0(\omega)\cos(\mathbf{k}\cdot\mathbf{r} - \omega t), \tag{1.54}$$

where ω is the angular frequency, $\mathbf{k} = \frac{\omega}{c}\hat{k}$ is the wave-number vector, and $A_0(\omega)$ describes the magnitude of this plane-wave component. Equation 1.51 requires the wave to be transverse, i.e., $\mathbf{k}\cdot\hat{\epsilon} = 0$. The intensity distribution (intensity per unit angular frequency range) is given by

$$I(\omega) = \frac{c}{8\pi}\omega^2 A_0^2(\omega). \tag{1.55}$$

Additionally, $I(\omega)$ is related to the number of photons $N(\omega)$ in a box with volume V by

$$I(\omega) = \frac{c\omega}{V} N(\omega). \tag{1.56}$$

In the Coulomb gauge, the Hamiltonian can be written as

$$H = -\frac{1}{2}\nabla^2 + V(r) - i\mathbf{A} \cdot \nabla + \frac{1}{2}A^2 = H_0 + H'(t), \tag{1.57}$$

where $H_0 = -\frac{1}{2}\nabla^2 + V(r)$ is the field-free Hamiltonian and $H'(t)$ is the time-dependent perturbation. Consider a weak field such that the A^2 term can be dropped. The perturbation becomes

$$H'(t) \approx -i\mathbf{A} \cdot \nabla. \tag{1.58}$$

According to first-order, time-dependent perturbation theory, transition probability from an initial state $|a\rangle$ to a final state $|b\rangle$ is given by

$$P_{ba}(t) = \left| \int_0^t \langle b|H'(t')|a\rangle e^{i\omega_{ba}t'} dt' \right|^2, \tag{1.59}$$

where $|a\rangle$ and $|b\rangle$ are eigenstates of the field-free Hamiltonian H_0 with energies E_a and E_b, respectively. $\omega_{ba} = E_b - E_a$ is the transition energy. For $E_b > E_a$, $\omega_{ba} > 0$, which is the case of photoabsorption, the transition probability due to a certain frequency component of the electromagnetic field can be calculated by rewriting Equation 1.54 into two exponential terms. After dropping the integral involving $e^{i(\omega+\omega_{ba})t'}$, the result is

$$P_{ba}(t) = A_0^2(\omega)|M_{ba}(\omega)|^2 \frac{\sin^2\left(\frac{\omega-\omega_{ba}}{2}t\right)}{(\omega - \omega_{ba})^2}. \tag{1.60}$$

The matrix element M_{ba} is given by

$$M_{ba}(\omega) = \langle b|e^{i\mathbf{k}\cdot\mathbf{r}}\hat{\epsilon} \cdot \nabla|a\rangle. \tag{1.61}$$

Assuming that the radiation is incoherent, the transition probability due to all of the frequency components is obtained by integrating Equation 1.60 over the frequency ω. When t is large,

$$\frac{\sin^2\left(\frac{\omega-\omega_{ba}}{2}t\right)}{(\omega - \omega_{ba})^2} \approx \frac{\pi t}{2}\delta(\omega - \omega_{ba}). \tag{1.62}$$

Then, the transition probability integrated over the frequency is

$$P_{ba}(t) = \frac{\pi}{2}A_0^2(\omega_{ba})|M_{ba}(\omega_{ba})|^2 t \tag{1.63}$$

and the transition rate for photoabsorption is

$$W_{ba} = \frac{\pi}{2}A_0^2(\omega_{ba})|M_{ba}(\omega_{ba})|^2 = \frac{4\pi^2}{c\omega_{ba}^2} I(\omega_{ba})|M_{ba}(\omega_{ba})|^2. \tag{1.64}$$

An integrated absorption cross-section σ_{ba}, which is the rate of absorption of energy $\omega_{ba}W_{ba}$ divided by $I(\omega_{ba})$, can also be defined as

$$\sigma_{ba} = \frac{4\pi^2}{c\omega_{ba}}|M_{ba}(\omega_{ba})|^2. \tag{1.65}$$

When $E_b < E_a$, $\omega_{ba} < 0$, one can obtain the transition rate and cross-section integrated over frequency for stimulated emission in a similar way:

$$W_{ba} = \frac{4\pi^2}{c|\omega_{ba}|^2} I(|\omega_{ba}|)|\tilde{M}_{ba}(|\omega_{ba}|)|^2, \tag{1.66}$$

$$\sigma_{ba} = \frac{4\pi^2}{c|\omega_{ba}|}|\tilde{M}_{ba}(|\omega_{ba}|)|^2. \tag{1.67}$$

The transition-matrix elements between photoabsorption and photoemission are related by

$$\tilde{M}_{ba}(\omega) = \langle b|e^{-i\mathbf{k}\cdot\mathbf{r}}\hat{\epsilon}\cdot\nabla|a\rangle = -M^*_{ab}(\omega). \tag{1.68}$$

The stimulated emission and photoabsorption are in detailed balance, i.e., $W_{ba} = W_{ab}$ and $\sigma_{ba} = \sigma_{ab}$.

First-order perturbation theory can also be used for photoionization from an initial bound state $|i\rangle$ to continuum states $|E\rangle$. The ionization rate per unit energy is

$$\frac{dW_{ion}}{dE} = \frac{4\pi^2}{c\omega_{Ei}^2} I(\omega_{Ei})|M_{Ei}(\omega_{Ei})|^2\rho(E), \tag{1.69}$$

where $\omega_{Ei} = E - E_i$, $M_{Ei}(\omega) = \langle E|e^{i\mathbf{k}\cdot\mathbf{r}}\hat{\epsilon}\cdot\nabla|i\rangle$ and $\rho(E)$ is the number of states per unit energy. If $|E\rangle$ is energy normalized, $\rho(E) = 1$. The photoionization cross-section (per unit energy) is

$$\sigma_{ion} = \frac{4\pi^2}{c\omega_{Ei}}|M_{Ei}(\omega_{Ei})|^2\rho(E). \tag{1.70}$$

The semiclassical theory does not include spontaneous emission. According to quantum electrodynamics, when there is no external electromagnetic field, the transition from $|a\rangle$ to $|b\rangle$ ($E_a > E_b$) can still happen, and thus can emit a photon with momentum $\mathbf{k}$ and polarization $\hat{\epsilon}_\lambda$. Here, λ can be one or two to specify two independent polarizations perpendicular to $\mathbf{k}$. The spontaneous-emission rate for a single mode of photon is

$$W^s_{ba} = \frac{4\pi^2}{V\omega}|\tilde{M}^\lambda_{ba}(\omega)|^2\delta(\omega - |\omega_{ba}|). \tag{1.71}$$

The matrix element $\tilde{M}^\lambda_{ba}$ is given in Equation 1.68 with $\hat{\epsilon}$ replaced by $\hat{\epsilon}_\lambda$.

It is desirable to calculate the total rate for emitting photons with all possible energy ω, propagation direction $\hat{k}$, and polarization $\hat{\epsilon}$. Using box normalization, the number of photon modes is given by

$$dn = \frac{V}{(2\pi)^3}k^2 dk d\Omega = \frac{V}{(2\pi c)^3}\omega^2 d\omega d\Omega_k, \tag{1.72}$$

where Ω_k is the solid angle. Integrating over $d\omega$, $d\Omega_k$ and summing over two polarizations, the total spontaneous emission rate is

$$W^s_{ba} = \frac{|\omega_{ba}|}{2\pi c^3}\int d\Omega_k \sum_{\lambda=1,2}|\tilde{M}^\lambda_{ba}(|\omega_{ba}|)|^2. \tag{1.73}$$

Electric-Dipole Transition, Selection Rules, and Oscillator Strength

In the matrix element $M_{ba} = \langle b|e^{i\mathbf{k}\cdot\mathbf{r}}\hat{\epsilon}\cdot\nabla|a\rangle$, the exponential term can be expanded to

$$e^{i\mathbf{k}\cdot\mathbf{r}} = 1 + i\mathbf{k}\cdot\mathbf{r} + \frac{1}{2}(i\mathbf{k}\cdot\mathbf{r})^2 + \cdots \tag{1.74}$$

The first-order approximation is to replace $e^{i\mathbf{k}\cdot\mathbf{r}}$ by 1, which is valid provided that the characteristic distance of the atomic wavefunction is much smaller than the wavelength of the electromagnetic radiation, i.e., $kr \ll 1$. This approximation is known as the dipole approximation, which amounts to neglecting the spatial variation of the electromagnetic field across the atom. In the dipole approximation, the vector potential **A** and the electric field **E** depend only on t and the magnetic field vanishes.

Within the dipole approximation, the matrix element

$$M_{ba}^{(1)} = \hat{\epsilon}\cdot\langle b|\nabla|a\rangle = i\hat{\epsilon}\cdot\langle b|\mathbf{p}|a\rangle. \tag{1.75}$$

Applying the Heisenberg equation of motion

$$\mathbf{p} = \dot{\mathbf{r}} = -i[\mathbf{r}, H_0], \tag{1.76}$$

leads to the result

$$M_{ba}^{(1)} = -\omega_{ba}\hat{\epsilon}\cdot\langle b|\mathbf{r}|a\rangle = \omega_{ba}\hat{\epsilon}\cdot\mathbf{D}_{ba}. \tag{1.77}$$

This introduces the one-electron, electric-dipole moment operator $\mathbf{D} = -\mathbf{r}$. Its matrix element can take the length, velocity, or acceleration form

$$\mathbf{D}_{ba}^{L} = -\langle b|\mathbf{r}|a\rangle, \tag{1.78}$$

$$\mathbf{D}_{ba}^{V} = \frac{i}{\omega_{ba}}\langle b|\mathbf{p}|a\rangle, \tag{1.79}$$

$$\mathbf{D}_{ba}^{A} = -\frac{1}{\omega_{ba}^2}\langle b|\nabla V|a\rangle. \tag{1.80}$$

Provided that $|a\rangle$ and $|b\rangle$ are exact eigenstates of H_0, these three forms give identical results.

In the dipole approximation, the frequency-integrated transition rate for photo-absorption or stimulated emission becomes

$$W_{ba} = W_{ab} = \frac{4\pi^2}{c}I(\omega_{ba})|\hat{\epsilon}\cdot\mathbf{D}_{ba}|^2. \tag{1.81}$$

For unpolarized radiation, the polarization direction $\hat{\epsilon}$ is random. By defining Θ as the angle between $\hat{\epsilon}$ and $\mathbf{D}_{ba}$ and taking the average over all solid angles, the transition rate takes the form

$$W_{ba} = W_{ab} = \frac{4\pi^2}{c}I(\omega_{ba})|\mathbf{D}_{ba}|^2\langle\cos^2\Theta\rangle = \frac{4\pi^2}{3c}I(\omega_{ba})|\mathbf{D}_{ba}|^2, \tag{1.82}$$

where

$$|\mathbf{D}_{ba}|^2 = |x_{ba}|^2 + |y_{ba}|^2 + |z_{ba}|^2. \tag{1.83}$$

Similarly, the ionization rate per unit energy from an initial state $|i\rangle$ to continuum states $|E\rangle$ in the dipole approximation reads

$$\frac{dW_{ion}}{dE} = \frac{4\pi^2}{c} I(\omega_{Ei}) |\hat{\epsilon} \cdot \mathbf{D}_{Ei}|^2 \rho(E), \tag{1.84}$$

where $\omega_{Ei} = E - E_i$, $\mathbf{D}_{Ei} = -\langle E|\mathbf{r}|i\rangle$ and $\rho(E)$ is the density of states.

Starting from Equation 1.73, the spontaneous-emission rate from a higher state $|b\rangle$ to a lower state $|a\rangle$ in the dipole approximation can be calculated as

$$\begin{aligned} W^s_{ab} &= \frac{\omega_{ba}^3}{2\pi c^3} \int d\Omega_k \sum_{\lambda=1,2} |\hat{\epsilon}_\lambda \cdot \mathbf{D}_{ba}|^2 = \frac{\omega_{ba}^3}{\pi c^3} \int d\Omega_\epsilon |\hat{\epsilon} \cdot \mathbf{D}_{ba}|^2 \\ &= \frac{\omega_{ba}^3}{\pi c^3} |\mathbf{D}_{ba}|^2 \int \cos^2 \Theta d\Omega_\epsilon = \frac{4\omega_{ba}^3}{3c^3} |\mathbf{D}_{ba}|^2. \end{aligned} \tag{1.85}$$

The transition within the dipole approximation is said to be an electric-dipole transition. If $\mathbf{D}_{ba}$ is nonvanishing, then the transition between states $|a\rangle$ and $|b\rangle$ is allowed. Otherwise, the transition is said to be forbidden. If the transition is forbidden, one has to consider higher-order effects such as multipole and multiphoton transitions.

The selection rules of electric-dipole transitions are closely related to the matrix elements x_{ba}, y_{ba}, and z_{ba}. Here, another set of coordinates is introduced as

$$D^1 = -\frac{1}{\sqrt{2}}(x + iy) = -\frac{1}{\sqrt{2}} r \sin\theta e^{i\phi} = \sqrt{\frac{4\pi}{3}} r Y_{1,1}(\theta, \phi), \tag{1.86}$$

$$D^{-1} = \frac{1}{\sqrt{2}}(x - iy) = \frac{1}{\sqrt{2}} r \sin\theta e^{-i\phi} = \sqrt{\frac{4\pi}{3}} r Y_{1,-1}(\theta, \phi), \tag{1.87}$$

$$D^0 = z = r\cos\theta = \sqrt{\frac{4\pi}{3}} r Y_{1,0}(\theta, \phi), \tag{1.88}$$

$$|\mathbf{D}_{ba}|^2 = |D^0_{ba}|^2 + |D^1_{ba}|^2 + |D^{-1}_{ba}|^2. \tag{1.89}$$

The states $|a\rangle$ and $|b\rangle$ are eigenstates of the one-electron, field-free Hamiltonian, so they can be labeled by (Elm) and $(E'l'm')$, respectively. The matrix element D^q_{ba} $(q = 0, \pm 1)$ becomes

$$D^q_{ba} = \sqrt{\frac{4\pi}{3}} \langle u_{E'l'}|r|u_{El}\rangle \langle Y_{l'm'}|Y_{1q}|Y_{lm}\rangle, \tag{1.90}$$

where the angular part

$$\langle Y_{l'm'}|Y_{1q}|Y_{lm}\rangle = \sqrt{\frac{3(2l+1)}{4\pi(2l'+1)}} \langle l100|l'0\rangle \langle l1mq|l'm'\rangle. \tag{1.91}$$

Here, $\langle l_1 l_2 m_1 m_2|lm\rangle$ are the well-known Clebsch–Gordan (C–G) coefficients. The first C–G coefficient $\langle l100|l'0\rangle$ is nonvanishing only if $l' = l \pm 1$. Then

$$\langle l100|l+1,0\rangle = \sqrt{\frac{l+1}{2l+1}}, \tag{1.92}$$

$$\langle l100|l-1,0\rangle = -\sqrt{\frac{l}{2l+1}}. \tag{1.93}$$

Note that the parity of an eigenstate $|Elm\rangle$ is $(-1)^l$. This selection rule implies that the initial and final states should have opposite parities. Furthermore, getting a nonvanishing C–G coefficient $\langle l1mq|l'm'\rangle$ requires $m' = m + q$.

For dipole transitions, it is common to introduce a dimensionless quantity f_{ka} called the oscillator strength

$$f_{ka} = \frac{2\omega_{ka}}{3}|\mathbf{D}_{ka}|^2, \tag{1.94}$$

with $\omega_{ka} = E_k - E_a$. Thus $f_{ka} > 0$ is for photoabsorption and $f_{ka} < 0$ is for emission. Here, the final state $|k\rangle$ is confined as a bound state. For continuum final states $|E\rangle$, the oscillator strength per unit energy can be introduced as

$$\frac{df_{Ea}}{dE} = \frac{2\omega_{Ea}}{3}|\mathbf{D}_{Ea}|^2\rho(E). \tag{1.95}$$

The oscillator strength obeys the Thomas–Kuhn–Reiche sum rule

$$\sum_k f_{ka} + \int \frac{df_{Ea}}{dE}dE = 1. \tag{1.96}$$

Because the magnetic levels are degenerate for an one-electron atom, one can define the average oscillator strength for the transition $nl \rightarrow n'l'$ by averaging over the initial magnetic states and summing over all the final magnetic states, i.e.,

$$\bar{f}_{n'l',nl} = \frac{1}{2l+1}\sum_{m=-l}^{l}\sum_{m'=-l'}^{l'} f_{n'l'm',nlm}. \tag{1.97}$$

The spontaneous transition rate from a state $|b\rangle$ to a state $|a\rangle$ in the dipole approximation can be given in terms of the oscillator strength by

$$W^s_{ab} = \frac{2\omega_{ab}^2}{c^3}|f_{ab}|. \tag{1.98}$$

If b and a stand for levels that can be labeled by n and l (m-degenerate), the average spontaneous transition rate is given in terms of the averaged oscillator strength

$$\bar{W}^s_{ab} = \frac{2\omega_{ab}^2}{c^3}|\bar{f}_{ab}|. \tag{1.99}$$

The lifetime of level b is

$$\tau_b = \left(\sum_a \bar{W}^s_{ab}\right)^{-1}. \tag{1.100}$$

The sum in Equation 1.100 is over all the dipole-allowed final levels a that are lower than b. For typical optical transitions in atoms, the lifetime is in the order of nanoseconds. For hydrogenic atoms, the lifetime of a level with principal quantum number n is proportional to n^3, and therefore highly excited states will have longer lifetimes.

According to the uncertainty principle, the lifetime of a level b is inversely proportional to its natural width. For example, the lifetime of the $2p$ level in atomic hydrogen is 1.6 ns given a width of 3.32×10^{-3} cm^{-1}, or about 100 MHz. The lineshape takes the Lorentzian form for an isolated atom. The spectral width of a level can get additional width from pressure broadening, which is understood as due to the collisional de-excitation of the excited state from the surrounding medium and from Doppler broadening when the emitting atoms have a velocity distribution in accordance to the Maxwellian law.

Photoelectron Angular Distributions

Consider the photoionization of an one-electron system with model potential $V(r) = -Z_c/r + V_{sr}(r)$. Without loss of generality, it is assumed that the light is polarized along the z-direction. In the dipole approximation, according to Equation 1.84, the photoionization cross-section (per unit energy) is given by

$$\sigma = \frac{4\pi^2}{c}\omega|\langle E|z|i\rangle|^2 \tag{1.101}$$

in which $\omega = E - E_i$. Here, the continuum states $|E\rangle$ are normalized with respect to energy such that $\rho(E) = 1$. In spherical coordinates the continuum eigenstates of the field-free Hamiltonian can be written as

$$\langle \mathbf{r}|ELM\rangle = \frac{u_{EL}(r)}{r} Y_{LM}(\theta_r, \phi_r), \tag{1.102}$$

where u_{EL} represents the real radial wavefunctions normalized per unit energy and Y_{LM} represents the spherical harmonics. The spatial vector $\mathbf{r}$ is expressed in spherical coordinates (r, θ_r, ϕ_r). The initial state is given by

$$\langle \mathbf{r}|i\rangle = \frac{u_{nl}(r)}{r} Y_{lm}(\theta_r, \phi_r) \tag{1.103}$$

with well-defined angular momentum.

If $|ELM\rangle$, which has well-defined quantum numbers L and M, is chosen to be the continuum state $|E\rangle$ in Equation 1.101, then σ is the photoionization cross-section for this continuum partial wave. Alternatively, $|E, \theta, \phi\rangle$ can be chosen to describe the photoelectrons emitted along a particular direction (θ, ϕ). This state can be constructed by partial waves as

$$\langle \mathbf{r}|E, \theta, \phi\rangle = \sum_{L=0}^{\infty} \sum_{M=-L}^{L} i^L e^{-i\eta_{EL}} Y^*_{LM}(\theta, \phi) Y_{LM}(\theta_r, \phi_r) \frac{u_{EL}(r)}{r}. \tag{1.104}$$

Here, $\eta_{EL} = \arg[\Gamma(L + 1 - iZ_c/k)] + \delta_{EL}$ and δ_{EL} is the phase shift due to the short-range potential $V_{sr}(r)$ (see Equation 1.22). In the asymptotic region, this wave-function is the sum of a plane wave and an incoming spherical wave. If the continuum state $|E\rangle$ in Equation 1.101 is expressed by $|E, \theta, \phi\rangle$, then the photoionization differential cross-section is obtained as

$$\frac{d\sigma}{d\Omega} = \frac{16\pi^3}{3c}\omega \left| \sum_{L=0}^{\infty} \sum_{M=-L}^{L} a^*_{LM}(\theta, \phi) S_L \langle Y_{LM}|Y_{10}|Y_{lm}\rangle \right|^2, \tag{1.105}$$

where $a_{LM}(\theta,\phi) = i^L e^{-i\eta_{EL}} Y^*_{LM}(\theta,\phi)$ and $S_L = \langle u_{EL}|r|u_{nl}\rangle$.

The matrix element $\langle Y_{LM}|Y_{10}|Y_{lm}\rangle$ is nonvanishing only when $L = l \pm 1$ and $M = m$. They can be written explicitly as

$$A_{lm} = \langle Y_{l+1,m}|Y_{10}|Y_{lm}\rangle = \sqrt{\frac{3(l+m+1)(l-m+1)}{4\pi(2l+1)(2l+3)}}, \tag{1.106}$$

$$B_{lm} = \langle Y_{l-1,m}|Y_{10}|Y_{lm}\rangle = \sqrt{\frac{3(l+m)(l-m)}{4\pi(2l+1)(2l-1)}}. \tag{1.107}$$

Then the differential cross-section is

$$\begin{aligned}\frac{d\sigma}{d\Omega} &= \frac{16\pi^3}{3c}\omega\left|a^*_{l+1,m}(\theta,\phi)S_{l+1}A_{lm} + a^*_{l-1,m}(\theta,\phi)S_{l-1}B_{lm}\right|^2 \\ &= \frac{16\pi^3}{3c}\omega\Big\{|Y_{l+1,m}(\theta,\phi)|^2 S^2_{l+1}A^2_{lm} + |Y_{l-1,m}(\theta,\phi)|^2 S^2_{l-1}B^2_{lm} \\ &\quad - 2S_{l+1}S_{l-1}A_{lm}B_{lm}Y_{l+1,m}(\theta,\phi)Y^*_{l-1,m}(\theta,\phi)\cos(\eta_{E,l+1} - \eta_{E,l-1})\Big\}.\end{aligned} \tag{1.108}$$

When the initial level (nl) has m-degeneracy, one must average over the initial states with different quantum number m. Therefore

$$\begin{aligned}\frac{d\sigma}{d\Omega} = \frac{16\pi^3}{3c(2l+1)}\omega\Bigg\{&S^2_{l+1}\sum_{m=-l}^{l} A^2_{lm}|Y_{l+1,m}(\theta,\phi)|^2 + S^2_{l-1}\sum_{m=-l}^{l} B^2_{lm}|Y_{l-1,m}(\theta,\phi)|^2 \\ &-2S_{l+1}S_{l-1}\cos(\eta_{E,l+1} - \eta_{E,l-1}) \\ &\sum_{m=-l}^{l} A_{lm}B_{lm}Y_{l+1,m}(\theta,\phi)Y^*_{l-1,m}(\theta,\phi)\Bigg\}.\end{aligned} \tag{1.109}$$

After some algebraic calculations, the result can be simplified to the form

$$\frac{d\sigma}{d\Omega} = \frac{\sigma_{total}}{4\pi}[1 + \beta P_2(\cos\theta)] = \frac{\sigma_{total}}{4\pi}\left[1 + \beta\frac{(3\cos^2\theta - 1)}{2}\right], \tag{1.110}$$

where σ_{total} is the total ionization cross-section integrating over all solid angles. The β-parameter is given by

$$\beta = \frac{l(l-1)S^2_{l-1} + (l+1)(l+2)S^2_{l+1} - 6l(l+1)S_{l+1}S_{l-1}\cos(\eta_{E,l+1} - \eta_{E,l-1})}{(2l+1)[lS^2_{l-1} + (l+1)S^2_{l+1}]}. \tag{1.111}$$

The value β ranges from -1 to 2. For $\beta = 2$, the differential cross-section has a $\cos^2\theta$ dependence with the peak in the direction of the light polarization. Note that if $l = 0$ then $\beta = 2$. For $\beta = -1$, the differential cross-section has a $\sin^2\theta$ dependence with the peak at 90 degrees with respect to the polarization axis.

1.2 Two-Electron Atoms

1.2.1 Schrödinger Equation and Exchange Symmetry of Two-Electron Atoms

For the two-electron helium atom, the nonrelativistic Schrödinger equation for the spatial part of the wavefunction is given by

$$H\psi(\mathbf{r}_1,\mathbf{r}_2) = \left[-\frac{1}{2}\nabla_1^2 - \frac{1}{2}\nabla_2^2 - \frac{Z}{r_1} - \frac{Z}{r_2} + \frac{1}{r_{12}}\right]\psi(\mathbf{r}_1,\mathbf{r}_2) = E\psi(\mathbf{r}_1,\mathbf{r}_2), \quad (1.112)$$

where $\mathbf{r}_1$ and $\mathbf{r}_2$ are the coordinates of the two electrons with respect to the nucleus, and $r_{12} = |\mathbf{r}_1 - \mathbf{r}_2|$ is the distance between the two electrons. This total wavefunction cannot be separated into products of wavefunctions of the two electrons due to the electron–electron interaction $1/r_{12}$. Thus, the two electrons are said to be correlated. Equation 1.112 is invariant under the interchange of labels 1 and 2. Thus, the spatial wavefunction should be even or odd under particle exchange, i.e.,

$$\psi(\mathbf{r}_1,\mathbf{r}_2) = \pm\psi(\mathbf{r}_2,\mathbf{r}_1). \quad (1.113)$$

Since electrons are fermions and each electron has spin but spin interaction is not included in Equation 1.112, the total wavefunction can be written as the product of a spatial part and a spin part for the two electrons as

$$\Psi(q_1,q_2) = \psi(\mathbf{r}_1,\mathbf{r}_2)\chi(1,2). \quad (1.114)$$

The total spin of the two electrons can either be the symmetric triplet $S = 1$ or the antisymmetric singlet $S = 0$ under particle exchange. Since the total wavefunction should be antisymmetric, singlet states of helium have symmetric spatial wavefunctions while the spatial wavefunction is antisymmetric for triplet states.

1.2.2 Shell Model and Level Scheme of Two-Electron Atoms

Since Equation 1.112 is not separable, it is desirable to rewrite the Hamiltonian in Equation 1.112 as

$$H = H_0 + H', \quad (1.115)$$

where H_0 is given by

$$H_0 = -\frac{1}{2}\nabla_1^2 - \frac{1}{2}\nabla_2^2 + V(r_1) + V(r_2), \quad (1.116)$$

and

$$H' = -\frac{Z}{r_1} - \frac{Z}{r_2} + \frac{1}{r_{12}} - V(r_1) - V(r_2). \quad (1.117)$$

The $V(r)$ is a central potential that is chosen to make the effect of H' small. This means that a portion of the electron–electron interaction is now included in H_0. As shown in Section 1.3, this approach can be generalized to a many-electron system. Eigensolutions of H_0 can be expressed in terms of the products of one-electron wavefunctions. For a helium atom, the eigensolutions of H_0 that are symmetric or antisymmetric under electron exchange are then given by

$$\psi_{\pm}^{(0)}(\mathbf{r}_1, \mathbf{r}_2) = \frac{1}{\sqrt{2}}[\psi_{n_1 l_1 m_1}(\mathbf{r}_1)\psi_{n_2 l_2 m_2}(\mathbf{r}_2) \pm \psi_{n_2 l_2 m_2}(\mathbf{r}_1)\psi_{n_1 l_1 m_1}(\mathbf{r}_2)]. \quad (1.118)$$

The energy is degenerate with respect to particle exchange

$$E^{(0)} = E_{n_1 l_1} + E_{n_2 l_2}. \quad (1.119)$$

The quantum numbers n_1, l_1, n_2, l_2 together with the total orbital and spin quantum numbers L, M_L, S, M_S can be used to describe an eigenstate of H_0. This state also has well-defined parity $\pi = (-1)^{l_1+l_2}$. $(n_1 l_1 n_2 l_2)$ is used to define electron configurations such as $1s^2$, $1s2s$, or $1s2p$. In spectroscopic language $^{2S+1}L^{\pi}$ is used to denote a term. The eigenstates of H_0 can be expressed by $(n_1 l_1 n_2 l_2)\ ^{2S+1}L^{\pi}$, which forms a complete set of the whole Hilbert space. However, they are not eigenstates of the total Hamiltonian H. The good quantum numbers for H are L, M_L, S, M_S, and the parity π within the Hamiltonian given in Equation 1.112. Since H and H_0 are in the same Hilbert space, the eigenstate of H with quantum numbers L, M_L, S, M_S, and π can be expanded in terms of

$$|\Psi\rangle = \sum_{\alpha} c_{\alpha} |\alpha\rangle. \quad (1.120)$$

Here, α spans the complete set of $(n_1 l_1 n_2 l_2)$ for the given L, M_L, S, M_S, and π. This procedure for finding a more accurate wavefunction of H is usually called the *configuration interaction method*. The expansion Equation 1.120, in general, is seriously truncated and only configurations that have energies near the state of interest are included. If one coefficient c_{α} is dominant in Equation 1.120, then that specific $\alpha = (n_1 l_1 n_2 l_2)$ is used to designate this eigenstate of H, and, in this case, (n_1, l_1, n_2, l_2) are approximate, good quantum numbers.

Here, the level scheme of helium is described. The ground state is a singlet state $(1s)^2\ ^1S$ with energy -79.0 eV. The discrete levels are comprised of singly excited states $1snl\ ^{2S+1}L$ in which $L = l$, and thus these states can be denoted by $n\ ^{2S+1}l$ for short. They are divided into singlets (para-helium) and triplets (ortho-helium). If the outer electron becomes free, the remaining ion $He^+(n = 1)$ has energy -54.4 eV. Therefore the first ionization threshold of He is 24.6 eV. There are also doubly excited states ($n_1 \geq 2$ and $n_2 \geq 2$) that lie in the continuum with energies higher than that of $He^+(n = 1)$.

If relativistic effects and spin interactions are included, then L and S are not good quantum numbers when considered separately. It is important to consider the total angular momentum $\mathbf{J}$. In this case, J, M_J, and parity are good quantum numbers. The angular momentum coupling schemes are:

$L - S$ coupling	$\mathbf{J} = (\mathbf{l}_1 + \mathbf{l}_2) + (\mathbf{s}_1 + \mathbf{s}_2) = \mathbf{L} + \mathbf{S}$
$j - j$ coupling	$\mathbf{J} = (\mathbf{l}_1 + \mathbf{s}_1) + (\mathbf{l}_2 + \mathbf{s}_2) = \mathbf{j}_1 + \mathbf{j}_2$

For low-Z atoms, the $L - S$ coupling scheme is a better approximation and the term $^{2S+1}L_J$ can be used to describe an energy level. The $j - j$ coupling is better for high-Z atoms. The admixture of $L - S$ and $j - j$ coupling is called *intermediate coupling*.

These coupling schemes work best for valence orbitals and low-lying excited states. For Rydberg states where the outer electron is farther away from the inner electron, a better coupling scheme is $\mathbf{J} = (\mathbf{l}_1 + \mathbf{s}_1) + \mathbf{l}_2 + \mathbf{s}_2$ where the inner electron's j couples with the orbital angular momentum and then to the spin of the outer electron, called the *J–L coupling scheme*. The approximate coupling scheme depends on the relative energies of the various interactions.

1.2.3 Doubly Excited States, Autoionization, and Fano Resonances

Based on the shell model, there are doubly excited states in which both electrons occupy excited orbitals. These states can be designated as $n_1 l_1 n_2 l_2\ {}^{2S+1}L$ where $n_1, n_2 \geq 2$. They lie above the first ionization threshold and are therefore embedded in the continuum. In photoabsorption or photoelectron spectra they are seen in the form of sharp resonances superimposed on the smooth continuum background. For helium, these doubly excited states exist at about 60–79 eV above the ground state. Since the ionization potential for helium is only 24.6 eV, these doubly excited states are degenerate with $1s\epsilon l$ continuum states. Thus the eigenstate of the Hamiltonian clearly should be the linear combination of the bound doubly excited state and continuum states. This problem was addressed by Fano [1]. Take the $2s2p\ {}^1P$ doubly excited state as an example and denote it with $|\alpha\rangle$. It is degenerate with the $1sEp\ {}^1P$ continuum states, which is denoted by $|\beta_E\rangle$. The Hilbert space consists of two channels. The closed channel has one eigenstate $|\alpha\rangle$. The open channel has continuum states $|\beta_E\rangle$. Between the two channels there is a coupling term that causes the doubly excited state to decay to the continuum states. Thus the Hamiltonian can be written as

$$\langle\alpha|H|\alpha\rangle = E_r, \tag{1.121}$$

$$\langle\beta_E|H|\beta_{E'}\rangle = E\delta(E - E'), \tag{1.122}$$

$$\langle\beta_E|H|\alpha\rangle = V_E. \tag{1.123}$$

The eigenstate of this Hamiltonian can be expressed as

$$|\psi_E\rangle = a_E|\alpha\rangle + \int b_{EE'}|\beta_{E'}\rangle dE'. \tag{1.124}$$

If one assumes that $V_E = V$ is constant in the vicinity of $|\alpha\rangle$, then the coefficients can be solved analytically

$$a_E = \frac{\sin\Theta}{\pi V}, \tag{1.125}$$

$$b_{EE'} = \frac{\sin\Theta}{\pi(E - E')} - \delta(E - E')\cos\Theta, \tag{1.126}$$

$$\Theta = -\arctan\frac{\pi V^2}{E - E_r}. \tag{1.127}$$

Given a transition operator T that takes the atom from the initial state $|g\rangle$ to the final state $|\psi_E\rangle$, the ratio of the resonance with respect to the smooth background is

$$\frac{|\langle\psi_E|T|g\rangle|^2}{|\langle\beta_E|T|g\rangle|^2} = \frac{(\epsilon + q)^2}{1 + \epsilon^2}, \tag{1.128}$$

where

$$\epsilon = \frac{E - E_r}{\Gamma/2}, \tag{1.129}$$

$$\Gamma = 2\pi V^2, \tag{1.130}$$

$$q = \frac{\langle \alpha | T | g \rangle}{\pi V \langle \beta_E | T | g \rangle}. \tag{1.131}$$

Equation 1.128 demonstrates that the cross-section in the vicinity of the resonance can be expressed as

$$\sigma(E) = \sigma_0 \frac{(\epsilon + q)^2}{1 + \epsilon^2}. \tag{1.132}$$

In this equation, a resonance is characterized by its position E_r, width Γ, and shape parameter q. The profile varies with q and for $q \to \infty$, the profile becomes Lorentzian. The parametrization in Equation 1.132 provides a simple way to characterize the shape of a resonance, known as a Fano lineshape. Note that the Fano resonance is also called the Feshbach resonance since the formulation given above was used earlier by Feshbach. Section 1.1.3 mentioned that a shape resonance can also be parametrized, like a Fano resonance. A shape resonance is a single-channel phenomenon and, in general, lies close to the threshold. Feshbach resonances usually lie closer to the threshold of the closed channel. However, such a distinction is not always possible in a real physical problem. The formulation of a Fano resonance is quite general. It happens whenever a confined standing wave is coupled with an external dissipative medium. Figure 1.1 shows examples of some Fano resonances observed in rare gas atoms.

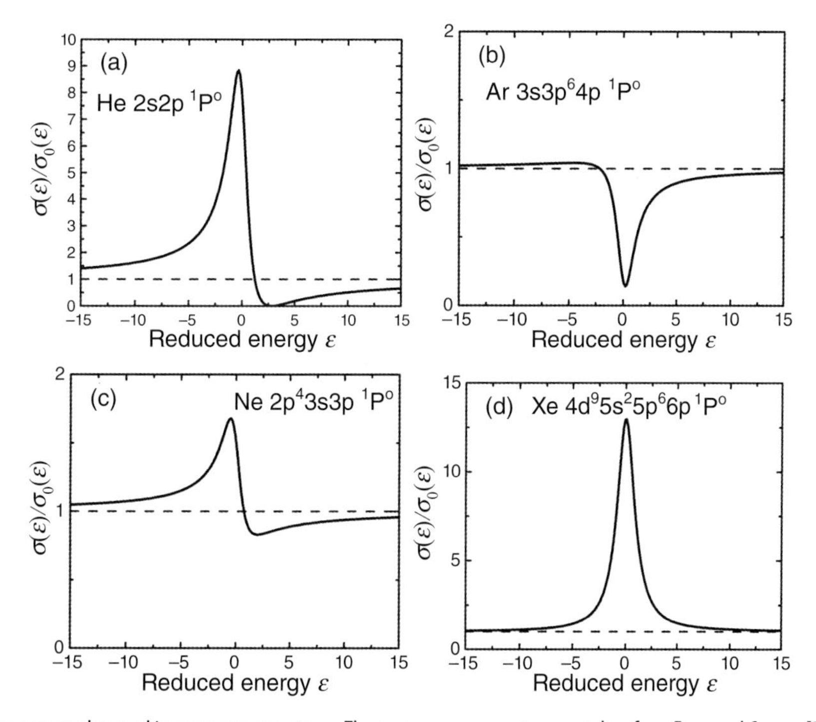

Figure 1.1 Fano resonances observed in some rare gas atoms. The resonance parameters are taken from Fano and Cooper [2].

Doubly excited states can also decay to other bound states via radiative transitions. Since the autoionization lifetime estimated from the resonance width is on the order of a few tens or a few femtoseconds, it is much shorter than the typical nanoseconds-long lifetime for radiative decay. Thus, for valence electrons, autoionization is the main pathway for the decay of a resonance.

Doubly excited states are examples where electron correlation is large such that shell-model description fails completely. Alternative descriptions of such strongly correlated systems in atoms can be found, for example, in Lin [3].

1.2.4 Description of Continuum States of Helium and Electron-He$^+$ Collisions

The Schrödinger equation that governs electron-He$^+$ collisions is the same as the one for the helium atom

$$H\Psi(\mathbf{r}_1,\mathbf{r}_2) = \left[-\frac{1}{2}\nabla_1^2 - \frac{1}{2}\nabla_2^2 - \frac{2}{r_1} - \frac{2}{r_2} + \frac{1}{r_{12}}\right]\Psi(\mathbf{r}_1,\mathbf{r}_2) = E\Psi(\mathbf{r}_1,\mathbf{r}_2) \tag{1.133}$$

except that the total energy of the system $E = -2.0 + k_1^2/2$ where $E_1 = k_1^2/2$ is the incident energy of the electron. For sufficient incident energy E_1, some or all of the following processes may occur:

$$\begin{aligned} e^- + \mathrm{He}^+(1s) &\rightarrow e^- + \mathrm{He}^+(1s) \quad \text{elastic scattering} \\ &\rightarrow e^- + \mathrm{He}^+(2s,2p) \quad \text{excitations} \\ &\quad \ldots \\ &\rightarrow e^- + e^- + \mathrm{He}^{++} \quad \text{impact ionization} \\ &\rightarrow \mathrm{He} + \hbar\omega \quad \text{photorecombination.} \end{aligned}$$

Note that the charge state of helium changes for impact ionization and radiative recombination processes. To isolate a particular process, the charge state of the target, the energy, and/or the momentum of the electron are measured.

If the incident energy is higher than the first excitation threshold, inelastic transitions can occur. For the He$^+$ target, the minimum energy to excite He$^+$ from $n = 1$ to $n = 2$ is 40.8 eV. Below this energy lie the doubly excited states. Similarly, for electron energy above 48.4 eV, there are processes where He$^+$ is excited to $n = 3$ states and below such a threshold there are many doubly excited states as well. All of the energetically possible processes are called open channels. In contrast, doubly excited states belong to closed channels. Clearly, many more open channels become possible as the collision energy is increased. Thus, when the total energy is sufficiently above the ground state, the two-electron wavefunction for a given energy E can be expressed in the general form

$$\Psi(\mathbf{r}_1,\mathbf{r}_2) = A\sum_i F_i(r_1)\Phi_i(\hat{r}_1,\mathbf{r}_2), \tag{1.134}$$

where A is the antisymmetrization operator, F_i is the radial function of the outer electron, and $\Phi_i(\hat{r}_1,\mathbf{r}_2)$ is the channel function, including the angular variables of the outer electron. In other words, one has to deal with multichannel problems. In the past few decades, many sophisticated theoretical tools and computational packages have been developed. These are very technical in nature and will not be discussed in this book.

1.3 Many-Electron Atoms

1.3.1 The Hierarchy of Approximate Description of Many-Electron Atoms

No one can solve the structure of a many-electron atom exactly. The description of a many-electron system relies on different degrees of approximations that depend on the problems at hand. In a many-electron system one may need to take into account the following interactions seen by the electron:

(1) Coulomb interactions,
(2) spin-orbit interactions,
(3) spin-spin interactions, other relativistic corrections, mass polarization, and QED effects.

This section focuses on the already monumental task of dealing with Coulomb interactions. The approximated Hamiltonian reads:

$$H = \sum_{i=1}^{N} \left(-\frac{1}{2}\nabla_i^2 - \frac{Z}{r_i} \right) + \sum_{i<j=1}^{N} \frac{1}{r_{ij}}, \tag{1.135}$$

where $\mathbf{r}_i$ is the coordinate of the electron i from the nucleus and r_{ij} is the distance between the two electrons i and j. The many-electron wavefunction satisfies the Schrödinger equation

$$H\Psi(q_1, q_2, \ldots, q_N) = E\Psi(q_1, q_2, \ldots, q_N), \tag{1.136}$$

where q_i denotes the spatial coordinates $\mathbf{r}_i$ and the spin of the electron i. The wavefunction Ψ should be constructed such that it is an eigenstate of L^2, S^2, and parity. Moreover, the Pauli exclusion principle requires that the total wavefunction be antisymmetric under the interchange of any two electrons. Once such a wavefunction is obtained, the inclusion of the smaller interactions (2) and (3) can be done using first-order perturbation theory.

1.3.2 The Central Field Approximation and Shell Model of Many-Electron Atoms

In the central field approximation, each electron moves in an effective potential, which represents the attraction of the nucleus and the average effect of the repulsive interactions between this electron and the $(N-1)$ other electrons. This effective potential is assumed to be spherically symmetric such that

$$V(r) = -\frac{Z}{r} + S(r), \tag{1.137}$$

where Z is the charge of the nucleus and $S(r)$ describes the screening effect due to other electrons. With the effective central potential, Equation 1.135 can be rewritten as

$$H = \sum_{i=1}^{N} \left(-\frac{1}{2}\nabla_i^2 + V(r_i) \right) + \left[\sum_{i<j=1}^{N} \frac{1}{r_{ij}} - \sum_{i=1}^{N} S(r_i) \right] = H_c + H_1. \tag{1.138}$$

The first part H_c is separable since it is the sum of one-electron Hamiltonians

$$h_i = -\frac{1}{2}\nabla_i^2 + V(r_i). \tag{1.139}$$

The second part H_1 contains the remaining spherical and all the nonspherical parts of the inter-electron repulsion, which can be treated as a perturbation as it is much smaller than H_c.

The eigenfunction of the one-electron Hamiltonian h_i can be represented by a spin-orbital

$$h_i u_\alpha(q_i) = h_i R_{nl}(r_i) Y_{lm_l}(\theta_i, \phi_i) \chi_{1/2,m_s} = E_{nl} u_\alpha(q_i), \tag{1.140}$$

where $\alpha = (n, l, m_l, m_s)$. Clearly, $u_{\alpha_1}(q_1)u_{\alpha_2}(q_2)\ldots u_{\alpha_N}(q_N)$ is an eigensolution of H_c with the total energy

$$E_c = E_{n_1 l_1} + E_{n_2 l_2} + \cdots + E_{n_N l_N}. \tag{1.141}$$

However, to make sure that the wavefunction of the N-electron atom is fully antisymmetric, the many-electron wavefunction should be constructed in terms of the Slater determinant

$$\Psi_c(q_1, q_2, \ldots, q_N) = \frac{1}{\sqrt{N!}} \begin{vmatrix} u_{\alpha_1}(q_1) & u_{\alpha_2}(q_1) & \cdots & u_{\alpha_N}(q_1) \\ u_{\alpha_1}(q_2) & u_{\alpha_2}(q_2) & \cdots & u_{\alpha_N}(q_2) \\ \vdots & \vdots & \ddots & \vdots \\ u_{\alpha_1}(q_N) & u_{\alpha_2}(q_N) & \cdots & u_{\alpha_N}(q_N) \end{vmatrix}. \tag{1.142}$$

Such Slater determinants are combined such that L,S and their projections, M_L, M_S are good quantum numbers. After this is done, the basis states can be represented by $|\Psi_c\rangle = |n_1 l_1 n_2 l_2, \ldots, {}^{2S+1}L^{\pi}\rangle$. The set $\{n_i, l_i\}$ is called the *electron configuration*, and ${}^{2S+1}L$ is called a *term*. Note that the basis states have well-defined parity given by $\pi = (-1)^{\sum_i l_i}$. Both the configuration and the term can be used to designate exact eigenstates of the atom. For example, the ground state of Be is $|1s^2 2s^2\ {}^1S^e\rangle$ and an excited state of Be is $|1s2s^2 4p\ {}^3P^o\rangle$. The latter is said to have a hole in the K-shell. The residual interaction term H_1 and spin-orbit interaction term can then be calculated perturbatively.

The discussion so far has not addressed how the central field potential $V(r)$ is obtained. If the spin-orbitals are chosen variationally such that the total energy is minimized, then the method is called the Hartree–Fock (HF) method. Using such an approach, the spin-orbitals $u_\alpha(q_i)$ can be shown to satisfy a set of coupled integral-differential equations. The details of this will not be discussed here.

One can also construct a better trial wavefunction with more than one Slater determinant, that is

$$\Psi = \sum_i c_i \Psi_i, \tag{1.143}$$

where Ψ_i are Slater determinants that differ in the choice of the occupied spin-orbitals and therefore correspond to different configurations. The coefficients c_i can be obtained variationally. This approach is known as the multi-configuration HF (MCHF) method.

1.3.3 Density Functional Theory

The HF theory can, in principle, be extended to large systems like molecules and solids. However, as the systems become large, the HF method becomes too complicated. The density functional theory (DFT) is an alternative approach in which the electron number density plays the central role instead of the many-electron wavefunctions. From Equation 1.135, the total (nonrelativistic) Hamiltonian of an N-electron atom can be separated into

$$H = T + W + H_2, \tag{1.144}$$

where

$$T = \frac{1}{2}\sum_{i=1}^{N} \nabla_i^2 \tag{1.145}$$

is the total kinetic energy. The term

$$W = \sum_{i=1}^{N} U(\mathbf{r}_i) \tag{1.146}$$

in which

$$U(\mathbf{r}_i) = -\frac{Z}{r_i} \tag{1.147}$$

is the interaction energy between electrons and the nucleus and

$$H_2 = \sum_{i<j=1}^{N} \frac{1}{r_{ij}} \tag{1.148}$$

is the interaction energy between electron pairs. If $\Psi(\mathbf{r}_1, \mathbf{r}_2, \ldots, \mathbf{r}_N)$ is the exact many-electron wavefunction of the ground state, including exchange and correlation effects, the single-electron number density can be defined by

$$\rho(\mathbf{r}) = \int d\mathbf{r}_2 d\mathbf{r}_3 \ldots d\mathbf{r}_N |\Psi(\mathbf{r}, \mathbf{r}_2, \mathbf{r}_3, \ldots, \mathbf{r}_N)|^2. \tag{1.149}$$

The DFT was first proposed by Hohenberg and Kohn, who proved that every observable quantity of a stationary, many-electron system like the ground-state energy is uniquely determined by the ground-state number density $\rho(\mathbf{r})$, and that a functional $E[\rho]$ exists such that the minimum value of this functional is the ground-state energy and the number density that yields this minimum is the exact single-electron number density of the ground state.

It was then shown by Kohn and Sham [4] that it is possible to replace the many-electron problem by an exactly equivalent set of self-consistent, one-electron equations. The total energy functional can be written as

$$E[\rho] = T[\rho] + \frac{1}{2}\int d\mathbf{r}d\mathbf{r}'\frac{\rho(\mathbf{r})\rho(\mathbf{r}')}{|\mathbf{r}-\mathbf{r}'|} + \int d\mathbf{r}U(\mathbf{r})\rho(\mathbf{r}) + E_{xc}[\rho]. \tag{1.150}$$

On the right-hand side of Equation 1.150, the T term is supposedly the kinetic energy, the second term is the expectation value of the electron–electron interaction, the third term is the electron–nucleus interaction, and the E_{xc} term gives the many-body correction, which is called the exchange-correlation energy.

If the system is represented by single-electron orbitals $u_i(\mathbf{r})$, the single-electron number density of this system is given by

$$\rho(\mathbf{r}) = \sum_{i=1}^{N}|u_i(\mathbf{r})|^2, \tag{1.151}$$

and the expectation value of the kinetic energy is

$$T[\rho] = \frac{1}{2}\sum_{i=1}^{N}\int d\mathbf{r}|\nabla u_i(\mathbf{r})|^2. \tag{1.152}$$

Note that the true kinetic-energy operator cannot be written in this way, but when the orbitals are optimized, Equation 1.152 turns out to be a good approximation.

By minimizing $E[\rho]$ with respect to $\rho(\mathbf{r})$, which is equivalent to finding the optimal $u_i(\mathbf{r})$, one can obtain the Kohn–Sham equation

$$\left[-\frac{1}{2}\nabla^2 + V_{eff}(\mathbf{r})\right]u_i(\mathbf{r}) = E_i u_i(\mathbf{r}), \tag{1.153}$$

where E_i is the Lagrange multiplier that ensures normalization as in the HF theory. The effective potential V_{eff} is given by

$$V_{eff}(\mathbf{r}) = U(\mathbf{r}) + \int d\mathbf{r}'\frac{\rho(\mathbf{r}')}{|\mathbf{r}-\mathbf{r}'|} + V_{xc}(\mathbf{r}). \tag{1.154}$$

The last term

$$V_{xc}(\mathbf{r}) = \frac{\delta E_{xc}}{\delta\rho} \tag{1.155}$$

is called the exchange-correlation potential, which is not known but can be approximated in different ways. Given an approximation for V_{xc}, the Kohn–Sham equation can be solved self-consistently.

The performance of the DFT method depends on finding an adequate approximation for E_{xc}. The simplest one is the local-density approximation (LDA) through which more advanced DFT theories have been proposed. The advantage of the DFT is that it reduces the problem to that of three degrees of freedom (say, x, y, and z), rather than $3N$ degrees of freedom. DFT is in use widely, especially for large systems such as big molecules or solid-state systems, and has contributed to the development of many computational packages.

1.3.4 Photoionization of Rare Gas Atoms

Section 1.2.4 briefly mentioned the theoretical tool for describing continuum states of helium atoms, or more precisely, the e+He$^+$ system after photoionization. This theoretical method can be generalized to describe an N-electron atom if there is only one continuum electron. For example, Equation 1.134 can be generalized to

$$\Psi(\mathbf{r}_1, \mathbf{r}_2, \ldots, \mathbf{r}_N) = A \sum_i F_i(r_1)\Phi_i(\hat{r}_1, \mathbf{r}_2, \mathbf{r}_3, \ldots, \mathbf{r}_N), \tag{1.156}$$

where A is the antisymmetrization operator, r_1 is the radial distance of the continuum electron, and Φ_i is the eigenstate of the (N-1)-electron target ion coupled with the angular momentum of the continuum electron to form eigenstates of L^2, S^2, and L_z, S_z (if the spin interaction is neglected) of the whole N-electron atom. The summation in this equation is over all the channels that satisfy $E = IP_i + \varepsilon_i$ where IP_i is the ionization potential and $\varepsilon_i = \frac{1}{2}k_i^2$ is the kinetic energy of the photoelectron for the ith channel.

Solutions of the time-dependent Schrödinger equation are complicated and different computer packages have been developed. These solutions will not be discussed here except to summarize the information obtained experimentally. In a typical photoionization experiment where photoelectrons are measured, one can separate subshell photoionization cross-sections. As an example, consider the photoionization of Ne. The left panel of Figure 1.2 shows the subshell cross-sections from accurate theoretical calculations and data tabulated from various experiments. There is only one channel open (the fine structure due to spin interaction is neglected) above the $2p$ but below the $2s$ ionization threshold. Above the $2s$ threshold, there are two channels; one group of photoelectrons is associated with the $2p$ hole and another one with the $2s$ hole. Similarly, at the photon energy where the $1s$ hole is created, three channels are open. Thus, three groups of electrons with different energies can be detected. In the simple discussion up to now, each continuum electron can

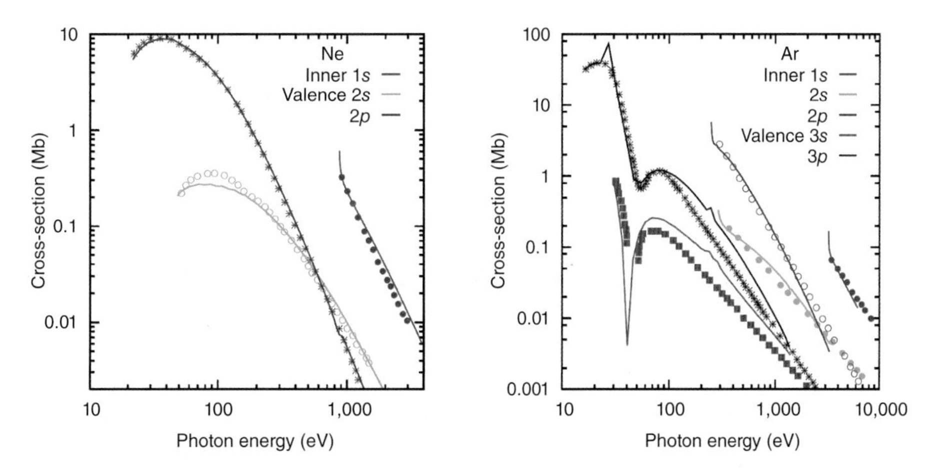

Figure 1.2 Subshell photoionization cross-sections of Ne and Ar. Symbols are from experiments and lines are from theoretical calculations (Figure adopted with permission from A. S. Kheifets et al., *Phys. Rev. A*, **92**, 063422 (2015) [5]. Copyrighted by the American Physical Society.)

be associated with one hole in the Ne^+ ion. At this elementary level, electron correlation is essentially neglected and subshell cross-sections can be reasonably calculated using the one-electron model potential. Note that ionization from the inner shell becomes dominant as the photon energy increases.

On the other hand, consider the subshell cross-sections of Ar as shown in the right panel of Figure 1.2. For $3p$ ionization, it is clear that there is a minimum in the cross-section, known as the Cooper minimum. It has a simple origin and can be predicted using the one-electron model. The minimum occurs at the energy when the dipole-matrix element changes sign. There is also a very pronounced minimum in the $3s$ subshell cross-section. This minimum has a different origin. It has been traced to intershell interaction and observed often only in the weak channel when coupled to a strong channel. While the electron can be removed directly from the $3s$ shell to the continuum, the final state can also be reached by a two-step process: first, the electron escapes from the $3p$, this electron then interacts with the $3s$ electron and knocks it into the $3p$ hole, resulting in a $3s$ hole, and a continuum electron emerges with the same kinetic energy as the one from the direct process from the 3s. The interference between these two processes results in a minimum in the $3s$ subshell cross-section.

In concluding this brief summary of atoms, it is important to emphasize that the shell model serves as a good starting point for describing what can happen in an experiment. However, deviation from this model is expected in general since interaction among the electrons is only treated approximately. Corrections to the simple shell model are generally attributed to electron-correlation effect.

1.4 General Concepts and Structure of Diatomic Molecules

Molecules are much more complicated than atoms. At first glance, the main difficulty stems from the fact that one has to take into account the motion of electrons and nuclei. Fortunately, due to their differences in mass, electronic and nuclear motions occur on very different time scales. This leads to an approximate separation of motions, each of which is also associated with a different energy scale (see Figure 1.3). Nuclear motion can be further separated into vibrational and rotational motions. Typical time and energy scales are summarized in Table 1.1.

1.4.1 The Born-Oppenheimer Approximation and Beyond

The Born-Oppenheimer (BO) approximation is the most important concept for understanding molecular structure. It provides a framework for separated treatments of electronic and nuclear degrees of freedom, and therefore simplifies the description of molecules. As discussed above, due to the different time and energy scales associated with various types of motion, one expects that the molecular wavefunction can be approximately written as a product of an electronic part and a nuclear part. The nuclei in a molecule are much slower than the electrons so it can be reasonably assumed that the nuclei are nearly fixed

Table 1.1 Typical time and energy scales associated with different types of motion in molecules

Type of motion	Time scale (fs)	Energy scale (eV)
Electronic	0.1	10
Vibrational	10	0.1
Rotational	1,000	0.001

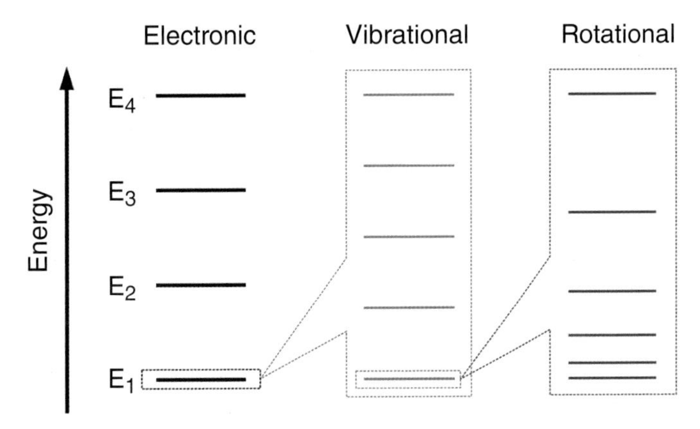

Figure 1.3 Schematic of molecular energy levels showing clear separation of electronic, vibrational, and rotational energies.

during the electrons' motion. In the BO approximation, the original task of solving the molecular-structure problem is broken into two steps. The first step solves for the electronic wavefunction that depends parametrically on the nuclear coordinates (see Equation 1.160). This task can be carried out for a range of nuclear coordinates to obtain the potential energy surface (PES). The second step solves for the nuclear motion along the PESs (see Equation 1.163).

Formal derivation and detailed discussion can be found in many textbooks on molecular quantum mechanics. Here, the main points are briefly summarized in order to introduce the concept, equations, and notations referred to in the rest of the book.

The Hamiltonian of a molecule can be written (in atomic units) as

$$\hat{H} = -\sum_{A} \frac{1}{2M_A}\nabla_A^2 - \frac{1}{2}\sum_{i}\nabla_i^2 - \sum_{A,i}\frac{Z_A}{r_{Ai}} + \sum_{i>j}\frac{1}{r_{ij}} + \sum_{A>B}\frac{Z_A Z_B}{R_{AB}}, \tag{1.157}$$

or, in short, as

$$\hat{H} = \hat{T}_N(\mathbf{R}) + \hat{T}_e(\mathbf{R}) + V_{eN}(\mathbf{r},\mathbf{R}) + V_{ee}(\mathbf{r}) + V_{NN}(\mathbf{R}), \tag{1.158}$$

where **R** (or **r**) is a shorthand notation for the set of nuclear (or electronic) coordinates, indexes A, B (or i, j) refer to nuclei (or electrons), and Z_A is the nuclear charge of nucleus A. Here, it is assumed that relativistic effect and spin-orbit interaction are negligible. Formally, the treatment here is only adequate for molecules with light atoms although extension to heavy molecules is straightforward. For fixed nuclei, the nuclear kinetic energy term goes to zero and the Hamiltonian is simplified to

$$\hat{H}_{el} = \hat{T}_e(\mathbf{R}) + V_{eN}(\mathbf{r}, \mathbf{R}) + V_{ee}(\mathbf{r}) + V_{NN}(\mathbf{R}). \tag{1.159}$$

In principle, the V_{NN} term in this electronic Hamiltonian can also be omitted and included in the nuclear Hamiltonian in the second step. To simplify the following equations, assume that this term is included in the equation above. For any fixed **R**, this term only gives a constant shift in electronic energies.

Assume that the first step can solve the eigenvalue problem for the electronic part

$$\hat{H}_{el}\Phi_k(\mathbf{r}; \mathbf{R}) = U_k\Phi_k(\mathbf{r}; \mathbf{R}), \tag{1.160}$$

where the nuclear coordinate **R** acts as a parameter. The wavefunction is determined up to a phase that, in most cases, can be chosen so the wavefunction is real. For any fixed **R**, the electronic wavefunctions form an orthonormal and complete basis set. The molecular wavefunction can be expanded in this basis set as

$$\Psi(\mathbf{r}, \mathbf{R}) = \sum_k \Phi_k(\mathbf{r}; \mathbf{R})\chi_k(\mathbf{R}). \tag{1.161}$$

Here, $\chi_k(\mathbf{R})$ is the nuclear wavefunction that serves as the expansion coefficients. After inserting the above equation into the Hamiltonian (Equation 1.157) and doing some straightforward mathematics, the following set of coupled equations is obtained:

$$\left[\hat{T}_N + T'_{kk} + T''_{kk} + U_k - E\right]\chi_k(\mathbf{R}) = -\sum_{k'\neq k}\left[T'_{kk'} + T''_{kk'}\right]\chi_{k'}(\mathbf{R}), \tag{1.162}$$

where

$$\begin{aligned}
T'_{kk'}(\mathbf{R}) &= \sum_A \frac{-1}{M_A}\mathcal{P}^{(A)}_{kk'}(\mathbf{R})\cdot\nabla_A \\
T''_{kk'}(\mathbf{R}) &= \sum_A \frac{-1}{2M_A}\mathcal{Q}^{(A)}_{kk'}(\mathbf{R}) \\
\mathcal{P}^{(A)}_{kk'}(\mathbf{R}) &= \langle\Phi_k(\mathbf{r}; \mathbf{R})|\nabla_A|\Phi_{k'}(\mathbf{r}; \mathbf{R})\rangle \\
\mathcal{Q}^{(A)}_{kk'}(\mathbf{R}) &= \langle\Phi_k(\mathbf{r}; \mathbf{R})|\nabla_A^2|\Phi_{k'}(\mathbf{r}; \mathbf{R})\rangle.
\end{aligned}$$

The brackets in the above equations denote integration over electronic coordinates **r**.

Equation 1.162 is still formally exact as long as the expansion in Equation 1.161 is infinite. The terms T' and T'' are called *nonadiabatic couplings*. In general, the second-derivative term T'' is weaker than the first derivative term T' so it can usually be neglected. Furthermore, after taking a derivative of $\langle\Phi_k(\mathbf{r}; \mathbf{R})|\Phi_k(\mathbf{r}; \mathbf{R})\rangle = 1$ with respect to nuclear coordinates, the diagonal term of T' vanishes if the electronic wavefunction is taken to be real. In certain cases, it is desirable to minimize the non-diagonal term $T'_{kk'}$ on the right-hand side of Equation 1.162 by using a diabatic basis set instead of an adiabatic basis set expansion as shown in Equation 1.161.

The couplings on the right-hand side of Equation 1.162 are quite small in most cases where state k is energetically well separated from other states k'. If they can be neglected, the BO-approximation equation for the nuclear motion on a single PES can be obtained as

$$\left[\hat{T}_N + U_k\right]\chi_k(\mathbf{R}) = E\chi_k(\mathbf{R}). \tag{1.163}$$

The BO approximation basically corresponds with the limit when only a single term of the expansion in Equation 1.161 is retained. Based on the small m/M ratio, this approximation is expected to be quite accurate.

Within the BO approximation framework, one first solves Equation 1.160 for electrons for a range of $\mathbf{R}$ to obtain the PES. In the second step, the PES is used in Equation 1.163, which one needs to solve with respect to nuclear coordinates $\mathbf{R}$. The nuclear motions are further divided into vibrational and rotational motions. The following sections address each of these steps.

There are certain cases where the BO approximation breaks down. This happens when the right-hand side in Equation 1.162 cannot be neglected. In fact, by using the Hellmann–Feynman theorem it can be shown that

$$\langle\Phi_k|\nabla\Phi_{k'}\rangle = \frac{\langle\Phi_k|\nabla\hat{H}_{el}|\Phi_{k'}\rangle}{U_{k'} - U_k} \quad \text{for} \quad k \neq k'. \tag{1.164}$$

Term $T'_{kk'}$ could be non-negligible when the electronic energy gap between state k and k' is small. Near a degeneracy of two PESs, a nuclear wave packet on one PES initially can spread to the other one without the emission of a photon, resulting in a radiationless transition. This typically happens near a conical intersection. In such cases, the treatment should be extended to include a few relevant terms in the expansion Equation 1.161. Instead of using an adiabatic basis, the practical treatment typically involves the use of a diabatic representation, as mentioned above.

1.4.2 Electronic Energy in Diatomic Molecules

HF Equation and Classification of Molecular Orbitals

Now, the solution of the electronic Equation 1.160 at fixed nuclei will be discussed. As with atoms, it is quite productive to start with the independent particle model. Within this approach, each electron in a molecule is assumed to move independently in a field created by other electrons and nuclei. As with atoms, the simplest electronic wavefunction (for both spatial and spin coordinates) can be written as a Slater determinant in order to satisfy the Pauli principle. The Slater determinant is constructed from single-electron wavefunctions, which are called *molecular orbitals* (MOs). This approximation to Equation 1.160 leads to the HF equation, which needs to be solved self-consistently to get the MOs. The available electrons fill these MOs according to the Aufbau principle. To go beyond the independent-particle model, one can add more Slater determinants to describe electrons more accurately and to account for the so-called *correlation energies*. This is typically done within the configuration interaction (CI) approach, MCHF method, or many-body perturbation theory.

Formally, everything up to this point still looks quite similar to atoms. However, in the case of molecules, the main difficulty in solving the HF equation stems from the molecules' multicenter nature, which makes the central-field approximation not applicable. To understand the complicated electronic structures and facilitate the calculations one therefore uses symmetry properties of the molecule. Here, only diatomic molecules that

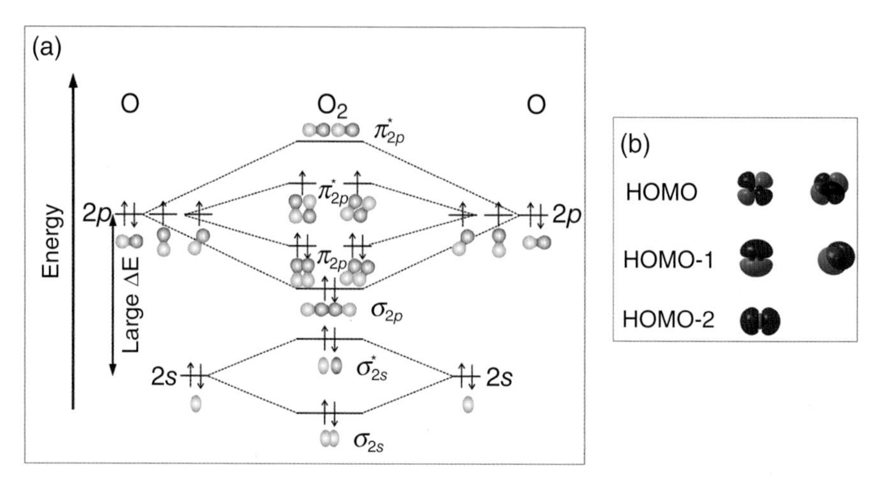

Figure 1.4 (a) Molecular orbitals diagram of O_2 schematically illustrating how MOs are formed from atomic orbitals. Up and down arrows indicate different spin states. (b) The MOs calculated with Gaussian quantum chemistry software are shown for the highest occupied molecular orbital (HOMO), HOMO-1, and HOMO-2.

possess cylindrical symmetry are considered. Polyatomic molecules will be treated in Section 1.4.3.

Due to cylindrical symmetry, the MOs of diatomic molecules can be classified according to the projection of the orbital-angular momentum on the molecular axis (taken as the symmetry axis). Namely, MOs are denoted by $\sigma, \pi, \delta, \ldots$, which corresponds with $|m| = 0, 1, 2, \ldots$, respectively. In homonuclear diatomic molecules, there is a center of symmetry at the center of mass. In such cases the MOs are further classified according to their parity into g (or *gerade*, which is German for even) or u (or *ungerade*, which is German for odd), e.g., $\sigma_g, \pi_u, \ldots$. It is also customary to use * to denote antibonding orbitals (see, for example, Figure 1.4).

MOs: Illustrative Example of O_2

Although the solution to the HF equation can be performed routinely with quantum chemistry software such as Gaussian, Gamess, Molpro, and many other packages, it is still instructive to see how MOs are formed based on a simple linear combination of atomic orbitals (LCAO). This is illustrated for O_2, as shown in Figure 1.4(a). Note the atomic orbitals of the two isolated oxygen atoms shown in the left and right sides of the figure. In this case, $2s$, $2p_x$, $2p_y$, and $2p_z$ merge together in specific ways to form MOs of different symmetries when they approach each other to form a chemical bond (shown in the middle part of the figure). Here, the different colors (red or blue) denote the different signs (positive or negative) of the MOs. Also note the different orientations of the orbitals. The MOs obtained from this heuristic approach compare nicely with more accurate results obtained using the Gaussian quantum chemistry package shown in Figure 1.4(b).

Molecular Terms

Due to cylindrical symmetry, the projection of the total orbital angular momentum of the molecule to the molecular axis will be a constant of motion. Much like atoms, the molecular term can then be classified as ${}^{2S+1}\Lambda^{(+/-)}_{\Omega,(g/u)}$, where $\Lambda = |M|$ is now the total orbital angular momentum of electrons about the molecular axis and S is the total electron spin. $2S+1$ is called *spin multiplicity*. The g/u subscript labels the symmetry of the total electronic wavefunction with respect to inversion through a center of symmetry and is only applicable to molecules with a center of symmetry. The $+/-$ superscript labels the symmetry of the wavefunction with respect to the reflection in a plane containing the nuclei. Similar to the notation for MOs, the term is denoted as $\Sigma, \Pi, \Delta, \ldots$ for $\Lambda = 0, 1, 2, \ldots$. When spin-orbit splitting is important, an extra subscript $\Omega = \Lambda + M_s$, the projection of total (orbital and spin) electronic angular momentum on the molecular axis, is added to the term symbol.

First, let us take N_2 as an example. This molecule has the ground-state configuration of $(1\sigma_g)^2(1\sigma_u^*)^2(2\sigma_g)^2(2\sigma_u^*)^2(1\pi_u)^4(3\sigma_g)^2$. Here, all the electrons are paired in closed shells so the total orbital momentum $\Lambda = 0$ and spin $S = 0$. Therefore the term symbol is ${}^1\Sigma_g^+$. However, much like atomic terms, one open-shell electronic configuration in a molecule gives rise to different molecular electronic terms. To illustrate this point, consider an O_2 molecule that has the ground-state configuration $(1\sigma_g)^2(1\sigma_u^*)^2(2\sigma_g)^2(2\sigma_u^*)^2(3\sigma_g)^2(1\pi_u)^4(1\pi_g^*)^2$ (see Figure 1.4). It is only necessary to consider partially filled shells since full shells are of ${}^1\Sigma^+$ symmetry (or ${}^1\Sigma_g^+$) as shown in the N_2 example. The two electrons can be filled in to the degenerate π_g^* orbitals as shown schematically in Figure 1.5(a–c). Total Λ and S can be easily understood from the figure.

Potential energy curves of O_2 are shown in Figure 1.6, where the three lowest curves can be identified as these three terms together with a few other excited states.

Apart from molecular-term symbol notations based on the angular momenta (or the group representation in general), alternative notations are used in the literature. For diatomic molecules, the ground state is labeled X. Electronic excited states are labeled in order of increasing energies as $A, B, C, \ldots$ if their total electronic spin is the same as the ground states, or as $a, b, c, \ldots$ otherwise. To avoid possible confusion with symmetry labels based on group representation (see, for example, Table 1.4), a tilde is usually added to those labels (e.g., $\tilde{A}, \tilde{a}$) for polyatomic molecules. The use of both kinds of notations can be seen in the example for O_2 in Figure 1.6.

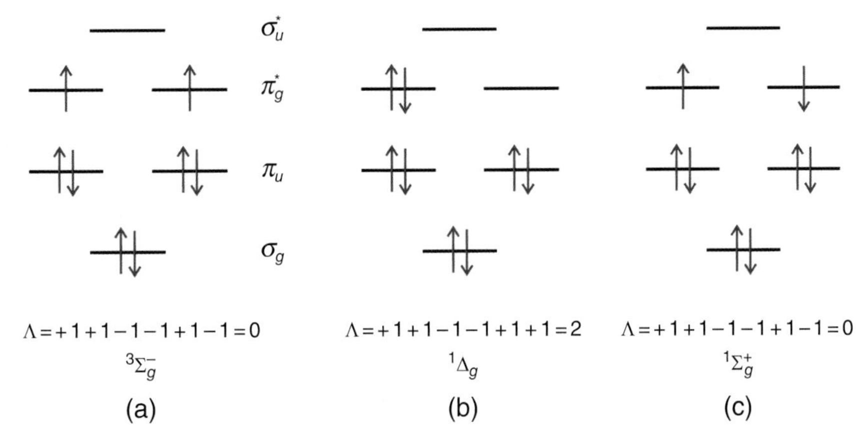

Figure 1.5 Schematic of MO energies and the three lowest terms in O_2. Electron occupations from HOMO-2 and higher are shown, although one only needs to take care of the two π_g^* electrons.

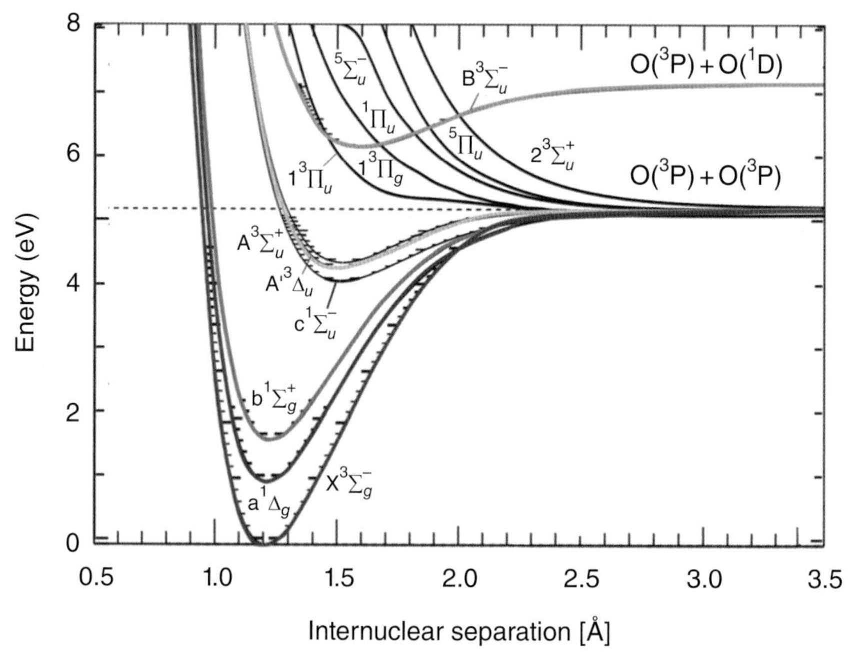

Figure 1.6 O_2 potential energy curves. (Adopted from Farooq Zahid et al., *Phys. Chem. Chem. Phys.* **16**, 3305 (2014) [6]. Copyrighted by Royal Society of Chemistry.)

1.4.3 Nuclear Motion in Diatomic Molecules

Within the BO approximation, the next step is to solve the nuclear motion in Equation 1.163. After separating the center of mass motion, Equation 1.163 can be rewritten for a diatomic molecule in a specific electronic state n as

$$\left[-\frac{1}{2\mu}\nabla_R^2 + U_n(R)\right]\chi^{(n)}(\mathbf{R}) = E^{(n)}\chi^{(n)}(\mathbf{R})$$

or

$$\left[-\frac{1}{2\mu}\left(\frac{\partial^2}{\partial R^2} + \frac{2}{R}\frac{\partial}{\partial R}\right) + \frac{1}{2\mu R^2}\hat{\mathbf{J}}^2 + U_n(R) - E^{(n)}\right]\chi^{(n)}(\mathbf{R}) = 0. \tag{1.165}$$

Here, $\mathbf{R}$ is the internuclear vector, μ is the reduced mass, and $\hat{\mathbf{J}}$ is the nuclear-rotational, angular-momentum operator

$$\hat{\mathbf{J}}^2 = -\frac{1}{\sin\theta}\frac{\partial}{\partial\theta}\sin\theta\frac{\partial}{\partial\theta} - \frac{1}{\sin^2\theta}\frac{\partial^2}{\partial\phi^2},$$

where $\{\theta, \phi\}$ gives the orientation of $\mathbf{R}$ in space. The motion described by the equation above can be separated into two independent motions: one is rotational motion in angles θ and ϕ and the other is vibrational motion in R such as

$$\chi^{(n)}(R,\theta,\phi) = \frac{F(R)}{R}Y_{JM}(\theta,\phi). \tag{1.166}$$

Rotational Energy

The rotational motion is described by the Y_{JM} functions. These functions are the spherical harmonics that satisfy the following equations:

$$\hat{\mathbf{J}}^2 Y_{JM}(\theta,\phi) = J(J+1)Y_{JM}(\theta,\phi),$$
$$\hat{J}_z Y_{JM}(\theta,\phi) = MY_{JM}(\theta,\phi).$$

Therefore, the associated rotational energy is

$$E_{\text{rot}}(J) = \frac{1}{2\mu R_e^2}J(J+1) = B_e J(J+1). \tag{1.167}$$

Here, the internuclear distance is fixed at the equilibrium (the *rigid rotor approximation*). B_e is called the *rotational constant.*

The above discussion is strictly valid only for molecules with electrons in closed shells when the total electronic angular momentum is zero. For open-shell molecules, the electronic angular momentum is coupled with the nuclear part. Such situations are described by Hund coupling cases (a–d). This complication will not be addressed in this book.

Vibrational Energy

For each rotational state with quantum number J, the equation for vibrational motion can be written as

$$\left[-\frac{1}{2\mu}\frac{\partial^2}{\partial R^2} + \frac{1}{2\mu R^2}J(J+1) + U_n(R)\right]F(R) = E^{(n)}F(R). \tag{1.168}$$

In principle, this one-dimensional (1D) equation can be solved numerically to obtain vibration–rotation energies. Nevertheless, vibrations typically occur near equilibrium, so that further simplification can be made. In fact, one can expand the potential energy term $U_n(R)$ up to the second order around the equilibrium $R = R_e + \rho$ to get

$$\left[-\frac{1}{2\mu}\frac{\partial^2}{\partial \rho^2} + \frac{1}{2}k\rho^2\right]F(\rho) = \left[E^{(n)} - U_n(R_e) - E_{\text{rot}}(J)\right]F(\rho). \tag{1.169}$$

Here, the wavefunction is now written as a function of small deviation ρ from R_e. The rotational energy is also approximated with the rigid rotor approximation (see "Rotational Energy"). Equation 1.169 is for the quantum harmonic oscillator, which can be readily solved as

$$\left[-\frac{1}{2\mu}\frac{\partial^2}{\partial \rho^2} + \frac{1}{2}k\rho^2\right]F(\rho) = \omega\left(v + \frac{1}{2}\right)F(\rho), \tag{1.170}$$

with $\omega = \sqrt{k/\mu}$ and $v = 0, 1, 2, \ldots$. After comparing the right-hand side of the last two equations, it becomes clear that the total energy is written as a sum of electronic, vibrational, and rotational energies

$$E^{(n)} = U_n(R_e) + \omega\left(v + \frac{1}{2}\right) + B_e J(J+1). \tag{1.171}$$

The harmonic oscillator approximation is valid only for small-amplitude vibrations. Clearly, the harmonic oscillator is not adequate for large v. Indeed, molecules do not support an infinite number of vibrational states as in quantum harmonic oscillators – they eventually dissociate into two separate atoms at large distance R. To have a slightly more realistic description for vibrational states, a Morse potential is often used in order to include the anharmonicity effect

$$V(R) = D_e \left(1 - e^{-\beta(R-R_e)}\right)^2, \tag{1.172}$$

where $V(R_e) = 0$ and $V(R \to \infty) = D_e$ with D_e being the dissociation energy. For a more detailed discussion, the interested reader is referred to standard textbooks such as Bransden and Joachain [Bransden, Joachain].

1.5 Structure of Polyatomic Molecules

1.5.1 Brief Introduction to Molecular Symmetry

General Remarks

In general, symmetry simplifies the description significantly. It is fair to say that understanding polyatomic molecules is nearly impossible without a basic knowledge of molecular symmetry groups. Fortunately, most of the concepts in the symmetry groups in application to molecular physics are quite intuitive. Therefore, the following sections introduce and illustrate the basic concepts with a few, brief examples.

More specifically, in molecular physics, symmetry is used to

- classify electronic, vibrational wavefunctions;
- predict and understand some physical properties (e.g., dipole moment) and selection rules;
- construct (although current quantum chemistry software can handle this easily) and understand MOs.

Molecular Symmetry Operations

Basic symmetry elements, operations, and standard notations are given in Table 1.2. Some illustrations of these symmetry elements are also shown in Figure 1.7.

Molecular Symmetry Group and Group Multiplication

The complete set of symmetry operations of a molecule forms a group in the abstract mathematical sense. This book will not go into detail about the abstract properties of group theory. Instead, it accepts that every molecule belongs to a certain symmetry group (or a

Table 1.2 Symmetry operations and notations

Element	Operation	Symbol
Identity	Identity	E
Symmetry plane	Reflection of the plane	σ
Inversion center	Inversion of a point x, y, z to $-x, -y, -z$	i
Proper axis	Rotation by $360°/n$	C_n
Improper axis	Rotation by $360°/n$ followed by reflection in plane perp. to rotation axis	S_n

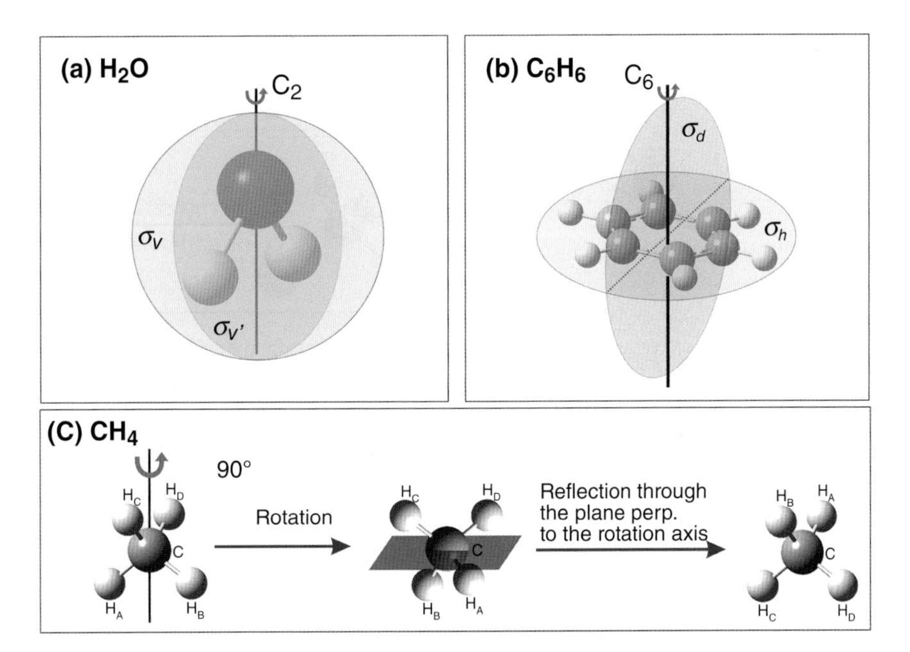

Figure 1.7 (a) H_2O: C_2 axis together with two vertical planes containing the principal axis, σ_v and $\sigma_{v'}$. (b) Benzene (C_6H_6): C_6 symmetry axis together with σ_h (*h*: horizontal plane) and σ_d (*d*: dyhedral plane). (c) Methane (CH_4): Rotation- reflection or improper rotation S_n.

molecular point group). A molecule belongs to a symmetry point group if it is unchanged under all the symmetry operations of this group.

As the multiplication of symmetry operations is at the heart of group theory, this concept is illustrated for CH_2Cl_2 in Figure 1.8. A product of two operations $A \times B$ (or simply AB) is defined as the consecutive actions of B and A. As shown in the figure, the product of two elements $\sigma(yz)\sigma(xz)$ is equal to C_2 and $\sigma(xz)C_2 = \sigma(yz)$. Other products can be found in the same fashion. Clearly, a product of two operations is another operation. This is summarized in Table 1.3, which is a multiplication table. In this particular case, four operations $E, C_2, \sigma(xz)$, and $\sigma(yz)$ form a point group called the C_{2v} group. H_2O, as illustrated in Figure 1.7(a), also belongs to the same C_{2v} symmetry group. The elements $\sigma(xz)$ and $\sigma(yz)$ are denoted by σ_v and σ_v' in Figure 1.7(a).

Table 1.3 Multiplication table for C_{2v} group

C_{2v}	E	C_2	$\sigma(xz)$	$\sigma(yz)$
E	E	C_2	$\sigma(xz)$	$\sigma(yz)$
C_2	C_2	E	$\sigma(yz)$	$\sigma(xz)$
$\sigma(xz)$	$\sigma(xz)$	$\sigma(yz)$	E	C_2
$\sigma(yz)$	$\sigma(yz)$	$\sigma(xz)$	C_2	E

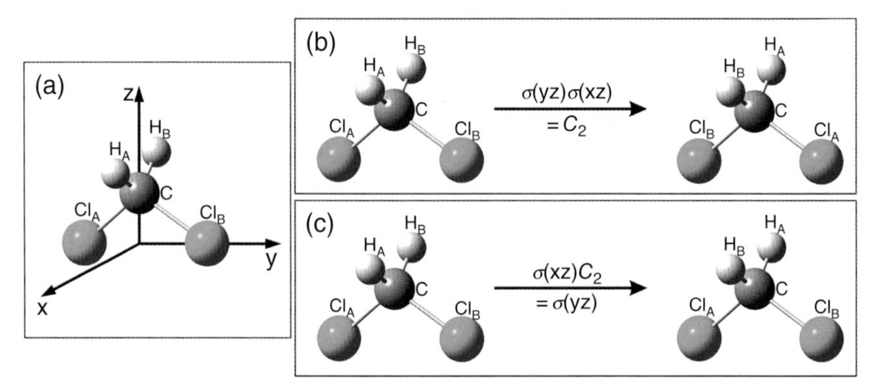

Figure 1.8 (a) CH_2Cl_2 molecule and coordinates in standard orientation. (b) The product of $\sigma(yz)$ and $\sigma(xz)$ is shown to be the same as the C_2 operation. (c) The product of $\sigma(xz)$ and C_2 is shown to be the same as $\sigma(yz)$.

Mulliken Notations and Symbols

The symmetry notations and labels used in textbooks and literature can be somewhat confusing due to different conventions and/or molecular orientations. To avoid ambiguity, this book follows the standard conventions given by Mulliken [7], the main elements of which are:

- lowercase letters are used for orbitals and normal vibrational modes, e.g., a_g;
- uppercase letters are used for electronic states (terms), vibrational and vibronic, e.g., A_g, of the whole molecule;
- character symbols and their definitions are used exactly as in the textbook by Herzberg [Herzberg, 1945] with the exception that the T (and t) symbol is used instead of F (and f) for triply degenerate states.

Mulliken notations for the irreducible representation (or simply "irrep") can be summarized in Table 1.4 and Table 1.5. According to the above conventions, all of these tables are valid if the uppercase letters for the Mulliken symbols are replaced by lowercase letters.

Character Tables

To illustrate the character table, the C_{2v} symmetry group is considered in Figure 1.9. Here, C_{2v} is the Schoenflies symbol for the symmetry group. Note that, although the

Table 1.4 Main symbol and dimension of the irreducible representations

Symbol	Dimension
A,B	1
E	2
T	3

Table 1.5 Subscripts and superscripts indices, reflecting additional classifications

	Rotation C_n	$\sigma_v(\sigma_d)$ or C_2 perp. to C_n	σ perp. to C_n or σ_h	Inversion
Symmetric	A	1	$'$	g
Antisymmetric	B	2	$''$	u

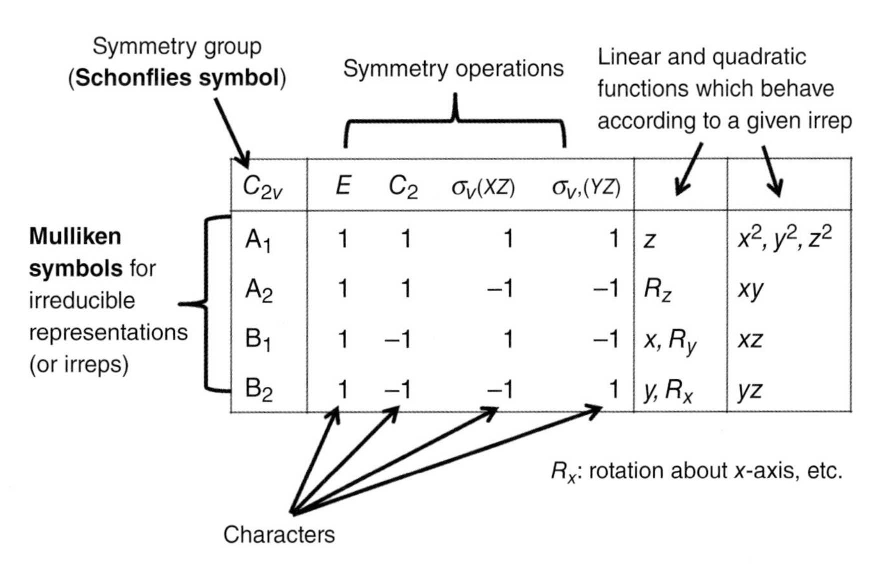

Figure 1.9 Character table for the C_{2v} symmetry group.

Mulliken symbols here (left-most column) are written with uppercase letters for the whole molecular terms, they are applicable for lowercase letters. The next four columns are for characters of the four symmetry operations of this group in their irreducible representations (to be denoted as irreps). The two right-most columns contain linear and quadratic-basis functions, respectively, for the irreps. Here, R_α (with $\alpha = x, y, z$) stands for the rotation about the axis α. These last two columns show how the dipole moment and dipole polarizability tensor transform with respect to the symmetry operations.

The condensed character table for the C_{3v} symmetry group in which two elements of C_3 class and three elements of σ_v class are combined together is also shown in Table 1.6.

Table 1.6 Character table for C_{3v} symmetry group

C_{3v}	E	$2C_3$	$3\sigma_v$		
A_1	1	1	1	z	x^2+y^2, z^2
A_2	1	1	-1	R_z	
E	2	-1	0	$(x,y),(R_x,R_y)$	$(x^2-y^2,xy)(xz,yz)$

Table 1.7 Product of A_2 and B_1 of symmetry group C_{2v}

C_{2v}	E	C_2	$\sigma_v(xz)$	$\sigma_v'(yz)$
A_2	1	1	-1	-1
B_1	1	-1	1	-1
$A_2 \times B_1$	1	-1	-1	1

Table 1.8 Product tables for C_{2v} group representation

C_{2v}	A_1	A_2	B_1	B_2
A_1	A_1	A_2	B_1	B_2
A_2	A_2	A_1	B_2	B_1
B_1	B_1	B_2	A_1	A_2
B_2	B_2	B_1	A_2	A_1

Character tables for other molecular symmetry groups can be found in standard textbooks on molecular symmetry (see, for example, appendix A in [Harris, Bertolucci]).

Product Table of Representation

The product rules for symmetry species are very useful when determining the molecular terms and characterizing the transition-dipole moments. If some molecular property A is a product of other properties B and C, the character A is a product of B and C characters, and thus may be determined from the character product table. For example, consider the direct product of A_2 and B_1 representations in the C_{2v} group. The last row of Table 1.7 is obtained by multiplying the characters in the first two rows. By comparing the last row with the character table for C_{2v} (see Figure 1.9), one confirms that $A_2 \times B_1 = B_2$. Other products can be found in the same manner. For future reference, these products are conveniently summarized in Table 1.8. Similar tables can be found for other symmetry groups in the standard textbooks on molecular symmetry (see for example, appendix B in [Harris, Bertolucci]).

There is a simpler way to obtain the table above by applying the following rules:

$$S \otimes S = S, A \otimes A = S, S \otimes A = A, A \otimes S = A \tag{1.173}$$

with S standing for symmetric and A for antisymmetric. These rules can be expanded as given in Table 1.9.

Table 1.9 General product rules for representations

$A\&B$	$A \otimes A = A$	$B \otimes B = A$	$A \otimes B = B \otimes A = B$
1&2 subscripts	$1 \otimes 1 = 1$	$2 \otimes 2 = 1$	$1 \otimes 2 = 2 \otimes 1 = 2$
$g\&u$ subscripts	$g \otimes g = g$	$u \otimes u = g$	$g \otimes u = u \otimes g = u$
$'$ and $''$	$(') \otimes (') = (')$	$('') \otimes ('') = (')$	$(') \otimes ('') = ('') \otimes (') = ('')$

Irreducible and Reducible Representations

Irreducible representations in a character table form a set of orthogonal vectors that span the complete space, such as

$$\sum_R \chi^{(i)}(R) \cdot \chi^{(j)}(R) = h\delta_{ij}. \tag{1.174}$$

Here, h is the order of the group (the total number of its elements) and the sum is carried out over all elements. Any reducible representation can be written as a linear combination of irreducible representations as follows:

$$\Gamma^{\text{red}} = \sum_k c_k^{\text{red}} \Gamma^k, \tag{1.175}$$

where

$$c_k^{\text{red}} = \frac{1}{h} \sum_R \chi^{\text{red}}(R)^* \chi^k(R). \tag{1.176}$$

The use of symmetry groups will be illustrated in Sections 1.5.2 and 1.6.2.

1.5.2 Electronic Energy

In comparison to diatomic molecules, the degree of complexity in the description of polyatomic molecules increases greatly due to the presence of many atomic centers. The point-group symmetry summarized in the previous section can be used to simplify the solutions of Equation 1.160 and the classification of molecular orbitals. The MOs of polyatomic molecules can be classified according to their point-group symmetry. Many older textbooks devote much time to the description of how appropriate MOs can be formed according to the symmetry of the molecule. Today, MOs can be calculated routinely in practice with modern quantum chemistry software packages. Therefore, it is important to take a more pragmatic approach and try to understand the implications of the calculations already carried out by any of the quantum chemistry packages.

As an illustration, the output from Gaussian G03 for H_2O obtained within the HF approximation for Equation 1.160 is analyzed. In Gaussian, MOs are constructed with Gaussian-type orbitals (GTOs) centered at each atom. For simplicity, the minimum basis set, STO-3G, is used. A part of the Gaussian G03 output shown in Figure 1.10 has been annotated with comments to make the succeeding points more clear.

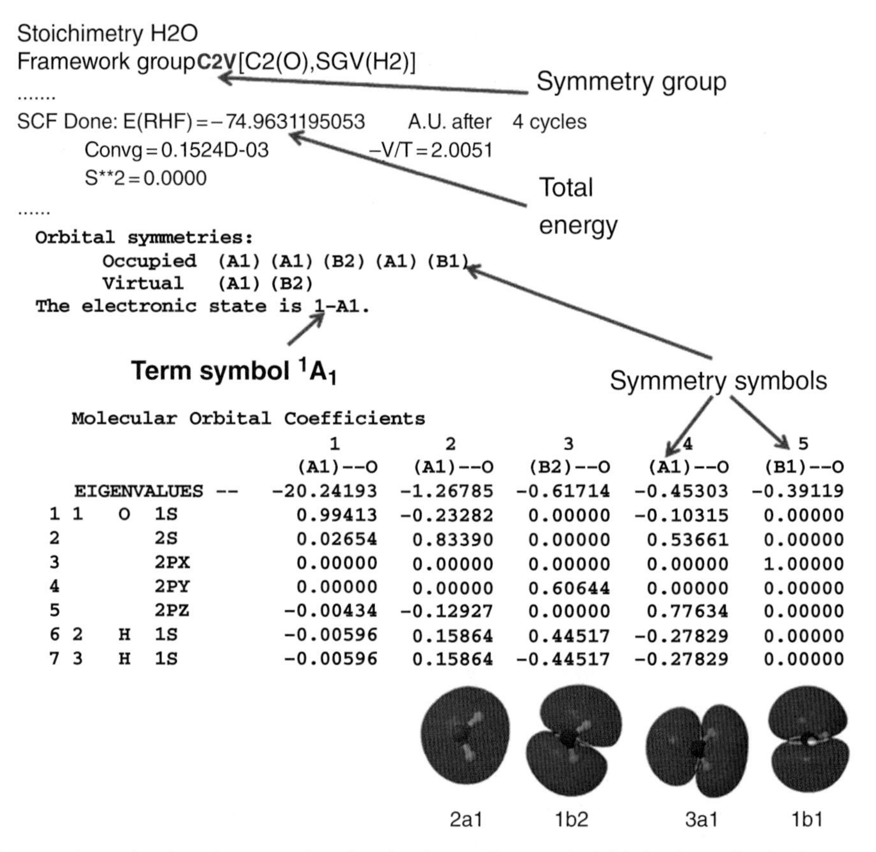

				1	2	3	4	5
				(A1)--O	(A1)--O	(B2)--O	(A1)--O	(B1)--O
EIGENVALUES --				-20.24193	-1.26785	-0.61714	-0.45303	-0.39119
1	1	O	1S	0.99413	-0.23282	0.00000	-0.10315	0.00000
2			2S	0.02654	0.83390	0.00000	0.53661	0.00000
3			2PX	0.00000	0.00000	0.00000	0.00000	1.00000
4			2PY	0.00000	0.00000	0.60644	0.00000	0.00000
5			2PZ	-0.00434	-0.12927	0.00000	0.77634	0.00000
6	2	H	1S	-0.00596	0.15864	0.44517	-0.27829	0.00000
7	3	H	1S	-0.00596	0.15864	-0.44517	-0.27829	0.00000

Figure 1.10 (a) Annotated text taken from G03 output for H_2O within the HF/STO-3G method. (b) The shape of each MO.

First, notice that G03 has correctly identified C_{2v} as the symmetry group. The total electronic energy $E_{HF} = -74.963$ au (within the HF approximation) is also given. The MOs in terms of the basis-set expansion at each atom are shown at the bottom of Figure 1.10. In the minimum STO-3G basis set, the oxygen atom is represented by five orbitals, $1s$, $2s$, $2p_x$, $2p_y$, and $2p_z$, while each hydrogen atom is represented by a $1s$ orbital (see left-most column).

There are a total of 10 electrons. They occupy five MOs (each with a spin-up and a spin-down electron). These electrons are shown in five columns labeled by $1, 2, \ldots, 5$. Only the occupied MOs (indicated by O next to the symmetry symbol in each column) are shown. Here, each MO is already classified according to its symmetry property as A_1, B_2, or B_1 of C_{2v} group (see character table for C_{2v} in Table 1.10). According to standard convention, these symbols should be written with lowercase letters a_1, b_2, and b_1. The HF energy (in au here) of each MO is also given in the line below the symmetry symbol and indicated by the "EIGENVALUES" keyword.

For each MO, the expansion coefficients in the basis set are given in the corresponding column. For example, the HOMO is given by column 5 with symmetry symbol B_1. This MO is formed entirely from the $O(2p_x)$ orbital (with a coefficient of 1 in the column); thus,

Table 1.10 Character table for C_{2v}

C_{2v}	E	C_2	$\sigma_v(xz)$	$\sigma_{v'}(yz)$		
A_1	1	1	1	1	z	x^2, y^2, z^2
A_2	1	1	-1	-1	R_z	xy
B_1	1	-1	1	-1	x, R_y	xz
B_2	1	-1	-1	1	y, R_x	yz

it is a nonbonding orbital. This MO shape is shown in Figure 1.10(b). In addition, checking that this MO behaves in accordance with B_1 symmetry is easy. For example, C_2 rotation around the molecular axis (along the z-axis) would change the sign of this MO, and thus its character is -1 for C_2. Reflection on the $\sigma_v(xz)$ plane would leave this MO unchanged (i.e., its character is 1), while reflection on the $\sigma_{v'}(yz)$ plane would change the sign of this MO (i.e., its character is -1). Comparison to the character table confirms that all these behaviors are in accordance with B_1.

Similar analysis can be done for other MOs. For example, the HOMO-1 (column 4) has a totally symmetric A_1 character, and its shape remains unchanged with respect to all symmetry operations of C_{2v}. This can be easily understood because, according to the expansion coefficients of column 4, this MO is constructed from a symmetric combination of H($1s$) and O($2p_z$) as well as O($2s$), and a little contribution from O($1s$).

The above results from G03 can be interpreted in the following way. Inspecting the central atom shows that O($2p_x$) and O($2p_y$) belong to B_1 and B_2, respectively, while O($1s$), O($2s$), and O($2p_z$) belong to totally symmetric A_1. Each of the non-central hydrogen atoms provides a $1s$ orbital, and together they can either form a symmetric combination (which behaves like A_1) or an antisymmetric combination (which behaves like B_2). Each of these combinations mixes with the oxygen orbital of the same symmetry to form an MO. Note that there is no combination of $1s$ orbitals from two hydrogen atoms that behaves like B_1; thus, the HOMO (B_1 symmetry) is entirely formed from the O($2p_x$) orbital. For more complex molecules, the construction of the symmetry-adapted LCAO can be conveniently carried out by using projection operators within each symmetry group.

Molecular-term symbols for polyatomic molecules are written in a similar form to linear molecules but with Λ replaced by the corresponding symmetry symbol of the molecular symmetry group. For H_2O, G03 has identified the term as $^1A_1$, i.e., singlet A_1 (see Figure 1.10). Note that for H_2O in the ground electronic state each MO is filled by two electrons (with spin-up and spin-down), and so, in agreement with the rule for molecules with closed shells, it has totally symmetric A_1 symmetry.

Equation 1.160 has to be solved for different geometries. In general, there are $3N - 6$ vibrational degrees of freedom so the PES will be a 3D hyper-surface in the case of H_2O in the ground electronic state. Similar PESs exist for excited electronic states. Once a PES is found, the solution of the nuclear motions of Equation 1.163 can be taken. Sections 1.5.3–1.5.4 will be devoted to the solution of Equation 1.163 for polyatomic molecules.

1.5.3 Rotational Motion

It is desirable to separate the rotational motion of the molecule as a whole from Equation 1.163. In the first approximation, the rigid-rotor model is assumed. With a higher-level theory, one also can account for centrifugal distortion, vibration-rotational coupling, and electronic-angular momentum as treated by Hund's cases. Fortunately, such levels of accuracy are not needed in strong-field physics and they will be avoided in the following. The Hamiltonian for molecular rotation as a rigid rotor is written in the body-fixed frame as

$$\hat{H}_{\text{rot}} = \frac{\hat{J}_a^2}{2I_a} + \frac{\hat{J}_b^2}{2I_b} + \frac{\hat{J}_c^2}{2I_c}, \tag{1.177}$$

where I_a, I_b, and I_c are the moments of inertia of the molecules with respect to the principal axes a, b, and c, and $\hat{J}_a$, $\hat{J}_b$, and $\hat{J}_c$ are the corresponding angular momentum components. Note that, in general, the moment of inertia is a second-rank tensor with nine components. Nevertheless, with a proper choice of principal axes, this tensor becomes diagonal, as chosen, for example, in Equation 1.177. The principal axes are normally labeled in such a way that

$$I_a \leq I_b \leq I_c. \tag{1.178}$$

The rotational constants A, B, and C are defined as

$$A = \frac{\hbar^2}{2I_a}, \tag{1.179}$$

and similar equations hold for B and C. With the convention above, one has $A \geq B \geq C$.

To describe the rotation of the molecule, the space-fixed-coordinates frame XYZ and the body-fixed-coordinates frame abc will be distinguished. These two frames are related by a rotation characterized by Euler angles $(\alpha\beta\gamma)$. The direction of the angular momentum can be specified by two projections. The eigenstates are $|JKM\rangle$, which are Wigner D-matrix $D^J_{MK}(\alpha\beta\gamma)$. The meaning of each of the quantum numbers is:

- J as the angular quantum number;
- M (which takes the values as $-J, -J+1, \ldots, J$) as the quantum number associated with the projection of J on the space-fixed Z-axis;
- K (which takes the values as $-J, -J+1, \ldots, J$) is similar to M, but for the projection on the body-fixed frame (normally chosen as the symmetry axis, for example, c-axis).

Spherical Top Molecules (with $I_a = I_b = I_c$)

Examples of spherical top molecules are CH_4 or SF_6. The Hamiltonian is as simple as

$$\hat{H}_{\text{rot}} = \frac{\hat{J}^2}{2I} \tag{1.180}$$

for which the solution can be written as

$$\hat{H}_{\text{rot}}\Psi_{JKM} = E_J\Psi_{JKM} \tag{1.181}$$

with the rotational energies

$$E_J = \frac{\hbar^2}{2I} J(J+1) = BJ(J+1) \tag{1.182}$$

where B is the rotational constant. Equation 1.182 is quite similar to linear molecules. However, since the energy does not depend on M and K, there is a degeneracy factor $g = (2J+1)^2$ for each J instead of $2J+1$ as in linear molecules.

Oblate Symmetric Top Molecules ("Disc" Shape, with $I_a = I_b < I_c$)

Examples of oblate symmetric top molecules are benzene C_6H_6 or ammonia NH_3. For convenience $a \to x$, $b \to y$ and $c \to z$ were chosen. By rearranging the Hamiltonian it can be shown that

$$\begin{aligned} \hat{H}_{\text{rot}} \Psi_{JKM} &= \left[\frac{1}{2I_b} (\hat{J}_x^2 + \hat{J}_y^2) + \frac{1}{2I_c} \hat{J}_z^2 \right] \Psi_{JKM} \\ &= \left[\frac{1}{2I_b} (\hat{J}^2 - \hat{J}_z^2) + \frac{1}{2I_c} \hat{J}_z^2 \right] \Psi_{JKM} \\ &= \left\{ BJ(J+1) + (C-B)K^2 \right\} \Psi_{JKM}. \end{aligned} \tag{1.183}$$

Here, $B > C$ (since $I_b < I_c$) are the rotational constants. Clearly, the rotational energy not only depends on J, but also on $|K|$. Since $B > C$, a state with larger $|K|$ has a lower energy than a state with smaller $|K|$ if both have the same J. For $K = 0$, the degeneracy factor is $g = 2J + 1$ due to the degeneracy in M. For $K \neq 0$, due to additional degeneracy of $\pm K$, the degeneracy factor $g = 2(2J+1)$.

Prolate Symmetric Top Molecules ("Cigar" Shape, with $I_a < I_b = I_c$)

An example of such a molecule is CH_3I. For convenience $a \to x$, $b \to y$ and $c \to z$ were chosen. By proceeding similarly to the case above, one gets

$$\begin{aligned} \hat{H}_{\text{rot}} \Psi_{JKM} &= \left[\frac{1}{2I_b} (\hat{J}^2 - \hat{J}_z^2) + \frac{1}{2I_c} \hat{J}_z^2 \right] \Psi_{JKM} \\ &= \left\{ BJ(J+1) + (A-B)K^2 \right\} \Psi_{JKM}. \end{aligned} \tag{1.184}$$

Here, $A > B$ (since $I_a < I_b$) are the rotational constants. Since $A > B$, a state with larger $|K|$ has a higher energy than a state with smaller $|K|$ if both have the same J. The degeneracy factor is similar to the oblate symmetric top.

Linear Molecules (with $I_a = 0 < I_b = I_c$)

Linear molecules are a special case of prolate symmetric top for which $K = 0$. This case is the same for diatomic molecules. Since the energy does not depend on M (which can take any value as $-J, -J+1, \ldots, J$), the degeneracy is $g = 2J + 1$.

Asymmetric Top Molecules (with $I_a < I_b < I_c$)

Most polyatomic molecules belong to asymmetric top molecules. In this case, while J and M are still good quantum numbers, K is not well defined since there is no longer a symmetry axis. Therefore, $|JKM\rangle$ are not the eigenfunctions of the general rotational Hamiltonian. However, they can still serve as the basis set through which one can express the rotational wavefunction as

$$\Psi_{J\tau M} = \sum_{K=-J}^{J} c_{\tau,K} |JKM\rangle. \tag{1.185}$$

The rotational energies of an asymmetric top and its wavefunction can be obtained by direct matrix diagonalization for each fixed J. Note that for a qualitative understanding of an asymmetric top molecule, one can think of it as an intermediate case between two extreme cases of oblate and prolate symmetric tops.

1.5.4 Vibrational Motion

In principle, Equation 1.163 has to be solved numerically for nuclear motions. However, if only the vibrational motions near the nuclear equilibrium geometry are of interest, the analysis can be simplified significantly. The description of vibrations is quite similar to that of diatomic molecules. However, now $f = 3N - 6$ (or $3N - 5$ for linear molecules) nuclear degrees of freedom must be dealt with instead of $f = 1$ (as in diatomic molecules). Therefore the notations will be more complicated.

Assume one must solve Equation 1.163 for which the nuclear Hamiltonian is written in the following form:

$$\hat{H} = \sum_{i=1}^{3N} \frac{\hat{p}_i^2}{2m_i} + U(x_1, \ldots, z_N), \tag{1.186}$$

where U is the PES of the electronic state of interest. Considering small oscillations about the nuclear equilibrium, one can introduce mass-weighted displacements defined as

$$q_i = \sqrt{m_i}\Delta x_i = \sqrt{m_i}(x_i - x_i^e). \tag{1.187}$$

The PES can be expanded in a Taylor series near equilibrium as

$$U = U_0 + \sum_{i=1}^{3N} \left.\frac{\partial U}{\partial q_i}\right|_e q_i + \frac{1}{2}\sum_{i=1}^{3N}\sum_{j\leq i}^{3N} \left.\frac{\partial^2 U}{\partial q_i \partial q_j}\right|_e q_i q_j + \cdots, \tag{1.188}$$

where the second term vanishes at the equilibrium while the first term is a constant that can be set to $U_0 = 0$ as a reference energy for convenience. The index e in Equations 1.187 and 1.188 indicates that the expressions are taken at equilibrium. The Hamiltonian then becomes a quadratic form in $\{q\}$. To further simplify, one uses normal coordinates as appropriate linear combinations of displacements. This is shown in the following:

$$Q_j = \sum_{i=1}^{3N} C_{ij} q_i \qquad j = 1, 2, \ldots, 3N - 6. \tag{1.189}$$

Here, it is assumed that the coordinates corresponding to overall translation and rotation motions have been separated; thus, there are only $3N-6$ (or, for linear molecules, $3N-5$) normal coordinates. The normal coordinates are defined with a suitable choice of C_{ij} such that the Hamiltonian is written as a sum of Hamiltonians for $3N-6$ independent harmonic oscillators as

$$\hat{H}=\sum_{i=1}^{3N-6}\left(\frac{1}{2}\dot{Q}_i^2+\frac{1}{2}k_iQ_i^2\right). \tag{1.190}$$

The solution of the Schrödinger equation for vibrational motions can then be readily written as a product of wavefunctions for the $3N-6$ harmonic oscillators as

$$\Psi_{v_1,v_2,\ldots,v_{3N-6}}=\Psi_{v_1}(Q_1)\Psi_{v_2}(Q_2)\ldots\Psi_{v_{3N-6}}(Q_{3N-6}). \tag{1.191}$$

Therefore, energy is the sum of vibrational energy due to $3N-6$ harmonic oscillators as

$$E_{v_1,v_2,\ldots,v_{3N-6}}=\sum_{i=1}^{3N-6}\omega_i\left(v_i+\frac{1}{2}\right). \tag{1.192}$$

As in the case of diatomic molecules, extension can be done to include higher-order terms in the Taylor expansion of the PES near equilibrium. This would lead to an anharmonic-oscillator model for vibrations in polyatomic molecules. The Morse potential, in particular, can also be used instead of the harmonic oscillator model.

1.6 Molecular Spectra

1.6.1 Spectra of Diatomic Molecules

This section provides a brief overview of the basic concepts in radiative transitions that happen when a diatomic molecule absorbs or emits a photon. These weak-field (one-photon) radiative transitions lie at the heart of the traditional molecular spectroscopy and are still relevant to strong-field phenomena.

Vibrational–Rotational Transitions

This subsection considers the transitions when both the initial and final states belong to the same electronic state. In such cases, the transition intensity within BO approximation can be written as

$$I\propto|\langle\psi_{\mathrm{v}}'|\mu(Q)|\psi_{\mathrm{v}}''\rangle|^2, \tag{1.193}$$

where $\mu(Q)=\langle\psi_e|\hat{\mu}|\psi_e\rangle$ is the molecular dipole moment of the given electronic state. For diatomic molecules, the internuclear separation R plays the role of the only normal coordinate Q. For homonuclear diatomic molecules, the dipole moment vanishes due to the symmetry. Therefore, these molecules do not have dipole-allowed vibrational–rotational spectra. For heteronuclear diatomic molecules, one can expand the dipole moment near equilibrium R_e in a Taylor series as

$$\mu(R) = \mu(R_e) + \left.\frac{\partial \mu}{\partial R}\right|_{R_e} (R - R_e) + \frac{1}{2} \left.\frac{\partial^2 \mu}{\partial R^2}\right|_{R_e} (R - R_e)^2 + \cdots. \tag{1.194}$$

Now, two cases are distinguished. First, consider a pure rotational transition that occurs between different rotational states within the same vibrational state. The spectra typically lie in the microwave region. In this case, one can limit oneself to the first term in the above expansion. Because of the symmetry, the dipole vector lies along the molecular axis. The transition-dipole matrix element becomes proportional to

$$\int Y^*_{J'M'_J}(\Theta, \Phi) \mathbf{e}_R \cdot \mathbf{e} Y_{JM_J}(\Theta, \Phi) \sin\Theta d\Theta d\Phi, \tag{1.195}$$

where $\mathbf{e}$ is the unit polarization vector and $\mathbf{e}_R$ is the unit vector along the molecular axis. Therefore, the selection rules are analogous to those from atomic transitions, namely

$$\begin{aligned} \Delta J &= \pm 1 \\ \Delta M &= 0, \pm 1. \end{aligned} \tag{1.196}$$

The above selection rules can be understood because the absorbed or emitted photon carries one unit of angular momentum. The transition energy between states $(J \to J + 1)$ is

$$E = B(J+1)(J+2) - BJ(J+1) = 2B(J+1). \tag{1.197}$$

Thus spectral lines in a pure rotational spectrum all have the same separation of $2B$.

Now, consider a transition that occurs between different vibrational states. The spectra typically lie in the infrared to mid-infrared regions. In this case, the first term in the above expansion does not contribute to the transition dipole due to the orthogonality of the vibrational wavefunctions of the initial and final states. Now the main contribution comes from the second term

$$\mathbf{D}_v = \left(\frac{d\mu}{dR}\right)_{R_e} \int \psi'^*_v x \psi''_v dx, \tag{1.198}$$

where $x = R - R_e$. By using harmonic oscillator wavefunctions, the following selection rules can be derived such that the second factor of the Equation 1.198 does not vanish:

$$\Delta v = \pm 1. \tag{1.199}$$

If vibrational transitions are also accompanied by rotational transitions, the selection rules are

$$\begin{aligned} \Delta J &= \pm 1 \quad &\text{if} \quad \Lambda = 0 \\ \Delta J &= 0, \pm 1 \quad &\text{if} \quad \Lambda \neq 0. \end{aligned} \tag{1.200}$$

The upper state is normally denoted as (v', J'), while the lower state is normally denoted as (v'', J''). The transition with $\Delta J = J' - J'' = -1$ forms the so-called P branch in the lower part the spectrum, while the transition with $\Delta J = J' - J'' = 1$ forms the R branch in the higher part of the spectrum. The transition energy is

$$\begin{aligned} E &= E(v+1, J-1) - E(v,J) = E_{v0} - 2BJ \qquad &\text{with } J = 1,2,\ldots \\ E &= E(v+1, J+1) - E(v,J) = E_{v0} + 2B(J+1) \quad &\text{with } J = 0,1,\ldots \end{aligned} \tag{1.201}$$

for the P and R branches, respectively. E_{v0} is called the *band origin* and lies in the middle of the two branches. The spectral lines in both branches are separated by a constant energy of $2B$.

For molecules with electronic states different from Σ (i.e., with electronic-angular momentum different from zero in, for example, a nitric oxide (NO) molecule), transitions with $\Delta J = 0$ are also allowed. This could happen because the change in nuclear-angular momentum is accompanied by an opposite change in electronic-angular momentum such that the total molecular-angular momentum does not change. These transitions form the so-called Q branch. Within the rigid rotor approximation, the Q branch is reduced to a single line as

$$E = E(v+1, J) - E(v,J) = E_{v0} \quad \text{with} \quad J \geq \Lambda. \tag{1.202}$$

Most molecules in the gas phase at room temperature are in the vibrational ground state with $v'' = 0$. Thus, the most dominant transitions are from $v'' = 0 \to v' = 1$. Furthermore, the intensity of each spectral line not only depends on the transition dipole, but also on the occupancies (therefore, also on the degeneracy) of the states involved. The latter typically follows the Boltzmann distribution at a given temperature. Schematics of the transitions and the corresponding absorption spectrum are shown in Figure 1.11.

Electronic Transitions

Now consider the case where initial and final electronic states are different. Since the energy gap between different electronic states is much larger than that within the same electronic state, the electronic spectra lie in the visible or ultraviolet regions.

Selection Rules for Electronic Transitions

First, consider a pure electronic transition with the motion of nuclei neglected for the time being. The electronic-transition dipole can be written as

$$\mathbf{D}_e = \int \psi_e'^* \hat{\mu} \psi_e'' d\tau_e, \tag{1.203}$$

where τ_e is the volume element of the electronic coordinates. For the transition to occur, the electronic-transition dipole has to be nonzero. Much like electronic transitions in atoms, certain selection rules apply. Some basic rules are:

- $\Delta\Lambda = 0, \pm 1$. For example, transitions $\Sigma - \Sigma, \Pi - \Sigma$ are allowed. When $\Delta\Lambda = 0$, the transition is called *parallel* (that is, with the transition dipole along the molecular axis), whereas for $\Delta\Lambda = \pm 1$ it is called *perpendicular transition*.

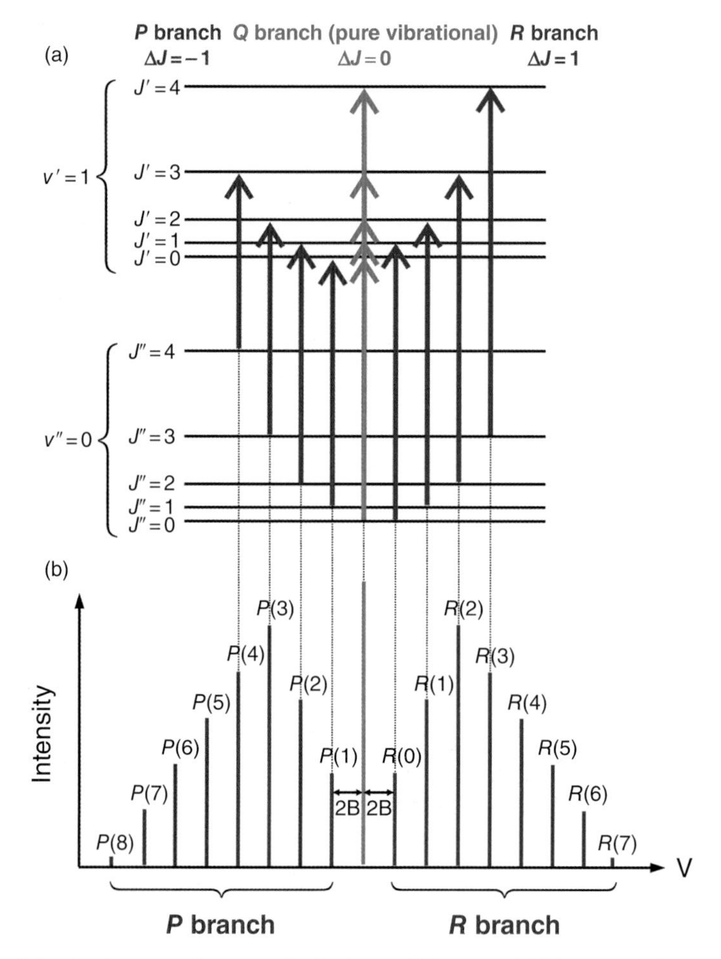

Figure 1.11 (a) Schematic of vibrational–rotational transition for the *P*, *Q*, and *R* branches. (b) Schematic of vibrational–rotational absorption spectrum.

- Transition occurs between $g \leftrightarrow u$ (called the *Laporte rule*).
- $\Sigma^+ - \Sigma^+$ and $\Sigma^- - \Sigma^-$ are allowed.
- $\Delta S = 0$. In other words, the transition between singlet and triplet states is forbidden since the electric-dipole operator does not depend on spin coordinates.

Vibronic Transitions and Franck–Condon Principle

Electronic transitions are expected to occur in a very short time scale as compared to vibrational motion. In fact, according to the Franck–Condon principle, the nuclei can be considered to be frozen during an electronic transition.

More quantitatively, within the BO approximation the transition dipole can be written as

$$\mathbf{D}_{ev} = \int\int \psi_e'^* \psi_v'^* \hat{\boldsymbol{\mu}} \psi_e'' \psi_v'' d\tau_e dR. \tag{1.204}$$

Following the Franck–Condon principle, one assumes that $\mathbf{D}_e$ (see Equation 1.203) is independent of internuclear distance such that

$$\mathbf{D}_{ev} = \int \psi_v'^* \mathbf{D}_e \psi_v'' dR = \mathbf{D}_e \int \psi_v'^* \psi_v'' dR, \tag{1.205}$$

where the last factor is called the Franck–Condon factor. This result indicates that, within the electronic-band spectra, the strongest transition occurs between the vibrational states that have the largest overlap. Note that there is no selection rule for vibrational quantum number v since the vibrational wavefunctions belong to different PESs.

Rovibronic Transitions

Much like vibrational–rotational transitions within the same electronic state, vibronic transitions are accompanied by certain rotational transitions. In fact, similar selection rules, in addition to selection rules listed above for pure electronic transitions, apply here. Namely,

$$\begin{aligned} \Delta J &= \pm 1 \quad \text{if } \Lambda = \Lambda' = 0 \\ \Delta J &= 0, \pm 1 \quad \text{if either } \Lambda \text{ or } \Lambda' \text{ are not } 0. \end{aligned} \tag{1.206}$$

Spectral lines belonging to the P branch are given as

$$E = E_v + B'J(J-1) - BJ(J+1) \text{ for P branch } (\Delta J = -1) \tag{1.207}$$

and so on for the $Q(\Delta J = 0)$ and $R(\Delta J = 1)$ branches. Note that rotational constants B and B' are generally different since they belong to different electronic states.

Raman Scattering

In addition to absorption (or transmission) and emission spectra discussed in the above subsections, the measurements of radiation scattered from molecules are also possible. In Raman scattering theory, this radiation scattering corresponds to a two-photon process that involves the absorption of a photon of energy ω and momentum k by a molecule to a virtual intermediate state and the emission of a different photon of energy ω' and momentum k'. If $\omega' = \omega$, the process is called *Rayleigh scattering*. If $\omega' < \omega$, the spectral line is called the *Stokes line* while $\omega' > \omega$ is called the *anti-Stokes line*.

The Raman transition dipole is proportional to the matrix element of the polarizability tensor. The polarizability near equilibrium can be expanded in a Taylor series in a similar fashion to the infrared transition in Equation 1.194. Retaining the second term of the Taylor expansion gives (cf. Equation 1.198)

$$\mathbf{D}_v = \left(\frac{d\alpha}{dR}\right)_{R_e} \boldsymbol{E}_0 \int \psi_v'^* x \psi_v'' dx, \tag{1.208}$$

where $x = R - R_e$ and $\boldsymbol{E}_0$ is the amplitude of the oscillating electric field. Therefore, the selection rules are the same as those for the infrared vibrational transitions studied earlier (see Equation 1.199), namely,

$$\Delta v = \pm 1. \tag{1.209}$$

One can further derive the selection rules in the case of rotational Raman (two-photon) transitions as

$$\Delta J = 0, \pm 2. \tag{1.210}$$

Therefore the vibrational–rotational spectrum for Raman scattering contains three branches: S (with $\Delta J = 2$), $Q(\Delta J = 0)$, and $O(\Delta J = -2)$.

Raman scattering does not require molecules to have permanent dipoles as in the case of the infrared vibrational–rotational transitions. Indeed, according to Equation 1.208, Raman scattering intensity depends on the change in dipole polarizability α instead of the change in dipole moment.

Section 3.6 discusses will discuss molecular alignment by intense laser fields, which can be interpreted as due to coherent Raman excitation of rotational states. Similarly, efficient Raman excitation has been used to create coherent vibrational wave packets in various strong-field pump-probe experiments and theoretical simulations.

Nuclear Spin

The rotational Raman spectra of some molecules show interesting features. For example, there are no transitions from states with even J in O_2. The absence of these rotational transitions can generally be understood as due to the Pauli principle. The Pauli principle is relevant here since nuclei rotation might lead to the interchange of identical nuclei. This book only briefly summarizes the main results and discusses a few simple examples on homonuclear diatomic molecules. The interested reader is referred to more detailed discussions in standard textbooks such as Bransden and Joachain [Bransden, Joachain] or Atkins and Friedman [Atkins, Friedman] as well as more advanced treatment in the monograph by Herzberg [Herzberg, 1950].

The total molecular wavefunction can be written as

$$\Psi = \Phi_{el}\psi_n\chi_{ns}, \tag{1.211}$$

where Φ_{el}, ψ_n, and χ_{ns} are the electronic, nuclear, and nuclear-spin wavefunctions, respectively. For the interchange $\hat{P}_n$ of nuclei, the following must be true:

$$\hat{P}_n\Psi = \pm\Psi, \tag{1.212}$$

depending on whether the identical nuclei are bosons (with + sign) or fermions (with − sign). It can be shown that the interchange of the nuclei is equivalent to the following combination (see [Atkins, Friedman]):

$$\hat{P}_n\Psi = \hat{\sigma}_h\hat{\imath}\Phi_{el}(-1)^J\psi_n\hat{P}_{ns}\chi_{ns} \equiv \pm\Psi. \tag{1.213}$$

Here, $\hat{\sigma}_h$ is the reflection operation of electronic wavefunction with respect to a plane through the molecular axis, $\hat{\imath}$ is the inversion of the electronic coordinates, $\hat{P}_{ns}$ is the interchange (or permutation) of two nuclei spins, and the appearance of the factor $(-1)^J$ is due to the rotation by 180° (that is C_2) of ψ_n around an axis perpendicular to the molecular axis. Equation 1.213 imposes certain restrictions on angular momentum J and nuclear-spin wavefunction. Note that, for most cases, the ground electronic state is totally symmetric

Σ_g^+ so the electronic part does not affect Equation 1.213. However, this is not always the case (see, for example, O_2, below). The consequences of this will be analyzed in a few simple examples.

First, consider an O_2 molecule. Since an ^{16}O nucleus is a boson (with nuclear spin $I = 0$), the total wavefunction is symmetric with respect to the interchange of two oxygen nuclei. From two nuclei with $I = 0$, one can only construct a symmetric- singlet, nuclear-spin function χ_{ns} that is $\hat{P}_{ns}\chi_{ns} = \chi_{ns}$. Recall that the ground-state electronic wavefunction of O_2 is $^3\Sigma_g^-$, and thus $\hat{\sigma}_h\hat{\iota}\Phi_{el} = -\Phi_{el}$. That means that the factor $(-1)^J$ has to give -1, or J can only take odd values. This is consistent with the experimentally observed rotational Raman spectra as mentioned earlier.

Similar arguments can be used for CO_2 with respect to interchange of two oxygen nuclei. However, since the ground-state electronic wavefunction of CO_2 is $^1\Sigma_g^+$, one has $\hat{\sigma}_h\hat{\iota}\Phi_{el} = \Phi_{el}$ as opposed to $-\Phi_{el}$ for O_2. This means that J can only be even.

For N_2 the ground-state electronic wavefunction is $^1\Sigma_g^+$ so only the nuclei part in Equation 1.213 has to be considered. ^{14}N nucleus has a spin $I = 1$ (boson) so $\hat{P}_n\Psi = +\Psi$. From two nuclei with spin $I = 1$ we can construct one singlet χ_{ns} (with total nuclear spin $I_{tot} = 0$, symmetric), one triplet ($I_{tot} = 1$, antisymmetric), and one quintuplet ($I_{tot} = 2$, symmetric). Note that these results can be easily obtained using Clebsch–Gordan tables. By using Equation 1.213, it is clear that the symmetric nuclear spin (singlet and quintuplet) should have an even J, while the antisymmetric one (triplet) should have an odd J. Due to the degeneracy of $2I_{tot} + 1$, the ratio of an even J to an odd J is (1+5):3, or 2:1.

Finally, it can also be shown that the ratio of symmetric to antisymmetric states is $(I + 1)/I$ (see [Atkins, Friedman]).

1.6.2 Spectra of Polyatomic Molecules

Polyatomic molecules have very complicated spectra. Nevertheless, one can understand their basic features by extending the concepts learned from diatomic molecules. For that purpose, we use methods from point-group symmetry extensively.

For a transition to occur, the transition dipole

$$\mathbf{D}_{ev} = \int \psi_e'^* \psi_v'^* \hat{\mu} \psi_e'' \psi_v'' d\tau_e d\tau_v \tag{1.214}$$

should not vanish. This means that, according to point-group theory, the integrand should be totally symmetric with respect to all operations of the symmetry group of the molecule. Now consider a few cases.

Vibrational Transitions within the Same Electronic State

In this case, by following the same method as for diatomic molecules, one can expand the permanent electric dipole near equilibrium in a Taylor series and show that the transition dipole can be written as

$$\mathbf{D}_v = \int \psi_v'^* \mu \psi_v'' d\tau_v \approx \sum_{n=1}^{3N-6} \left(\frac{\partial \mu}{\partial Q_n}\right)_0 \int \psi_v'^* Q_n \psi_v'' d\tau_v. \tag{1.215}$$

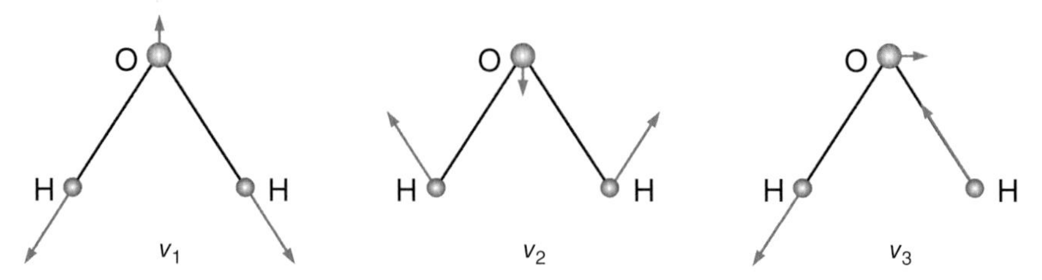

Figure 1.12 Normal modes of H_2O. Both ν_1 and ν_2 belong to A_1 symmetry, while ν_3 belongs to B_2 symmetry of the C_{2v} group.

Here, the sum is over all $3N - 6$ normal modes. By using the harmonic oscillator approximation, it can be shown that the integral $\int \psi_v'^* Q_n \psi_v'' d\tau_v$ is different from zero when $v_n' = v_n'' \pm 1$ and $v_m' = v_m''$ for $m \neq n$. Therefore the selection rules for each mode ν_n are

$$\Delta v_n = \pm 1. \tag{1.216}$$

Here, v_n is used to denote a normal mode or its quantum number, respectively. Clearly, for the transition to occur in mode ν_n, the factor $\left.\frac{\partial \mu}{\partial Q_n}\right|_0$ should be nonvanishing. In other words, the dipole has to change during the vibration of the given mode. In that case, the mode is called *infrared active*. This is illustrated by the case of H_2O vibrations shown in Figure 1.12. After inspection, it becomes clear that the dipole moment changes during the vibration of all three modes. More precisely, for ν_1 and ν_2 modes (both with A_1 symmetry), the dipole moment changes in magnitude along the z-axis (molecular axis), while for ν_3 (with B_2 symmetry), the dipole moment changes its direction during those vibrations. Therefore, all three modes are infrared active.

The above example is simple but difficult to carry out for more complicated cases. For a general method, a nonvanishing transition dipole requires that the integrand $\psi_v'^* \mu \psi_v''$ contains the totally symmetric irreducible representation, namely

$$\Gamma_{\text{vib}}' \otimes \Gamma_\alpha \otimes \Gamma_{\text{vib}}'' \supset A \tag{1.217}$$

for at least one component $\alpha = x, y, z$. Here, A stands for the totally symmetric irreducible representation although, for certain groups, it might be labeled as A_1, $\Sigma \ldots$. Note that in the above equation, Γ_{μ_α} has been replaced with Γ_α since dipole μ_α transforms as a vector. Irreducible representations of $\{x, y, z\}$ can be found in the character table for any given symmetry group. Here, this method is illustrated in the same example of H_2O. Consider a transition from the ground vibrational state $(\nu_1, \nu_2, \nu_3) = (0, 0, 0)$, which has A_1 symmetry, to an excited state of symmetric stretch (1, 0, 0) also of A_1 symmetry. According to the character table of C_{2v} group (see Table 1.10), $\{x, y, z\} = \{B_1, B_2, A_1\}$. By using product tables (see Table 1.3), the direct products of these representations are

$$\Gamma_{\text{vib}}' \otimes \Gamma_\alpha \otimes \Gamma_{\text{vib}}'' = A_1 \otimes \begin{Bmatrix} B_1 \\ B_2 \\ A_1 \end{Bmatrix} \otimes A_1 = \begin{Bmatrix} B_1 \\ B_2 \\ A_1 \end{Bmatrix} \tag{1.218}$$

so the allowed transition is due to z-component of the dipole, which agrees with the result above.

The above analysis can be extended to Raman scattering in polyatomic molecules. Here, the dipole polarizability tensor is dealt with instead of the dipole moment. For Raman scattering to occur in a particular vibrational mode, at least one component of the dipole polarizability should change during the vibration near the equilibrium. In such cases, the mode is called *Raman active*. Following a similar derivation as Equation 1.215, the same selection rules are obtained for a Raman vibrational transition $\Delta v_n = \pm 1$ as for infrared transitions. Similarly, instead of Equation 1.217, for Raman scattering there is

$$\Gamma'_{\text{vib}} \otimes \Gamma_{\alpha\beta} \otimes \Gamma''_{\text{vib}} \supset A \tag{1.219}$$

where $\alpha\beta$ are $xx, xy, \ldots$. Irreducible representation $\Gamma_{\alpha\beta}$ can be found in the character table for any symmetry group.

Electronic Transitions

Finally, consider the transition when the initial and final electronic states are different and assume that nuclei are fixed. For the electronic-transition dipole

$$\mathbf{D}_e = \int \psi_e'^* \mu \psi_e'' d\tau_e \tag{1.220}$$

to be different from zero, one should have

$$\Gamma'_{\text{el}} \otimes \Gamma_{\alpha} \otimes \Gamma''_{\text{el}} \supset A \tag{1.221}$$

for at least one of the components α. The symmetry representations for electronic wavefunctions can be obtained using the methods presented in Section 1.5.2 so further analysis for symmetry-allowed transitions can be done in the same manner as those for vibrational transitions discussed in the previous subsection. Since the electric-dipole operator does not depend on spin coordinates, there is the spin selection rule $\Delta S = 0$. This selection rule is valid for light atoms as long as the spin-orbit interaction is small.

If the electronic transition is allowed, the nuclear motion should be taken into account. In fact, electronic transitions are typically accompanied by vibrational transitions. In the spirit of the Franck–Condon principle, it is assumed that the electronic-transition dipole $\mathbf{D}_e$ is independent of the nuclear coordinates near the equilibrium such that the transition dipole $\mathbf{D}_{ev}$ in Equation 1.214 can be factored out as

$$\mathbf{D}_{ev} = \mathbf{D}_e \int \psi_v'^* \psi_v'' d\tau_v. \tag{1.222}$$

Therefore the vibronic-transition dipole is proportional to the Franck–Condon factor, i.e., the overlap integral between two vibrational states belonging to two different electronic states in Equation 1.222. For the transition to occur, this factor must be nonvanishing, or

$$\Gamma'_{\text{vib}} \otimes \Gamma''_{\text{vib}} \supset A. \tag{1.223}$$

This means that transitions occur only between the same vibrational symmetry species in the two electronic states.

Note that even when a transition is electronic forbidden, it may still be weakly observed. This is due to the coupling between electronic and vibrational degrees of freedom, or the so-called *vibronic coupling*. In particular, the electronic-transition dipole in Equation 1.220 is strictly not a constant, but depends on nuclear coordinates. Therefore, the vibronic-transition dipole $\mathbf{D}_{ev}$ in Equation 1.214 is simply required to be nonvanishing, or

$$\Gamma'_{\text{vib}} \otimes \Gamma'_{\text{el}} \otimes \Gamma_{\alpha} \otimes \Gamma''_{\text{el}} \otimes \Gamma''_{\text{vib}} \supset A. \tag{1.224}$$

1.7 Propagation of a Laser Pulse in Free Space

1.7.1 Maxwell's Equations

This section summarizes the propagation of the electric field of a laser beam in free space. Before a laser beam arrives in the interaction region it has to undergo free propagation and pass various optical materials like mirrors, lenses, or plates. Maxwell's equations (in SI units), which govern the coupling of the electric-field vector $\mathbf{E}(E_x, E_y, E_z)$ and the magnetic-field vector $\mathbf{H}(H_x, H_y, H_z)$ are

$$\begin{aligned}
\nabla \times \mathbf{E} &= -\mu \frac{\partial \mathbf{H}}{\partial t}, \\
\nabla \times \mathbf{H} &= \varepsilon \frac{\partial \mathbf{E}}{\partial t}, \\
\nabla \cdot \varepsilon \mathbf{E} &= 0, \\
\nabla \cdot \mu \mathbf{H} &= 0.
\end{aligned} \tag{1.225}$$

In Equation 1.225, both vectors are functions of position and time, represented by their Cartesian components and the time derivative. μ is the magnetic permeability and ε is the electric permittivity of the propagation medium. There is no free charge. Taking the curl of the first two time-derivative equations of Equation 1.225 results in the wave equations

$$\begin{aligned}
\nabla^2 \mathbf{E} - \frac{n^2}{c^2} \frac{\partial^2 \mathbf{E}}{\partial^2 t} &= 0, \\
\nabla^2 \mathbf{H} - \frac{n^2}{c^2} \frac{\partial^2 \mathbf{H}}{\partial^2 t} &= 0.
\end{aligned} \tag{1.226}$$

In the derivation the medium is taken to be linear, isotropic, nondispersive, and homogeneous such that μ and ε are constants. Define μ_0 and ε_0 as the constants in vacuum, the refractive index is $n = (\varepsilon/\varepsilon_0)^{1/2}$, and the velocity of propagation in vacuum is $c = 1/\sqrt{\varepsilon_0 \mu_0}$.

Both the electric and magnetic fields, as well as their components, satisfy the wave equations. For example, the x-component of the electric field satisfies

$$\nabla^2 E_x - \frac{n^2}{c^2} \frac{\partial^2 E_x}{\partial^2 t} = 0. \tag{1.227}$$

Equation 1.227 can be generalized to any scalar field component such as $u(x, y, z, t)$ in the following:

$$\nabla^2 u(x, y, z, t) - \frac{n^2}{c^2} \frac{\partial^2 u(x, y, z, t)}{\partial^2 t} = 0. \tag{1.228}$$

Consider a laser beam propagating along the z direction that is axially symmetric. The wave number k is given by $k = 2\pi/\lambda$ where λ is the central wavelength. The frequency in the homogeneous dielectric medium is ν and the wavelength is defined as $\lambda = c/n\nu$. One solution that represents a traveling wave is

$$u(x,y,z,t) = \widetilde{u}(x,y,z)\exp(i2\pi\nu t), \tag{1.229}$$

in which $\widetilde{u}(x,y,z)$ satisfies the time-independent Helmholtz equation

$$(\nabla^2 + k^2)\widetilde{u}(x,y,z) = 0. \tag{1.230}$$

1.7.2 Paraxial Wave Equation

This section looks at an optical beam primarily propagating along the z direction. The spatial dependence of $\widetilde{u}(x,y,z)$ has the $\exp(-ikz)$ variation, which gives a spatial period of one wavelength λ along z. For any beam of practical interest, its amplitude and phase generally have transverse variation in x and y that specifies the beam's transverse profile. The transverse amplitude and phase change slowly with distance z compared to the variation of the plane wave $\exp(-ikz)$ in the z direction for a reasonably well-collimated beam. Thus, each relevant vector component of the field (such as E_x or E_y) can be written in the form

$$\widetilde{u}(x,y,z) \equiv \widetilde{E}(x,y,z)e^{-ikz}, \tag{1.231}$$

where $\widetilde{E}(x,y,z)$ is a complex scalar wave amplitude that describes the transverse profile of the beam. By substituting $\widetilde{u}(x,y,z)$ into Equation 1.230, the amplitude in Cartesian coordinates satisfies

$$\frac{\partial^2\widetilde{E}}{\partial x^2} + \frac{\partial^2\widetilde{E}}{\partial y^2} + \frac{\partial^2\widetilde{E}}{\partial z^2} - 2ik\frac{\partial\widetilde{E}}{\partial z} = 0. \tag{1.232}$$

By factoring out $\exp(-ikz)$, the remaining slow z-varying amplitude $\widetilde{E}(x,y,z)$ is expressed mathematically by the paraxial approximation

$$\left|\frac{\partial^2\widetilde{E}}{\partial z^2}\right| \ll \left|2k\frac{\partial\widetilde{E}}{\partial z}\right| \text{ or } \left|\frac{\partial^2\widetilde{E}}{\partial x^2}\right| \text{ or } \left|\frac{\partial^2\widetilde{E}}{\partial y^2}\right|. \tag{1.233}$$

Dropping the second partial derivative in z reduces Equation 1.232 to the paraxial-wave equation

$$\frac{\partial^2\widetilde{E}}{\partial x^2} + \frac{\partial^2\widetilde{E}}{\partial y^2} - 2ik\frac{\partial\widetilde{E}}{\partial z} = 0. \tag{1.234}$$

This paraxial-wave equation can be written as

$$\nabla_\perp^2\widetilde{E}(\mathbf{r},z) - 2ik\frac{\partial\widetilde{E}(\mathbf{r},z)}{\partial z} = 0, \tag{1.235}$$

where $\mathbf{r}$ refers to the transverse coordinates $\mathbf{r} \equiv (x,y)$ or $\mathbf{r} \equiv (r,\theta)$ depending on the selected coordinate system (Cartesian or cylindrical). In this equation, $\nabla_\perp^2$ is the Laplacian operator.

1.7.3 Huygens' Integral under the Paraxial Approximation

The paraxial-wave propagation can be obtained using an integral approach that employs Huygens' principle in the Fresnel approximation. This section introduces spherical waves, Fresnel approximation, and Huygens' integral.

Spherical Waves

A uniform spherical wave $\widetilde{u}(\mathbf{r};\mathbf{r}_0)$ emanating from a source point $\mathbf{r}_0 = (x_0, y_0, z_0)$ to an observation point $\mathbf{r} = (x, y, z)$ can be written in the form

$$\widetilde{u}(\mathbf{r};\mathbf{r}_0) = \frac{\exp[-ik\rho(\mathbf{r},\mathbf{r}_0)]}{\rho(\mathbf{r},\mathbf{r}_0)}. \tag{1.236}$$

The distance $\rho(\mathbf{r},\mathbf{r}_0)$ is given by

$$\rho(\mathbf{r},\mathbf{r}_0) = \sqrt{(x-x_0)^2 + (y-y_0)^2 + (z-z_0)^2}. \tag{1.237}$$

Note that a spherical wavefunction in this form is an exact solution of the full scalar wave equation Equation 1.230.

Fresnel Approximation

If the source point and the observation point are not too far away from the z-axis, one can expand the distance $\rho(\mathbf{r},\mathbf{r}_0)$ in a power series

$$\rho(\mathbf{r},\mathbf{r}_0) = z - z_0 + \frac{(x-x_0)^2 + (y-y_0)^2}{2(z-z_0)} + \cdots. \tag{1.238}$$

By dropping out all terms greater than the quadratic ones in the phase-shift factor $\exp[-ik\rho(\mathbf{r},\mathbf{r}_0)]$ and replacing $\rho(\mathbf{r},\mathbf{r}_0)$ by $z - z_0$ in the denominator $1/\rho$, a "paraxial-spherical wave" is obtained

$$\widetilde{u}(x,y,z) \approx \frac{1}{z-z_0}\exp\left[-ik(z-z_0) - ik\frac{(x-x_0)^2 + (y-y_0)^2}{2(z-z_0)}\right]. \tag{1.239}$$

This gives

$$\widetilde{E}(x,y,z) \approx \frac{1}{z-z_0}\exp\left[-ik\frac{(x-x_0)^2 + (y-y_0)^2}{2(z-z_0)}\right]. \tag{1.240}$$

In these expressions, a spherical wave has a quadratic phase variation as observed on a transverse plane (x, y) located at a distance $z - z_0$ away from the source point along the z axis, as shown in Figure 1.13.

Huygens' Integral

Huygens' integral was initially proposed as an intuitive physical principle. It was then put into a more rigorous mathematical form first by Fresnel and Kirchoff and later by Rayleigh and Sommerfeld. Given an incident field distribution $\widetilde{u}_0(x_0, y_0, z_0)$ over a closed surface

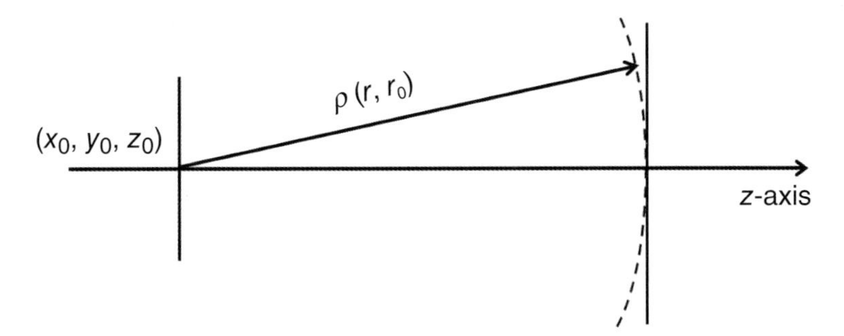

Figure 1.13 Fresnel approximation to the spherical wave.

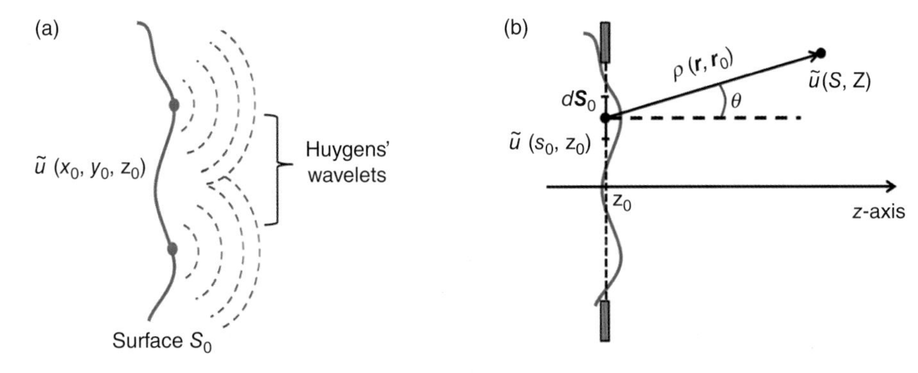

Figure 1.14 (a) Illustration of Huygens' principle. (b) Geometry for evaluating Huygens' integral in the Fresnel approximation.

S_0, the field at each point on that surface can be considered to be the source for a uniform spherical wave, or the *Huygens' wavelet*, as illustrated in Figure 1.14(a). Huygens' wavelets should be viewed as a spherical wave with a form like Equation 1.236. The total field at any other point inside or beyond the surface S_0 can then be calculated by summing over all of these Huygens' wavelets emanating from the surface S_0. This leads to Huygens' integral equation in the form

$$\tilde{u}(\mathbf{s}, z) = \frac{i}{\lambda} \int\int_{S_0} \tilde{u}(\mathbf{s}_0, z_0) \frac{\exp[-ik\rho(\mathbf{r}, \mathbf{r}_0)]}{\rho(\mathbf{r}, \mathbf{r}_0)} \cos\theta(\mathbf{r}, \mathbf{r}_0) d\mathbf{S}_0, \tag{1.241}$$

where **s** denotes the surface elements, $d\mathbf{S}_0$ is an incremental element of the surface area, and $\cos\theta(\mathbf{r}, \mathbf{r}_0)$ is an "obliquity factor" that depends on the angle $\theta(\mathbf{r}, \mathbf{r}_0)$ between the line element $\rho(\mathbf{r}, \mathbf{r}_0)$ and the direction normal to the surface element $d\mathbf{S}_0$.

In both the Kirchoff–Fresnel and Rayleigh–Sommerfeld approaches to diffraction theory, slightly different forms for the obliquity factor in Equation 1.241 were predicted. However, in either situation, this factor goes to unity if the angle θ is limited to small values. The normalization factor of i/λ in front of Huygens' integral comes out of a more detailed approach to the theory and is necessary in order to obtain the correct near-field and far-field dependence from Huygens' integral.

Fresnel Approximation to Huygens' Integral

Consider how the output beam profile $\widetilde{u}(x, y, z)$ across a plane at a distance $z = z_0 + L$ can be accurately calculated within the paraxial approximation (see Figure 1.14(b)). On the input plane $z = z_0$, it is assumed that the field distribution of a paraxial optical beam $\widetilde{u}_0(x_0, y_0, z_0)$ is given. Huygens' integral is used to calculate the field at a point (x, y, z)

$$\widetilde{u}(x,y,z) = \frac{ie^{-ik(z-z_0)}}{(z-z_0)\lambda} \int\int \widetilde{u}_0(x_0, y_0, z_0) \exp\left[-ik\frac{(x-x_0)^2 + (y-y_0)^2}{2(z-z_0)}\right] dx_0 dy_0. \tag{1.242}$$

In Equation 1.242, the distance $z - z_0 = L$ between the input and output planes is assumed to be large such that the obliquity factor $\cos\theta$ in Equation 1.241 can be approximated by unity.

Similarly, the reduced electric field $\widetilde{E}(x, y, z)$ can be written as

$$\widetilde{E}(x,y,z) = \frac{i}{L\lambda} \int\int \widetilde{E}_0(x_0, y_0, z_0) \exp\left[-ik\frac{(x-x_0)^2 + (y-y_0)^2}{2L}\right] dx_0 dy_0 \tag{1.243}$$

where the integration is over the input plane. Huygens' integral in this form is most often used with plane-wave phase shift $\exp(-ikL)$ omitted since the primary interest usually is the transverse variation of $\widetilde{E}(x, y)$.

1.7.4 Gaussian Beam

In optics, a Gaussian beam is a beam of monochromatic electromagnetic radiation whose transverse magnetic- and electric-field amplitude profiles are described by Gaussian functions. This fundamental (or TEM_{00}) transverse Gaussian mode describes the intended output of most (but not all) lasers such that a beam can be focused onto a concentrated spot. In this section, the expression of the Gaussian beam will first be derived from the paraxial-wave equation, then the properties of the Gaussian beams will be discussed.

Complex Amplitude

For simplicity, it is assumed that the laser beam is axially symmetric and propagates in free space along the z direction. In the cylindrical coordinate system, the Helmholtz equation can be written as

$$\frac{1}{r}\frac{\partial}{\partial r}\left[r\frac{\partial}{\partial r}\widetilde{E}(r,z)\right] - 2ik\frac{\partial}{\partial z}\widetilde{E}(r,z) = 0. \tag{1.244}$$

The solution of this equation is the paraboloidal wave

$$\widetilde{E}(r,z) = \frac{E_1}{z}\exp\left(-ik\frac{r^2}{2z}\right), \tag{1.245}$$

where E_1 is a constant.

By a simple transformation in which z in Equation 1.245 is replaced by $z - \xi$ with a constant ξ, another solution to Equation 1.244 is obtained

$$\widetilde{E}(r,z) = \frac{E_1}{q(z)} \exp\left[- ik\frac{r^2}{2q(z)} \right], \quad q(z) = z - \xi. \tag{1.246}$$

This solution represents a paraboloidal wave centered at $z = \xi$ instead of $z = 0$. Equation 1.246 remains a solution to Equation 1.244, even when ξ is a complex constant. However, the solution acquires dramatically different properties. In particular, when ξ is purely imaginary, say, $\xi = -iz_R$, where z_R is real, Equation 1.246 gives the complex envelope of the Gaussian beam

$$\widetilde{E}(r,z) = \frac{E_1}{q(z)} \exp\left[- ik\frac{r^2}{2q(z)} \right], \quad q(z) = z + iz_R. \tag{1.247}$$

The quantity $q(z)$ is called the q-parameter of the beam and the parameter z_R is known as the Rayleigh range.

By defining two new real functions, $R(z)$ and $w(z)$, the complex function $1/q(z)$ can be written as

$$\frac{1}{q(z)} = \frac{1}{R(z)} - i\frac{\lambda}{\pi w^2(z)}. \tag{1.248}$$

Here, $w(z)$ and $R(z)$ are measures of the beam width and wavefront curvature, respectively. Subsequently, Equation 1.247 can be expressed as

$$\widetilde{E}(r,z) = E_0\frac{w_0}{w(z)} \exp\left[- \frac{r^2}{w^2(z)} \right] \exp\left[- ik\frac{r^2}{2R(z)} + i\phi(z) \right], \tag{1.249}$$

where

$$w(z) = w_0\sqrt{1 + \left(\frac{z}{z_R}\right)^2}, \tag{1.250}$$

$$R(z) = z\left[1 + \left(\frac{z_R}{z}\right)^2\right], \tag{1.251}$$

$$\phi(z) = \tan^{-1}\left(\frac{z}{z_R}\right), \tag{1.252}$$

$$w_0 = \sqrt{\frac{\lambda z_R}{\pi}}. \tag{1.253}$$

A new constant $E_0 = E_1/iz_R$ has been defined for convenience.

The propagation factor $\exp(-ikz)$ in the Gaussian beam is not included in Equation 1.249 since we start from Equation 1.235. The expression for the complex amplitude of the Gaussian beam in Equation 1.249 is described by two independent parameters, E_0 and z_R, which are determined from the boundary conditions.

Properties of a Gaussian Beam

As shown in Figure 1.15, the geometry and behavior of a Gaussian beam are governed by a set of beam parameters. The properties of a Gaussian beam are as follows.

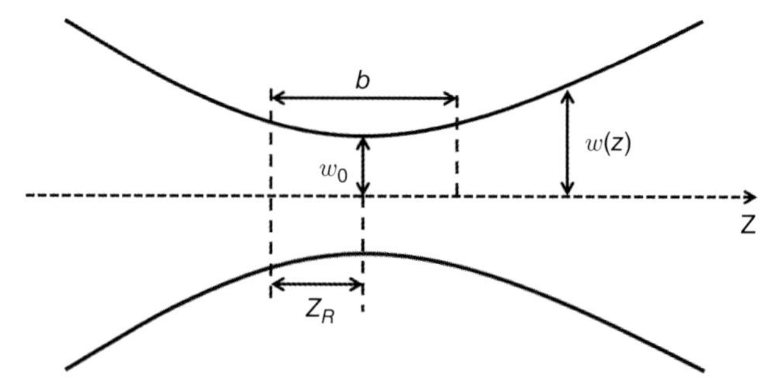

Figure 1.15 Schematic diagram of a Gaussian beam. Beam width $w(z)$ is a function of the axial distance z; w_0: beam waist; b: confocal parameter, which is twice the Rayleigh range z_R.

Intensity. The optical intensity $I(r,z) = |\widetilde{E}(r,z)|^2$ as a function of the axial and radial positions, z and r, respectively, can be expressed as

$$I(r,z) = I_0 \left(\frac{w_0}{w(z)} \right)^2 \exp\left[-\frac{2r^2}{w^2(z)} \right], \tag{1.254}$$

where $I_0 = |E_0|^2$. At any value z, the intensity is a Gaussian function of the radial distance r. The Gaussian function has its peak on the z-axis (at $r = 0$), and decreases monotonically as r increases. The beam width $w(z)$ of the Gaussian distribution increases with the axial distance z as shown in Figure 1.15.

On the beam axis ($r = 0$), the intensity reduces to

$$I(0,z) = I_0 \left(\frac{w_0}{w(z)} \right)^2 = \frac{I_0}{1 + (z/z_R)^2}. \tag{1.255}$$

Power. The total power carried by the Gaussian beam can be calculated from

$$P = \int_0^\infty I(r,z) 2\pi r dr, \tag{1.256}$$

which gives

$$P = \frac{1}{2} I_0 (\pi w_0^2). \tag{1.257}$$

As expected, the beam power is independent of z and is half the peak intensity multiplied by the beam area. At position z, the ratio of the power within a radial distance of r_0 in the transverse plane to the total power is

$$\frac{1}{P} \int_0^{r_0} I(r,z) 2\pi r dr = 1 - \exp\left[-\frac{2r_0^2}{w^2(z)} \right]. \tag{1.258}$$

As calculated above, the power contained within a radial distance $r_0 = w(z)$ is approximately 86% of the total power. About 99% of the power is contained within a circle of radius 1.5 $w(z)$.

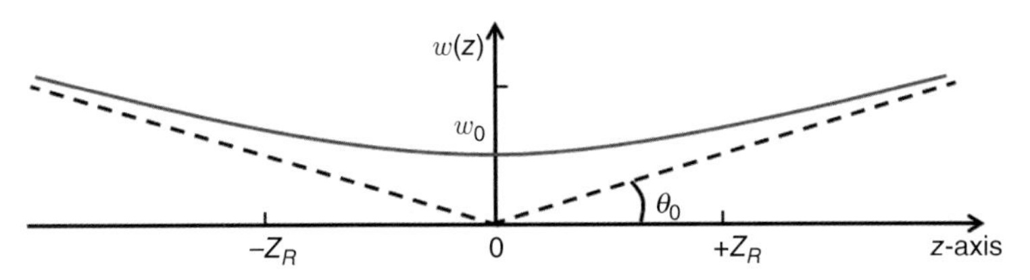

Figure 1.16 The beam radius $w(z)$ has its minimum value w_0 at the waist ($z = 0$). It reaches $\sqrt{2}w_0$ at $z = \pm z_R$ and increases linearly with z for large z.

Beam width. At any transverse plane, the beam intensity has its peak value on the beam axis and decreases by a factor of $1/e^2 \approx 0.135$ at the radial distance $r = w(z)$. Since within a circle of radius $w(z)$, 86% of the power is carried, $w(z)$ is regarded as the beam radius (or beam width). The dependence of the beam width $w(z)$ on z is governed by Equation 1.250. Its value w_0 at $z = 0$ is known as *beam waist* or *waist radius*. The waist diameter $2w_0$ is called the *spot size*. From Equation 1.250 it can be seen that the beam width increases monotonically with z and is $\sqrt{2}w_0$ at $z = \pm z_R$.

Beam divergence. For $z \gg z_R$, the first term of Equation 1.250 can be neglected. The following is obtained:

$$w(z) \approx \frac{w_0}{z_R} z = \theta_0 z. \tag{1.259}$$

By using Equation 1.253, the half-angle of the beam divergence is

$$\theta_0 = \frac{w_0}{z_R} = \frac{\lambda}{\pi w_0}. \tag{1.260}$$

Approximately 86% of the beam power is confined within this cone (see Figure 1.16).

After rewriting the Equation 1.260 in terms of the spot size, the angular divergence of the beam becomes

$$2\theta_0 = \frac{4}{\pi} \frac{\lambda}{2w_0}. \tag{1.261}$$

From Equation 1.261, the divergence angle is directly proportional to the wavelength λ and inversely proportional to the spot size $2w_0$. Therefore, squeezing the spot size would lead to increased beam divergence. Making a beam with fat waist at a short wavelength would lead to a highly directional beam.

Depth of focus. The beam achieves its best focus (or minimum) at $z = 0$. The depth of focus is twice the Rayleigh range

$$b = 2z_R = \frac{2\pi w_0^2}{\lambda}. \tag{1.262}$$

Therefore the depth of focus is proportional to the square of the beam waist w_0 and inversely proportional to the wavelength λ. Thus, a beam focused to a small spot size has a short depth of focus, making it harder to locate the plane of focus. Small spot size and long depth of focus can only be attained simultaneously with short-wavelength lasers.

Phase. The phase of a Gaussian beam from Equation 1.249 is

$$\varphi(r,z) = -\phi(z) + \frac{kr^2}{2R(z)}. \tag{1.263}$$

On the beam axis ($r = 0$) the phase becomes

$$\varphi(0,z) = -\phi(z). \tag{1.264}$$

This represents a phase retardation $\phi(z)$ that ranges from $-\pi/2$ at $z = -\infty$ to $\pi/2$ at $z = \infty$. Compared to a plane wave or a spherical wave, this phase retardation means an excess delay of the wavefront. Thus, when a wave travels from $z = -\infty$ to $z = \infty$, the total accumulated excess retardation is π. This phenomenon is known as the *Gouy effect*, and $\phi(z)$ is called the *Gouy phase*.

1.8 Matrix Optics

1.8.1 Introduction

Matrix optics is a method that can be used to trace paraxial rays. The rays are assumed to travel only within a single plane such that the formalism is applicable to systems with planar geometry.

A ray can be described by its position and angle with respect to the optical axis. These parameters vary as the ray travels through an optical system. In the paraxial approximation, the position and angle at the input and output planes are connected by two linear algebraic equations. As a result, the optical system is described by a 2×2 matrix, which is called the *ray-transfer matrix*.

Because the ray-transfer matrix of a cascade of optical components is a product of the ray-transfer matrices of the individual components, it is convenient to use matrix methods. Therefore matrix optics provides a formal means for describing a complex optical system in the paraxial approximation.

1.8.2 The Ray-Transfer Matrix

First, consider a circularly symmetric optical system that consists of a succession of refracting and reflecting surfaces all centered about the same optical axis. The z-axis lies along the optical axis and points in the ray's traveling direction. Assume the rays lie on a plane containing the optical axis, say, the $x - z$ plane. A ray crossing the transverse plane at z is fully described by the x coordinate of its crossing point and the angle θ.

An optical system consisting of a set of optical components is placed between two transverse planes at z_0 and z. These planes are referred to as the input and output planes, respectively. The system is completely determined by its effect on an incoming ray of arbitrary (x_0, θ_0). It steers the ray so that it has a new (x, θ) on the output plane (see Figure 1.17).

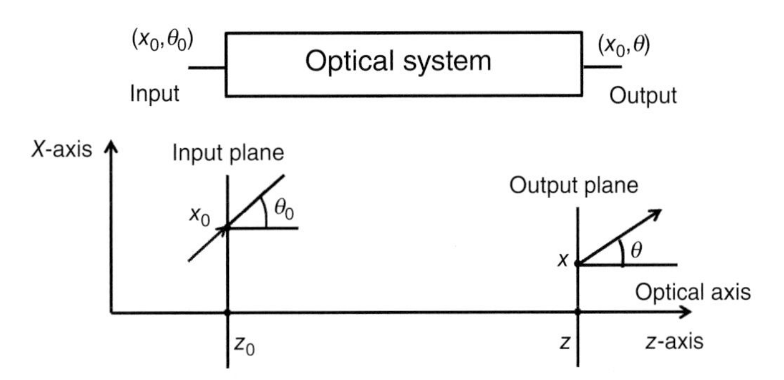

Figure 1.17 A ray enters an optical system at position x_0 and angle θ_0 and leaves at position x and angle θ.

In the paraxial approximation when all angles are sufficiently small so $\sin\theta \approx \theta$, the relation between (x, θ) and (x_0, θ_0) is linear and can be written as

$$x = Ax_0 + B\theta_0, \tag{1.265}$$

$$\theta = Cx_0 + D\theta_0, \tag{1.266}$$

where A, B, C, and D are real numbers. In matrix form

$$\begin{pmatrix} x \\ \theta \end{pmatrix} = \begin{pmatrix} A & B \\ C & D \end{pmatrix} \begin{pmatrix} x_0 \\ \theta_0 \end{pmatrix}. \tag{1.267}$$

The matrix **M** in Equation 1.267, with its elements A, B, C, and D, completely characterizes the optical system. This matrix can determine $(x,\ \theta)$ for any $(x_0,\ \theta_0)$ and is known as the ray-transfer matrix.

1.8.3 Matrices of Simple Optical Components

A complicated optical system consists of simple optical components, and each component can be characterized by its own ray-transfer matrix. The following demonstrates the $ABCD$ matrices for commonly used optical systems.

Free-space propagation. As shown in Figure 1.18(a), the rays travel along straight lines in a medium with a uniform refractive index such as in free space; thus, a ray traversing a distance d is altered in accordance with $x = x_0 + \theta_0 d$ and $\theta = \theta_0$. The ray-transfer matrix can be written as

$$\mathbf{M} = \begin{pmatrix} 1 & d \\ 0 & 1 \end{pmatrix}. \tag{1.268}$$

Refraction at a planar boundary. As shown in Figure 1.18(b), at a planar boundary between two media of refractive indices n_0 and n, the ray angle changes in accordance with Snell's law $n_0 \sin\theta_0 = n \sin\theta$. In the paraxial approximation, $n_0\theta_0 = n\theta$ and $x = x_0$. The ray-transfer matrix is

$$\mathbf{M} = \begin{pmatrix} 1 & 0 \\ 0 & n_0/n \end{pmatrix}. \tag{1.269}$$

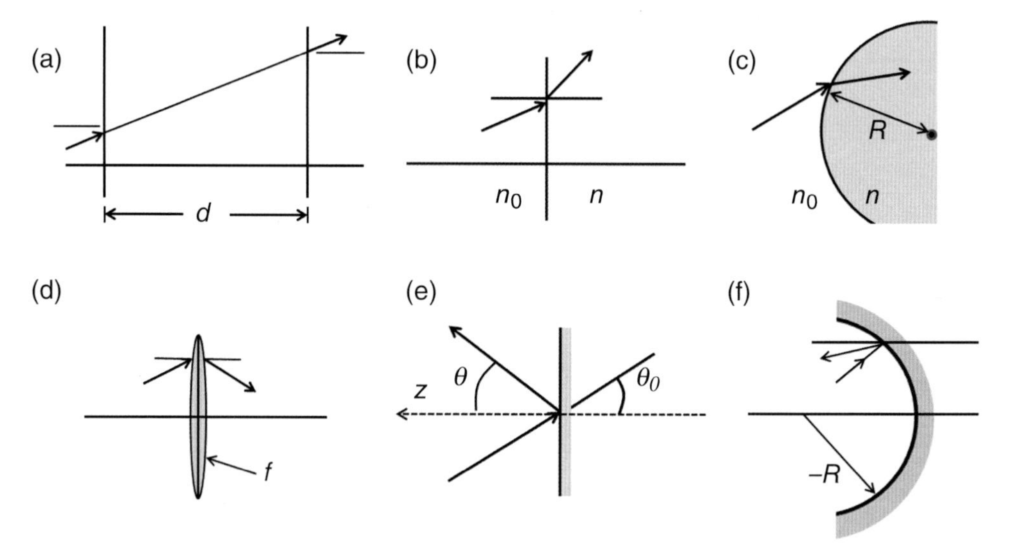

Figure 1.18 Illustration of *ABCD* matrix construction for different optical systems.

Refraction at a spherical boundary. For paraxial rays refracted at a spherical boundary between two media, the relation between θ_0 and θ is provided by

$$\theta \simeq \frac{n_0}{n}\theta_0 - \frac{n - n_0}{nR}x, \tag{1.270}$$

where R is the radius of curvature of the spherical surface. The ray height is not altered, $x \approx x_0$ (see Figure 1.18(c)). The ray-transfer matrix is

$$\mathbf{M} = \begin{pmatrix} 1 & 0 \\ -(n - n_0)/nR & n_0/n \end{pmatrix}. \tag{1.271}$$

Transmission through a thin lens. For paraxial rays transmitted through a thin lens of focal length f, shown in Figure 1.18(d), the relation between θ_0 and θ is given by

$$\theta = \theta_0 - \frac{x}{f}. \tag{1.272}$$

The height remains unchanged ($x = x_0$) so

$$\mathbf{M} = \begin{pmatrix} 1 & 0 \\ -1/f & 1 \end{pmatrix}. \tag{1.273}$$

Reflection from a planar mirror. As shown in Figure 1.18(e), the ray-transfer matrix is the identity matrix

$$\mathbf{M} = \begin{pmatrix} 1 & 0 \\ 0 & 1 \end{pmatrix}. \tag{1.274}$$

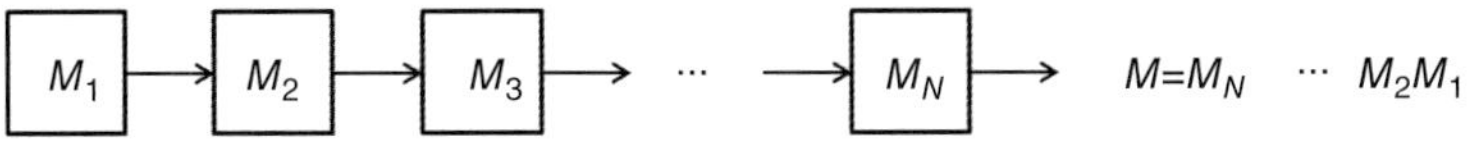

$M_1 \to M_2 \to M_3 \to \cdots \to M_N \to \quad M{=}M_N \cdots M_2M_1$

Figure 1.19 Illustration of a cascaded matrix.

Reflection from a spherical mirror. As shown in Figure 1.18(f), the ray-transfer matrix

$$\mathbf{M} = \begin{pmatrix} 1 & 0 \\ 2/R & 1 \end{pmatrix}. \tag{1.275}$$

Detailed derivation of *ABCD* matrices can be found in [Saleh, Teich].

1.8.4 Matrices of Cascaded Optical Components

A cascade of N optical components or systems whose ray-transfer matrices are $\mathbf{M}_1, \mathbf{M}_2, \ldots, \mathbf{M}_N$ is equivalent to a single optical system of ray-transfer matrix as shown in Figure 1.19

$$\mathbf{M} = \mathbf{M}_N \cdots \mathbf{M}_2\mathbf{M}_1. \tag{1.276}$$

Note that the matrix of the system that is crossed by the rays is first placed to the right. A sequence of matrix multiplication is not, in general, commutative.

1.9 Huygens' Integral through the General *ABCD* System

1.9.1 Huygens' Integral in Free Space

From Equation 1.242, Huygens' integral in one transverse dimension for an optical ray traveling over a distance L in free space can be written in the form

$$\begin{aligned} \widetilde{u}(x) &= e^{-ikL} \int_{-\infty}^{\infty} \widetilde{K}(x, x_0)\widetilde{u}_0(x_0)dx_0 \\ &= \sqrt{\frac{i}{L\lambda}} \int_{-\infty}^{\infty} \widetilde{u}_0(x_0) \exp[-ik\rho(x, x_0)]dx_0, \end{aligned} \tag{1.277}$$

where the path length $\rho(x, x_0)$ from the initial position x_0 on plane z_0 to the final position x on plane $z = z_0 + L$ is given in the paraxial approximation by

$$\rho(x, x_0) = \sqrt{L^2 + (x - x_0)^2} \approx L + \frac{(x - x_0)^2}{2L}. \tag{1.278}$$

This approximate optical path length $\rho(\mathbf{r}, \mathbf{r}_0)$ involved in Huygens' integral is sometimes called the *eikonal function* of an optical source point at $\mathbf{r}_0$ and a field measurement point at $\mathbf{r}$. Thus, the Huygens–Fresnel kernel for wave propagation in free space is expressed as

$$\widetilde{K}(x, x_0) = \sqrt{\frac{i}{L\lambda}} \exp\left[-i\frac{\pi(x - x_0)^2}{L\lambda}\right] \tag{1.279}$$

in one transverse dimension. As discussed in later sections, this form can be extended to two transverse dimensions using a product of two such kernels.

1.9.2 Huygens' Integral through a General *ABCD* Matrix

Let the total optical path length through the paraxial system from plane z_0 to plane z for a ray traveling exactly on the axis be denoted by L_0. This length is given by a sum

$$L_0 = \sum_i n_i L_i, \tag{1.280}$$

where each individual element has physical thickness L_i and index of refraction n_i.

Huygens' integral for an optical ray traveling through the paraxial system from plane z_0 to plane z can be written in the same general form as the free-space situation

$$\widetilde{u}(x) = e^{-ikL_0} \int_{-\infty}^{\infty} \widetilde{K}(x, x_0)\widetilde{u}_0(x_0)dx_0, \tag{1.281}$$

with the kernel given by

$$\begin{aligned}\widetilde{K}(x, x_0) &= \sqrt{\frac{i}{B\lambda_0}} \exp\left[-i\frac{\pi}{B\lambda_0}(Ax_0^2 - 2x_0x + Dx^2)\right] \\ &= \sqrt{\frac{ik}{2\pi B}} \exp\left[-\frac{ik}{2B}(Ax_0^2 - 2x_0x + Dx^2)\right],\end{aligned} \tag{1.282}$$

where λ_0 is the optical wavelength in free space and $k = 2\pi/\lambda_0$. The derivation of Equation 1.282 can be found in section 20.1 of [Siegman]. The scale factor $\sqrt{i/B\lambda_0}$ in front of the kernel is inserted in order to conserve power and make the general Huygens' kernel agree with the free-space result. The matrix element B plays the same role in this kernel as the length $L = z - z_0$ plays in the free-space Huygens' integral in Equation 1.279.

This form of Huygens' integral shows that, within the paraxial approximation, an arbitrary optical wave can be propagated through an optical system by using only the knowledge of the overall *ABCD* coefficients of the system.

1.9.3 Imaging of Coherent Fields through Lenslike Systems

From Equations 1.281 and 1.282, Huygens' integral in two transverse dimensions can be written as

$$\begin{aligned}\widetilde{u}(x, y) = \frac{ik}{2\pi B} \exp(-ikL_0) \int\int \exp\left\{\left(-\frac{ik}{2B}\right)\right. \\ \left.\left[A(x_0^2 + y_0^2) - 2x_0x - 2y_0y + D(x^2 + y^2)\right]\right\}\widetilde{u}_0(x_0, y_0)dx_0dy_0.\end{aligned} \tag{1.283}$$

By rearranging and manipulating the exponent, the Equation 1.283 can be expressed as

$$\widetilde{u}(x,y) = \left(\frac{ik}{2\pi B}\right)\exp(-ikL_0)\exp\left[\left(-\frac{ik}{2B}\right)\left(D-\frac{1}{A}\right)(x^2+y^2)\right]$$
$$\int\int \exp\left\{\left(-\frac{ik}{2B}\right)\left[A\left(x_0-\frac{x}{A}\right)^2+A\left(y_0-\frac{y}{A}\right)^2\right]\right\}\widetilde{u}_0(x_0,y_0)dx_0dy_0. \tag{1.284}$$

Using the identity

$$\lim_{B\to 0}\sqrt{\frac{i}{2\pi B}}\exp\left(-i\frac{x^2}{2B}\right)=\delta(x), \tag{1.285}$$

where $\delta(x)$ is the Dirac delta function. In the limit $B \to 0$ one can get

$$\widetilde{u}(x,y) = \frac{\exp(-ikL_0)}{A}\widetilde{u}_0\left(\frac{x}{A},\frac{y}{A}\right)\exp\left[-i\frac{k(DA-1)}{2AB}(x^2+y^2)\right]. \tag{1.286}$$

For a lossless system, $AD - BC = 1$, Equation 1.286 is rewritten as

$$\widetilde{u}(x,y) = \frac{\exp(-ikL_0)}{A}\widetilde{u}_0\left(\frac{x}{A},\frac{y}{A}\right)\exp\left[-i\frac{kC}{2A}(x^2+y^2)\right]. \tag{1.287}$$

Equation 1.287 shows that when $B = 0$ the output is an exact scaled replica of the input field with the exception of a quadratic phase factor. The image magnification is A, and thus the generalized imaging condition is $B = 0$.

1.9.4 Huygens' Integral in Cylindrical Coordinates

If both of the electric fields $\widetilde{u}_0$ and $\widetilde{u}$ on the input and output planes have cylindrical symmetry, Equation 1.283 can be written as

$$\widetilde{u}(r) = \frac{ik}{2\pi B}\exp(-ikL_0)\int_0^\infty \exp\left[\left(-\frac{ik}{2B}\right)(Ar_0^2-2r_0r+Dr^2)\right]\widetilde{u}_0(r_0)2\pi r_0dr_0. \tag{1.288}$$

By using the relation $J_0(x) = \exp(ix)$ where J_0 is the zero-order Bessel function of the first kind, Equation 1.288 can be rewritten as

$$\widetilde{u}(r) = \frac{ik}{2B}\exp\left[-ik\left(L_0+\frac{Dr^2}{2B}\right)\right]\int_0^\infty \exp\left(-\frac{ikAr_0^2}{2B}\right)J_0\left(\frac{krr_0}{B}\right)\widetilde{u}_0(r_0)r_0dr_0. \tag{1.289}$$

When $B = 0$, Equation 1.287 in cylindrical coordinates becomes

$$\widetilde{u}(r) = \frac{1}{A}\exp\left[-ik\left(L_0+\frac{Cr^2}{2A}\right)\right]\widetilde{u}_0\left(\frac{r}{A}\right). \tag{1.290}$$

In Section 1.9.5, examples will be given for cases where the electric field $\widetilde{u}_0(r_0)$ on the input plane is replaced by a truncated Bessel or truncated Gaussian beam.

1.9.5 Examples of *ABCD* Matrices for Two Optical Systems

In focused-laser experiments there are two kinds of commonly used laser beams. One is the truncated Bessel beam generated on the exit plane of a gas-filled, hollow-core fiber (HCF) where the fiber is used to compress the incident light to produce few-cycle pulses. The other is the truncated Gaussian beam generated by applying an aperture into a laser beam directly from the amplifier. These beams in general are guided with lenses and/or mirrors before they are used for applications.

For a truncated Bessel beam its electric field can be approximately given by $\widetilde{E}_1(r_0) = E_0 J_0(2.405 r_0/a_1)$ with $r_0 \le a_1$, where r_0 is the radial coordinate, E_0 is the on-axis peak electric field, a_1 is the capillary radius, and J_0 is the zero-order Bessel function of the first kind. For a truncated Gaussian beam its electric field is described by a Gaussian function, and an integration over r_0 in Equation 1.289 should be terminated at a_2 where a_2 is the radius of the aperture.

The following demonstrates how to obtain the electric field of a truncated Bessel beam through an optical system used in experiments by constructing its *ABCD* matrix. The primary propagation factor $\exp(-ikL_0)$ can be taken out of $\widetilde{u}(r)$ (similar to Equation 1.243). With a truncated Bessel beam on the input plane, Equations 1.289 and 1.290 can be explicitly written as

$$\widetilde{E}_{TB}(r,\xi) = E_0\left(\frac{ik}{2B}\right)\exp\left(-\frac{ikDr^2}{2B}\right)\int_0^{a_1}\exp\left(-\frac{ikAr_0^2}{2B}\right)J_0\left(\frac{krr_0}{B}\right) \times J_0\left(2.405\frac{r_0}{a_1}\right)r_0 dr_0, \tag{1.291}$$

and

$$\widetilde{E}_{TB}(r,\bar{\xi}) = \frac{E_0}{A}\exp\left(-\frac{ikCr^2}{2A}\right)J_0\left(2.405\frac{r}{a_1 A}\right), \tag{1.292}$$

when $B = 0$ ($\xi = \bar{\xi}$).

Consider the setup of an optical system used in the experiment of Nisoli et al. [8], as depicted in Figure 1.20. The radius of the capillary is $a_1 = 0.25$ mm, the focal length of the focus mirror is $f = 250$ mm, and ξ and $\bar{\xi}$ of the focus plane are shown in Figure 1.20. A laser pulse emerging from the HCF is propagated in free space for a distance $d = 2{,}000$ mm to the focusing mirror where it is further propagated for a distance ξ after the mirror to reach the output plane. The laser pulse is also compressed by chirped mirrors, but they are not included in the *ABCD* matrix. For this optical system, the *ABCD* matrix is found to be

$$\begin{aligned} A(\xi) &= 1 - \xi/f, \\ B(\xi) &= d + \xi(1 - d/f), \\ C &= -1/f, \\ D &= 1 - d/f. \end{aligned} \tag{1.293}$$

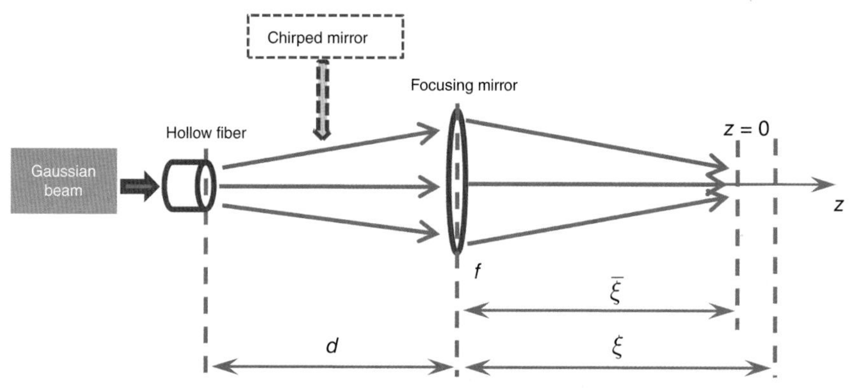

Figure 1.20 Sketch of the experimental setup (see [8]); (Reprinted from Cheng Jin and C. D. Lin, *Phys. Rev. A*, **85**, 033423 (2012) [9]. Copyrighted by the American Physical Society.)

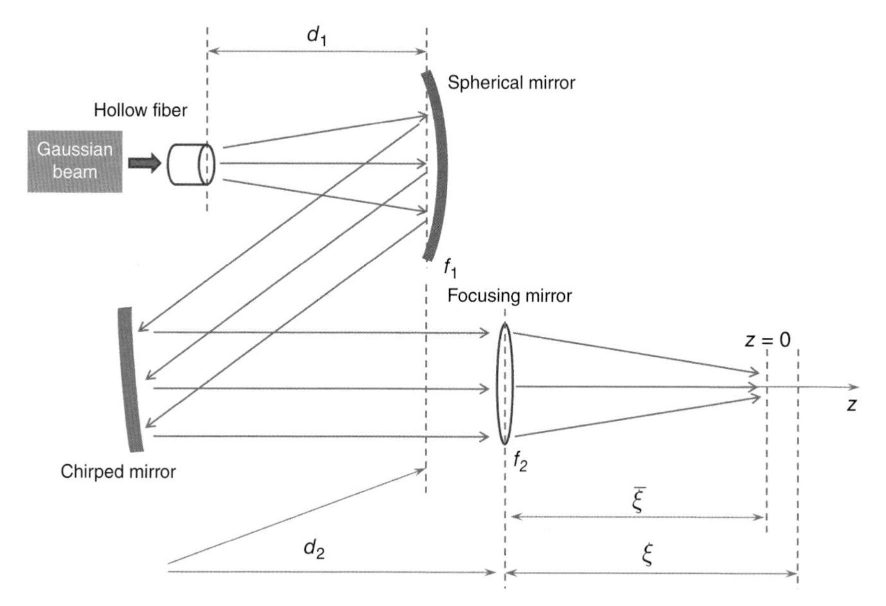

Figure 1.21 Sketch of the experimental setup used in Wörner et al. [10]. (Reprinted from Cheng Jin and C. D. Lin, *Phys. Rev. A*, **85**, 033423 (2012) [9]. Copyrighted by the American Physical Society.)

Next, consider the setup used in the experiment of Wörner et al. [10], as depicted in Figure 1.21. The HCF is similar to Nisoli et al. [8]. The beam that comes out of the HCF (with radius $a_1 = 0.125$ mm) is divergent. It is re-collimated by a spherical mirror (focal length $f_1 = 1{,}000$ mm) placed 1 m after the output of the HCF ($d_1 = 1{,}000$ mm). The beam is then reflected on chirped mirrors eight times and propagated for a distance of 2 m from the spherical mirror ($d_2 = 2{,}000$ mm) to a focusing mirror (focal length $f_2 = 500$ mm). It further propagates for a distance ξ after the mirror to the output plane. Without considering the effect of the chirped mirrors, the *ABCD* matrix for this optical system can be found to be

$$A(\xi) = \left(1 - \frac{d_2}{f_1}\right)\left(1 - \frac{\xi}{f_2}\right) - \frac{\xi}{f_1},$$
$$B(\xi) = \left(d_1 + d_2 - \frac{d_1 d_2}{f_1}\right)\left(1 - \frac{\xi}{f_2}\right) - \xi\left(\frac{d_1}{f_1} - 1\right),$$
$$C = -\frac{1}{f_1} - \frac{1}{f_2} + \frac{d_2}{f_1 f_2},$$
$$D = -\frac{d_1}{f_2} + \left(1 - \frac{d_1}{f_1}\right)\left(1 + \frac{d_2}{f_2}\right). \tag{1.294}$$

If the input is a truncated Gaussian beam, then for the same optical system one can use the same *ABCD* matrix, and Equations 1.291 and 1.292 can also be used except that the electric field on the input plane is replaced by the truncated Gaussian beam.

Notes and Comments

Sections 1.1, 1.2, and 1.3 were adapted from Bransden and Joachain [Bransden, Joachain]. Atomic spectra can be found at the database provided from the National Institute of Standards and Technology (NIST) website. It is desirable to be able to solve the radial wavefunctions of bound states or continuum states for an electron in a given model potential numerically (see Equations 1.19 and 1.21). Such programs are available by directly contacting the authors (atle@phys.ksu.edu). Sections 1.4, 1.5, and 1.6 were mostly extracted from Bransden and Joachain [Bransden, Joachain], Demtröder [Demtröder], and Herzberg [Herzberg, 1945, Herzberg, 1950]. Section 1.5.1 was mostly adopted from the book of Harris and Bertolucci [Harris, Bertolucci]. Sections 1.7.2, 1.7.3, 1.9.1, and 1.9.2 were adapted from Siegman [Siegman], and Sections 1.7.4 and 1.8 were adapted from Saleh and Teich [Saleh, Teich].

Exercises

1.1 Review one-electron atoms in parabolic coordinates. Consider a hydrogen atom in a static electric field $F\hat{z}$, and plot the effective potentials in the ξ and η coordinates. Choose the ground state of hydrogen and take $F = 1$.

1.2 The model potential $V(r)$ for the Li atom can be found in Schweizer and Fassbinder, *Atom. Data Nucl. Data* **72**, 33 (1999). It is given in the form

$$V(r) = -\frac{1}{r}\left[Z_c + (Z - Z_c)\exp(-a_1 r) + a_2 r \exp(-a_3 r)\right], \tag{1.295}$$

with parameters $a_1 = 3.395$, $a_2 = 3.212$, $a_3 = 3.207$, nuclear charge $Z = 3$, and $Z_c = 1$. Calculate the binding energies of 2s, 2p, 3s, 4f states. Compare the calculated values with the data tabulated from NIST's atomic database.

1.3 Use the same model potential $V(r)$ in Problem 1.2 to calculate the differential cross-sections for electron and Li^+ collisions at incident energies of 5, 20, and 40 eV, respectively.

1.4 A model potential for Ar has been tabulated in Tong and Lin, *J. Phys. B*, **38**, 2593 (2005). Calculate the differential cross-sections for e-Ar^+ collisions at the incident energy of 20 eV over the whole angular range. Zoom in to look at the distributions for the angular range of 100–180 degrees.

1.5 Repeat the calculations in Problem 1.4 for different incident energies and compare your results with the results in figure 15 of Chen et al., *Phys. Rev. A*, **79**, 033409 (2009).

1.6 Use the model potential of Ar in Problem 1.4 to calculate photoionization cross-sections of Ar for photon energy from 20 to 80 eV. Assume the electric field is polarized along the z-axis. Consider the forward electron ($\theta = 0$) only. If you replace the continuum wavefunction of the electron by a plane wave, show the corresponding cross-section. Compare your results with figure 1(a) of Le et al., *Phys. Rev. A*, **78**, 023814 (2008).

1.7 Calculate the lifetimes of the 3p and 3d states of atomic hydrogen, of He^+ and of Ne^{9+}. How do they scale with nuclear charge Z?

1.8 Use the simple model potential of Ar in Problem 1.4 to calculate the β-parameter (Equation 1.111) for photoionization from the 3p state of the Ar for photon energies at 40 and 50 eV.

1.9 For two-electron atoms, assume $H_0 = -\frac{1}{2}\nabla_1^2 - \frac{1}{2}\nabla_2^2 - \frac{Z}{r_1} - \frac{Z}{r_2}$ and $H' = \frac{1}{r_{12}}$. Use the first-order perturbation method to calculate the ground-state energy of helium and compare it with the exact energy.

1.10 (a) The $2s2p\ ^1P$ state of helium is a doubly excited state. When it autoionizes, what is the angular momentum of the continuum electron? What are the Fano resonance parameters and the autoionization lifetime (look them up)?
(b) In principle, the $2s2p\ ^1P$ state can also decay radiatively. What is the most likely state it will decay to? Estimate the lifetime due to the radiative decay.
(c) Can the $2p^2\ ^3P$ state decay by autoionization? Will it decay radiatively? If it does, to what state?

1.11 (a) Estimate the first three rotational energies of $^{14}N_2$ and $^{16}O_2$ by using the rigid rotor model with the equilibrium distance $R_e = 1.097$ Å and $R_e = 1.21$ Å for these molecules, respectively.
(b) Taking into account the nuclear-spin statistics discussed in Section 1.6.1, estimate the molecule's angular momentum distribution at a temperature of 100 K.

1.12 Consider N_2, O_2, and F_2 molecules at the equilibrium geometries. Which of these molecules can be expected to be stabilized by removing an electron from the HOMO? How does the bond length change in each case? How does the situation change when an electron is added to the lowest unoccupied molecular orbital (LUMO)?

1.13 Identify the point-group symmetry of the CF_3I molecule and the symmetry for each of the normal vibration modes. Which of the modes is IR active? Which one is Raman active?

1.14 Ethylene (C_2H_4) belongs to the D_{2h} point group. Its ground state electronic configuration is $\ldots, (b_{1u})^2(b_{2g})^0$. By using the D_{2h} product table, identify the symmetry terms for the ground state and the first excited states, which can be generated by exciting an electron from the HOMO (b_{1u}) to the LUMO (b_{2g}). To which of these excited states is the transition from the ground state expected to be strongest or weakest? Why?

1.15 (a) The HOMO (and LUMO) of ethylene (C_2H_4) can be well approximated by a linear symmetric (and antisymmetric) combination of the $2p_z$ of carbon atoms separated by a distance of 1.34 Å along the y-axis. By using the Slater-type orbital of the form

$$\psi_{nlm}(r, \theta, \varphi) = Nr^{n-1} \exp(-\zeta r) Y_{lm}(\theta, \varphi), \tag{1.296}$$

with $\zeta = 1.63$ au^{-1}, plot the HOMO and LUMO wavefunctions as functions of y, z at $x = 0$.

(b) Using the results from (a), calculate the transition dipole for excitation from the HOMO to the LUMO.

1.16 Read section 20.1 in Siegman [Siegman], and derive the Huygens–Fresnel kernel in Equation 1.282 for wave propagation through a general $ABCD$ matrix.

1.17 Using the matrices of simple optical components in Section 1.8.3, calculate the $ABCD$ matrices in Equations 1.293 and 1.294 for the optical systems shown in Figures 1.20 and 1.21, respectively.

1.18 Using the analytical formulae in Equations 1.291 and 1.292, plot the intensity and phase of truncated Bessel beams for the two examples given in Section 1.9.5, and compare your results in figures 13 and 15 of Jin and Lin, *Phys. Rev. A*, **85**, 033423 (2012). [E_0 in Equations 1.291 and 1.292 needs to be adjusted to obtain the same intensities at the focus in Figures 1.13 and 1.15.]

References

[1] U. Fano. Effects of configuration interaction on intensities and phase shifts. *Phys. Rev.*, **124**:1866–1878, Dec. 1961.

[2] U. Fano and J. W. Cooper. Spectral distribution of atomic oscillator strengths. *Rev. Mod. Phys.*, **40**:441–507, Jul. 1968.

[3] C. D. Lin. Doubly excited states, including new classification schemes. *Adv. At. Mol. Phys.*, **22**:77–142, 1986.

[4] W. Kohn and L. J. Sham. Self-consistent equations including exchange and correlation effects. *Phys. Rev.*, **140**:A1133–A1138, Nov. 1965.

[5] A. S. Kheifets, S. Saha, P. C. Deshmukh, D. A. Keating, and S. T. Manson. Dipole phase and photoelectron group delay in inner-shell photoionization. *Phys. Rev. A*, **92**:063422, Dec. 2015.

[6] Z. Farooq, D. A. Chestakov, B. Yan, G. C. Groenenboom, W. J. van der Zande, and D. H. Parker. Photodissociation of singlet oxygen in the UV region. *Phys. Chem. Chem. Phys.*, **16**:3305–3316, 2014.

[7] R. S. Mulliken. Report on notation for the spectra of polyatomic molecules. *J. Chem. Phys.*, **23**(11):1997–2011, 1955.

[8] M. Nisoli, E. Priori, G. Sansone, et al. High-brightness high-order harmonic generation by truncated bessel beams in the sub-10-fs regime. *Phys. Rev. Lett.*, **88**:033902, 2002.

[9] C. Jin and C. D. Lin. Comparison of high-order harmonic generation of Ar using truncated Bessel and Gaussian beams. *Phys. Rev. A*, **85**:033423, 2012.

[10] H. J. Wörner, H. Niikura, J. B. Bertrand, P. B. Corkum, and D. M. Villeneuve. Observation of electronic structure minima in high-harmonic generation. *Phys. Rev. Lett.*, **102**:103901, 2009.

2 Basic Formulation of Interactions between an Intense Laser Pulse and Atoms

2.1 The Formal Theory

2.1.1 Choice of Gauges of the Electromagnetic Fields

Consider a many-electron atom exposed to an ultrashort laser pulse. The basic equation governing the motion of all the electrons in the system is given by the time-dependent Schrödinger equation (TDSE)

$$i\frac{\partial}{\partial t}\Psi(\mathbf{x},t) = H(t)\Psi(\mathbf{x},t), \tag{2.1}$$

$$H(t) = H_0 + H_{\text{int}}(t). \tag{2.2}$$

Here, H_0 is the Hamiltonian of the target, $\mathbf{x}$ represents all the degrees of freedom of the atom and $H_{\text{int}}(t)$ is its interaction with the classical laser field. In principle, if the time-dependent wavefunction $\Psi(\mathbf{x},t)$ is solved exactly, all the experimentally observable parameters can be calculated. In practice, so far such calculations can only be accurately carried out for effective, one-electron atomic systems. Our goal is to develop models for complex many-electron systems. But to calibrate the model we test the results from the model against TDSE results for simple one-electron atoms. Thus, for simplicity, our formulation starts with a one-electron atom where

$$H_0 = \frac{p^2}{2} + V(r), \tag{2.3}$$

with $V(r)$ being a central-field model potential seen by the electron. Since laser interaction does not change the spin of the electron, within the nonrelativistic theory we further ignore its spin. Within this model, the Hilbert space of the atom is generated by the complete set of eigenfunctions $|u_i\rangle$ of H_0:

$$H_0|u_i\rangle = \varepsilon_i|u_i\rangle, \tag{2.4}$$

where ε_i is the eigenenergy. In classical electrodynamics the Hamiltonian of an electron in electromagnetic fields is given by (in the Coulomb gauge $\nabla \cdot \mathbf{A} = 0$)

$$H = \frac{1}{2}(\mathbf{p}+\mathbf{A})^2 + V \tag{2.5}$$

$$= \frac{p^2}{2} + V + \mathbf{A}\cdot\mathbf{p} + \frac{A^2}{2} \tag{2.6}$$

$$= H_0 + H_{\text{int}}. \tag{2.7}$$

Thus

$$H_{\text{int}} = \mathbf{A} \cdot \mathbf{p} + \frac{A^2}{2}. \tag{2.8}$$

For typical infrared laser wavelengths, the vector potential $\mathbf{A}$ is independent of the position coordinates. Thus, the A^2 term can be eliminated by a unitary transformation

$$\Psi(\mathbf{x}, t) = \Psi^V(\mathbf{x}, t) e^{-i \int^t \frac{1}{2} A^2(t') dt'} \tag{2.9}$$

such that

$$i \frac{\partial}{\partial t} \Psi^V(\mathbf{x}, t) = [H_0 + \mathbf{A} \cdot \mathbf{p}]\, \Psi^V(\mathbf{x}, t). \tag{2.10}$$

The $\Psi^V(\mathbf{x}, t)$ is expressed in the velocity gauge since the interaction is given in terms of the momentum operator. If we choose the unitary transform

$$\Psi(\mathbf{x}, t) = e^{-i\mathbf{A}\cdot\mathbf{x}} \Psi^L(\mathbf{x}, t), \tag{2.11}$$

then we obtain

$$i \frac{\partial}{\partial t} \Psi^L(\mathbf{x}, t) = [H_0 + \mathbf{F}(t) \cdot \mathbf{x}]\, \Psi^L(\mathbf{x}, t). \tag{2.12}$$

This equation is expressed in the length gauge since the dipole operator is proportional to $\mathbf{x}$ where the electric field $\mathbf{F} = -\partial \mathbf{A}/\partial t$. Finally, by defining

$$\Psi^V(\mathbf{x}, t) = e^{-i\boldsymbol{\alpha}\cdot\mathbf{p}} \Psi^A(\mathbf{x}, t), \tag{2.13}$$

where

$$\boldsymbol{\alpha}(t) = \int^t \mathbf{A}(t') dt', \tag{2.14}$$

the TDSE is expressed in the acceleration gauge

$$i \frac{\partial}{\partial t} \Psi^A(\mathbf{x}, t) = \left[\frac{p^2}{2} + V(\mathbf{x} + \boldsymbol{\alpha}(t)) \right] \Psi^A(\mathbf{x}, t). \tag{2.15}$$

In the acceleration gauge, the electron is moving with the electric field $\mathbf{F}(t)$. It sees a time-dependent potential $V(\mathbf{x} + \boldsymbol{\alpha}(t))$ from the atomic core. This frame is also called the *Kramers–Henneberger (K–H) frame*. For a free electron in a linearly polarized monochromatic field $F(t) = F_0 \sin \omega t$, $\alpha(t) = -\alpha_0 \sin \omega t$, where $\alpha_0 = F_0/\omega^2$ is the excursion amplitude of the electron in the laser field. In atomic units, α has the units of displacement, the vector potential has the units of velocity, and the electric field has the units of acceleration. Classically, a free electron in a sinusoidal field $F(t) = F_0 \sin \omega t$ executes a simple harmonic oscillation. The cycle-averaged kinetic energy is $U_p = F_0^2/4\omega^2$. U_p is called the *cycle-averaged quiver energy* of an electron in the laser field, or the *ponderomotive energy*. Note that U_p is proportional to $I\lambda^2$. In the infrared wavelength region,

$$U_p(\text{eV}) = 9.33 \times I(10^{14}\ \text{W/cm}^2)\lambda^2(\mu\text{m}). \tag{2.16}$$

For an 800 nm laser at a peak intensity of 10^{14} W/cm^2, U_p is 6 eV, which is close to the typical binding energy of an atom or molecule.

Note that wavefunctions in different gauges are related to each other by a unitary transformation. Physical observables calculated with the exact numerical solution of the TDSE should be independent of the gauge chosen. However, if the TDSE equation is solved approximately, the final results depend on the gauge. This holds true for the conventional, single-photon ionization processes, the nonlinear multiphoton ionization processes, or in high-order harmonic generation.

2.1.2 Volkov Wavefunction

In the absence of an atomic potential, the TDSE of a free electron in an oscillatory laser field can be solved analytically. The solution is called the *Volkov wavefunction*, and its specific expression depends on the gauge chosen. Starting from Equation 2.10, write $\Psi^V(\mathbf{x},t) = e^{i\mathbf{k}\cdot\mathbf{x}}\varphi(t)$ to pull out the plane-wave part, then $\varphi(t)$ satisfies

$$i\frac{d}{dt}\varphi(t) = \left[\frac{k^2}{2} + \mathbf{k}\cdot\mathbf{A}(t)\right]\varphi(t), \tag{2.17}$$

which can be integrated to

$$\varphi(t) = Ce^{-ik^2t/2 - i\mathbf{k}\cdot\boldsymbol{\alpha}(t)}. \tag{2.18}$$

Thus, the Volkov state in the velocity gauge is given by

$$\Psi^V(\mathbf{x},t) = Ce^{i\mathbf{k}\cdot(\mathbf{x}-\boldsymbol{\alpha}(t)) - iEt} \tag{2.19}$$

with $E = k^2/2$. From Equation 2.9, the Volkov wavefunction for the original interaction Hamiltonian Equation 2.8 can be readily obtained.

A similar procedure can be used to derive the Volkov wavefunctions from the TDSE Equations 2.12 and 2.15 for the length gauge and the acceleration gauge, respectively.

$$\Psi^L(\mathbf{x},t) = Ce^{i\left\{[\mathbf{k}+\mathbf{A}(t)]\cdot\mathbf{x} - \frac{1}{2}\int_{-\infty}^{t}[\mathbf{k}+\mathbf{A}(t')]^2dt'\right\}}, \tag{2.20}$$

$$\Psi^A(\mathbf{x},t) = Ce^{i(\mathbf{k}\cdot\mathbf{x} - Et)}. \tag{2.21}$$

In all of these equations, C depends on the convention of normalization for the continuum states. If $\langle\Psi_{\mathbf{k}}|\Psi_{\mathbf{k}'}\rangle = \delta(\mathbf{k}-\mathbf{k}')$, then $C = (2\pi)^{-3/2}$. If it is energy normalized, $\langle\Psi_{\mathbf{k}}|\Psi_{\mathbf{k}'}\rangle = \delta(E-E')\delta(\hat{k},\hat{k}')$ is chosen, then $C = (2\pi)^{-3/2}\sqrt{k}$ where $k = |\mathbf{k}|$ (see Equations 1.10 and 1.11).

2.2 Formulation of the Solution of the TDSE

In this section, two direct numerical methods for solving the TDSE are given. Two other methods based on the perturbation-expansion methods are also outlined. Approximate solutions will then be discussed in Sections 2.4 and 2.5.

2.2.1 Numerical Solution of TDSE

For a given model potential $V(r)$ and known laser pulse $F(t)$, the TDSE can be solved numerically. The length gauge is chosen and the laser is linearly polarized along the z-axis. In terms of the complete set of eigenstates $f_{nl}(r)Y_{lm}(\hat{\mathbf{r}})$ of $H_0 = -\nabla^2/2 + V(r)$, the time-dependent wavefunction $\Psi(\mathbf{r}, t)$ can be expanded as

$$\Psi(\mathbf{r}, t) = \sum_{nl} C_{nl}(t) f_{nl}(r) Y_{lm}(\hat{\mathbf{r}}). \tag{2.22}$$

Here, the Y_{lm} functions are the spherical harmonics. The radial eigenfunctions $f_{nl}(r)$ are often expanded in terms of some basis set, like the discrete variable representation (DVR) functions. After projecting the angular part and confining the radial wavefunctions within a finite range $r \in [0, r_{\max}]$, i.e., within a box, the time-dependent coefficients $C_{nl}(t)$ can be solved by the split-operator method

$$\begin{aligned} C_{nl}(t+\Delta t) \approx \sum_{n'l'} \{&\exp(-iH_0\Delta t/2)\exp[-iH'(t+\Delta t/2)\Delta t] \\ &\times \exp(-iH_0\Delta t/2)\}_{nl,n'l'} C_{n'l'}(t), \end{aligned} \tag{2.23}$$

where H' is the interaction term. The matrix elements are efficiently evaluated using the DVR quadrature. For linearly polarized lights, the interaction system has cylindrical symmetry; thus, m is a good quantum number where the polarization axis is taken to be the quantization axis. If the ground state of the atom is not an s state, then each magnetic substate can be solved separately.

Once the time-dependent wavefunction $\Psi(\mathbf{r}, t)$ has been solved, the angular distribution (or the momentum distribution) of the photoelectron can be extracted by projecting the final wavefunction at the end of the pulse onto eigenstates of the target. For an electron emitted with energy E and angle θ (no φ dependence due to cylindrical symmetry), the eigenfunction with well-defined momentum vector $\mathbf{p}$ is given by

$$\Phi_{\mathbf{p}}^{-} = \sum_{l} \sum_{m=-l}^{l} i^l e^{-i\eta_l} R_{pl}(r) Y_{lm}^{*}(\hat{\mathbf{p}}) Y_{lm}(\hat{\mathbf{r}}). \tag{2.24}$$

This eigenfunction satisfies the time-independent Schrödinger equation

$$\left[-\frac{1}{2}\nabla^2 + V(r)\right]\Phi_{\mathbf{p}}^{-} = E\Phi_{\mathbf{p}}^{-}. \tag{2.25}$$

The solution of the continuum wavefunction was discussed in Equation 1.104. Doubly differential probability distributions of the photoelectron are then given by

$$\frac{\partial^2 P}{\partial E \partial \theta} = \left| \sum_{l} (-i)^l e^{i\eta_l} Y_{lm}(\hat{\mathbf{p}}) \sum_{n} C_{nl}(\tau) \langle R_{pl}(r) | f_{nl}(r) \rangle \right|^2 2\pi p \sin\theta, \tag{2.26}$$

where τ is the end of the laser pulse, and the normalization $\langle \Phi_{\mathbf{p}}^{-} | \Phi_{\mathbf{p}'}^{-} \rangle = \delta(\mathbf{p} - \mathbf{p}')$ is chosen. By integrating over the scattering angle, the electron energy spectra

$$\frac{\partial P}{\partial E} = \int \frac{\partial^2 P}{\partial E \partial \theta} d\theta \tag{2.27}$$

can be calculated.

The derivation above is limited to linearly polarized laser pulses. The specification of the laser field will be addressed in later sections. If circularly or elliptically polarized lights are used, the expansion Equation 2.22 has to include summation over m.

2.2.2 Weak-Field Perturbation Expansion Method

Consider $H(t) = H_0 + H_i(t)$ again, where H_0 is the atomic Hamiltonian and $H_i(t)$ is the interaction with the electromagnetic field in the length gauge. When $H_i(t)$ is small, one can solve

$$i\frac{\partial}{\partial t}|\Psi(t)\rangle = H(t)|\Psi(t)\rangle \tag{2.28}$$

by treating $H_i(t)$ as a small perturbation, using the time-dependent perturbation theory. Define a unitary time-evolution operator $U(t, t_0)$ where

$$|\Psi(t)\rangle = U(t, t_0)|\Psi(t_0)\rangle, \tag{2.29}$$

$$i\frac{\partial}{\partial t}U(t, t_0) = H(t)U(t, t_0) = (H_0 + H_i)U(t, t_0). \tag{2.30}$$

Similarly, one can define a time-evolution operator $U_0(t, t_0)$ for H_0 where

$$i\frac{\partial}{\partial t}U_0(t, t_0) = H_0 U_0(t, t_0). \tag{2.31}$$

Clearly,

$$U_0(t, t_0) = e^{-iH_0(t-t_0)} = \sum_k e^{-iE_k(t-t_0)}|u_k\rangle\langle u_k|, \tag{2.32}$$

where $|u_k\rangle$ is the eigenstate of H_0 with energy E_k.

Equation 2.30 can be solved iteratively for small $H_i(t)$

$$U(t, t_0) = U_0(t, t_0) - i\int_{t_0}^{t} dt_1 U_0(t, t_1)H_i(t_1)U(t_1, t_0). \tag{2.33}$$

Write

$$U(t, t_0) = U_0(t, t_0) + \sum_{n=1}^{\infty} U^{(n)}(t, t_0) \tag{2.34}$$

then

$$U^{(n)}(t, t_0) = (-i)^n \int_{t_0}^{t} dt_1 \int_{t_0}^{t_1} dt_2 \ldots \int_{t_0}^{t_{n-1}} dt_n \times U_0(t, t_1)H_i(t_1)U_0(t_1, t_2)\ldots U_0(t_{n-1}, t_n)H_i(t_n)U_0(t_n, t_0) \quad n \geq 1. \tag{2.35}$$

Let the atom be in the initial state at $t = t_0$:

$$|\phi_i(t_0)\rangle = |u_i\rangle e^{-iE_i t_0}, \tag{2.36}$$

the state at time t is

$$|\Psi_i(t)\rangle = U(t, t_0)|\phi_i(t_0)\rangle = \sum_{n=0}^{\infty} |\Psi_i^{(n)}(t)\rangle \tag{2.37}$$

where $|\Psi_i^{(0)}(t)\rangle = |u_i\rangle e^{-iE_it}$. For $n \geq 1$

$$|\Psi_i^{(n)}(t)\rangle = (-i)^n \int_{t_0}^{t} dt_1 \int_{t_0}^{t_1} dt_2 \ldots \int_{t_0}^{t_{n-1}} dt_n \times U_0(t,t_1)H_i(t_1)U_0(t_1,t_2)\ldots U_0(t_{n-1},t_n)H_i(t_n)|\phi_i(t_n)\rangle. \quad (2.38)$$

The transition amplitude to a final state $|\phi_f(t)\rangle$ at time t is

$$a_{fi}(t) = \langle\phi_f(t)|\Psi_i(t)\rangle = \sum_{n=1}^{\infty}\langle\phi_f(t)|\Psi_i^{(n)}(t)\rangle = \sum_{n=1}^{\infty} a_{fi}^{(n)}(t), \quad (2.39)$$

where

$$a_{fi}^{(n)}(t) = (-i)^n \int_{t_0}^{t} dt_1 \int_{t_0}^{t_1} dt_2 \ldots \int_{t_0}^{t_{n-1}} dt_n \times \langle\phi_f(t_1)|H_i(t_1)U_0(t_1,t_2)\ldots U_0(t_{n-1},t_n)H_i(t_n)|\phi_i(t_n)\rangle. \quad (2.40)$$

By inserting $U_0(t_p,t_q) = \sum_k e^{-iE_k(t_p-t_q)}|u_k\rangle\langle u_k|$,

$$a_{fi}^{(n)}(t) = (-i)^n \sum_{k_1,k_2,\ldots,k_{n-1}} \int_{t_0}^{t} dt_1 \int_{t_0}^{t_1} dt_2 \ldots \int_{t_0}^{t_{n-1}} dt_n \times e^{iE_ft_1}\langle u_f|H_i(t_1)|u_{k_1}\rangle e^{-iE_{k_1}(t_1-t_2)}\langle u_{k_1}|H_i(t_2)|u_{k_2}\rangle \times \ldots e^{-iE_{k_{n-1}}(t_{n-1}-t_n)}\langle u_{k_{n-1}}|H_i(t_n)|u_i\rangle e^{-iE_it_n} \quad (2.41)$$

is obtained. Specifically,

$$a_{fi}^{(1)}(t) = -i\int_{t_0}^{t} \langle u_f|H_i(t_1)|u_i\rangle e^{i\omega_{fi}t_1}dt_1, \quad (2.42)$$

$$a_{fi}^{(2)}(t) = (-i)^2\sum_{k_1}\int_{t_0}^{t} dt_1 \int_{t_0}^{t_1} dt_2 e^{iE_ft_1}\langle u_f|H_i(t_1)|u_{k_1}\rangle \times e^{-iE_{k_1}(t_1-t_2)}\langle u_{k_1}|H_i(t_2)|u_i\rangle e^{-iE_it_2}. \quad (2.43)$$

Consider a short light pulse expressed as

$$\mathbf{F}(t) = \hat{\epsilon}F_0(t)\cos(\omega t + \alpha). \quad (2.44)$$

In the length gauge $\mathbf{d} = -\mathbf{r}$

$$H_i(t) = -\frac{F_0(t)}{2}\left[e^{i(\omega t+\alpha)} + e^{-i(\omega t+\alpha)}\right]\hat{\epsilon}\cdot\mathbf{d}. \quad (2.45)$$

If $F_0(t) = F_0$, i.e., a monochromatic light, then the first-order theory reduces to the theory given in Section 1.1.4. For a finite pulse, the amplitude is

$$a_{fi}^{(1)}(t) = -iM_{fi}G(t), \tag{2.46}$$

$$G(t) = \int_{t_0}^{t} F_0(t_1)\cos(\omega t_1 + \alpha)e^{i\omega_{fi}t_1}\,dt_1, \tag{2.47}$$

where $M_{fi} = \langle u_f|z|u_i\rangle$ if the light is polarized along the z direction.

For the absorption process, the transition amplitude peaks at $\omega \approx \omega_{fi}$, with the bandwidth governed by the Fourier transform of $F_0(t)$. For the time being, the absorption process is considered in a monochromatic light wave. The first-order transition probability for absorption is given by

$$P_{fi}^{(1)} = |a_{fi}^{(1)}|^2 = 2\pi\frac{F_0^2}{4}|M_{fi}|^2 t\delta(E_f - E_i - \omega) \tag{2.48}$$

from which the absorption rate is obtained,

$$W_{fi}^{(1)} = 2\pi\frac{F_0^2}{4}|M_{fi}|^2\rho(E_f), \tag{2.49}$$

where $\rho(E_f)$ is the density of states. If the continuum wavefunction is normalized per unit energy, then $\rho(E_f) = 1$.

Similarly, the transition rate for two-photon absorption is

$$W_{fi}^{(2)} = 2\pi\left(\frac{F_0^2}{4}\right)^2|T_{fi}^{(2)}|^2\rho(E_f), \tag{2.50}$$

where

$$T_{fi}^{(2)} = \sum_k \frac{\langle u_f|\hat{\epsilon}\cdot\mathbf{d}|u_k\rangle\langle u_k|\hat{\epsilon}\cdot\mathbf{d}|u_i\rangle}{(E_i + \omega) - E_k}. \tag{2.51}$$

The summation is over all the intermediate states. This can be further generalized to the absorption of n photons

$$W_{fi}^{(n)} = 2\pi\left(\frac{F_0^2}{4}\right)^n|T_{fi}^{(n)}|^2\rho(E_f), \tag{2.52}$$

where the n-photon transition amplitude is

$$T_{fi}^{(n)} = \sum_{k_1,k_2,\ldots,k_{n-1}} \frac{\langle u_f|\hat{\epsilon}\cdot\mathbf{d}|u_{k_{n-1}}\rangle\ldots\langle u_{k_2}|\hat{\epsilon}\cdot\mathbf{d}|u_{k_1}\rangle\langle u_{k_1}|\hat{\epsilon}\cdot\mathbf{d}|u_i\rangle}{(E_i + (n-1)\omega - E_{k_{n-1}})\ldots(E_i + 2\omega - E_{k_2})(E_i + \omega - E_{k_1})}. \tag{2.53}$$

Additional discussion can be found in Peng et al. [1].

2.2.3 Strong-Field or S-Matrix Expansion Method

The total Hamiltonian can be separated in two different ways:

$$H(t) = \left(-\frac{1}{2}\nabla^2 + V(r)\right) + \mathbf{r}\cdot\mathbf{F}(t) = H_0 + H_i, \tag{2.54}$$

$$= \left(-\frac{1}{2}\nabla^2 + \mathbf{r}\cdot\mathbf{F}(t)\right) + V(r) = H_F + V. \tag{2.55}$$

The upper equation is used, as in Section 2.2.2, for perturbation expansion if H_i is small. However, if the laser field is strong, then the electron in the final state is mostly governed by the Hamiltonian H_F. The probability amplitude of detecting a photoelectron with momentum $\mathbf{p}$ can be written as

$$f(\mathbf{p}) = -i \lim_{t\to\infty} \int_{-\infty}^{t} dt' \langle \Psi_{\mathbf{p}}(t)|U(t,t')H_i(t')|\Psi_0(t')\rangle, \tag{2.56}$$

where $U(t,t')$ is the time-evolution operator that satisfies the Dyson equation

$$U(t,t') = U_F(t,t') - i \int_{t'}^{t} dt'' U_F(t,t'')VU(t'',t'), \tag{2.57}$$

where $U_F(t,t')$ is the time-evolution operator for the Hamiltonian of a free electron in the laser field H_F as given in Equation 2.55.

The eigenstates of $H_F(t)$ are the Volkov states, Equation 2.20,

$$|\chi_{\mathbf{p}}(t)\rangle = |\mathbf{p} + \mathbf{A}(t)\rangle \exp[-iS_{\mathbf{p}}(t)], \tag{2.58}$$

with the action

$$S_{\mathbf{p}}(t) = \frac{1}{2} \int_{-\infty}^{t} dt' [\mathbf{p} + \mathbf{A}(t')]^2. \tag{2.59}$$

The vector potential of the laser field $\mathbf{F}(t)$ is denoted by $\mathbf{A}(t)$, and $|\mathbf{k}\rangle$ is a plane-wave state

$$\langle \mathbf{r}|\mathbf{k}\rangle = \frac{1}{(2\pi)^{3/2}} \exp(i\mathbf{k}\cdot\mathbf{r}). \tag{2.60}$$

The Volkov time-evolution operator is

$$U_F(t,t') = \int d\mathbf{k} |\chi_{\mathbf{k}}(t)\rangle\langle\chi_{\mathbf{k}}(t')|. \tag{2.61}$$

By approximating $U(t'',t')$ on the right-hand side of Equation 2.57 by $U_F(t'',t')$, and $\langle\Psi_{\mathbf{p}}(t)|$ in Equation 2.56 by $\langle\chi_{\mathbf{p}}(t)|$, the ionization-probability amplitude may be expressed as

$$f = f^{(1)} + f^{(2)}, \tag{2.62}$$

where the first term

$$f^{(1)} = -i \int_{-\infty}^{\infty} dt \langle\chi_{\mathbf{p}}(t)|H_i(t)|\Psi_0(t)\rangle \tag{2.63}$$

corresponds to the standard strong-field approximation (SFA). This term will be called SFA1 from time to time when clarity is needed. The second-order term is SFA2

$$f^{(2)} = -\int_{-\infty}^{\infty} dt \int_{-\infty}^{t} dt' \int d\mathbf{k} \langle\chi_{\mathbf{p}}(t)|V|\chi_{\mathbf{k}}(t)\rangle\langle\chi_{\mathbf{k}}(t')|H_i(t')|\Psi_0(t')\rangle, \tag{2.64}$$

which accounts for the first-order correction by the atomic potential. This expression can be easily understood by reading it from the right side to the left. The electron is first released by the laser field at time t'. It then propagates in the laser field from t' to t, when it is rescattered by the atomic potential V into a state with momentum $\mathbf{p}$. Thus, this second-order term describes the rescattering of the electron with the ion core. Note that, in the SFA expansion, the final states are eigenstates of H_F while the initial state is the eigenfunction of H_0. The two sets of functions are not orthogonal.

It is important to point out that the basis functions used in the S-matrix formulation are not orthogonal. A similar situation occurs in the theoretical treatment of any rearrangement scattering problem such as a charge-transfer process in ion–atom collision, where the initial state is the eigenstate of the target while the final state is the eigenstate of the projectile. There is a large amount of literature on ion–atom collisions that addresses the different forms of perturbation-expansion series. The corresponding first-order term $f^{(1)}$ in charge transfer is called *Oppenheimer–Brinkman–Kramer* (OBK) [2, 3] theory, while the second-order term $f^{(2)}$ is called the *second-Born approximation* [4]. There are similarities in physical phenomena between strong-field ionization and charge transfer as well. For example, for charge-transfer collisions at high energies, the second-Born term is dominant over the first Born (the OBK theory). In laser–atom collisions, high-energy ATI spectra are dominated by the $f^{(2)}$ term, i.e., the back rescattering term, rather than by the direct ionization term $f^{(1)}$. Using the propagator $U_F(t'', t')$, it is straightforward to derive higher-order terms in the perturbation expansion. However, there has been no proof that the S-matrix expansion series converges in both examples. For more details on this method see Becker and Faisal [5].

2.2.4 Numerical Solution of the Time-Dependent Schrödinger Integral Equation

A more efficient way to include higher-order terms in the S-matrix formulation is to solve the TDSE in integral equation form (see Tong et al. [6]). Here, the time-dependent wavefunction from $T = -\infty$ to t is expressed as

$$\Psi(t) = -i \int_T^t U(t, \tau) H_i(\tau) U_0(\tau, T) \Phi_0 d\tau + U_0(t, T)\Phi_0 \tag{2.65}$$

with the propagator $U_0(t, \tau) = e^{-iH_0(t-\tau)}$ and $U(t, \tau) = e^{-i\int_\tau^t H dt'}$ of H_0 and $H = H_0 + H_i(t)$, respectively, and Φ_0 being the initial field-free wavefunction. The second term on the right of this expression gives the propagation of the initial state Φ_0 from T to t. The first term propagates by U_0 from T to τ, where it interacts with the laser field $H_i(\tau)$ and then propagates to time t under the full Hamiltonian. Note that if U is replaced by the propagator of the Hamiltonian H_F of a free electron in the laser field, $U_F(t, \tau)$, the SFA is recovered.

The propagation using the full $U(t, \tau)$ is equivalent to the full solution of the TDSE. To eliminate the reflection of the wave packet from the boundary in a typical numerical calculation, at each time t_i, the wavefunction is split into two regions, I and II

$$\Psi(t_i) = \Psi_{\text{I}}(t_i) + \Psi_{\text{II}}(t_i) = \Psi(t_i)[1 - F_s(R_c)] + \Psi(t_i)F_s(R_c), \tag{2.66}$$

where $F_s(R_c) = 1/[1 + e^{-(r-R_c)/\Delta}]$. Here, R_c separates the inner region $I(0 \to R_c)$ and the outer region $II(R_c \to R_{max})$ and Δ is the width of the crossover region. To simplify the calculation, $\Psi_{II}(t)$ is obtained using the propagator from only the laser field. The criterion to choose the $t_1, t_2, \ldots t_i$ is that the fastest electron velocity is smaller than $(R_{max} - R_c)/(t_i - t_{i-1})$. This is first done by defining $\Psi_{II}(t)$ in the momentum space

$$C(\mathbf{p}, t_i) = \frac{1}{(2\pi)^{3/2}} \int \Psi_{II}(t_i) e^{-i[\mathbf{p}+\mathbf{A}(t_i)]\cdot\mathbf{r}} d^3 r, \tag{2.67}$$

then it propagates with the propagator of the Hamiltonian H_F to infinity,

$$\Psi_{II}(\infty) = U_F(\infty, t_i) \int C(\mathbf{p}, t_i) \frac{e^{i\mathbf{p}\cdot\mathbf{r}}}{(2\pi)^{3/2}} d^3 p. \tag{2.68}$$

With $\bar{C}(\mathbf{p}, t_i) = U_F(\infty, t_i) C(\mathbf{p}, t_i)$, the angular distribution is then given by the coherent superposition of the amplitudes

$$\frac{dP}{dEd\Omega} = \sqrt{2E} \left| \sum_i \bar{C}(\mathbf{p}, t_i) \right|^2. \tag{2.69}$$

As in the S-matrix method, the integral approach is advantageous as subcycle ionization dynamics can be analyzed,

$$\Psi(\infty) = -i \sum_n \int_{t_{n-1}}^{t_n} U(\infty, \tau) H_i(\tau) U_0(\tau, -\infty) \Phi_0 d\tau + U_0(\infty, -\infty) \Phi_0. \tag{2.70}$$

This expression is useful for analyzing the interferences of photoelectron momentum distributions from the different pathways.

2.2.5 The Floquet Method

For a long laser pulse that can be approximated as a monochromatic laser field $F(t) = F_0 \cos(\omega t)$, the total Hamiltonian is periodic, $H(t + T) = H(t)$, where T is the period. Therefore the TDSE

$$i \frac{\partial \Psi(t)}{\partial t} = H(t) \Psi(t), \tag{2.71}$$

in which

$$H(t) = H_0 + \hat{V} e^{i\omega t} + \hat{V}^\dagger e^{-i\omega t} \tag{2.72}$$

can be expressed by introducing

$$\Psi(t) = e^{-i\varepsilon t} \Phi_\varepsilon(t). \tag{2.73}$$

The function $\Phi_\varepsilon(t)$ should satisfy

$$\left(H - i \frac{\partial}{\partial t} \right) \Phi_\varepsilon = \varepsilon \Phi_\varepsilon. \tag{2.74}$$

This equation appears as a time-independent eigenvalue problem with the new Hamiltonian

$$\hat{H} = H - i\frac{\partial}{\partial t}. \tag{2.75}$$

The eigenvalues of Equation 2.75 are called quasi-energies, and the eigenfunctions are called *quasi-energy states* or *Floquet states*. In fact, for each eigenvalue, there is a complete set of Floquet states $\Phi_\varepsilon e^{-in\omega t}$, with eigenvalues $\varepsilon - n\omega$, where $n = 0, \pm 1, \pm 2, \cdots$. Any eigenstates of Equation 2.75 can then be expressed as

$$\Phi_\varepsilon = \sum_n e^{-in\omega t}\Psi_{\varepsilon,n}. \tag{2.76}$$

According to Equations 2.72 and 2.74, the equations for $\Psi_{\varepsilon,n}$ can be obtained

$$H_0\Psi_{\varepsilon,n} + \hat{V}\Psi_{\varepsilon,n+1} + \hat{V}^\dagger\Psi_{\varepsilon,n-1} = (\varepsilon + n\omega)\Psi_{\varepsilon,n}, \tag{2.77}$$

with

$$\varepsilon_i = E_i + \Delta_i - i\Gamma_i/2. \tag{2.78}$$

The complex eigenvalues give the AC Stark shift Δ_i and the decay width Γ_i of each atomic line. The Floquet method has been widely used since the 1990s to study atoms and molecules in strong laser fields where the pulse durations are in the order of picoseconds. It is the most powerful method in the multiphoton ionization regime. A detailed description of the Floquet method can be found in chapter 4 of [Joachain, Kylstra, Potvliege].

2.2.6 Many-Electron Theories

The TDSE for a many-electron atom or molecule in an intense laser field can be obtained by generalizing from the one-electron atom theory. The main challenge, of course, is finding suitable methods to solve such a multidimensional TDSE, including all the degrees of freedom. The simplest direct-numerical approach is to discretize the many-electron wavefunction in the spatial grids (see chapter 8 of [Schultz, Vrakking]). Let $c_i(t)$ be the complex N-body wavefunction at time t. The TDSE can be rewritten in terms of a system of first-order, coupled, ordinary differential equations. The size of such equations will be enormous. So far, such reports of direct, many-particle TDSE calculations are mostly limited to a helium atom in a linearly polarized laser field. The accuracy of such calculations is difficult to affirm except in the few-photon UV to XUV ionization regimes where the nonlinearity is small.

The structures of atoms and molecules in the absence of external fields are not directly studied on the grid points, but are more conveniently formulated in terms of electron orbitals. Many quantum chemistry packages have been developed to calculate the electronic structure of a molecule such as the density functional theory (DFT), Hartree–Fock (HF) or multi-configuration HF (MCHF) theory, and the R-matrix method. The DFT is probably the most widely used method where the many-electron correlation effect is empirically incorporated into the so-called *exchange-correlation potential*. The time-dependent DFT (TDDFT) has been extended to study simple molecules in intense laser fields [7, 8]. The TDDFT is a straight extension of the independent electron model. For each electron, or more precisely, each spin-orbital $\{i\sigma\}$ where i is the orbital and σ is spin up or down, the time-dependent one-electron orbital satisfies

$$i\frac{\partial}{\partial t}\varphi_{i\sigma}(\mathbf{r},t) = \left[\frac{p^2}{2} + V_{eff}(\mathbf{r},t)\right]\varphi_{i\sigma}(\mathbf{r},t), \tag{2.79}$$

$$V_{eff}(\mathbf{r},t) = V_{ne}(\mathbf{r},t) + V_h(\mathbf{r},t) + V_{xc,\sigma}(\mathbf{r},t) + \mathbf{r}\cdot\mathbf{F}(t). \tag{2.80}$$

Here, $V_{ne}(\mathbf{r},t)$ is the electron-nucleus interaction potential, $V_h(\mathbf{r},t)$ is the Hartree potential, $V_{xc,\sigma}(\mathbf{r},t)$ is the exchange-correlation potential, and the last term is the interaction with the laser field. Various efficient numerical methods have been developed to solve these equations, including methods on how to choose nonuniform grid points. Different groups may use different exchange-correlation potentials.

Another approach is the time-dependent HF method (TDHF) [9, 10]. It is a direct generalization of the HF theory from the time-independent Schrödinger equation. For each spin orbital, it satisfies

$$i\frac{\partial}{\partial t}\varphi_i(\mathbf{r},t) = (h + \mathbf{r}\cdot\mathbf{F}(t))\varphi_i(\mathbf{r},t), \tag{2.81}$$

where

$$h\varphi_i(\mathbf{r},t) = \left[\frac{\mathbf{p}^2}{2} - \frac{Z}{r} + \sum_j \int d\mathbf{r}' \frac{|\varphi_j(\mathbf{r}',t)|^2}{|\mathbf{r}-\mathbf{r}'|}\right]\varphi_i(\mathbf{r},t) + \sum_{j>i}\int d\mathbf{r}' \frac{\varphi_j(\mathbf{r}',t)\varphi_i(\mathbf{r}',t)}{|\mathbf{r}-\mathbf{r}'|}\varphi_j(\mathbf{r},t). \tag{2.82}$$

The TDHF is complicated by the exchange potential term, which makes TDHF equations much more difficult to solve. The TDHF theory is based on representing the time-dependent, many-electron wavefunction in terms of a single determinant.

In structure theory, one can use the MCHF method to account for electron correlation. Thus multi-configuration, time-dependent HF (MCTDHF) theory has been formulated, and coupled electronic along with nuclear dynamics can also be included [11, 12].

Another common and powerful approach for molecular-structure calculation is the R-matrix theory, which relies on separating the configuration space into two regions: an inner region where electron–electron correlation is important and an outer region where correlation can be neglected. The inner region can be calculated using quantum chemistry codes. The time-dependent R-matrix theory has been developed using the same concept, where the outer region can be expanded in terms of Volkov states or eikonal–Volkov states.

For practical applications, all of these time-dependent, many-electron theories are extremely demanding computationally. There are no known standard systems where results from different methods are compared to check the convergency. As stated earlier, direct comparison between precise calculations with experimental data is extremely difficult since precise experimental laser parameters are not generally known and volume integration has to be carried out. For molecules, the effect of partial alignment/orientation also has to be included. Due to such numerical challenges, accurate total ionization probability in strong fields (not to mention photoelectron spectra or high-order harmonic generation) cannot be "routinely" and accurately calculated in spite of great computational efforts in recent years. Nevertheless, a number of such all-purpose computational codes are already available.

2.3 Ultrashort Femtosecond Lasers: Representation, Generation, and Characterization

2.3.1 Pulse Representation

In strong-field experiments of atoms with lasers, the laser field or the time-dependent electric field of the laser should be specified precisely. For a linearly polarized monochromatic laser field, the oscillating electric field **F** and its vector potential **A**, respectively, can be represented by

$$\mathbf{F}(t) = \hat{\varepsilon} F_0 \sin(\omega t + \delta), \tag{2.83}$$

$$\mathbf{A}(t) = \hat{\varepsilon} \left(\frac{F_0}{\omega}\right) \cos(\omega t + \delta). \tag{2.84}$$

For a short laser pulse, a laser field is often represented by

$$\mathbf{F}(t) = \hat{\varepsilon} F_0(t) \cos[\omega_0 t + \varphi(t)], \tag{2.85}$$

where ω_0 is the carrier frequency, $F_0(t)$ is the envelope function and $\varphi(t)$ is the time-dependent phase. The time-derivative of the phase gives the instantaneous frequency

$$\omega(t) = \frac{d}{dt}[\omega_0 t + \varphi(t)]. \tag{2.86}$$

If $\varphi(t) = \alpha t^2$, then $\omega(t) = \omega_0 + 2\alpha t$. Such a pulse is linearly chirped. If $\alpha > 0$, the pulse is positively chirped and the wave oscillation becomes faster from cycle to cycle. If $\alpha < 0$, the negatively chirped pulse oscillates more and more slowly as time increases. Figure 2.1 shows examples of non-chirped, positive-chirped, and negative-chirped pulses, respectively.

A short pulse can also be written as the real part of a complex field

$$\varepsilon(t) = E_0(t) \exp[i\omega_0 t + i\varphi(t)]. \tag{2.87}$$

If the time dependence of the pulse envelope is Gaussian, then the complex electric field can be expressed as

$$\varepsilon(t) = E_0 e^{-2\ln 2(t/\tau)^2} e^{i[\omega_0 t + \varphi(t)]}. \tag{2.88}$$

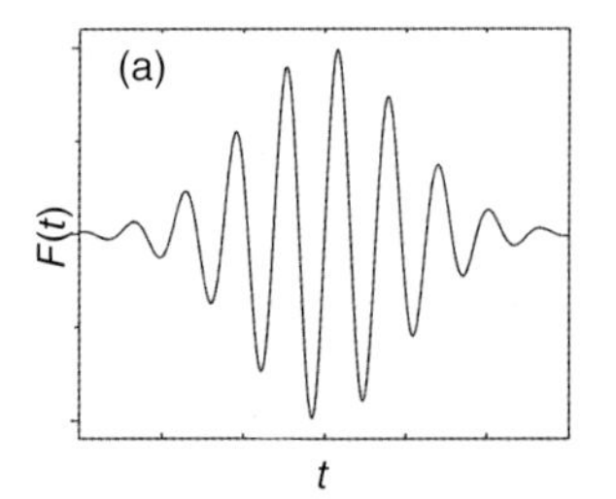

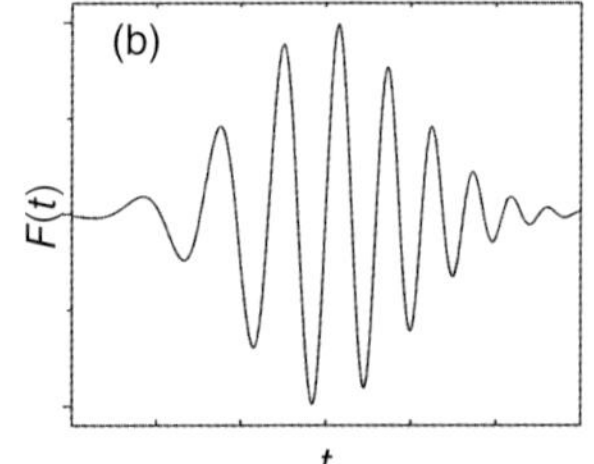

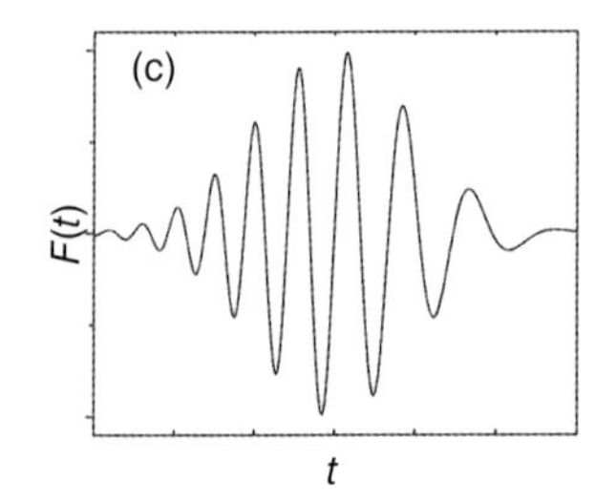

Figure 2.1 Illustration of non-chirped (left), positive-chirped (middle), and negative-chirped (right) pulses.

Here, τ is the full width at half maximum (FWHM) of the laser intensity

$$I(t) = I_0 e^{-4\ln 2(t/\tau)^2}, \tag{2.89}$$

with $I_0 = \epsilon_0 c E_0^2/2$, where ϵ_0 is the vacuum permittivity. In the frequency domain, the ultrafast pulse is obtained by taking the Fourier transform

$$\varepsilon(\omega) = \frac{1}{2\pi}\int_{-\infty}^{\infty} \varepsilon(t)e^{-i\omega t}dt = U(\omega)e^{i\Phi(\omega)}. \tag{2.90}$$

The spectral distribution of the laser intensity in the frequency domain $I(\omega) \propto U^2(\omega)$ gives the bandwidth of the laser pulse. The spectral phase $\Phi(\omega)$ plays a central role in determining the shape of a short pulse. In general, $\Phi(\omega)$ can be expanded into terms of a Taylor series about a central frequency ω_0. The constant term $\Phi_0^{(0)} = \Phi(\omega_0)$ is the carrier-envelope phase (CEP). The linear spectral phase term $\Phi_0^{(1)} = (d\Phi/d\omega)_{\omega_0}$ is a constant group delay (GD) time. It corresponds to a delay between the pulse and an arbitrary origin of time. The quadratic spectral phase term $\Phi_0^{(2)} = (d^2\Phi/d\omega^2)_{\omega_0}$ is the group delay dispersion (GDD). It determines the linear stretch of the pulse resulting from dispersion. This term can be reduced by passing the pulse through a well-designed dispersive medium. Consider a transform-limited (no-chirp) pulse with pulse duration τ. If its temporal distribution is Gaussian, the spectral distribution is also Gaussian with a spectral FWHM given by $\Gamma = 4\ln 2/\tau$. If the pulse duration is 1 fs, the bandwidth is about 1.8 eV. Thus, for a 1 ps pulse, the bandwidth is 1.8 meV and for a 10 as pulse, the bandwidth is 180 eV. A more detailed discussion of an electron wave packet in which the wave nature is identical to the wave nature of a laser pulse is given in Section 7.5.1.

2.3.2 Generation of Femtosecond Laser Pulses

Laser Oscillator

Figure 2.2(a) shows schematically the operational principle of a mode-locked laser oscillator. (The rest of this subsection is partly adapted from the chapter by Gallmann and Keller in [Quack, Merkt]). It consists of a cavity of length L, one highly reflected mirror and one output coupler together with a gain material and a saturable loss modulator. A light pulse makes a round-trip between the two end mirrors, taking time $T_R = 2L/v_g$, where v_g is the group velocity inside the cavity. Optical gain and loss change the energy contained within a pulse. The amplitude of the pulse gain is e^{gx}, where x is the length of the gain medium and g is the gain factor. However, the gain is saturated at a certain intensity. Optical loss, like reflection, is not saturable, but loss resulting from absorption is saturable due to the depletion of the ground state. The latter can act as a self-amplitude modulation. In a typical solid-state laser, the gain is saturated by the average intensity inside the cavity. If the saturable loss modulator is fast enough, the pulse can open up a short gain window over the duration. This mechanism is illustrated in Figure 2.2(b). The most common fast loss modulation is Kerr-lens mode-locking (KLM). It uses the self-focusing Kerr effect, where the index of refraction increases linearly with laser intensity. KLM is most frequently used

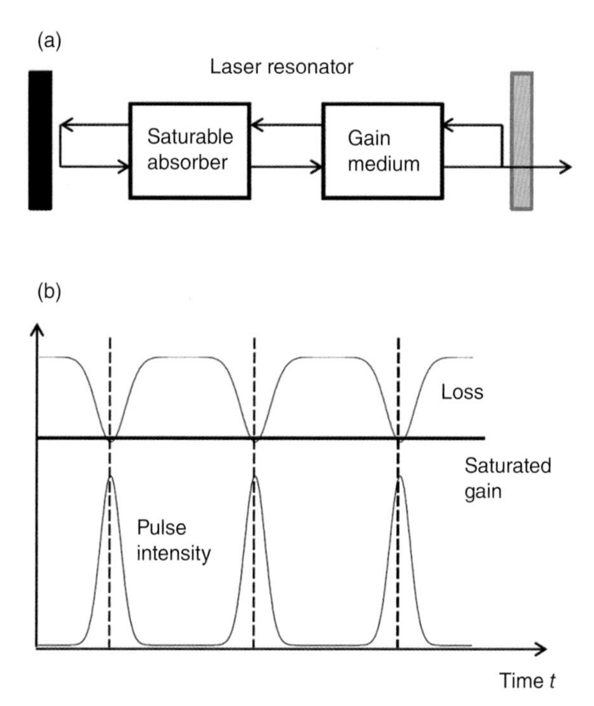

Figure 2.2 (a) Schematic operation principle of a mode-locked laser. A resonator contains a gain material and a saturable loss mechanism. In (b), the saturable loss gets saturated dynamically while the gain is saturated by the time-averaged intensity only. This allows for the generation of intense pulses.

in titanium–sapphire oscillators. For the common titanium–sapphire laser, the gain material is Ti^{3+} ion-doped sapphire crystal (Al_2O_3). It is pumped by Nd-based lasers. The laser transition is strongly phonon broadened, resulting in a spectrum from 650 nm to 1,000 nm with peak gain at about 780 nm. Typical output pulse energies from the oscillators are on the order of nanojoules and the repetition rate is about 1 MHz.

Amplifier

The amplification of femtosecond pulses to significantly higher energies has been made possible by the chirped-pulse amplification (CPA) technique [13]. Without CPA, the amplification is limited due to the damage of the amplifier medium. The CPA circumvents this problem by temporally stretching the femtosecond pulse from the oscillator to tens or hundreds of picoseconds. The energy of such a long pulse is amplified and then followed by a compressor to recompress the amplified pulse down to its femtosecond duration. Typically, the repetition rate of amplifiers is on the order of few to tens of kilohertz regime with pulse energy of few to tens of millijoules.

CEP

In a laser pulse, the envelope propagates in the oscillator cavity at group velocity, while the carrier wave propagates at phase velocity. The resulting phase difference is the CEP.

In the frequency domain, this phase difference manifests itself as an offset in the frequency comb from zero frequency, $f_m = mf_{rep} + f_0$. In frequency-domain measurements, exact knowledge of f_0 is essential. In time-domain applications, the CEP needs to be measured and controlled, especially for few-cycle pulses where the electric field varies from one half-cycle to the next. The measurement of f_0 can be carried out using the f-to-$2f$ method. It relies on a wave spanning over one octave in the spectral range. Using the second harmonic-generation crystal, the frequency train is doubled, $2f_m = 2mf_{rep} + 2f_0$. Comparing this frequency with $f_{2m} = 2mf_{rep} + f_0$ yields a difference of f_0. Therefore f_0 is the beating frequency between the second harmonic and the original wave that lies in the microwave range and can be measured electronically. The CEP can be controlled by modifying the difference in group and phase delays and/or adding a small power modulation to the oscillator. The change of the CEP can be tracked with the f-to-$2f$ method and locked to a stable microwave reference frequency. Today, CEP-stable femtosecond lasers are available commercially for ultrafast experiments, but the absolute value of the CEP still has to be determined in the experiment.

Pulse Compression

For the generation of single attosecond pulses and other applications, few-cycle laser pulses are needed. To shorten the pulse, the spectral range of the pulse should be extended and the spectral phase should be flat. The spectral range can be extended by sending an amplified laser pulse into a rare gas-filled, hollow-core optical fiber that is typically around one meter in length with a diameter of 200–300 μm and gas pressure of several hundreds of millibars. When a pulse propagates inside the fiber, because of the nonlinear self-phase modulation, its spectral content is expanded. Spectral broadening can also be carried out based on filamentation in a rare gas-filled cell. The nonlinear Kerr effect, plasma generation and interaction, etc., all contribute to the spectral broadening. Both methods can maintain CEP stabilization of the input pulse. To reach transform-limited short pulses, dispersion compensation is needed. The latter is often the limiting factor of getting the shortest pulse for a given spectral content of the compressed pulse.

Optical Parametric Amplification

Optical parametric sources are not only excellent, but are also sometimes the only choice for ultrafast pulses in special spectral regions not covered by ordinary lasers. These sources offer a high degree of flexibility since a large wavelength range can typically be covered with a single device. If an intense laser drives a noncentrosymmetric transparent material hard, its polarization essentially splits a higher-energy photon into two lower-energy photons. From energy conservation, the frequencies of the pump (p), the signal (s), and the idler (i) satisfy

$$\omega_p = \omega_s + \omega_i. \tag{2.91}$$

Parametric amplification is efficient only when phase matching or momentum conservation is fulfilled:

$$\mathbf{k}_p = \mathbf{k}_s + \mathbf{k}_i. \tag{2.92}$$

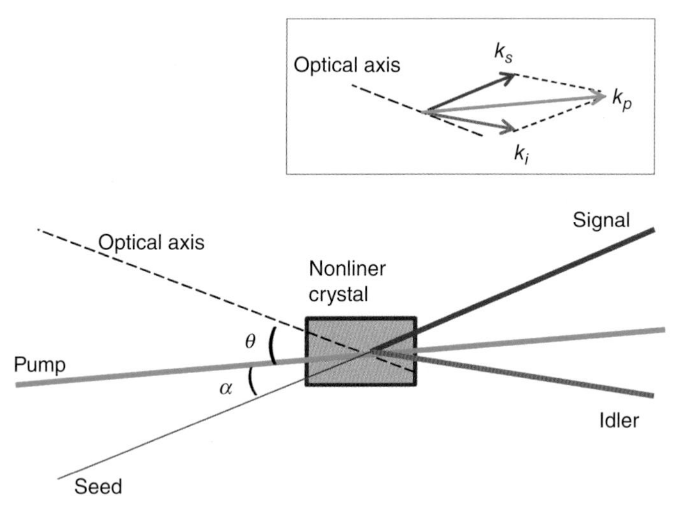

Figure 2.3 The schematic of optical parametric amplification. A weak seed signal is amplified by an intense pump. After the pulses go through a nonlinear crystal, a signal and an idler pulse are generated when phase-matching conditions are satisfied.

The amplification of a weak-seed signal by an OPA system in a difference-frequency scheme via a nonlinear medium is described in standard nonlinear optics books. Briefly, let $A_{p,s,i}$ be the amplitudes of the pump, signal, and idler fields, respectively. In the schematic diagram Figure 2.3, both the signal and the idler gain significantly as the pump pulse goes through the second-harmonic generation material without absorption. The amplitudes satisfy the equations (see chapter 2 of [Boyd])

$$\frac{dA_i}{dz} = 4\pi i \frac{\chi^{(2)}\omega_i}{n_i c} A_p A_s^* e^{i(\Delta\mathbf{k}\cdot\hat{z})z}, \tag{2.93}$$

$$\frac{dA_s}{dz} = 4\pi i \frac{\chi^{(2)}\omega_s}{n_s c} A_p A_i^* e^{i(\Delta\mathbf{k}\cdot\hat{z})z}, \tag{2.94}$$

$$\frac{dA_p}{dz} = 4\pi i \frac{\chi^{(2)}\omega_p}{n_p c} A_s A_i e^{-i(\Delta\mathbf{k}\cdot\hat{z})z}. \tag{2.95}$$

Here, n_i, n_s, and n_p are the index of refraction at each wavelength. $\chi^{(2)}(\omega_p, \omega_s, \omega_i)$ is the second-order susceptibility of the material. $\Delta\mathbf{k} = \mathbf{k}_p - \mathbf{k}_s - \mathbf{k}_i$ is the phase mismatch. As the intense pump pulse goes through the material, the amplitude A_p can be treated as a constant; then the intensities for the signal and idler can be calculated as

$$I_s(z) = I_s(0)\cosh^2 gz, \tag{2.96}$$

$$I_i(z) = I_s(0)\frac{n_s\omega_i}{n_i\omega_s}\sinh^2 gz. \tag{2.97}$$

Here, $I_s(0)$ is the intensity of the weak signal at the beginning. For both equations to be simultaneously satisfied, the incident angle θ for the pump and α for the signal pulse should satisfy Equations 2.91 and 2.92, which leads to phase matching $\Delta\mathbf{k} = 0$. Otherwise, if the coherent length of the nonlinear medium is less than $\pi/\Delta k$, then phase matching can still be maintained. Thus, the phase-matching bandwidth is primarily determined by the dispersion properties of the nonlinear medium. For femtosecond OPA, the pump is usually a titanium–sapphire amplifier system at the 800 nm wavelength or its second harmonic at about 400 nm. The pulse energy reached in such a system is typically a few microjoules.

The widely used titanium–sapphire laser typically operates at repetition rates from 10 Hz to 10 kHz, with a pulse duration between 20 and 40 fs. This laser is also limited to a wavelength of about 800 nm. With OPAs driven by titanium–sapphire lasers, new, intense-driving laser wavelengths can be extended from visible to the mid-infrared spectral region. These long-wavelength lasers can generate soft X-ray harmonics or attosecond pulses. However, OPAs driven by titanium–sapphire lasers cannot reach high repetition rates or high average power.

In recent years, an alternative to titanium–sapphire-driven OPA called *optical parametric chirped-pulse amplification* (OPCPA) has been developed. Here, an ultrashort seed pulse is stretched and amplified via nonlinear mixing with an intense picosecond pump pulse in a nonlinear crystal. Much like difference-frequency generation but with the powerful Yb-doped chirp-pulse-amplified picosecond pump lasers, high-averaged power and high-repetition-rate pulses covering a broad wavelength region can be produced. Today, several different Yb-doped OPCPA systems with repetition rates ranging from kHz to MHz have been used for the nonlinear generation of high harmonics. Nonlinear physics with such laser systems will be discussed in later chapters.

2.3.3 Characterization of Ultrashort Pulses

Autocorrelation

The duration of a short pulse can be estimated most easily using autocorrelation. A test pulse is split into two replicas with a relative delay and both replicas are focused inside a second harmonic generation (SHG) crystal. The SHG signal is proportional to

$$S(\tau) \sim \int_{-\infty}^{\infty} I(t)I(t-\tau)dt. \tag{2.98}$$

When the two pulses do not overlap in time, the signal goes to zero. This gives an estimate of the pulse duration only. The profile of the pulse and its phase are not available.

For short pulses, there are many different methods for complete pulse characterization. By measuring $\overline{E}(\omega) = |\overline{E}(\omega)|e^{i\Phi(\omega)}$ completely, $E(t)$ can be obtained through an inverse Fourier transform. $|\overline{E}(\omega)|$ can be best measured using a spectrometer, where the output

$$\widetilde{E}_{out}(\omega) = H(\omega)\overline{E}(\omega), \tag{2.99}$$

$$I_{out}(\omega) = |H(\omega)|^2|\overline{E}(\omega)|^2. \tag{2.100}$$

Here, it is assumed that $H(\omega)$ is known. The remaining task is to measure $\Phi(\omega)$. Below, two common methods for full characterization of short pulses, FROG and SPIDER, will be summarized. Both methods rely on nonlinear optics and self-reference.

FROG: Frequency-Resolved Optical Gating

The most widely used FROG implementation for measuring the phase is the SHG–FROG. The gate is a time-delayed replica of the test pulse itself. The signal reads

$$S(\omega, \tau) = \left| \int_{-\infty}^{\infty} E(t)E(t-\tau)e^{i\omega t}dt \right|^2. \tag{2.101}$$

The SHG–FROG method is considered a type of amplitude gating since the conversion efficiency increases with laser intensity. Figure 2.4(a) shows a schematic setup of SHG–FROG. The time delay is usually introduced by the Michelson interferometer. The generated 2D signal, or FROG trace (see Figure 2.4(b)), is analyzed by an iterative algorithm to retrieve the phase. The sampling covers both the frequency and time domain. Since the signal is the slow-varying intensity, the retrieval is sensitive to noise and the

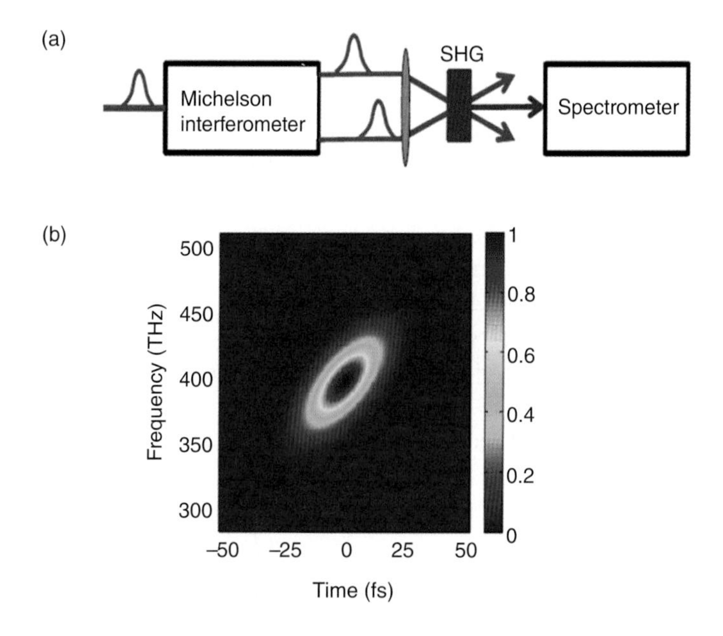

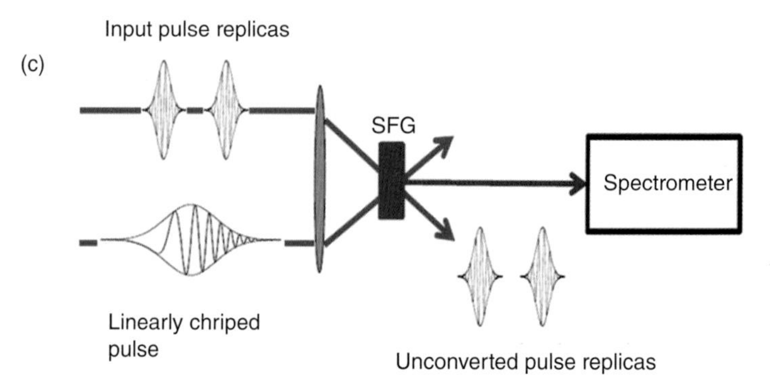

Figure 2.4 (a) The working principle of FROG. (b) A typical FROG trace. (c) The working principle of SPIDER. See the text for detail.

uniformity of the detector. For complex pulses, the number of data points grows quickly and convergence becomes slow in the iterative algorithm. There are many other varieties of FROG that depend on other types of gate functions. An extension of FROG is used to characterize single attosecond pulse (see Section 7.5.2).

SPIDER: Spectral Phase Interferometry for Direct Electric-Field Reconstruction

SPIDER is based on the spectral interference between a test pulse and its replica pulse, which is shifted both in time and frequency. Using spectral interferometry, the interference between $\overline{E}(\omega)$ and a delayed $\overline{E}(\omega-\Omega)$ gives

$$I(\omega,\tau) = |\overline{E}(\omega) + \overline{E}(\omega-\Omega)e^{i\omega\tau}|^2 \tag{2.102}$$

$$= |\overline{E}(\omega)|^2 + |\overline{E}(\omega-\Omega)|^2 + 2|\overline{E}(\omega)||\overline{E}(\omega-\Omega)| \times \cos[\Phi(\omega) - \Phi(\omega-\Omega) - \omega\tau]. \tag{2.103}$$

By taking the Fourier transform of $I(\omega,\tau)$ at each ω,

$$I_\tau(\omega) \sim \cos[\Phi(\omega) - \Phi(\omega-\Omega) - \omega\tau] \tag{2.104}$$

or

$$\theta(\omega) = \Phi(\omega) - \Phi(\omega-\Omega) - \omega\tau \tag{2.105}$$

can be extracted. This directly gives the spectral phase difference between ω and $\omega-\Omega$. To obtain the spectral phase of a single pulse, one generates a spectral shear between the carrier frequencies of two replicas of this pulse. The spectral shear is obtained by sum or difference-frequency generation in a nonlinear optical crystal between the two replicas and a strongly chirped pulse (see Figure 2.4(c)). The spectral phase of the test pulse is obtained by concatenation or integration with the exception of a linear term, which does not affect the shape of the pulse. Note that the spectral interference method is linear, but to get spectral shift (spectral shearing), nonlinear optics is needed. The phase-retrieval method for SPIDER is non-iterative.

Determination of Laser Intensity

Accurate knowledge of the intensity of a focused ultrashort laser pulse is crucial to correctly interpreting experimental results in nonlinear strong-field physics. Typically, experimentalists measure the average power P, the pulse duration τ, the radius w of a Gaussian beam, and the repetition rate ν of the laser. The laser peak intensity is then estimated through

$$I = \frac{2P}{(\pi w^2)\nu\tau}, \tag{2.106}$$

where $P = U\nu$ with U being the pulse energy. Typically, the laser power and pulse duration are accurate to about 5% and the focused area at the center deduced from beam-waist determination is accurate to about 10%. Thus, the total uncertainty is about 10%. Since there is no simple way to measure the beam profile, the estimation may still be optimistic. Because of this, accurate, direct determination of laser intensity in an experiment for a focused ultrafast laser pulse is very difficult. Several approaches have been suggested for *in situ* intensity calibration based on the comparison of experimental data and theoretical calculations. Unfortunately, "exact" theoretical calculations for a known laser pulse can only be carried out for atomic hydrogen by solving the TDSE [14]. However, few laboratories have the setup to generate atomic hydrogen targets, and the result can still be dependent on the beam profile in time and space as it propagates to the scattering chamber. While intensity accuracy better than 10% may not be needed in applications, anything much worse than 10% is not desirable. Various methods have been suggested for estimating laser intensities based on the measured ionization yields, photoelectron spectra, or high-order harmonic spectra. These methods all have their own difficulties.

2.4 Tunnel Ionization Theory

2.4.1 Static or Adiabatic Ionization Model

When a hydrogen-like atom is exposed to a static electric field, an electron can tunnel ionize through the combined potential $V(r) = -Z/r + Fz$, where Z is the charge of the ionic core and F is the static electric field (along the z-axis). The time-independent Schrödinger equation is separable in parabolic coordinates (see Section 1.1.1). By expressing

$$\psi(\xi,\eta,\phi) = \frac{f(\xi)g(\eta)}{\sqrt{\xi\eta}}\frac{e^{im\phi}}{2\pi}, \tag{2.107}$$

the two functions $f(\xi)$ and $g(\eta)$ satisfy

$$\left(-\frac{1}{2}\frac{d^2}{d\xi^2} + V_1(\xi)\right)f(\xi) = \frac{E}{4}f(\xi), \tag{2.108}$$

$$\left(-\frac{1}{2}\frac{d^2}{d\eta^2} + V_2(\eta)\right)g(\eta) = \frac{E}{4}g(\eta), \tag{2.109}$$

where the effective potentials are

$$V_1(\xi) = \frac{m^2-1}{8\xi^2} - \frac{\nu_1}{2\xi} + \frac{F}{8}\xi, \tag{2.110}$$

$$V_2(\eta) = \frac{m^2-1}{8\eta^2} - \frac{\nu_2}{2\eta} - \frac{F}{8}\eta. \tag{2.111}$$

Here, $\nu_1 + \nu_2 = Z$. For $F > 0$, $V_1(\xi) \to \infty$ as $\xi \to \infty$, the electron is bounded in the small ξ region. On the other hand, $V_2(\eta) \to -\infty$ as $\eta \to \infty$. The potential in $V_2(\eta)$ has a barrier. Let $E = -\kappa^2/2$, while the asymptotic wavefunction for large r in spherical coordinates is

$$\psi_{as} \to C_l r^{(Z/\kappa)-1} e^{-\kappa r} Y_{lm}(\theta,\phi). \tag{2.112}$$

Since $F > 0$, the electron will be ionized along the $-z$ direction. Consider $\theta \rightarrow \pi$,

$$Y_{lm}(\theta,\phi) \approx Q(l,m)\frac{\sin^{|m|}\theta}{2^{|m|}|m|!}\frac{e^{im\phi}}{\sqrt{2\pi}}, \tag{2.113}$$

$$Q(l,m) = (-1)^{l+m}(-1)^{(m+|m|)/2}\sqrt{\frac{2l+1}{2}\frac{(l+|m|)!}{(l-|m|)!}}, \tag{2.114}$$

$$\sin\theta \approx 2\sqrt{\xi/\eta}. \tag{2.115}$$

Thus

$$\psi_{as} \rightarrow B_{lm}\frac{2^{-(Z/\kappa)+1}}{|m|!}\xi^{|m|/2}e^{-\kappa\xi/2}\eta^{Z/\kappa-|m|/2-1}e^{-\kappa\eta/2}\frac{e^{im\phi}}{\sqrt{2\pi}}, \tag{2.116}$$

where $B_{lm} = C_l Q(l,m)$.

Since the asymptotic solution for ξ should satisfy Equation 2.108,

$$f(\xi) = \sqrt{\xi}\xi^{|m|/2}e^{-\kappa\xi/2} \tag{2.117}$$

is obtained and from Equation 2.108, $\nu_1 = \kappa(|m|+1)/2$. Recall $\nu_2 = Z - \nu_1$.

To calculate the ionization rate, one needs to integrate the probability current density in the $-z$ direction over a surface orthogonal to it. For a large value of negative $z = -\eta/2$, the flux is

$$w = \int j_z dS. \tag{2.118}$$

Here, j_z is the current density. In parabolic coordinates, this is simplified to

$$w = \frac{i}{2}\frac{|m|!}{\kappa^{|m|+1}}\left[g\frac{dg^*}{d\eta} - g^*\frac{dg}{d\eta}\right]. \tag{2.119}$$

To obtain analytical expressions for ionization, semiclassical theory is used to solve $g(\eta)$. The momentum, from Equation 2.109, is

$$p = \left[-\left(\frac{\kappa}{2}\right)^2 + \frac{\nu_2}{\eta} + \frac{F}{4}\eta - \frac{m^2-1}{4\eta^2}\right]^{1/2}. \tag{2.120}$$

The outer turning point for small F occurs at

$$\eta_0 \sim \kappa^2/F. \tag{2.121}$$

From the Wentzel–Kramers–Brillouin (WKB) formula, the wavefunction for $\eta < \eta_0$ is

$$g(\eta) = \frac{D}{\sqrt{|p|}}e^{-\int_{\eta_0}^{\eta}|p|d\eta} \tag{2.122}$$

and for $\eta > \eta_0$

$$g(\eta) = \frac{D}{\sqrt{|p|}}e^{i(\int p d\eta - \pi/4)}. \tag{2.123}$$

By using the above equation for $\eta > \eta_0$ and combining it with Equation 2.117, the following can be obtained:

$$\psi = D\kappa^{-1/2} 2^{-Z/\kappa+|m|/2+1} \left(\frac{F}{2\kappa^2}\right)^{Z/\kappa-|m|/2-1/2} e^{\kappa^3/3F} \times \xi^{|m|/2} e^{-\kappa\xi/2} \eta^{Z/k-|m|/2-1} e^{-\kappa\eta/2} \frac{e^{im\phi}}{\sqrt{2\pi}}. \quad (2.124)$$

The relation between D and B_{lm} can be obtained by comparing Equation 2.116 and Equation 2.124

$$\frac{D}{B_{lm}} = \kappa^{1/2} \frac{1}{2^{|m|/2}|m|!} \left(\frac{2\kappa^2}{F}\right)^{Z/\kappa-|m|/2-1/2} e^{-\kappa^3/3F}. \quad (2.125)$$

From Equaton 2.119

$$w = \frac{|m|!}{\kappa^{|m|+1}} |D|^2 \quad (2.126)$$

is obtained.

With Equation 2.125, the final expression for static ionization rate is given by

$$w = \frac{C_l^2 \frac{(2l+1)(l+|m|)!}{2(l-|m|)!}}{2^{|m|}|m|!} \frac{1}{\kappa^{2Z/\kappa-1}} \left(\frac{2\kappa^3}{F}\right)^{2Z/\kappa-|m|-1} e^{-2\kappa^3/3F}. \quad (2.127)$$

Note that the static ionization rate depends on the ionization potential through κ, the magnetic quantum number $|m|$, and C_l, which, from Equation 2.112, is related to the amplitude of the electron wavefunction in the tunnel-ionization direction. The derivation presented here follows Bisgaard and Madsen [15]. The coefficients C_l for some atomic targets have been tabulated in Tong et al. [16].

2.4.2 Ionization from an Intense Low-Frequency Field

In the adiabatic approximation where the ionization time (measured with respect to the inverse of ionization rate) is short compared to the optical cycle of the oscillating field, one can replace F by $|F(t)|$ in Equation 2.127. To obtain the cycle-averaged ionization rate from a linearly polarized light, take $F(t) = F_0 \cos \omega t$. The cycle-averaged ionization rate is

$$\bar{w} = \frac{1}{T} \int_{-T/2}^{T/2} w(t)dt. \quad (2.128)$$

Since $w(t)$ depends exponentially on $F(t)$, one can expand $1/\cos \omega t \sim 1 + (\omega t)^2/2$ and obtain the cycle-averaged ionization rate

$$\bar{w} = \left(\frac{3F_0}{\pi\kappa^3}\right)^{1/2} w_{stat}(F_0). \quad (2.129)$$

For adiabatic-tunneling ionization by infrared laser fields, the ionization rate thus calculated is often called the ADK model, after Ammosov, Delone, and Krainov [17]. The original ADK static-ionization rates for atoms are given by

$$w_{ADK} = |C_{n^*l^*}|^2 G_{lm} I_p \left(\frac{2E_0}{F}\right)^{2n^*-|m|-1} e^{-\frac{2E_0}{3F}}, \quad (2.130)$$

where the effective principal quantum number $n^* = Z/\sqrt{2I_p}$, the effective orbital quantum number $l^* = n^* - 1$, the parameter $E_0 = (2I_p)^{3/2}$, and the two coefficients are

$$|C_{n^*l^*}|^2 = \frac{2^{2n^*}}{n^*\Gamma(n^*+l^*+1)\Gamma(n^*-l^*)}, \tag{2.131}$$

$$G_{lm} = \frac{(2l+1)(l+|m|)!}{2^{|m|}|m|!\,(l-|m|)!}. \tag{2.132}$$

By comparing Equations 2.127 and 2.130, it is clear that $C_l^2 = |C_{n^*l^*}|^2\kappa^{2Z/\kappa+1}$.

For a circularly polarized light, $\bar{w} = w_{stat}(F_0)$, where F_0 is the magnitude of the electric field. For a slightly elliptically polarized light, $\mathbf{F} = F_0[\cos\omega t\,\hat{x} + \varepsilon\sin\omega t\,\hat{y}]$, the cycle-averaged rate is

$$\bar{w} = \left(\frac{3F_0}{\pi(1-\varepsilon^2)\kappa^3}\right) w_{stat}(F_0), \tag{2.133}$$

where ε is the eccentricity.

Within the ADK model, the ionization probability P_{ion} by a laser pulse $F(t)$ can be calculated from

$$P_{ion} = 1 - e^{-\int_{-\infty}^{\infty} w[|F(t)|]dt}, \tag{2.134}$$

or by summation over the ionization probability from each cycle using Equation 2.129.

Besides tunneling ionization, the position of the tunnel exit is also of interest. It can be easily calculated from the 1D model, $z_0 = \frac{I_p}{F} = \frac{\kappa^2}{2F}$, where $I_p = \kappa^2/2$. The initial transverse-velocity distribution at tunneling has been derived in Delone and Krainov [18]. It has the Gaussian form

$$e^{-\frac{\sqrt{2I_p}}{F}v_\perp^2} = e^{-\frac{\kappa}{F}v_\perp^2}. \tag{2.135}$$

However, recent experiments [19] and theory [20] indicate that the lateral width should be about 30% wider.

The ADK theory given above was derived based on the physical model in which the typical timescale of the electron in an atom (150 as for hydrogen atoma) is much shorter than the optical period of the low-frequency laser (2.6 fs for 800 nm laser and 13 fs for 4 μm laser). Thus the ADK model is expected to work better for long-wavelength lasers. Deviation of tunnel ionization from the prediction of the ADK model is included in some nonadiabatic ionization theories, which will be addressed in Section 2.5.7. For high intensities, the initial state may lie above the top of the potential barrier formed by the Coulomb field and the instantaneous static field of the laser such that the electron can be released through the over-the-barrier ionization (OBI) mechanism. In this case, the tunneling model is not valid. For an electron with binding energy given by $I_p = \kappa^2/2$, the threshold for OBI in the 1D model is easily calculated to be $F_{OBI} = \frac{1}{16}\kappa^4$. When it is derived in 3D, the OBI threshold is given by $F_{OBI} = \frac{\kappa^4}{8(2-\kappa)}$ [21]. The exact ionization rate of an atom (in the one-electron-model potential approximation) in a static field can be calculated by solving the time-independent Schrödinger equation using the so-called *complex scaling method*, which consists of multiplying the coordinates $\mathbf{r}_i$ in the

Hamiltonian by a complex factor $\exp[i\theta]$. The resulting non-Hermitian Hamiltonian has complex eigenvalues whose imaginary parts are used to calculate the ionization rates. In the limit of small static fields, the ionization rate obtained from the numerical solution agrees with the prediction of the tunneling model. However, at high fields, the tunneling model would overestimate the rate. A simple empirical formula was proposed in Tong and Lin [22], where the OBI rate is given by

$$W_{OBI}(F) = W_{ADK}(F)e^{-\alpha(Z_c^2/I_p)(F/\kappa^3)}, \tag{2.136}$$

where the second factor on the right is to correct the error of the tunnel-ionization expression. A single empirical parameter α was obtained by fitting the OBI rate to the one calculated from numerical solution using the complex scaling method.

In deriving the ADK-ionization rate, it was assumed that the laser field is adiabatic with respect to the motion of the electron. Since the laser pulse is an oscillating field, it cannot formally be taken to the limit of a DC static field even when the wavelen gth is substantially increased. At any wavelength there is a nonadiabatic correction to the ADK rate. To identify the nonadiabatic effect, a perturbative expansion approach starting with the Schrödinger equation is needed. This theory is called weak-field asymptotic theory (WFAT) and will be touched upon in Section 3.7.3.

2.4.3 Classical Theory of Electron in a Laser Field and the Recollision Model

The tunneling-ionization model describes the release rate of an electron from the potential well. Consider a laser field given by $F(t) = F_0 \cos\omega t$ and assume that the electron is removed far away from the core. Its motion is then governed only by the laser field. The classical equation of motion for the electron is

$$\ddot{z} = -F_0 \cos\omega t, \tag{2.137}$$

$$\dot{z} = -\frac{F_0}{\omega}\sin\omega t + v_0, \tag{2.138}$$

$$z = +\frac{F_0}{\omega^2}\cos\omega t + v_0 t + z_0. \tag{2.139}$$

Assume the electron is liberated at $t = t_0$ with $z = 0$ and $\dot{z} = 0$. Equations 2.137, 2.138, and 2.139 are solved to obtain

$$\dot{z}(t) = \frac{F_0}{\omega}(\sin\omega t_0 - \sin\omega t), \tag{2.140}$$

$$z(t) = \frac{F_0}{\omega^2}(\cos\omega t - \cos\omega t_0) + \frac{F_0}{\omega}(t - t_0)\sin\omega t_0. \tag{2.141}$$

If the electron goes directly to the detector where the vector potential is zero, it will have zero velocity if ionization occurs at the peak of the laser field. It will have maximal velocity if it is ionized near the zero field. This simple model predicts that the electron energy distribution will drop quickly from zero to a maximal value of $2U_p$. On the other hand, an electron will return to the origin at $t = t_r$ if the following condition is satisfied

$$\cos\omega t_r - \cos\omega t_0 + \omega(t_r - t_0)\sin\omega t_0 = 0. \tag{2.142}$$

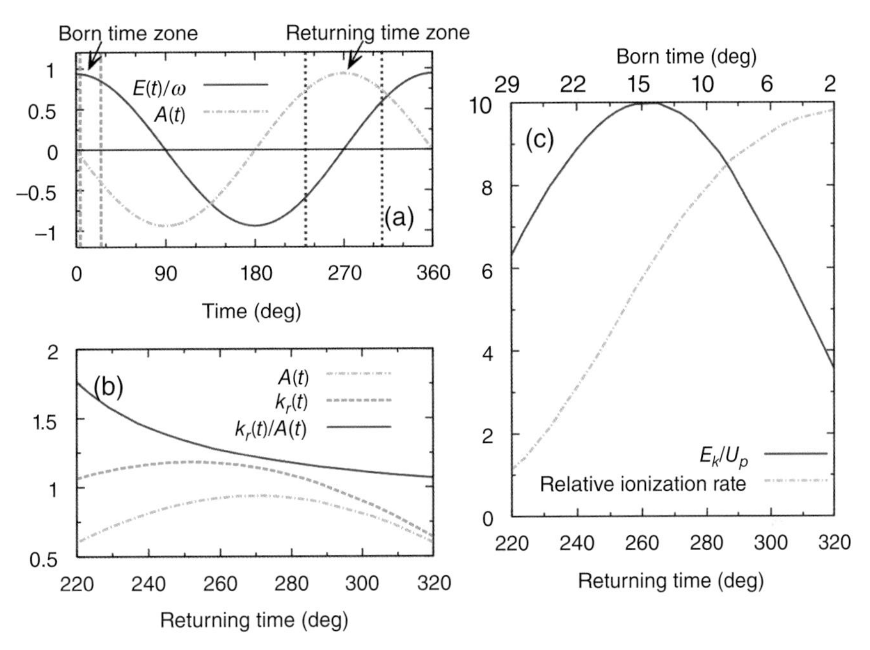

Figure 2.5 Classical model of a 1D electron in a monochromatic laser field. (a) Electric field and vector potential for a laser at an intensity of 1.0×10^{14} W/cm^2 and a wavelength of 800 nm. The born time and returning time zones are marked. (b) The electron velocity k_r, vector potential A, and their ratio at the time of return $t = t_r$ within the returning time zone. (c) Photoelectron energy after it has been backscattered by an angle 180° against the return time (bottom horizontal axis). Also shown is the relative ionization rate for electrons released with respect to the born time (top horizontal axis). (Reprinted from Zhangjin Chen et al., *Phys. Rev. A*, **79**, 033409 (2009) [23]. Copyrighted by the American Physical Society.)

Consider the first return. The momentum k_r at the time of return at the origin is

$$k_r \equiv \dot{z}(t_r) = -\frac{F_0}{\omega}(\sin \omega t_r - \sin \omega t_0). \tag{2.143}$$

When $\omega t_0 = 17°$, the electron returns with a maximum kinetic energy of $3.17U_p$. At $\omega t_0 = 13°$, the electron returns at $\omega t_r = 270°$. Figure 2.5(a) shows the born time within $4° < \omega t_0 < 25°$. The corresponding return time is in the range $231° < \omega t_r < 309°$.

Next, assume that the returning electron undergoes an elastic scattering with the ion by an angle θ_r with respect to its incident direction. For $t > t_r$, the velocity components parallel and perpendicular to the polarization axis are

$$\dot{z}(t) = -\frac{F_0}{\omega}[\sin \omega t - \sin \omega t_r + \cos \theta_r(\sin \omega t_r - \sin \omega t_0)], \tag{2.144}$$

$$\dot{y}(t) = -\frac{F_0}{\omega}\sin \theta_r(\sin \omega t_r - \sin \omega t_0). \tag{2.145}$$

From the equations above, the photoelectron energy E_k measured by the detector outside the laser field is obtained by subtracting the ponderomotive energy U_p from the averaged kinetic energy

$$E_k = 2U_p[\sin^2 \omega t_0 + 2 \sin \omega t_r(1 - \cos \theta_r)(\sin \omega t_r - \sin \omega t_0)]. \tag{2.146}$$

This equation shows that at $\omega t_0 = 14°$, for which $\omega t_r = 265°$, the photoelectron will have maximal kinetic energy of $10U_p$ for $\theta_r = 180°$. Figure 2.5(c) shows the relative tunneling ionization rate and the photoelectron energy after it has been backscattered by 180°. This figure also shows that below $10U_p$ there are two sets of born-return times that would give the same kinetic energy except for $\omega t_0 = 14°$ ($\omega t_r = 265°$). Electrons that were born before 14° would return after 265°, while those born after 14° would return before 265°. The former belong to long-trajectory electrons, and the latter belong to short-trajectory electrons. If the electrons are scattered in the forward direction, they would emerge as low-energy electrons with energies below $2U_p$.

So far, the derivation considers only the so-called *first return*, i.e., the scattering occurs within the same optical cycle of the born time. For a multi-cycle pulse, higher returns are possible. Furthermore, electrons can also be born in the next half cycle after the field has reached the peak. Classically, all of these electrons generated in each half cycle contribute to the electron yield. In quantum mechanics, for each given final electron momentum, contributions from these different paths interfere coherently. This results in very complicated photoelectron angular distributions.

In this subsection, the electron has been treated as a classical particle. Later, in quantum theory of the motion of the electron, these classical results can be qualitatively obtained under the semiclassical approximation.

2.5 SFA and Its Modifications

2.5.1 The KFR Theory

The KFR theory for strong-field ionization was first derived by Keldysh [24], Faisal [25], and Reiss [26]. In the SFA, the ionization probability amplitude for an electron ejected with momentum $\mathbf{p}$ is

$$M(\mathbf{p}) = -i \int_{-\infty}^{+\infty} \left\langle \Psi_{\mathbf{p}} \left| H_{\text{int}} \right| \Psi_0 \right\rangle dt, \tag{2.147}$$

from which the differential ionization probability is given as

$$\frac{dW(\mathbf{p})}{d^3p} = |M(\mathbf{p})|^2. \tag{2.148}$$

The precise expression for the interaction H_{int} and the final state wavefunction $\Psi_{\mathbf{p}}$ depends on the gauge used. In the conventional (non-transformed) gauge,

$$H_{\text{int}}(\mathbf{r}, t) = -i\mathbf{A}(t) \cdot \nabla + \frac{\mathbf{A}^2(t)}{2} \tag{2.149}$$

and the final continuum state is taken to be the Volkov wavefunction

$$\Psi_{\mathbf{p}}(\mathbf{r}, t) = \frac{1}{(2\pi)^{3/2}} \exp\left\{ i \left[\mathbf{p} \cdot \mathbf{r} - \frac{1}{2} \int_{-\infty}^{t} \mathbf{v}_{\mathbf{p}}^2(t')\, dt' \right] \right\}, \tag{2.150}$$

where $\mathbf{v_p}(t) = \mathbf{p} + \mathbf{A}(t)$. For ionization by a monochromatic laser field with optical period T, the relation

$$\langle \Psi_{\mathbf{p}} |H_{\text{int}}| \Psi_0 \rangle (t+T) = \langle \Psi_{\mathbf{p}} |H_{\text{int}}| \Psi_0 \rangle (t) \exp\left(iT \left[\frac{p^2}{2} + I_p + U_p \right] \right) \quad (2.151)$$

holds between two neighboring cycles since, by averaging over one cycle,

$$\begin{aligned} \left\langle \frac{\mathbf{v}_{\mathbf{p}}^2(t)}{2} \right\rangle_T &= \frac{1}{2} \left\langle (\mathbf{p} + \mathbf{A}(t))^2 \right\rangle_T \\ &= \frac{p^2}{2} + \frac{1}{2} \left\langle \mathbf{A}^2 \right\rangle_T \equiv \varepsilon_p + U_p. \end{aligned} \quad (2.152)$$

Using the identity

$$\sum_{k=-\infty}^{+\infty} e^{2i\pi kx} = \sum_{n=-\infty}^{+\infty} \delta(n - x), \quad (2.153)$$

the differential ionization rate can be expressed as

$$dW(\mathbf{p}) = \sum_n \delta\left(\varepsilon_p + I_p + U_p - n\omega\right) R(\mathbf{p})\, d\varepsilon_p d\Omega_{\mathbf{p}}, \quad (2.154)$$

where $\varepsilon_p = \frac{p^2}{2}$ and $R(\mathbf{p}) = \frac{\omega^2 p}{2\pi} \left| M_K(\mathbf{p}) \right|^2$.

The delta function in Equation 2.154 is the consequence of energy conservation

$$\frac{p_n^2}{2} = n\omega - I_p - U_p. \quad (2.155)$$

It imposes the absorption of a minimum number of N photons given by

$$N_{\min} = \left[(I_p + U_p)/\omega \right] + 1 \quad (2.156)$$

in order for ionization to occur. Here, the square bracket denotes the integer value of the quantity inside. Thus the allowed kinetic energies for photoelectrons are governed by Equation 2.155 with successive peaks separated by ω. The resulting peaks are called *above-threshold ionization (ATI) peaks*. Note that, in this strong-field model, the threshold for ionization is shifted by the ponderomotive energy U_p.

The partial transition amplitude per optical cycle is defined by

$$M_K(\mathbf{p}) = -i \int_0^T dt \langle \Psi_{\mathbf{p}} |H_{\text{int}}(t)| \Psi_0 \rangle, \quad (2.157)$$

which can be evaluated analytically as

$$M_K(\mathbf{p}) = 2\pi i \frac{I_p + p^2/2}{\omega} \Psi_0(\mathbf{p}) J_n\left(\frac{\mathbf{F}_0 \cdot \mathbf{p}}{\omega^2}, \frac{U_p}{2\omega} \right), \quad (2.158)$$

where $\Psi_0(\mathbf{p})$ is the momentum wavefunction of the ground state and the generalized Bessel function is defined by

$$J_n(a, b) = \sum_{k=-\infty}^{+\infty} J_{n+2k}(a) J_k(b). \quad (2.159)$$

Note that Equation 2.154 expresses the photoelectron spectrum as the sum of discrete peaks, but the derivation does not require the concept of photons. The peaks originate from the constructive interference of ionization-probability amplitudes from each optical cycle. However, the results are most conveniently understood using the photon concept, though the electron does acquire a ponderomotive energy that is absent in the conventional perturbative multiphoton theory.

The ionization amplitude given in Equation 2.147 is often called the *Keldysh ionization ansatz*. It is basically the first-order theory for laser–atom interaction, where the final continuum state is approximated by the Volkov wavefunction. This first-order theory has taken the name of the Keldysh theory, or KFR theory to include Faisal and Reiss, or the more neutral SFA. Since the predicted results from SFA are gauge dependent, there have been arguments over which gauge is better. However, the basic physics underlying these models is the same and the question of which gauge is better must rely on empirical evidence. The important thing is to improve the SFA (at least to the next order) as exemplified in the next few subsections.

The ionization of an atom by electromagnetic fields depends on how many photons the atom needs to absorb to promote the electron from the ground state (or the initial state in general) to the continuum. This can be represented by the parameter $K = I_p/\omega$. The other parameter is the degree of adiabaticity that can be represented by the Keldysh parameter $\gamma = (I_p/2U_p)^{1/2} = (2I_p)^{1/2}/A_0$, which is the ratio of the electron's characteristic momentum $(2I_p)^{1/2}$ to the field-induced momentum $A_0 = F_0/\omega$. For strong-field ionization of atoms and molecules from the ground state by intense infrared or mid-infrared lasers, K is typically much greater than one, and thus γ alone can be used to characterize the ionization mechanism. The tunneling model is applicable if $\gamma \ll 1$. However, if $\gamma \gg 1$ it belongs to the multiphoton ionization regime. The ADK theory discussed in Section 2.4 only works when the Keldysh parameter is less than one. On the other hand, the SFA theory applies formally to the whole range of γ.

2.5.2 Modification of SFA Theory for Strong-Field Ionization

As discussed in Section 2.2, the TDSE can be solved with the use of the evolution operator. Beginning with the Hamiltonian in the length gauge and introducing the Green function G_V, the wavefunction at a later time takes the form

$$\Psi(\mathbf{r},t) = \int_{-\infty}^{+\infty} dt_1 \int d^3r_1 G_V(\mathbf{r},t;\mathbf{r}_1,t_1) V(\mathbf{r}_1)\Psi(\mathbf{r}_1,t_1), \tag{2.160}$$

where $V(\mathbf{r})$ is the atomic potential and the Green function satisfies

$$\left[i\frac{\partial}{\partial t} + \frac{\nabla_{\mathbf{r}}^2}{2} - \mathbf{F}(t)\cdot\mathbf{r}\right] G_V(\mathbf{r},t;\mathbf{r}_1,t_1) = \delta(t-t_1)\,\delta(\mathbf{r}-\mathbf{r}_1) \tag{2.161}$$

with the initial condition $G_V(\mathbf{r},t;\mathbf{r}_1,t_1) = 0$ for t less than t_1. This equation can be solved analytically,

$$G_V(\mathbf{r},t;\mathbf{r}_1,t_1) = -i\int d^3p\Psi^*_{\mathbf{p}}(\mathbf{r}_1,t_1)\Psi_{\mathbf{p}}(\mathbf{r},t)$$
$$= -i\frac{1}{(2\pi)^3}\int d^3p\exp\left\{i\left[\mathbf{v_p}(t)\mathbf{r}-\mathbf{v_p}(t_1)\mathbf{r}_1-\frac{1}{2}\int_{t_1}^{t}\mathbf{v}^2_{\mathbf{p}}(t')dt'\right]\right\}. \quad (2.162)$$

By applying G_V and the initial condition in Equation 2.160, when the laser field is off, the wavefunction with t going to infinity takes the form

$$\Psi(\mathbf{r},t) = \frac{1}{(2\pi)^{3/2}}\int d^3pe^{i\,\mathbf{pr}-ip^2t/2}M(\mathbf{p}) \quad (2.163)$$

with

$$M(\mathbf{p}) = \frac{-i}{(2\pi)^{3/2}}\int_{-\infty}^{+\infty}dt_1\int d^3r_1e^{-i\mathbf{v_p}(t_1)\mathbf{r}_1-iS_0(\mathbf{p},t_1)}\mathbf{F}(t_1)\cdot\mathbf{r}_1\Psi_0(\mathbf{r}_1). \quad (2.164)$$

This is the ionization-probability amplitude in the length gauge. Recall that $\mathbf{v_p}(t) = \mathbf{p}+\mathbf{A}(t)$ and

$$S_0(\mathbf{p},t) = \int_{t}^{+\infty}\left[\frac{1}{2}\mathbf{v}^2_{\mathbf{p}}(t')+I_p\right]dt'. \quad (2.165)$$

While it is straightforward to carry out the integral in Equation 2.164 numerically, especially for short laser pulses of tens of femtoseconds or less, it is more illustrative to evaluate the integral using the saddle-point approximation. This has the advantage of offering a simpler semiclassical interpretation and thus can be modified to include "corrections" beyond the SFA – in particular, the effect of the Coulomb potential, such that the theory can be applied to atomic and molecular targets. The Coulomb correction is not present in strong-field detachment of negative ions where SFA has been found to describe those processes well.

2.5.3 Saddle Point Approximation

Define

$$M_K(\mathbf{p}) = \int_{-\infty}^{+\infty}dtP(\mathbf{p},t)e^{-iS_0(\mathbf{p},t)}, \quad (2.166)$$

$$P(\mathbf{p},t) = \frac{-i}{(2\pi)^{3/2}}\int d^3re^{-i\mathbf{s}\cdot\mathbf{r}}H_{\text{int}}(t)\Psi_0(\mathbf{r}) \quad (2.167)$$

with $\mathbf{s} = \mathbf{v_p}(t)$ in the length gauge. By introducing scaled quantities, $\varphi = \omega t$, $\mathbf{q} = \mathbf{p}/p_f$, $\mathbf{a} = \mathbf{A}/p_f$, where $p_f = F_0/\omega$ and $K_0 = I_P/\omega$, $z_F = U_p/\omega$, one can rewrite

$$S_0(\mathbf{p},\varphi) = -K_0\varphi + 2z_F\int_{\varphi}^{+\infty}(\mathbf{q}+\mathbf{a}(\varphi'))^2d\varphi'. \quad (2.168)$$

The saddle points t_s are the solution of $\partial S_0/\partial t = 0$, resulting in

$$\gamma^2 + (\mathbf{q} + \mathbf{a}(t_s))^2 = 0. \tag{2.169}$$

Clearly, its solution is complex. $t_s = t_0 + i\tau_0$. If $P(\mathbf{p}, t_s)$ is not singular, by saddle-point approximation,

$$M_K(\mathbf{p}) \approx \sum_\alpha \sqrt{\frac{2\pi}{iS''(t_{s\alpha})}} P(\mathbf{p}, t_{s\alpha}) e^{-iS_0(\mathbf{p}, t_{s\alpha})}, \tag{2.170}$$

where the summation is over all the solutions of the saddle-point equation (Equation 2.169). The real part of the solution gives the "time" for the emission of the electron, and the imaginary part is related to the rate of ionization. If the laser field is a linearly polarized monochromatic field such that $F(t) = F_0 \cos(\omega t)$, then

$$\omega t_s = \sin^{-1}\left(q_z \pm i\sqrt{\gamma^2 + q_\perp^2}\right), \tag{2.171}$$

where q_z and $q_\perp$ are the dimensional-momentum components parallel and perpendicular to the linear polarization direction. It is possible to carry out further calculations using Equation 2.171 to obtain the momentum distributions and total ionization rates. However, these calculations will be omitted here. In practical applications using short pulses, it is more efficient to numerically evaluate the integral in Equation 2.166.

2.5.4 The Validity of the SFA

The KFR or the SFA theory, according to Equation 2.154, predicts sharp ATI peaks that are separated in energy by $\hbar\omega$. Equation 2.155 predicts the position of the first ATI peak for a given laser intensity and also predicts that the position of each ATI peak shifts to lower energy as the laser intensity increases due to the increase of ponderomotive energy with laser intensity. These predictions are in good agreement with earlier experimental results. However, the total ionization probability calculated from SFA is not in good agreement with calculations obtained from solving the TDSE equations. In comparison, ionization probability calculated based on the ADK model has been shown to be in better agreement with TDSE results. The main reason for this is that, in SFA, the continuum electron is approximated by Volkov wavefunction that neglects the effect of the potential from the atomic ion core. Thus the first attempt to improve KFR or SFA is to introduce Coulomb correction to the theory.

2.5.5 Coulomb-Volkov Approximation

One intuitive approach to include the Coulomb field is to replace the plane wave $e^{i\mathbf{k}\cdot\mathbf{r}}$ in the final Volkov wavefunction by the continuum Coulomb function

$$\Psi_{\mathbf{k}}^{(-)}(\mathbf{r}) = \frac{1}{(2\pi)^{3/2}} e^{\pi/2k} \Gamma\left(1 + \frac{i}{k}\right) e^{-i\mathbf{k}\cdot\mathbf{r}} {}_1F_1\left(-\frac{i}{k}, 1, -i(kr + \mathbf{k}\cdot\mathbf{r})\right). \tag{2.172}$$

This approach is called Coulomb–Volkov approximation (CVA). There has been some success in CVA, where the Coulomb effect is important (like the momentum spectra at low energies), but the improvement is not uniform. It works in some cases but not in others. Since the CVA wavefunction is not the eigenfunction of the laser–atom system when the electron is far away from the core, the improvement is not guaranteed using this intuitive approach.

2.5.6 Coulomb-Corrected SFA

A method of incorporating the Coulomb field to modify the SFA model that has proven to be more successful is the Coulomb-corrected SFA (CCSFA). This method starts by recognizing that the derivative of the action $S_0(\mathbf{p}, t)$ with respect to $\mathbf{p}$ corresponds to the classical electron trajectory

$$\mathbf{r}_\mathbf{p}(t) = \frac{\partial S_0}{\partial \mathbf{p}} = \int_{t_s}^{t} \mathbf{v}_\mathbf{p}(t')\, dt' = \mathbf{p}(t - t_s) + \int_{t_s}^{t} \mathbf{A}(t')\, dt'. \tag{2.173}$$

The electron starts from the origin at $t = t_s$ with a complex velocity $\mathbf{v}_\mathbf{p}(t_s)$ (see Equation 2.169). In the SFA, Newton's equation of motion in the laser field is

$$\frac{d^2 \mathbf{r}_\mathbf{p}(t)}{dt^2} = -\mathbf{E}(t). \tag{2.174}$$

The trajectory from Equation 2.174 can have an imaginary part even in real time. These complex trajectories are often called *quantum orbits*. The most probable trajectory is real in real time. This means that $d(\text{Im } S_0)/d\mathbf{p} = 0$. By including the Coulomb field as a perturbation, the corrected trajectory is expressed as $\mathbf{r}_0(\mathbf{p}, t) + \mathbf{r}_1(\mathbf{p}, t)$, where the correction $\mathbf{r}_1(\mathbf{p}, t)$ is found from

$$\frac{d^2 \mathbf{r}_1(t)}{dt^2} = -\frac{Z\mathbf{r}_0}{r_0^3}. \tag{2.175}$$

The derivation to reach the final answer is quite technical, and so interested readers are advised to consult original publications [27–29]. Although the trajectories can be calculated perturbatively, since they occur in the exponential function the correction can be quite large. The CCSFA has been used in a number of applications, but its most important application has been in the calculation of ionization rate (see the Section 2.5.7).

The CCSFA can be used to calculate the photoelectron momentum distributions. The Coulomb correction does not only change the ionization rate, but it also induces additional trajectories to a given final electron momentum. In SFA, for a linearly polarized field, there are two well-known trajectories: the long and the short (see Section 2.4.3). Including the Coulomb field, up to four trajectories can be found per optical cycle. These new trajectories can generate the so-called *low-energy structure* in the photoelectron spectra and form additional side lobes in the momentum distributions [30] (see Section 3.4).

2.5.7 The PPT Theory for Ionization of Atoms by Intense Light Fields

The Coulomb correction for ionization can be done as in the Perelomov–Popov–Terentev (PPT) theory. The derivation of the PPT theory is rather complicated. Interested readers can go back to the original papers for details. PPT theory for the total ionization rate has been found to be quite accurate, as will be shown in Chapter 3. An improved PPT model with newly recommended Coulomb correction has been given for atoms in Popruzhenko et al. [28]. The cycle-averaged ionization rate can be expressed as

$$
\begin{aligned}
w_{PPT}(F_0,\omega) &= \left(\frac{3F_0}{\pi\kappa^3}\right)^{1/2} \frac{C_l^2}{2^{|m|}|m|!} \frac{(2l+1)(l+|m|)!}{2(l-|m|)!} \\
&\times \frac{A_m(\omega,\gamma)}{\kappa^{2Z_c/\kappa-1}}(1+\gamma^2)^{|m|/2+3/4}\left(\frac{2\kappa^3}{F_0}\right)^{2Z_c/\kappa-|m|-1} \\
&\times (1+2e^{-1}\gamma)^{-2Z_c/\kappa} e^{-(2\kappa^3/3F_0)g(\gamma)},
\end{aligned} \tag{2.176}
$$

where C_l is the structure parameter of the atom (see Equation 2.112), $e = 2.718\ldots$, $\kappa = \sqrt{2I_p}$, and γ is the Keldysh parameter. Here, F_0, ω, and Z_c are the laser's peak field strength, laser frequency, and asymptotic charge seen by the electron, respectively. In this equation, $g(\gamma)$ is

$$
g(\gamma) = \frac{3}{2\gamma}\left[\left(1+\frac{1}{2\gamma^2}\right)\sinh^{-1}\gamma - \frac{\sqrt{1+\gamma^2}}{2\gamma}\right]. \tag{2.177}
$$

In the limit of $\gamma \to 0$, $A_m(\omega,\gamma)$, $(1+\gamma^2)^{|m|/2+3/4}$, $g(\gamma)$ and $(1+2e^{-1}\gamma)^{-2Z_c/\kappa}$ all go to 1.0 and the ADK model is recovered.

For completeness, the complex expressions for $A_m(\omega,\gamma)$ and other symbols used in the PPT theory are:

$$
A_m(\omega,\gamma) = \frac{4}{\sqrt{3\pi}}\frac{1}{|m|!}\frac{\gamma^2}{1+\gamma^2}\sum_{n\geq\nu}^{\infty} e^{-\alpha(n-\nu)}\omega_m\left(\sqrt{\beta(n-\nu)}\right), \tag{2.178}
$$

$$
\omega_m(x) = \frac{x^{2|m|+1}}{2}\int_0^1 \frac{e^{-x^2 t}t^{|m|}}{\sqrt{1-t}}dt, \tag{2.179}
$$

$$
\beta(\gamma) = \frac{2\gamma}{\sqrt{1+\gamma^2}}, \tag{2.180}
$$

$$
\alpha(\gamma) = 2\left[\sinh^{-1}\gamma - \frac{\gamma}{\sqrt{1+\gamma^2}}\right], \tag{2.181}
$$

$$
\nu = \frac{I_p}{\omega}\left(1+\frac{1}{2\gamma^2}\right). \tag{2.182}
$$

In applications, for short pulses described by $F(t) = F_0 f(t)\cos(\omega t+\varphi)$, it is desirable to be able to calculate the subcycle ionization rate as in the adiabatic ADK model. Following the procedure of Yudin and Ivanov [31], the expression is

$$
\begin{aligned}
\Gamma(t) &= \frac{C_l^2}{2^{|m|}|m|!}\frac{(2l+1)(l+|m|)!}{2(l-|m|)!}\frac{A_m(\omega,\gamma)}{\kappa^{2Z_c/\kappa-1}}(1+\gamma^2)^{|m|/2+3/4}\left[\frac{3\alpha(\gamma)}{2\gamma^3}\right]^{1/2} \\
&\times \left[\frac{2\kappa^3}{F_0 f(t)}\right]^{2Z_c/\kappa-|m|-1}(1+2e^{-1}\gamma)^{-Z_c/\kappa}\exp\left[-\frac{F_0^2 f^2(t)}{\omega^3}\Phi(\gamma(t),\theta(t))\right].
\end{aligned} \tag{2.183}
$$

The functions $\Phi(\gamma(t), \theta(t))$, $\gamma(t)$, and $\theta(t)$ are explicitly defined in Yudin and Ivanov [31]. The expression for PPT is more complicated than the ADK theory, but it allows for more accurate evaluation of the ionization rates, especially in the multiphoton ionization regime. The ADK theory is an adiabatic theory and the ionization rate drops exponentially as the static electric field is decreased. The nonadiabatic PPT theory will provide a larger correction for electrons emitted when the electric field is small during the cycle, and thus it tends to enhance the yield for direct electrons that are closer to the $2U_p$ cutoff energy.

2.5.8 The QTMC and GQTMC Models

The CCSFA theory, with the Coulomb field effect included, has been shown to be a significant improvement over the KFR (or SFA) model. Formulated in terms of classical trajectories and using an imaginary time method, the CCSFA relies on complex ionization times and complex trajectories from the saddle points (t_s) that satisfied Equation 2.169. Calculation of the complex t_s is not easy, especially for short laser pulses where the electric field varies from one cycle to another.

A alternative method to include the Coulomb field effect is to start with the classical-trajectory Monte Carlo (CTMC) simulation to describe the motion of the electron in the combined laser field and the field from the atomic ion. In a semiclassical extension, this is done by attaching to each trajectory a semiclassical quantum phase following Feynman's path-integral approach. This is advantageous as the trajectories and time are real quantities and one only needs to solve Newton's equation of motion. This quantum-trajectory Monte Carlo (QTMC) method has been proven capable of interpreting strong-field ionization electron spectra in great detail [32]. The QTMC method does not describe the ionization step. In its implementation, the ionization rate is described by the ADK theory. For a linearly polarized laser, the electron is born at the tunnel exit with zero velocity along the polarization axis but with a perpendicular initial velocity derived by Delone and Krainov [18], based on the adiabatic-tunneling ionization theory. Thus, at each time t_0, Monte Carlo trajectory simulations are carried out with an initial condition weighted by the ADK ionization rate and the perpendicular velocity distribution

$$W_1\left(v_\perp^j\right) \propto \left[\sqrt{2I_p}/\left|E(t_0)\right|\right] \exp\left[\sqrt{2I_p}\left(v_\perp^j\right)^2/\left|E(t_0)\right|\right] \tag{2.184}$$

until the electron reaches a virtual detector where the momentum of the electron is sampled. For each trajectory arriving at the detector, a quantum phase derived from the Feynman path integral is calculated.

$$\Phi_j = \int_{t_0}^{\infty} \left\{\mathbf{v}^2(t)/2 - 1/\left|\mathbf{r}(t)\right| + I_p\right\} dt. \tag{2.185}$$

The probability of finding an electron with a momentum within a bin $(\delta p_z, \delta p_r)$ is sampled and calculated as

$$N(\mathbf{p}) = \left|\sum_j \sqrt{W\left(t_0, v_\perp^j\right)} \exp\left(-i\Phi_j\right)\right|^2, \tag{2.186}$$

where $W\left(t_0, v_\perp^j\right)$ is the product of the ADK ionization rate $W_0(t_0)$ and the perpendicular momentum distribution Equation 2.184 for the emission rate of an electron j ionized at t_0 with initial transverse velocity $v_\perp^j$. Since the ADK ionization theory is valid only in the $\gamma \ll 1$ regime, to extend the QTMC to the multiphoton ionization regime one can replace the ADK rate by the rate from PPT, Equation 2.183. This new version is called generalized QTMC (GQTMC).

Notes and Comments

The different gauges for the Volkov wavefunctions have been described in the book by Bransden and Joachain [Bransden, Joachain]. The Floquet theory is treated extensively in [Joachain, Kylstra, Potvliege]. A good introduction to ultrafast femtosecond lasers can be found in the chapter by Gallmann and Keller in the encyclopedia edited by Quack and Merkt [Quack, Merkt]. Full discussion of the phase retrieval of femtosecond pulses can be found in a recent PhD tutorial in Monmayrant et al. [33]. The history of the KFR theory and its modification is best presented in Popruzhenko's review article [27]. In recent years, there have been various attempts to develop computational methods that can be used to solve many-electron atomic and molecular systems in strong laser fields based on first principles. In the multiphoton ionization regime, there is some success when the system is not very complicated. However, it is typically too difficult to guarantee that convergence in the calculation has been achieved, especially in the tunnel-ionization regime. The theoretical approaches presented in this chapter will be extensively used later in this book.

Exercises

2.1 Show that the Volkov-state wavefunctions in length gauge, velocity gauge, and acceleration gauge given in Section 2.1.2 indeed satisfy the corresponding differential equations given in Section 2.1.1.

2.2 For a short laser pulse, where the electric field is given by Equation 2.44 with $\alpha = 0$, the first-order transition amplitude is given by Equations 2.46 and 2.47. Write the corresponding expressions for the second-order transition amplitude.

2.3 Calculate the ponderomotive energy of an 800 nm linearly polarized laser pulse with a peak intensity of 10^{14} W/cm^2. What will be the excursion amplitude of a free electron in such a field? For the excursion amplitude, express the distance in atomic units. For the ponderomotive energy, express it in electron volts.

2.4 For a Gaussian laser pulse given by Equation 2.88, show that τ is the full width at half maximum (FWHM) of the intensity in the time domain. Write $\varphi(t) = \varphi_0 + \frac{1}{2}\alpha t^2$, calculate the pulse in the frequency domain, and use Equation 2.90. Calculate the spectral intensity $|\varepsilon(\omega)|^2$ and find its FWHM spectral bandwidth.

2.5 Go to section 2.2. of *Nonlinear Optics* [Boyd] and go through the derivation of the sum-frequency generation formulation (pp. 72–76, 2nd edition).
2.6 Estimate the saturation intensities for the ionization of Ar and Xe by 800 nm laser pulses with durations of 25 fs and 5 fs (the CEP equals zero), respectively. Saturation intensity is defined as the peak intensity, where the total ionization probability exceeds 95% at the end of the pulse. Assume the pulse shape is Gaussian in the time domain. Use the ADK theory Equations 2.127 and 2.134 to do the calculation. The coefficient C_l for Ar is 2.44 and that for Xe is 2.57 (see Tong et al., *Phys. Rev. A*, **66**, 033402 (2002).
2.7 The ADK theory is not correct at higher laser intensity. Equation 2.136 provides an empirical improvement. Redo Problem 2.6 using Equation 2.136 to find the saturation intensity for Ar when this correction is taken into account. The parameter $\alpha = 9.0$ for Ar (see Tong et al., *J. Phys. B-At. Mol. Opt.*, **38**, 2593 (2005).
2.8 Calculate the ratio of the ionization probability of the $m = 1$ state with respect to the $m = 0$ state in Ar by an 800 nm, 20 fs laser at peak intensities from 0.5 to 1.5×10^{14} W/cm^2 in steps of 0.1×10^{14} W/cm^2 to show that ionization from $m = 0$ state is much larger. Assume the pulse is Gaussian in the time domain. Use the ADK theory from Equation 2.127.
2.9 Follow Section 2.4.3 and reproduce the plots shown in Figure 2.5.

References

[1] L.-Y. Peng, W.-C. Jiang, J.-W. Geng, W.-H. Xiong, and Q. Gong. Tracing and controlling electronic dynamics in atoms and molecules by attosecond pulses. *Phys. Rep.*, **575**:1–71, 2015.

[2] J. R. Oppenheimer. On the quantum theory of the capture of electrons. *Phys. Rev.*, **31**:349–356, Mar. 1928.

[3] H. C. Brinkman and H. A. Kramers. Zur Theorie der Einfangung von Elektronen durch α-Teilchen. *Proc. Acad. Sci. (Amsterdam)*, **33**:973, 1930.

[4] J. S. Briggs and L. Dube. The second Born approximation to the electron transfer cross section. *J. Phys. B-At. Mol. Opt.*, **13**(4):771, 1980.

[5] A. Becker and F. H. M. Faisal. Intense-field many-body S-matrix theory. *J. Phys. B-At. Mol. Opt.* **38**(3):R1, 2005.

[6] X. M. Tong, K. Hino, and N. Toshima. Phase-dependent atomic ionization in few-cycle intense laser fields. *Phys. Rev. A*, **74**:031405, Sep. 2006.

[7] X. Chu and M. McIntyre. Comparison of the strong-field ionization of N_2 and F_2: a time-dependent density-functional-theory study. *Phys. Rev. A*, **83**:013409, Jan. 2011.

[8] T. Otobe and K. Yabana. Density-functional calculation for the tunnel ionization rate of hydrocarbon molecules. *Phys. Rev. A*, **75**:062507, Jun. 2007.

[9] K. C. Kulander. Time-dependent Hartree-Fock theory of multiphoton ionization: helium. *Phys. Rev. A* **36**:2726–2738, Sep. 1987.

[10] B. Zhang, J. Yuan, and Z. Zhao. Dynamic core polarization in strong-field ionization of CO molecules. *Phys. Rev. Lett.*, **111**:163001, Oct. 2013.

[11] D. J. Haxton, K. V. Lawler, and C. W. McCurdy. Multiconfiguration time-dependent Hartree-Fock treatment of electronic and nuclear dynamics in diatomic molecules. *Phys. Rev. A*, **83**:063416, Jun. 2011.

[12] H. Miyagi and L. B. Madsen. Time-dependent restricted-active-space self-consistent-field theory for laser-driven many-electron dynamics. II. Extended formulation and numerical analysis. *Phys. Rev. A*, **89**:063416, Jun. 2014.

[13] D. Strickland and G. Mourou. Compression of amplified chirped optical pulses. *Opt. Commun.*, **56**(3):19–221, 1985.

[14] M. G. Pullen, W. C. Wallace, D. E. Laban, et al. Measurement of laser intensities approaching 10^{15} W/cm^2 with an accuracy of 1%. *Phys. Rev. A*, **87**:053411, May 2013.

[15] C. Z. Bisgaard and L. B. Madsen. Tunneling ionization of atoms. *Am. J. Phys.*, **72**(2):249–254, 2004.

[16] X. M. Tong, Z. X. Zhao, and C. D. Lin. Theory of molecular tunneling ionization. *Phys. Rev. A*, **66**:033402, Sep. 2002.

[17] M. V. Ammosov, N. B. Delone, and V. P. Krainov. Tunnel ionization of complex atoms and of atomic ions in an alternating electromagnetic field. *Sov. Phys. JETP*, **64**:1191, 1986.

[18] N. B. Delone and V. P. Krainov. Energy and angular electron spectra for the tunnel ionization of atoms by strong low-frequency radiation. *J. Opt. Soc. Am. B*, **8**(6):1207–1211, Jun. 1991.

[19] L. Arissian, C. Smeenk, F. Turner, et al. Direct test of laser tunneling with electron momentum imaging. *Phys. Rev. Lett.*, **105**:133002, Sep. 2010.

[20] I. Dreissigacker and M. Lein. Quantitative theory for the lateral momentum distribution after strong-field ionization. *Chem. Phys.*, **414**:69–72, 2013.

[21] L. Hamonou, T. Morishita, and O. I. Tolstikhin. Molecular Siegert states in an electric field. *Phys. Rev. A*, **86**:013412, Jul. 2012.

[22] X. M. Tong and C. D. Lin. Empirical formula for static field ionization rates of atoms and molecules by lasers in the barrier-suppression regime. *J. Phys. B-At. Mol. Opt.*, **38**(15):2593, 2005.

[23] Z. Chen, A.-T. Le, T. Morishita, and C. D. Lin. Quantitative rescattering theory for laser-induced high-energy plateau photoelectron spectra. *Phys. Rev. A*, **79**:033409, Mar. 2009.

[24] L. V. Keldysh. Ionization in the field of a strong electromagnetic wave. *Sov. Phys. JETP*, **20**:1307, 1965.

[25] F. H. M. Faisal. Multiple absorption of laser photons by atoms. *J. Phys. B-At. Mol. Opt.*, **6**(4):L89, 1973.

[26] H. R. Reiss. Effect of an intense electromagnetic field on a weakly bound system. *Phys. Rev. A*, **22**:1786–1813, Nov. 1980.

[27] S. V. Popruzhenko. Keldysh theory of strong field ionization: history, applications, difficulties and perspectives. *J. J. Phys. B-At. Mol. Opt.*, **47**(20):204001, 2014.

[28] S. V. Popruzhenko, V. D. Mur, V. S. Popov, and D. Bauer. Strong field ionization rate for arbitrary laser frequencies. *Phys. Rev. Lett.*, **101**:193003, Nov. 2008.

[29] S. V. Popruzhenko and D. Bauer. Strong field approximation for systems with Coulomb interaction. *J. Mod. Opt.*, **55**(16):2573–2589, 2008.

[30] T.-M. Yan, S. V. Popruzhenko, M. J. J. Vrakking, and D. Bauer. Low-energy structures in strong field ionization revealed by quantum orbits. *Phys. Rev. Lett.*, **105**:253002, Dec. 2010.

[31] G. L. Yudin and M. Yu. Ivanov. Nonadiabatic tunnel ionization: looking inside a laser cycle. *Phys. Rev. A*, **64**:013409, Jun. 2001.

[32] M. Li, J.-W. Geng, H. Liu, et al. Classical-quantum correspondence for above-threshold ionization. *Phys. Rev. Lett.*, **112**:113002, Mar. 2014.

[33] A. Monmayrant, S. Weber, and B. Chatel. A newcomer's guide to ultrashort pulse shaping and characterization. *J. Phys. B-At. Mol. Opt.*, **43**(10):103001, 2010.

3 Strong-Field Ionization and Low-Energy Electron Spectra of Atoms and Molecules

3.1 Total Ionization Yield

3.1.1 Preliminary Remarks

When an atom is exposed to an intense laser field, it can be excited or ionized. Ionization processes are easily detected by measuring the charge state of the resulting ions, employing time-of-flight mass spectrometers. The pressure in the gas chamber is usually maintained at about 10^{-9}–10^{-10} Torr. Ions are collected from the whole focused laser volume and usually only the relative yields are determined.

In general, the validity of a theory is best tested by comparison with experimental data. However, this method is difficult to implement in strong-field physics. Experimentalists are often unable to have full knowledge of their laser parameters, especially for focused intense laser pulses. A pulse may be given in terms of its central wavelength, pulse duration, peak intensity, polarization, and in the case of short pulses, its carrier-envelope phase (CEP). However, these parameters alone are not enough to specify the electric field $F(t)$ of the laser. Since ionization depends nonlinearly on laser intensity, this makes precise absolute comparison between theory and experiment difficult. Furthermore, in a focused laser beam the intensity is not constant within the focused volume. Thus, in order to compare with experimental data, the single-atom ionization probability should be integrated over the whole interaction volume. This amounts to needing repeated calculations over many tens to hundreds of individual laser intensities. This latter step is often called *volume integration*.

Let the spatial-intensity distribution of a focused laser beam be represented by a Gaussian distribution

$$I(\rho, z) = \frac{I_0 w_0^2}{w(z)^2} e^{-2\rho^2/w(z)^2} \tag{3.1}$$

(see Section 1.7.4 for the definition of the parameters). If $P_I(\mathbf{p})$ is the calculated electron-momentum distribution at the peak intensity I, the total electron spectra that can be compared to the experiment are calculated from

$$S(\mathbf{p}, I_0) = D \int_0^{I_0} P_I(\mathbf{p}) \left(-\frac{\partial V}{\partial I} \right) dI, \tag{3.2}$$

where I_0 is the peak intensity at the focus. For such a Gaussian pulse the differential volume per unit of change of laser intensity is given by [1]

$$-\frac{\partial V}{\partial I}dI = \frac{\pi w_0^2 z_R}{3}\frac{1}{I}\left(\frac{I_0}{I}+2\right)\sqrt{\frac{I_0}{I}-1}dI. \tag{3.3}$$

In this expression it is assumed that the density of atoms in the interaction region is constant. In fact, if the target is a gas jet, the gas density is not uniform, nor is the pressure well determined. Thus, in general, it is better to compare the relative total ionization yield in a strong-field experiment.

3.1.2 Numerical Solution of the Time-Dependent Schrödinger Equation

For a one-electron model atom, it is possible to solve the time-dependent Schrödinger equation (TDSE) numerically. The results calculated with known laser parameters can be treated as numerical "experimental" data for the calibration of the simpler theoretical models discussed in Chapter 2. For simplicity, consider a hydrogen atom exposed to a 10 fs, 800 nm laser pulse. The laser field is given by

$$\mathbf{F}(t) = F_0\hat{z}a(t)\cos(\omega t + \varphi). \tag{3.4}$$

This is for a transform-limited pulse, linearly polarized along the z-axis, where ω is the carrier frequency and φ is the CEP. In computations, the envelope function $a(t)$ may take the form

$$a(t) = \cos^2\left(\frac{\pi t}{\tau}\right) \tag{3.5}$$

for the interval of $(-\tau/2, \tau/2)$ and zero elsewhere. The pulse duration is then given by $\Gamma = \tau/2.75$. In principle, a Gaussian envelope is more realistic but its slow drop to zero field makes the computation more time consuming.

3.1.3 Results for Atomic Hydrogen: Dependence on Intensity, Wavelength, and the Role of Excitation

Consider an atomic hydrogen target, the total ionization probabilities by a 10 fs, 800 nm laser for intensity from 0.5×10^{14} W/cm^2 to 1.6×10^{14} W/cm^2, obtained by solving the TDSE, as shown in Figure 3.1(a), together with results obtained from the ADK theory and PPT theory. The over-the-barrier intensity is 1.4×10^{14} W/cm^2. The Keldysh parameter γ ranges from 1.51 to 0.84 in this intensity region. Here we take the TDSE results as experimental data. Note that the ADK model is a perturbative adiabatic theory, thus at high intensity it overestimates the ionization rate. The ADK also underestimates the ionization rate at low intensities, resulting in a factor of 10 too small at 0.5×10^{14} W/cm^2, or at $\gamma = 1.51$. It is important to realize that the quasi-static ADK theory is valid only in the tunneling regime, i.e., for $\gamma < 1$. On the other hand, the nonadiabatic tunnel ionization (NTI), which is similar to PPT, is a nonadiabatic ionization theory. The PPT maintains in fair agreement with the TDSE result down to much lower intensities.

Next, take a close look at ionization at higher intensities, from 1.0×10^{14} W/cm^2 to 2.4×10^{14} W/cm^2 (see Figure 3.1(b)) and compare with the various ionization models. One approach is the static ionization (SI) theory where ionization rate at a fixed static

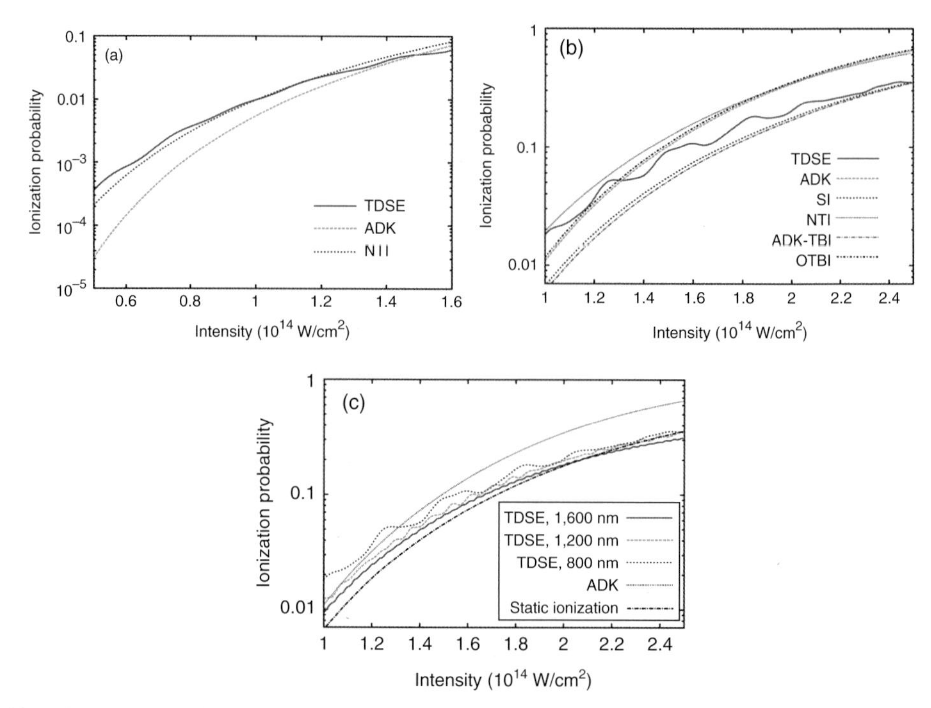

Figure 3.1 (a) Total ionization probability versus laser intensity for atomic hydrogen in a 10 fs, 800 nm laser pulse obtained from solving TDSE, ADK, and other models. (b) Comparison of different models in the high-intensity region. Note that the TDSE results show oscillatory probabilities. Excitation processes are included in TDSE but not in all the other, simpler theories. (c) TDSE results for three different wavelengths, 800, 1,200, and 1,600 nm, for the same peak intensity and pulse duration. Note that ADK is a perturbative static ionization model. (Figures adopted from Qianguang Li et al., *J. Phys. B-At. Mol. Opt.*, **47**, 204019 (2014) [3]. Copyrighted by the IOP publishing.)

electric field is calculated by solving the time-independent Schrödinger equation and the total ionization probability is obtained after integration over the duration of whole laser pulse. The SI agrees with TDSE at the higher-intensity end. The ADK–top-of-barrier-ionization (TBI) model was introduced in Tong et al. [2] such that the ADK rates at high intensities are reduced to the SI results. Other models shown in the figure are explained in Li et al. [3].

In Figure 3.1(c) the TDSE results for wavelength of 800, 1,200, and 1,600 nm lasers are shown. The probabilities calculated from the ADK theory and SI theory are independent of the wavelength since both are based on the adiabatic approximation. The 1,600 nm ionization probabilities are the closest to the SI model. This simply means that the adiabatic model becomes more accurate for longer-wavelength lasers.

The TDSE results for total ionization in the three figures above all display small modulations versus intensity. Such modulations do not appear in the other models. Figure 3.2 shows the modulations in ionization and excitation are mostly out of phase. The excitation probability is approximately 10% for an 800 nm laser, but it drops significantly to 1–3% for a 1,600 nm laser. It shows that excitation is fairly small in the tunnel ionization regime where total ionization probability increases smoothly with laser intensity.

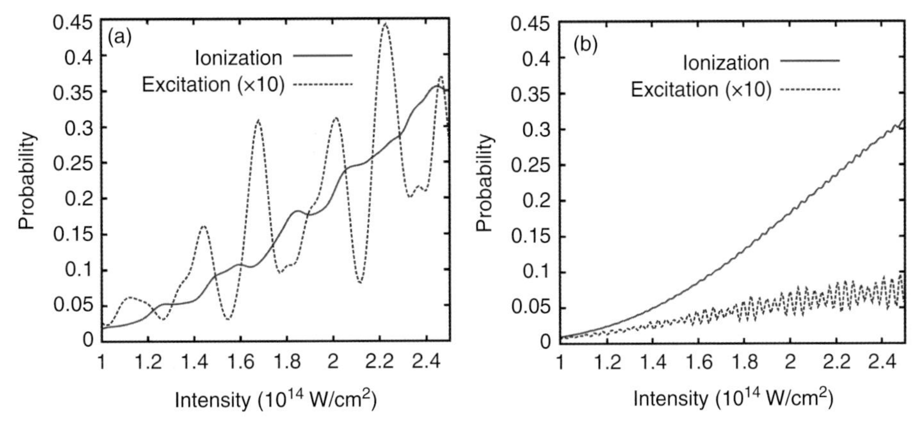

Figure 3.2 Comparison showing that excitation probability typically is a few percent of the total ionization probability for (a) 800 nm and (b) 1,600 nm laser pulses. The small oscillations in ionization and excitation probabilities are out of phase. Excitation diminishes quickly with the increase of wavelength. (Figures adopted from Qianguang Li et al., *J. Phys. B-At. Mol. Opt.*, **47**, 204019 (2014) [3]. Copyrighted by the IOP publishing.)

3.1.4 Role of Excited States in Strong-Field Ionization

Figure 3.2(a) shows the out-of-phase oscillation between excitation and ionization probabilities. For example, total excitation probability increases for intensity from 1.55×10^{14} W/cm^2 to a maximum at 1.70×10^{14} W/cm^2; from there it decreases to a minimum at about 1.80×10^{14} W/cm^2. Within this intensity range, ionization probability decreases to reach a minimum and then back to a maximum. This out-of-phase modulation is best understood by examining the ATI electron spectra. Figure 3.3 shows the calculated ATI spectra near the threshold region, but each curve has been extended to below the threshold, i.e., in the region of excited states. Normally the electron spectrum continuum is plotted as a probability density dP/dE while for the excited state n, it is given as a probability P_n. For each excited state n, define a probability density $dP_n/dE = P_n(dn/dE) = P_n n^3$, where $E = -(1/2n^2)$. For a non-hydrogenic atom, n is replaced by n^*. Then the probability-density distributions of the excited state and the continuum states can be plotted in the same graph, as in Figure 3.3.

The spectrum at each intensity in Figure 3.3 shows the typical ATI peaks for a short pulse. The ATI peaks, according to the multiphoton-ionization picture, are located at $E = n\omega - (I_p + U_p)$ for an electron to reach the continuum by absorbing n photons. As the laser intensity is increased, each ATI peak moves to the lower energy. This is clearly seen in Figure 3.3. When E reaches zero, the atom has to absorb $n + 1$ photons to reach the continuum where the new, first ATI peak will appear at an energy near ω above the threshold. At this particular intensity, the n-photon absorption for ionization is closed. This process is called *channel closing*. At channel closing the total ionization decreases with respect to a monotonically increasing curve. However, the atomic spectrum has discrete density of states below the threshold in the form of excited states. The n-photon "ATI peaks" then appear as excited states, including Rydberg states. Thus, as ionization yield

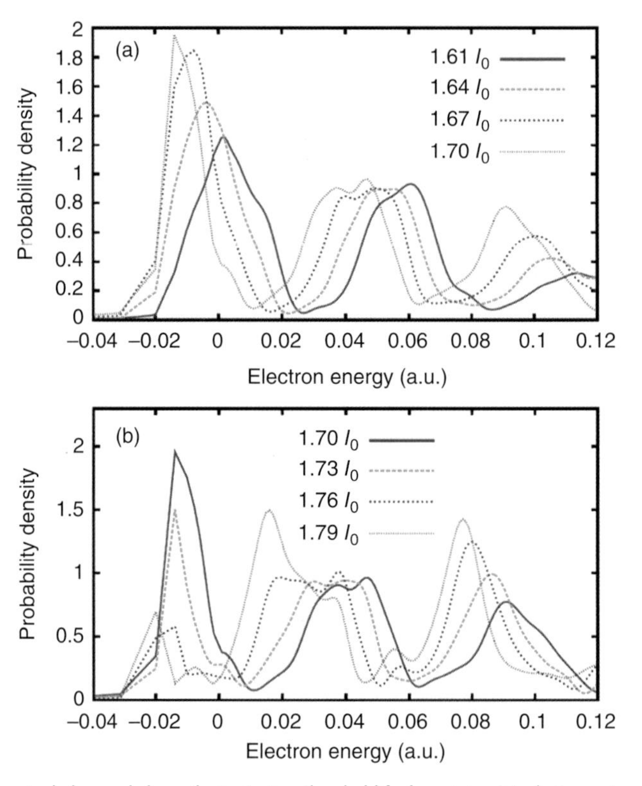

Figure 3.3 Detailed electron spectra below and above the ionization threshold for laser intensities between two channel-closing thresholds (15 and 16 photons). Below the ionization threshold, the excited-state probability is expressed in terms of probability density per unit energy. The figure shows that ATI peaks can then be extended to the below-threshold region. Each ATI peak is much broader than the bandwidth of the 10 fs pulse, as the result of contributions from Freeman resonances. Laser intensity $I_0 = 10^{14}$ W/cm^2. (Reprinted from Qianguang Li et al., *Phys. Rev. A*, **89**, 023421 (2014) [4]. Copyrighted by the American Physical Society.)

drops at the channel closing, the excitation yield increases, resulting in the out-of-phase modulations of their probabilities.

Compared to earlier experiments, the ATI peaks shown in Figure 3.3 are much broader. This is partly due to the short 10 fs laser pulses used in the calculation. For long pulses, the excited states play an important role in typical ATI spectra in the form of Freeman resonances [5]. These resonances are best understood using the multiphoton-ionization picture. In an intense laser field, absorption of n photons to reach an excited state $|i\rangle$ occurs when $(E_i - E_0) + U_p = n\omega$. Once in the excited states, they can further absorb one or more photons to reach the continuum. Since the continuum electrons also have ponderomotive shift U_p in the presence of the laser field, these electrons are located at energies $E_p = p\omega - |E_i|$ where p is the additional number of photons absorbed from the excited states. These peaks are called *Freeman resonances*. Unlike the nonresonant ATI peaks, Freeman resonances do not shift with laser intensity. In a typical spectrum, both resonant and nonresonant multiphoton processes are present, and their contributions to the ATI spectra should be added coherently. Clearly, Freeman resonances will be more prominent for long pulses than for short pulses. For short pulses, only a small fraction of the laser bandwidth can meet the resonant condition, so Freeman resonances are weaker. Therefore the bandwidth of the photoelectron spectrum in Figure 3.3 is due to the contribution of Freeman resonances and the bandwidth of the driving 10 fs laser.

3.1.5 Role of Resonant Excitation in Strong-Field Ionization

In the multiphoton-ionization regime, a specific, low-lying excited state may be in resonance with the absorption of a few photons. For long pulses where the bandwidth is very small, under the resonant condition, the population of the ground state and the excited state will execute Rabi oscillation and the probability is very sensitive to the detuning. For a short pulse, the bandwidth is broad. Thus resonance is less efficient such that a higher field is needed to significantly saturate the upper state. However, at high intensity other excited states as well as ionization may also become significant. The behavior of multiphoton ionization under such situations becomes complicated as it depends sensitively on the target's structure.

Figure 3.4(a) shows the ionization probability of Li atom from the $2s$ ground state, together with the populations of $2s$, $4f$, and Rydberg states, versus the input peak laser intensity from 1×10^{12} W/cm^2 to 14×10^{12} W/cm^2. The laser wavelength is 785 nm and duration is 30 fs. Such an experiment was carried out in Schuricke et al. [7], but only the TDSE results are used here for illustrations. First, the total ionization shows a few "bumps" between 1×10^{12} W/cm^2 and 5×10^{12} W/cm^2. The bumps are traced to the Rabi oscillation between $2s$ and $4f$. The latter is resonantly excited by three-photon absorption. Note that the ionization energy of Li is 5.39 eV and the $4f$ excitation energy is 4.54 eV, while three-photon absorption of a 785 nm pulse is 4.74 eV. At 3×10^{12} W/cm^2, the ponderomotive energy shift is about 0.2 eV, so the resonance condition can be met. Furthermore, although the input pulse is 30 fs, the $2s$ state is severely depleted before the pulse is over. Figures 3.4(c,d) show the time evolution of the probability distribution of the $2s$ state at a few input intensities. The population of the $2s$ displays damped Rabi oscillations. Figure 3.4(a) shows that when the $4f$ population grows, the ionization probability also increases rapidly as the intensity increases. It would take just one photon to ionize the Li from $4f$. Since depletion of Li from the $4f$ is large, the Rabi oscillation is severely damped.

3.1.6 High Rydberg States and Ionization Suppression at High Laser Intensities

Strong-field ionization of atoms grows quickly with laser intensity but the maximal ionization probability of an atom is 1.0. Take the model calculation of Li by an 800 nm laser discussed in Section 3.1.5. At a low laser intensity, ionization would first proceed by multiphoton ionization. For rare gas atoms ionized with 800 nm lasers, as the intensity is increased, ionization may go through tunnel ionization and then over-the-barrier ionization (OBI). Such a classification does not work for the ionization of Li by an 800 nm laser. As discussed in Section 2.4.2, the OBI threshold can be estimated by the simple (or the more accurate) model. For the present system, the threshold is located at an intensity of 3.4 (or 7.2) $\times 10^{12}$ W/cm^2 with Keldysh parameters of 3.6 (or 2.5). In fact, the adiabatic parameter K_0 introduced in Section 2.5.1 is 3.5 for this system. In this case there is no

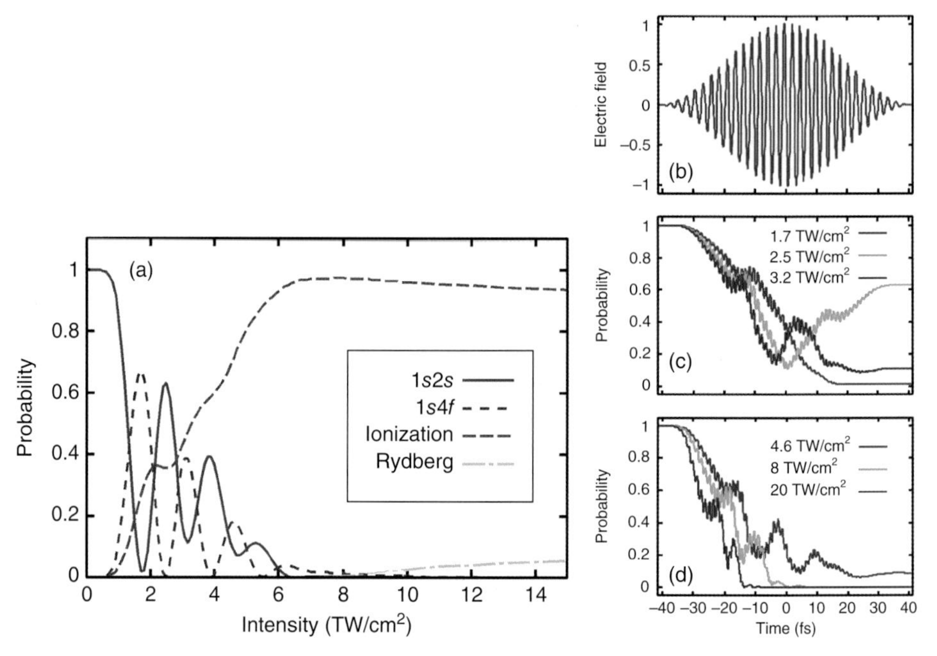

Figure 3.4 (a) Excitation and ionization probabilities obtained by solving the TDSE for a Li atom exposed to a 785 nm/30 fs laser pulse. Notable features are (1) three-photon Rabi oscillation between $2s$ and $4f$ levels, (2) full depletion of the $2s$ ground state for intensity beyond 6×10^{12} W/cm^2, and of the $4f$ state beyond 8×10^{12} W/cm^2, and (3) ionization suppression accompanied by growth of Rydberg states at intensities beyond 8×10^{12} W/cm^2. (b) Input laser's electric field. (c), (d) $2s$ survival probability as a function of time at different input intensities. The $2s$ electron is nearly fully removed before the pulse is over at the input intensities shown. (Figure adopted with permission from Toru Morishita and C. D. Lin, *Phys. Rev. A*, **87**, 063405 (2013) [6]. Copyrighted by the American Physical Society.)

tunneling ionization regime. At an intensity of 6×10^{12} W/cm^2, Figure 3.4 shows that $2s$ is fully depleted but the total ionization probability is still only about 0.95. As the laser intensity is further increased, the total ionization is actually suppressed. In addition, Figure 3.4 also shows that $4f$ is completely depleted at an intensity of about 9×10^{12} W/cm^2. Beyond 9×10^{12} W/cm^2, the total population of Rydberg states (n larger than 6) increases slightly with increasing laser intensity at the expense of decreasing ionization probability. Thus, at extremely high intensities, the atom is not fully ionized, and ionization decreases with increasing intensities. This is an example of ionization suppression.

In this example, ionization suppression is a consequence of the increasing population of Rydberg states. For Rydberg states, the timescale of the electron's "orbital" motion is much longer than the optical cycle of the 800 nm laser. Thus the static tunnel-ionization model cannot be applied. Instead, the fast optical oscillation of the laser exerts on average a zero impulse to the slow-moving electron. High Rydberg electrons, like free electrons, most of the time stay far away from the ion core. A free electron does not absorb a photon unless the electron is close to the ion core. Thus Rydberg electrons that have high orbital angular momenta are stable against ionization by infrared lasers. The stability of Rydberg states against infrared lasers has also been demonstrated by numerical calculations [8].

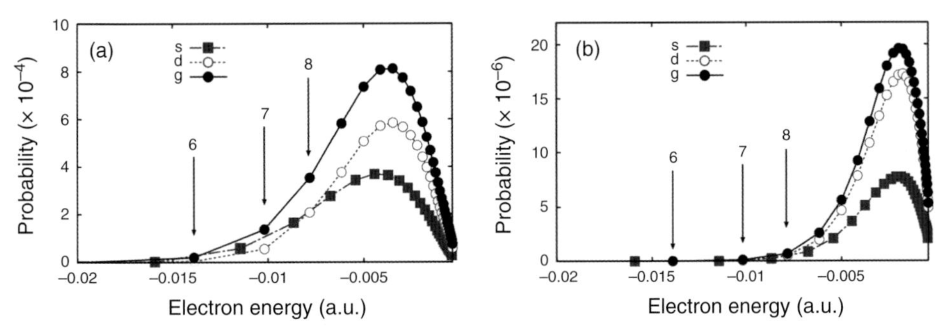

Figure 3.5 Distributions of high Rydberg states for Li exposed to a 30 fs/785 nm laser at an intensity of (a) 10 and (b) 7×10^{12} W/cm^2. The arrows and integers indicate the positions of the Rydberg states. (Figure adopted with permission from Toru Morishita and C. D. Lin, *Phys. Rev. A*, **87**, 063405 (2013) [6]. Copyrighted by the American Physical Society.)

Figure 3.5 shows the population of Rydberg states of Li calculated from solving the TDSE by 30 fs/800 nm laser at peak intensities of (a) 10×10^{12} W/cm^2 and (b) 7×10^{12} W/cm^2. Figure 3.4(d) shows that the ground state and lower excited states of Li are entirely depleted before the end of the laser pulse. Note that the total Rydberg-states probability is much higher at higher input intensities. The orbital angular momentum as well as the principal quantum numbers of the Rydberg states are indicated. These Rydberg states have high orbital quantum numbers, thus they are not readily ionized by the 800 nm lasers.

High Rydberg states have been widely reported in many strong-field experiments. In the literature, frustrated tunnel ionization (FTI) has been used to describe ionization suppression at high intensity [9] and the existence of high Rydberg states was attributed to recapture of continuum electrons by the ion core since it was observed that ionization suppression does not occur for circular or elliptically polarized lasers. This interpretation has been refuted [6]. The fact that Rydberg states are not populated by multiphoton absorption of circularly polarized light is well known [10]. Multiphoton ionization by circularly polarized light is difficult since it would have populated states with high magnetic quantum numbers that are associated with states of high orbital quantum numbers. These states are far away from the ion core, thus they do not absorb photons. If recapture were efficient then photoionization would be efficient as well. Classical simulation is consistent with this interpretation. It shows that these electrons are ionized by the laser in each half cycle just before the peak of the electric field is reached (see [11]). These electrons acquire very low energy from the laser field and are unable to escape to the continuum. Thus they end up in high Rydberg states.

3.2 Total Ionization Yields versus Laser Intensity and Wavelength: Experiment versus Theory

As shown earlier, the calculation of the total ionization probability of atoms by known intense laser fields within the single-active-electron model can be accurately carried out by solving the TDSE directly, especially for wavelengths on the order of 2 μm or less. However, in experiments, precise laser parameters, particularly the peak intensities, are not accurately known. In the meantime, absolute ionization yields are not generally determined in the experiment either. If ionization probability can be calculated accurately, then the calculated results can be used to extract laser information. In most experiments, electrons are collected from the whole focal volume and thus, volume integration should be carried out. For this purpose, having a simpler theoretical model that can obtain reasonably accurate ionization probabilities (comparable to those obtained from solving the TDSE) is desirable.

The simplest ionization model is the ADK theory, which is based on the static tunneling approximation. As the laser intensity and/or the adiabaticity decreases, the ADK theory is known to underestimate ionization probability severely. The PPT model, with the new Coulomb correction (see Section 2.5.7), has been found to reproduce TDSE results to about 50% or better over a broad range of laser intensities and wavelengths. Figure 3.6 shows the comparison of ionization probabilities calculated using ADK, PPT, and TDSE for rare gas atoms by 15 fs, 800 nm laser pulses over an intensity range of about one order of magnitude. Within this intensity range, the ionization probability increases by about six orders of magnitude. The comparison shows that ADK theory works quite well at an intensity where the Keldysh parameter γ (indicated in Figure 3.6) is close to 1.0. At lower intensities (or larger γ), the ADK model can be a few orders too small. On the other hand, the PPT model misses the whole intensity region by a factor of about 50% at most when compared to the TDSE result.

It is well known that ionization probability calculated using the strong-field approximation (SFA) gives reasonably correct laser-intensity dependence, but the absolute values are not correct. Formally, the PPT was obtained by "improving" the SFA. The total ionization yields calculated using ADK and PPT, like TDSE, do not involve an extra normalization factor, but this is not so if SFA is used.

Additional comparison between PPT and ADK versus TDSE for an atomic hydrogen target is shown in Figure 3.7 for wavelengths from 600 nm to 1,200 nm. The scaling factor needed for PPT to agree best with the TDSE result is about 1.5 or less and changes slightly with wavelength. Thus the PPT theory may be used to calibrate laser intensity. However, volume integration should be included.

The ionization yields obtained after volume integration using ADK and PPT are compared to the experiment in Figure 3.8 for Ar and Xe. The measured ionization yields versus the calculated PPT yields agree considerably once the experimental data is normalized to theory (where the two yields agree) at high intensity where the ionization curve is relatively flat. After lining up the experimental curve with the theory curve at higher intensities, the PPT theory and experimental curves are shown to overlap over a large intensity region. This method was suggested [12] for calibrating absolute laser intensity in the laboratory. Other common methods of estimating peak laser intensities include identification of the $2U_p$ cutoff from the electron spectra, the cutoff from the harmonic spectra, or from the focusing information with known input laser power. Both of the first two simple methods suffer the loss of distinct cutoff features after the volume integration in the experimental data.

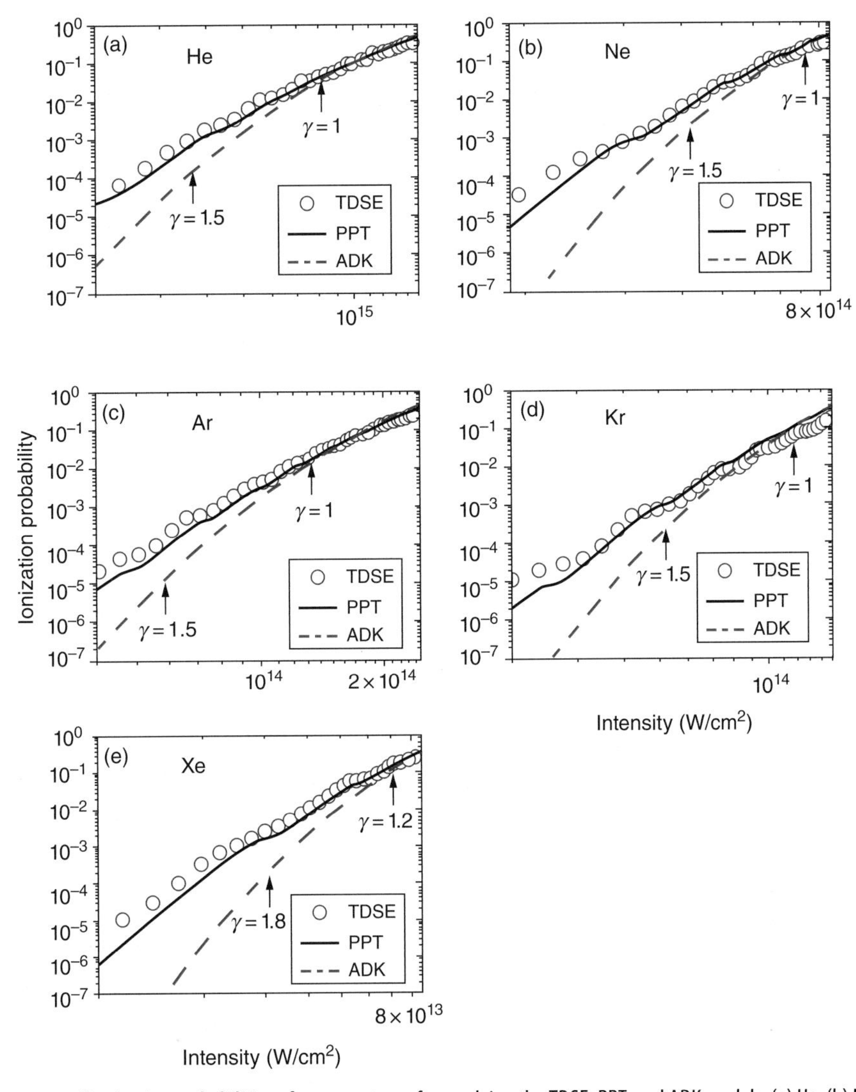

Figure 3.6 Comparison of ionization probabilities of rare gas atoms from solving the TDSE, PPT, and ADK models. (a) He; (b) Ne; (c) Ar; (d) Kr; (e) Xe. The laser is taken to be a Gaussian pulse with full width at half maximum (FWHM) of 15 fs. The central wavelength of the laser is 800 nm for Ar, Kr, and Xe, and 400 nm for He and Ne. (Reprinted from Song-Feng Zhao et al., *Phys. Rev. A*, **93**, 023413 (2016) [12]. Copyrighted by the American Physical Society.)

It is worth mentioning that target structure parameters enter the PPT and ADK models only through the binding energy, the orbital angular momentum, and the orbital wavefunction in the asymptotic region. If excited states are populated by near-resonant multiphoton excitation, then ionization yields will be severely modified, which cannot be accounted for by the simple PPT model. For few-cycle pulses, the ionization yield clearly depends on the CEP. If the CEP is not stabilized then the measured ionization yields should be compared to CEP-averaged calculations.

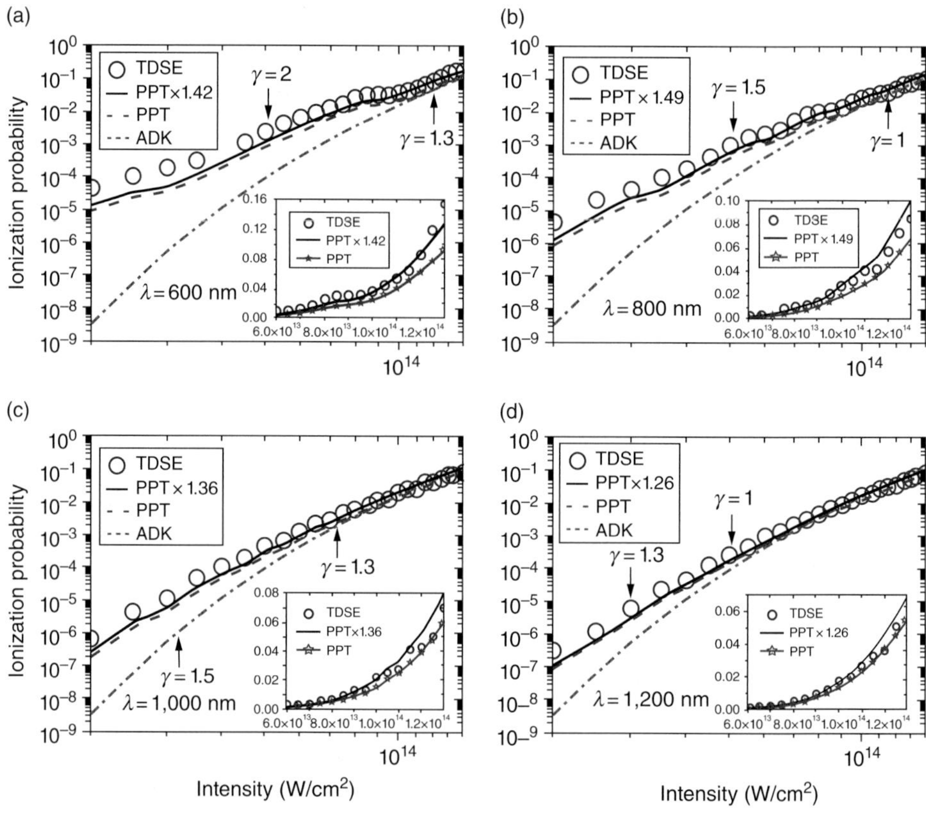

Figure 3.7 Comparison of ionization probabilities of H atom calculated from the TDSE, PPT, and ADK models for wavelengths from 600 to 1,200 nm. A Gaussian pulse with duration of 15 fs is used. Each inset zooms into the higher-intensity region to reveal the differences in the probabilities among the calculations. Based on these results, PPT theory can be used to calculate absolute ionization probabilities over a broad range of laser intensities, wavelengths, and atomic targets. (Reprinted from Song-Feng Zhao et al., *Phys. Rev. A*, **93**, 023413 (2016) [12]. Copyrighted by the American Physical Society.)

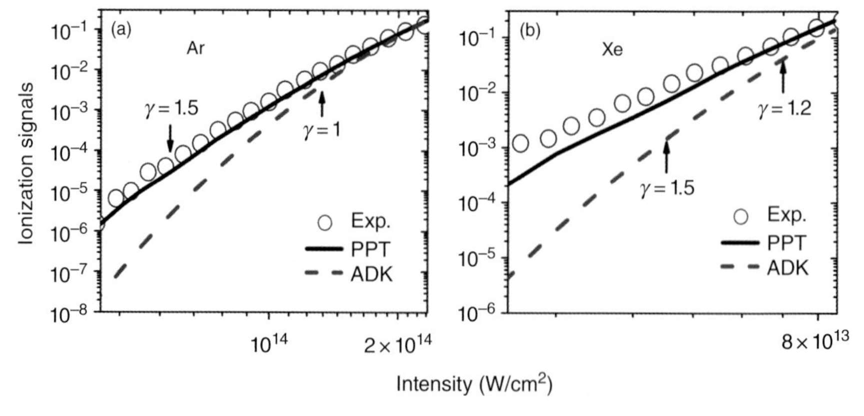

Figure 3.8 Comparison of ionization signals as a function of peak laser intensity. (a) Ar; (b) Xe. The laser field is a Gaussian pulse with a central wavelength of 800 nm and FWHM of 30 fs. The experimental data is from [13]. (Reprinted from Song-Feng Zhao et al., *Phys. Rev. A*, **93**, 023413 (2016) [12]. Copyrighted by the American Physical Society.)

The discussion for ionization so far assumes that strong-field ionization can be described by a one-electron model where a TDSE solution can be accurately carried out numerically. It is worth mentioning that many-electron effects in strong-field ionization have been considered in many theoretical papers, but the convergence of such calculations is difficult to check. In the tunneling-ionization regime, the electron wavefunction at large distance is more important, thus one may expect a many-electron effect to be less significant.

3.3 Low-Energy Electrons and 2D Momentum Spectra: Multiphoton Ionization Regime

3.3.1 TDSE versus SFA Models: Role of Coulomb Potential for Low-Energy Photoelectrons

Figures 3.9(a,b) show the 2D electron-momentum distributions for single ionization of argon in the one-electron model by a 10 fs, 600 nm laser pulse with peak intensity of 1.4×10^{14} W/cm^2 calculated by solving the TDSE and the simple SFA model. The 2D spectra are plotted for electron momentum along the laser-polarization direction against a direction perpendicular to it. Since the laser is linearly polarized, the system has cylindrical symmetry. In both theories the spectra show good similarity in terms of circular rings that correspond to different ATI orders. Also, in each figure, the first ring (the first ATI peak) shows a nodal structure at $p_{||} = 0$, while the second ATI peak shows an antinodal structure.

The two figures also exhibit pronounced differences. From the TDSE result, the first ATI peak shows a clear fan structure. The radial nodal lines mean that they are lines of constant polar angles. For the SFA, the major difference is that the nodal lines are perpendicular lines, implying that the electron wave packet is a direct product of the momentum along the polarization axis and a component perpendicular to it. The difference is attributed to the asymptotic Coulomb potential seen by an electron when the electron is stripped from a neutral atom. This can be confirmed by introducing a cutoff to the Coulomb potential. When the cutoff radius is close to the ion core, the 2D momentum spectra calculated from the TDSE begins to resemble the SFA result. Recall that, in the SFA model, after ionization the electron only sees the laser field, while in a real atom it would also see the Coulomb potential from the ionic core. The long-range Coulomb potential exerts an attractive force as the electron escapes to the detector. The Coulomb potential can also affect the electron-energy distributions. Figure 3.9(c) shows the electron-energy spectra from TDSE and SFA together with two TDSE calculations where the Coulomb potential cutoff was set at five and two atomic units. Clearly, the Coulomb force would pull the low-energy electron toward the lower-energy region compared to a situation where the Coulomb potential is weakened or turned off. Another pronounced effect of the Coulomb potential can be seen in the momentum distribution of photoelectrons in the perpendicular direction. The Coulomb force also shrinks the width of the perpendicular momentum distributions (see Figure 3.9(d)). These are all well-understood effects of the Coulomb potential on the low-energy electrons.

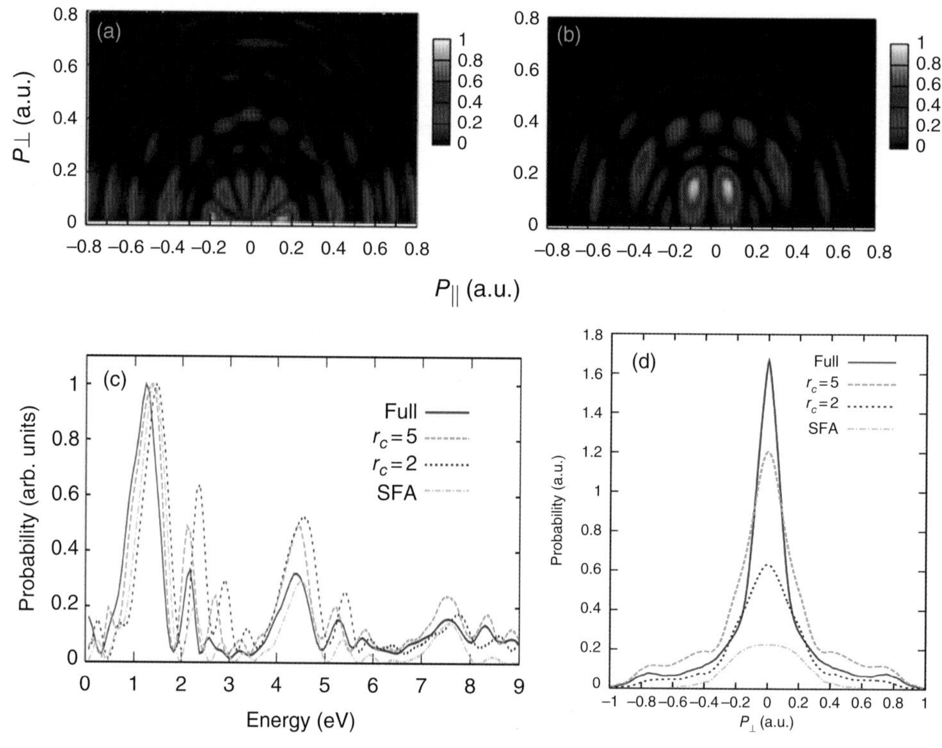

Figure 3.9 Typical 2D momentum distributions of low-energy photoelectrons for single ionization by a 10 fs, 600 nm laser. (a) From TDSE, showing fan structure near the origin. (b) From SFA, where parallel nodal lines are seen for the first ATI peak. (c) Distribution of low-energy photoelectrons obtained from solving the TDSE with calculations where the Coulomb potential is weakened at a cutoff radius of five and two atomic units or removed completely (the SFA). (d) Perpendicular momentum distribution of the photoelectrons. The effect of the Coulomb potential gives a cusp at the center. (Reprinted from Zhangjin Chen et al., *Phys. Rev. A*, **74**, 053405 (2006) [14]. Copyrighted by the American Physical Society.)

Extensive TDSE calculations and experimental data using lasers with wavelengths from 400 nm to 900 nm have shown that the TDSE results in Figure 3.9(a) are quite typical. The first ATI peak always exhibits a clear fan structure, which implies that the angular distribution is characterized by a relatively well-defined dominant angular momentum L. In other words, the angular distribution is well approximated by $P_{L0}^2(\cos\theta)$. The value L_0 was found to depend on the minimum number of photons the atom needs to absorb to reach the first ATI peak and the orbital angular momentum of the initial state.

An empirical model based on the multiphoton ionization picture was developed [14] to predict the value of L_0. Absorption of one photon from a level initially with angular momentum L will reach a level with angular momentum $L-1$ or $L+1$; thus, for example, if the initial state is an $L=0$ state, successive absorption of N photons will predominantly populate a final state with angular momentum $N/2$. However, it is known that the dipole-transition-matrix element from L to $L+1$ is always higher than that from L to $L-1$. Taking the ratio of these two transition dipoles to be a parameter to fit the dominant L_0, a ratio

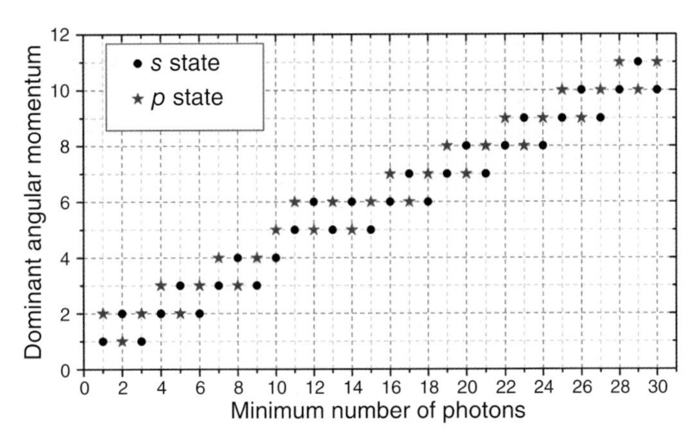

Figure 3.10 Dependence of the dominant L_0 on the minimum number of photons N needed to ionize the atom for an atom initially in the s or the p state, respectively. The dominant L_0 characterizes the number of angular nodal lines in the first ATI peak in the 2D momentum spectra. (Reprinted from Zhangjin Chen et al., *Phys. Rev. A*, **74**, 053405 (2006) [14]. Copyrighted by the American Physical Society.)

of 2:1 for the L to $L+1$ versus the L to $L-1$ transitions correctly predicts the dominant L_0. The results are shown in a universal graph in Figure 3.10. This graph shows that the dominant L_0 depends only on the minimum number of photons N needed to ionize the atom. Recall that the minimum number of photons needed is an integer N, which can make

$$E_e = N\hbar\omega - (I_p + U_P) \tag{3.6}$$

positive. The dominant L_0 also depends on whether the initial state is an s orbital or a p orbital. From Figure 3.10, the dominant L_0 does not change with the increase of laser intensity until a new channel closing appears. Thus clear nodal structure can be seen in the 2D experimental electron-momentum spectra. These nodal lines are not washed out after volume integration. Because the interpretation is based on the multiphoton picture, the model gradually fails with increasing laser wavelength toward the tunneling regime.

An alternative derivation of the dominant L_0 in the 2D momentum distribution for the first ATI peak was given in Arbó et al. [15] using a classical trajectory Monte Carlo method including tunnel effects (CTMC-T). The derived value is given by $L_0 = \sqrt{2\alpha}$ where $\alpha = F_0/\omega^2$ is the quiver amplitude of an electron in a laser field with field amplitude F_0 and frequency ω. Since angular momentum is quantized for an electron, the prediction of L_0 from CTMC-T lies within the uncertainty of $L_0 \pm 1$. Unlike the semiempirical multiphoton picture given in Maharjan et al. [14], which is based on the propensity rule in the multiphoton-absorption process, the L_0 given in Figure 3.11 depends not only on the laser parameter α, but also on the quantum nature of the initial state (s or p orbital) as well as the number of photons needed to ionize the atom in the presence of the driving field. The origin of the same (mostly) dominant L_0 is not clear. Note that in Figure 3.10 the L_0 for a given laser parameter always differs by one depending on s or p initial states, while the prediction in Arbó et al. [15] does not contain such information. Thus the latter model is incapable of predicting the nodal nature of the electron momentum spectrum at $p_{||} = 0$.

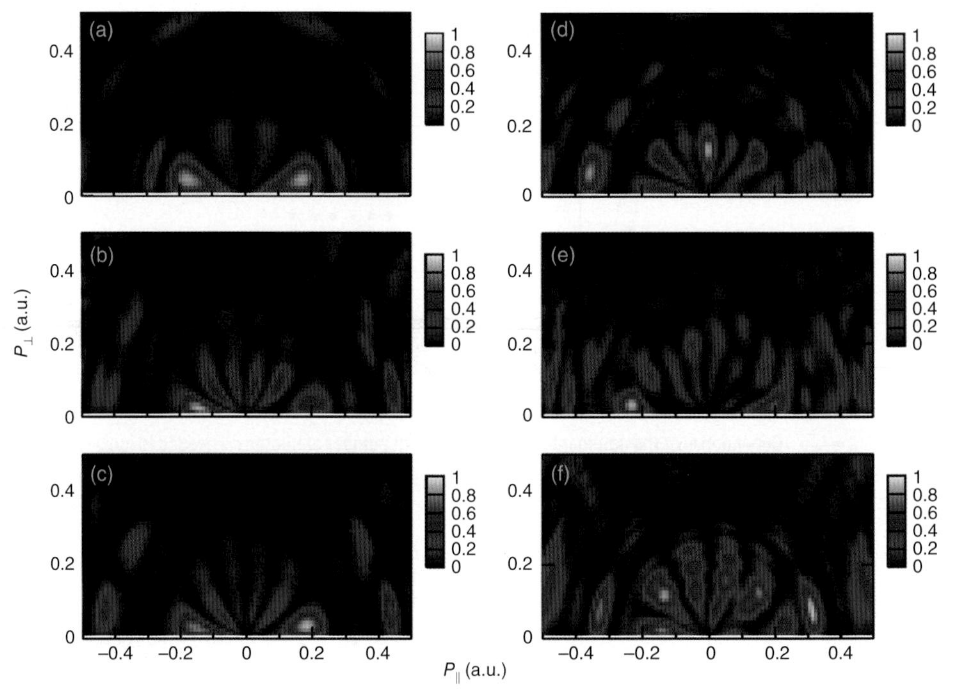

Figure 3.11 Low-energy, 2D photoelectron momentum spectra for single ionization of selective systems by 10 fs laser pulses. The parameters (target, wavelength (nm), peak intensity (10^{14} W/cm^2)) are (a) [Ar, 400, 1.7]; (b) [Ar, 590, 2.8]; (c) [Ne, 600, 2.1]; (d) [H, 800, 0.8]; (e) [H, 800, 2.03]; (f) [He, 400, 6.2]. The [Keldysh parameter, minimum number of photons needed for ionization, dominant angular momentum L_0 of the first ATI peak] are (a) [1.74, 6, 3]; (b) [0.9, 12, 5]; (c) [1.23, 14, 5]; (d) [1.19,12,6]; (e) [0.75, 17,7]; (f) [1.15,11,5]. The dominant L_0 is from Figure 3.10, which has been shown to agree with the number of nodal lines in the first ATI peak. (Reprinted from Zhangjin Chen et al., *Phys. Rev. A*, **74**, 053405 (2006) [14]. Copyrighted by the American Physical Society.)

Note that the most prominent structure of the 2D electron-momentum distributions, as exemplified in Figure 3.11, is the dominant L_0 of the first ATI peak. The number of approximate nodes (due to the contribution of the other, smaller angular-momentum components) for the first ATI peak is determined by L_0, which would predict that the electron momentum distribution at $p_{||} = 0$ is a node or an antinode. The electron angular distribution still peaks at 0° and 180° along the polarization axis. For the second ATI peak, its parity changes since its dominant angular momentum is $L_0 + 1$. On the other hand, the signal at 90° may become much weaker such that the second ATI peak becomes too feeble to be seen unless the distributions are displayed on a logarithmic scale.

The low-energy, 2D momentum spectra, according to the semiempirical rule for the dominant L_0, demonstrate that low-energy electron spectra in single ionization by an intense laser field, like PPT and ADK theory for ionization, depend little on the detailed structure of the atom except for the ionization potential and the orbital angular momentum of the electron from which ionization occurs. The core structure of the atom does not play an important role. This is "bad" since it implies that the abundant low-energy electrons in strong field ionization contain little structure information of the target.

3.3.2 Comparison of Experimental 2-D Electron Momentum Spectra with TDSE Simulations

2D electron momentum spectra of Ar ionized by 40 fs, 400–800 nm lasers have been reported using the cold target recoil ion momentum spectroscopy (COLTRIMS) detector in Maharjan et al. [16] and Li et al. [17] on Xe using an 800 nm, 25 fs laser, and in Marchenko

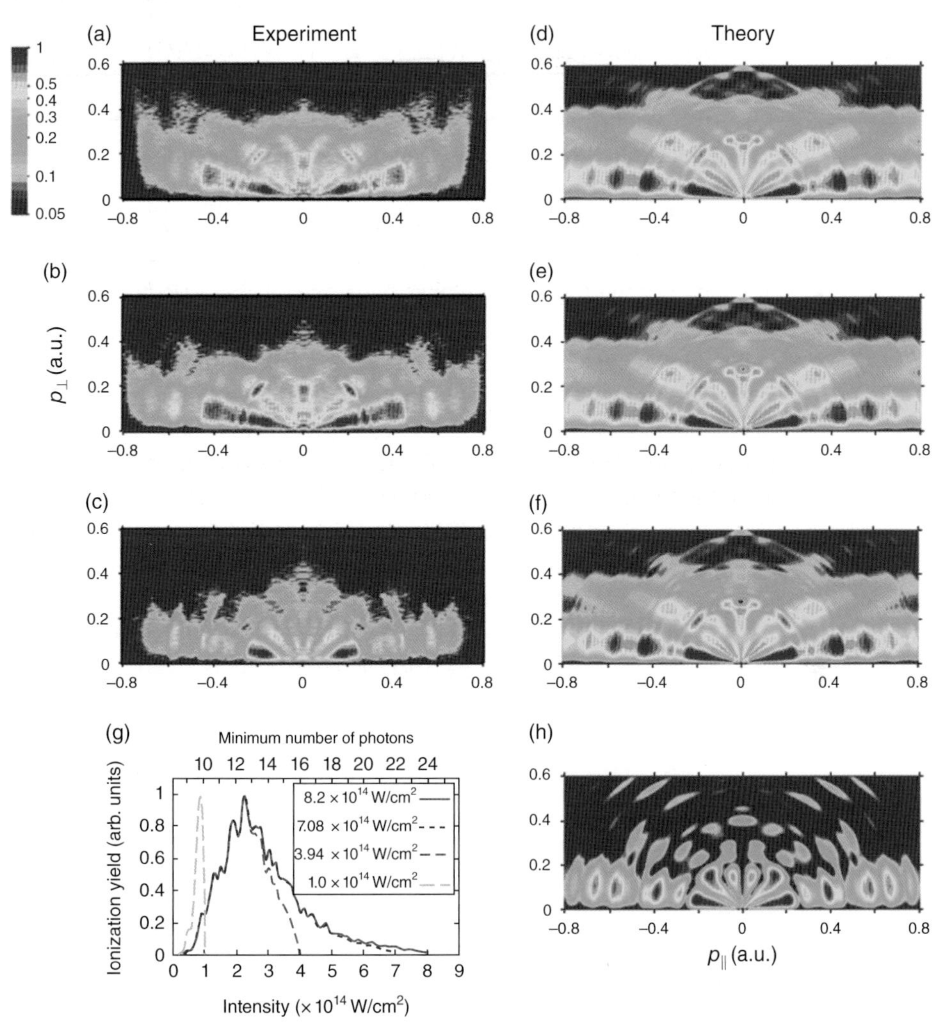

Figure 3.12 (a–c) Experimental 2D electron-momentum spectra of Ar ionized by a 40 fs, 640 nm laser pulse and the comparison with theoretical simulations (d,e,f). The peak intensity for the top row is 8.2×10^{14} W/cm^2; second row is 7.08×10^{14} W/cm^2, third row is 3.94×10^{14} W/cm^2. (h) Theoretical calculations for intensity at 1.0×10^{14} W/cm^2. Volume integration has been carried out in the theoretical calculation. (g) Effective peak laser-intensity distribution for each input laser pulse in the ionization of Ar including the focal volume effect. The minimum number of photons needed to ionize the atom is indicated on the upper horizontal axis. (Figure adopted with permission from Toru Morishita et al., *Phys. Rev. A*, **75**, 023407 (2007) [19]. Copyrighted by the American Physical Society.)

et al. [18] on Xe using 600–800 nm lasers and velocity map imaging (VMI) detectors. In the experiment in Maharjan et al. [16], the laser intensity was quite high such that ionization-saturation effect should be included carefully in the calculation together with volume integration. Figure 3.12 illustrates the degree of agreement between spectra reported from the experiment and from the theory. First, note that the experimental spectra (a) and (b) are essentially exactly identical even though the input intensity of (a) is 8.2×10^{14} W/cm^2 while that of (b) is 7.08×10^{14} W/cm^2. Although three different laser intensities were used in the calculation, the results shown in (d), (e), and (f) are all identical and in good agreement with the experimental data shown in (a) and (b). To get good agreement with the experimental data shown in (c), it was found that the theoretical input intensity must be reduced to 1.0×10^{14} W/cm^2 (see (h)), instead of the reported peak intensity of 3.94×10^{14} W/cm^2.

By taking into account the laser-intensity distribution in the focal volume, the equivalent effective laser-intensity distribution that contributes to the 2D electron momentum spectra can be calculated. The result is shown in Figure 3.12(g) [19] for the four input peak intensities at the laser focus. Note that, at the two highest input intensities, the effective intensity distributions in the interaction region are essentially identical. Thus, when saturation happens it does not matter whether higher laser power is used for the ionization (within the single-electron model). Even for the third intensity at 3.94×10^{14} W/cm^2, only a small portion of the high-intensity tail is slightly different. At the low intensity, where saturation does not occur, the 1.0×10^{14} W/cm^2 curve shows that ionization occurs only from a narrow intensity range. In studies of strong-field ionization, it is prudent to watch out for effects due to ionization saturation.

3.4 Surprising Features of Photoelectron Spectra for Ionization by Strong Mid-Infrared Lasers

3.4.1 Few eV to meV Peaks in Electron Spectra: Low-Energy Structures

In Section 3.3, the photoelectron spectra for atoms were examined for ionization by intense lasers with wavelengths of about 800 nm or less, where low-energy electron spectra are characteristic of multiphoton ionization. With optical parametric amplification (OPA) or optical parametric chirped-pulse amplification (OPCPA), intense long-wavelength lasers from 1.3 μm to 4.0 μm have been generated. For peak intensities of 10^{13} W/cm^2 or 10^{14} W/cm^2, the interaction of these mid-infrared lasers with atoms allows the study of ionization in the deep tunnel-ionization regime where the Keldysh parameter varies from less than one to about 0.2. One of the most unexpected results from these studies is the observation of low-energy electron peaks with kinetic energy of a few eVs down to a few meVs. Using a 2 μm driving laser, as first reported in Blaga et al. [20], low-energy electrons observed along a direction parallel to the linear polarization axis showed a pronounced peak. This peak was coined as "low-energy structure" (LES). Its location was found to be independent of the targets (Ar, N_2, H_2), see Figure 3.13(a). The peak is clearly absent in

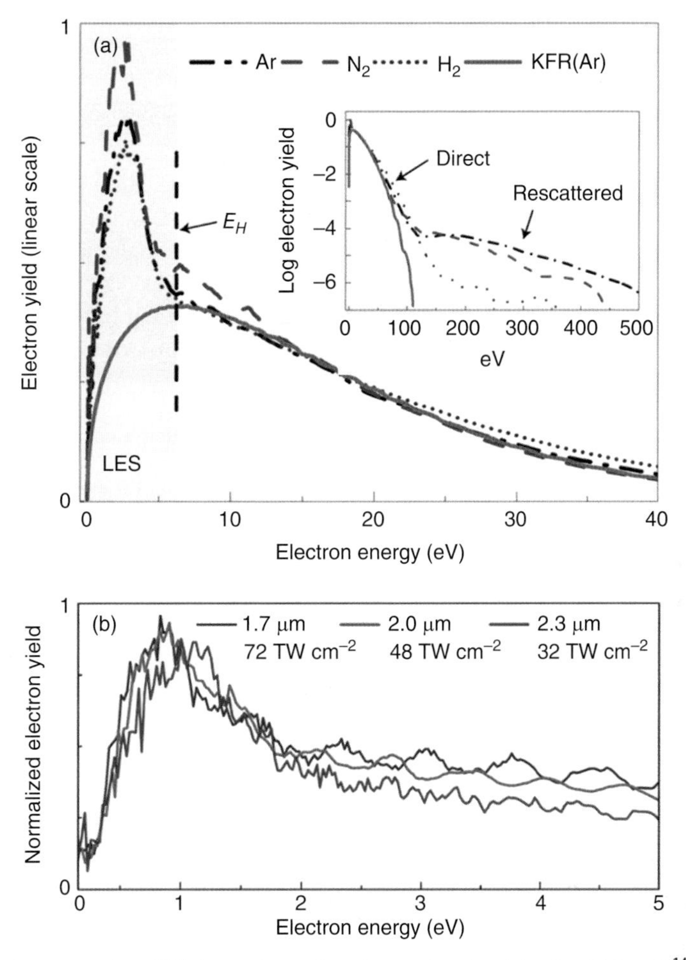

Figure 3.13 (a) Low-energy electron structure (LES) observed using 2 μm laser pulses at an intensity of 1.5×10^{14} for different targets. The inset shows the position of the LES versus the direct electrons and the high-energy rescattered electrons. (b) The position of the LES depends only on the ponderomotive energy or the Keldysh parameters γ. (Figure adopted with permission from C. I. Blaga et al., *Nat. Phys.*, **5**, 335 (2009) [20]. Copyrighted by Nature Publishing Group.)

the prediction of the first-order strong-field approximation (or the Keldysh–Faisal–Reiss (KFR) theory). Note that the peak was not observed for 800 nm lasers where the energy spectra show the familiar ATI peaks. The peak position has been found to be the same if the ponderomotive energy is the same, at about 1 eV for $U_p = 20$ eV, see Figure 3.13(b). The LES structure can be reproduced by solving the TDSE, and it was found that the long-range Coulomb potential plays an essential role, but the short-range ion core potential is not important.

Following Blaga et al. [20], the LES was also observed in Quan et al. [21], but another peak at an even lower energy was also seen. Subsequently, this same very low-energy peak was also observed in Wu et al. [22] at about 0.1–0.2 eV. This new peak is called the very low-energy structure (VLES). Later, the longitudinal momentum distribution was measured

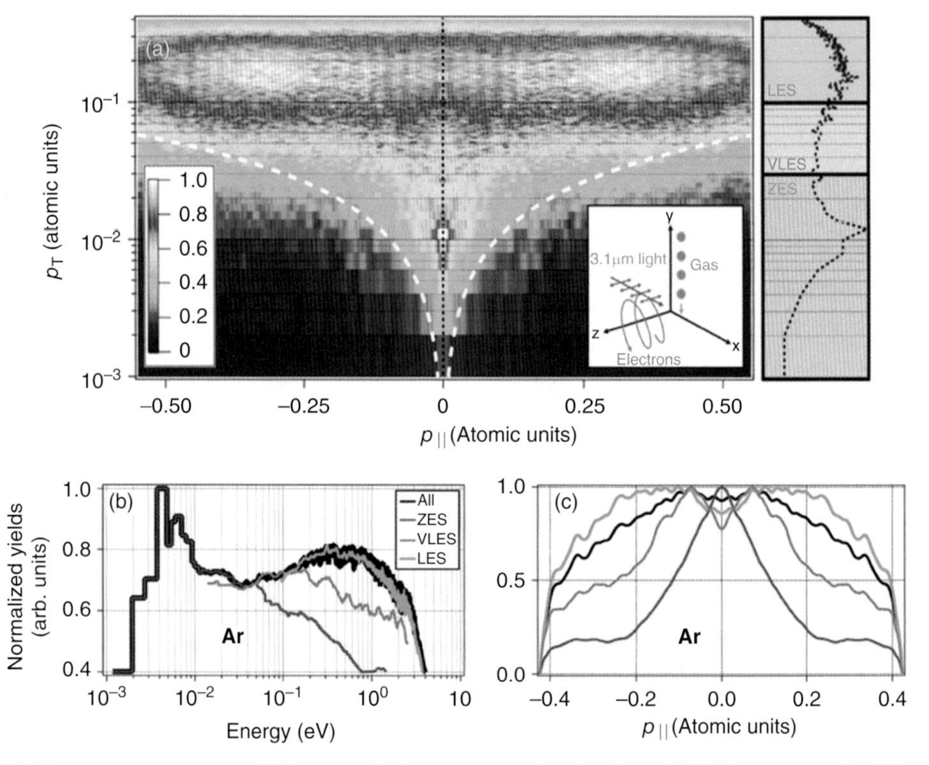

Figure 3.14 2D electron momentum spectra with the perpendicular momentum plotted on log scale. This figure was used to define the LES, VLES, and ZES spectra. On the right, the electron-energy spectra and the longitudinal-momentum spectra are shown. Both LES and VLES show depression at $p_{||} = 0$, but the depression for VLES is deeper. The ZES lies mostly perpendicular to the polarization axis with the peak energy at about 1 meV. (Figure adopted with permission from Michael Pullen et al., *J. Phys. B-At. Mol. Opt.*, **47**, 204010 (2014) [24]. Copyrighted by the IOP publishing.)

and the distribution has a minimum at $p_{||} = 0$ for the VLES along the polarization axis. The three experiments reported until 2012 used long-wavelength lasers up to 2 μm and the low-energy features have been observed for wavelengths as short as 1,320 nm. Another game-changing experiment was reported in 2013 in Dura et al. [23] using a 3,100 nm laser at an intensity where the ponderomotive energy was 95 eV and the Keldysh parameter was 0.3. With the new, high-resolution COLTRIMS detector, the electrons' fine features were observed down to a few meVs.

Figure 3.14 shows the 2D electron-momentum distributions with the perpendicular momentum plotted on a log scale. The measurement has extreme resolution of about 0.01 atomic units in both transverse and parallel momenta. By defining LES, VLES, and the zero energy structure (ZES) according to the magnitude of the perpendicular momenta, the energy distributions and the longitudinal momentum distributions of the LES, VLES, and ZES are shown on the two lower frames of Figure 3.14. The LES now is seen as a broad peak near about 0.5 eV, the VLES peaks at about 0.2 eV, and the ZES peaks at 4.5 meV. In the parallel longitudinal momentum distributions, the VLES is characterized by a deeper minimum than the LES at $p_{||} = 0$, while the ZES has a sharp peak at $p_{||} = 0$. The width of the longitudinal momentum is broadest for LES and smallest for ZES.

From the beginning it was realized that the LES is quite universal for laser wavelength greater than 1,000 nm. The LES, and to some extent the VLES, can be qualitatively reproduced by solving the TDSE. The LES is not reproduced in SFA (or KFR) theory. It is reproduced by TDSE even when the pulse is reduced to just one cycle but does not show up if an elliptical polarized laser is used. Based on these observations, it is believed that the LES is due to the "interference" between the direct scattering and rescattering that occurs within one optical cycle. Thus a change of pulse duration would not affect the main features. In Quan et al. [21], it is shown that classical theory can qualitatively interpret the observed peaks, but the Coulomb potential plays a major role. Further simulations in Liu and Hatsagortsyan [25] use classical theory and emphasize that multiple forward scattering is mainly responsible for the appearance of LES, and in Kästner et al. [26] the role of soft recollision is demonstrated, which can cause the bunching of photoelectrons through which a series of low-energy peaks emerges. These peaks do not require a long-range potential and can be derived classically.

According to the bunching model, one expects to find not just the low-energy LES and VLES, but a whole series. Is the observed VLES the limit of the series? The high-resolution spectra reported in Dura et al. [23] and Pullen et al. [24] did not raise this issue except to point out the new ZES. In a subsequent joint theory and experimental paper [24], by plotting the 2D data somewhat differently, there appears two LES peaks, LES1 and LES2 (see the top frame of Figure 3.15). The classical simulation, shown in the bottom frame, appears to reproduce these peaks, with the exception of the LES and VLES peaks, which all extend to quite low perpendicular momenta in the region where the ZES is located. In fact, the simulation in Pullen et al. [24] predicts up to LES4.

The new ZES was first observed in the experiment shown in Dura et al. [23], but not in the CTMC simulation, as can be seen by comparing the two pictures in the left column of Figure 3.15. It was then postulated that these ZESs are due to high-lying Rydberg states that were eventually ionized by the extracting field of the electron detector [27]. By including the extraction field of 1.5 V/cm in the simulation, the ZES then appears in the CTMC calculation (see the two figures in the right column of Figure 3.15). These results seem to support the suggestion that ZES is indeed due to the field ionization of Rydberg states. On the other hand, the simulation in Wolter et al. [27] performed the calculations up to a distance of 5,000 au, where excitation to high-Rydberg states and ionization were separated. Ideally, these separations should be performed by extracting the electron momentum at infinity.

Another simulation similar to [27] was performed in Xia et al. [28] by focusing on the ZES. In this work, the asymptotic momenta from the instantaneous positions and momenta are obtained using the three conserved quantities: total energy, angular momentum, and Runge–Lenz vector. Xia et al.'s results established that Coulomb potential does not play a significant role for LES, thus LES comes from multiple rescattering. For the VLES, the Coulomb potential does play a greater role, and below 10 meV, the Coulomb potential plays a significant role. In other words, these low-energy structures are the consequence of multiple rescattering of an electron in the combined laser and Coulomb fields. The transition from LES to VLES and to ZES is probably gradual, with stronger effect from the Coulomb potential, except for the "occasional" bunching of "orbitals" that results in sharp features in the 2D momentum spectra. According to classical dynamics, the final electron momentum depends on the time t_0 of ionization and its initial transverse velocity, $v_{\perp}$.

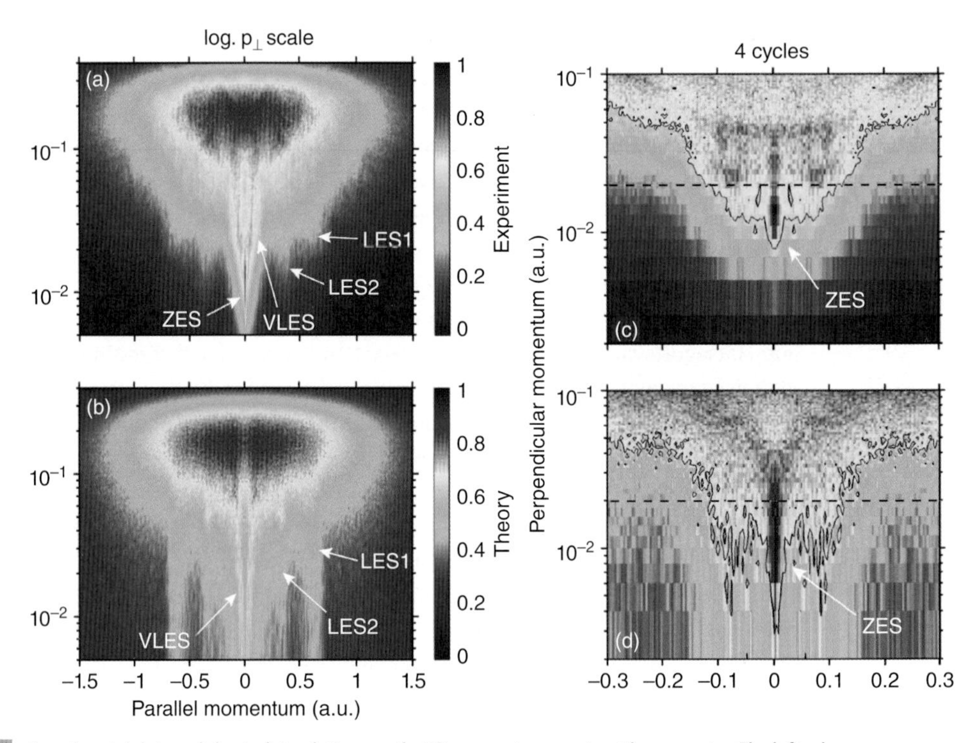

Figure 3.15 Experimental data and classical simulations on the 2D momentum spectra at low energies. The left column compares experiment (top) and CTMC calculations (bottom) without an extraction field. No ZES was found in the calculation. When a static extraction field is used, the ZES appears as shown in the right column. (Figure adopted with permission from Wolter Benjamin et al., *Phys. Rev. A*, **90**, 063424 (2014) [27]. Copyrighted by the American Physical Society.)

Bunching may be related to how the ionization probability depends on $(t_0, v_\perp)$. When the partial derivative with respect to t_0 and $v_\perp$ is zero, bunching occurs. A similar, simple example happens in rainbow scattering, which occurs when the derivative of the deflection function with respect to scattering angle goes to zero.

The simulation in Xia et al. [28] suggests that ZES can appear even without the external static field, and thus rules out field ionization of Rydberg states as the sole source of ZES for $\gamma = 0.3$, the intensity carried out in Dura et al. [23]. Interestingly, at a lower laser intensity (larger γ), the authors in Xia et al. [28] did not find ZES. Thus, one cannot rule out that ZES mostly originates from the ionization of Rydberg states. These low-energy structures are due to the intricate interplay between the laser field and the asymptotic Coulomb potential, and thus there is no need to study their dependence on the target. However, there is a need to study their dependence on laser parameters. In dealing with such low-energy electrons, the role of high-Rydberg states is important as well. While these Rydberg states have been "claimed" as due to the recapture of the electron after the laser field is over, it is generally believed that Rydberg electrons arise directly from the tunneling process with negative energy and large angular momentum. These electrons cannot be recaptured by the ion since the recapture process can only occur near the ion in order to conserve energy and momentum (except by radiative capture, a much weaker process).

To summarize, further experiments examining how the low-energy structures depend on the laser parameters are still needed. Despite the success of the semiclassical model in interpreting existing experiments, the model is still limited. Alternative approaches are clearly desirable but none are presently available. These low-energy structures remain one of the least understood phenomena in strong-field ionization.

3.4.2 Side Lobes, Spiders and Photoelectron Holography

Ionization of atoms in a strong laser field leads to the emission of electron wave packets that propagate in the combined laser and Coulomb fields. A plethora of interferences result from those that eventually acquire the same final momentum. The best known is the ATI peaks. For short pulses, many such interferences are intractable. However, when the number of pathways is limited, say, to only two, then clean interference patterns are visible in the photoelectron-momentum spectra if the amplitudes are comparable. Figure 3.16 shows examples. Using long-wavelength lasers, clear-interference fringes have been observed for electrons with large longitudinal momenta where the interference minima run nearly parallel to the polarization axis (see Figures 3.16(a,b)), or the fringes fan out like a spider. These fringes are interpreted as similar to holography, which result from the interference

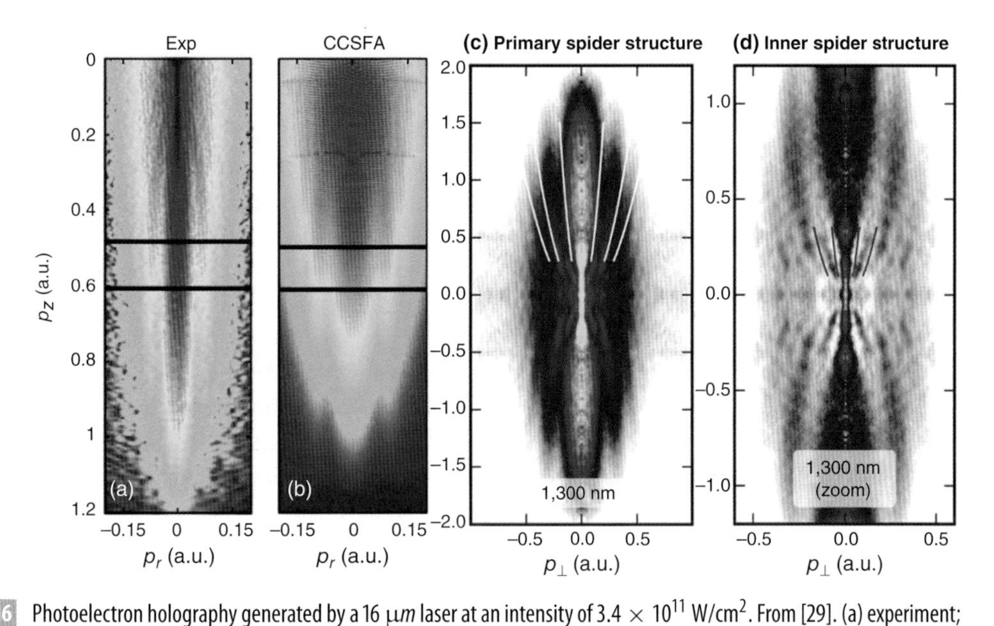

Figure 3.16 Photoelectron holography generated by a 16 μm laser at an intensity of 3.4×10^{11} W/cm^2. From [29]. (a) experiment; (b) calculations based on the CCSFA model. The fringes that are nearly parallel to the polarization axis are called sidelobes in Hickstein et al. [30]. The two frames on the right, (c) and (d), are similar spectra using a 40 fs, 1,300 nm laser with an intensity of 7.5×10^{13} W/cm^2 on Ar. The fringes are marked by white lines, but they are called primary spiders in (c) and inner spiders in (d). (Figures adopted with permission from Y. Huismans et al., *Phys. Rev. Lett.*, 109, 013002 (2012) [29] and from D. Hickstein et al., *Phys. Rev. Lett.*, **109**, 073004 (2012) [30]. Copyrighted by the American Physical Society.)

of a reference wave and a signal wave. The reference wave is from an electron with large perpendicular velocity (with respect to polarization) that makes a wide turn around the ion. The signal wave is from an electron with a small perpendicular velocity that has been driven back and scattered off of the ion. Based on the strong-field approximation, the phases accumulated from the two pathways can be calculated and the fringe can be reproduced qualitatively. The sidelobes (or the primary spider structure) (Figure 3.16(c)), are due to signal waves that undergo one rescattering only, while the inner spiders are signal waves undergoing two returns before rescattering. These direct and rescattered groups of electrons are born within the same quarter optical cycle.

However, to interpret the observed interferences, the simple model based on strong-field approximation is not enough since the effect of the Coulomb potential is not included. The CCSFA model (see Section 2.5.6) has some success (see Figure 3.16(b)), but large errors remain. The Coulomb effect and the quantum nature of electron waves make the description of the reference wave and the signal wave complicated, rendering it nearly impossible to pull out scattering information as opposed to optical holography [31] or photoelectron holography [32]. Thus theoretical studies on these interference features almost always depend on solving the TDSE, which does not allow the possibility of extracting the scattering amplitudes.

The QTMC method used to study low-energy structures has also been used to study photoelectron holography. However, it was found that the quantum-trajectory Monte Carlo (QTMC) method's cutoff energy is too low although it reproduces the interference fringes (see Figure 3.17(b) versus the experimental data in (a)). The error was traced to the deficiency of the adiabatic ADK tunnel-ionization theory used in QTMC. These interference fringes occur at scaled longitudinal momentum (with respect to the momentum of $2U_p$ in energy) of about 0.5 and higher. Thus the electrons are born at the laser phase away from the peak intensity, i.e., when the electric field is small. Figure 3.17(d) shows the subcycle ionization rate for the adiabatic ADK model versus the nonadiabatic PPT-tunneling model. Away from the peak, the ionization rate according to the PPT model is much larger. Thus, using the PPT model for ionization, the generalized QTMC (GQTMC) model was able to reproduce the interference fringes in better agreement with the experiment. This result provides a direct manifestation of nonadiabatic effect in tunnel ionization. In Figures 3.17(b,c) the volume integration is included. The simulation was carried out using the experimental parameters of Huismans et al. [33], but the laser intensity was raised by 30% to achieve the best visual fit to the experimental data. Figures 3.17(e,f) show that the interference fringes studied occur not only in the deep tunneling region ($\gamma = 0.55$), but also in the region where the Keldysh parameter $\gamma = 1.19$.

Before closing this section, it must be emphasized that, although the interferences discussed are dubbed "photoelectron holography," they are unlike optical holography where the signal wave contains target information. In photoelectron holography, the scattering is dominated by the Coulomb potential at a distance far away from the core, and thus it is unable to probe the structure of an object such as a molecule. In general, low-energy photoelectrons generated by intense lasers are very insensitive to the target structure. This will be further discussed in Section 3.5.

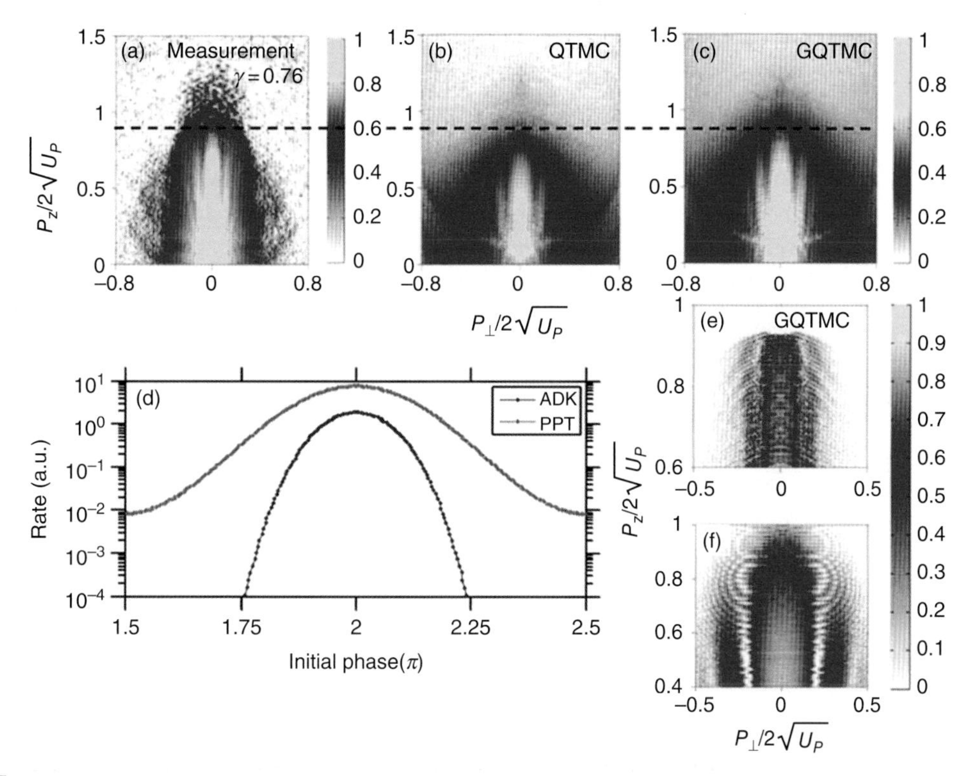

Figure 3.17 (a–c) Photoelectron holography observed in Huismans et al. [33] are studied using the QTMC and GQTMC methods. The sidelobes seen in the experiment were found to have a cutoff at lower parallel momenta using QTMC. The difference was traced to the use of the adiabatic ADK theory to describe tunnel ionization. Using the nonadiabatic tunneling model (PPT) the cutoff, as calculated in the GQTMC, agrees better with the experiment. (d) Comparison of sub-cycle ionization rate between ADK and PPT theories. (e,f) Simulation showing that sidelobes occur in the tunneling regime. (e) For $\gamma = 0.55$ in the multiphoton regime. (f) For $\gamma = 1.19$ (the separation line is arbitrarily set at $\gamma = 1$). (Figures reprinted from Xiaohong Song et al., *Sci. Rep.*, **6**, 28392 (2016) [34].)

3.5 Probing Atoms and Molecules with Low-Energy Photoelectrons by Strong-Field Ionization

Photoelectron spectroscopy by absorption of one photon is a powerful tool for probing the structure of atoms and molecules. The probability of ionization by an intense femtosecond laser is very large and the electron spectrum extends mostly from the threshold to about $2U_p$. If such laser pulses are capable of revealing the structure of the target, then strong-field photoelectron spectroscopy would offer a powerful tool for studying dynamic systems with femtosecond temporal resolution. Unfortunately, while the majority of the electrons are emitted from threshold to $2U_p$, there is very limited structural information in this energy region. Chapter 4 shows that most of the structural information of the target is embedded in the rescattering region for electron energies from $4U_p$ to about $10\,U_p$. Although LES and photoelectron holography (discussed in Section 3.4) contain contributions from rescattering, those contributions are from rescattering at large distances from the ion core where the Coulomb interaction is dominant, thus they offer little information about the target.

Section 3.3 showed that 2D electron-momentum spectra depend mostly on the ionization potential and the orbital symmetry of the initial-state wavefunction. From the tunneling model, the electron momentum gained from the laser field after ionization is given by the vector potential $A(t_i)$ at the ionization time t_i. The tunnel exit is located far outside the ion core, and thus electron-momentum distribution is mostly determined by the laser field only. As an example, consider the ionization of Xe and Kr by 1,320 nm, 35–40 fs lasers. The photoelectron momentum distributions for the two atoms at the same γ are shown in Figure 3.18. There is no discernible difference between the two at first glance. These spectra are not sensitive to the target, and thus low-energy electrons are not good tools for probing the structure.

Next, atoms and molecules that have comparable ionization potential, Ar versus N_2, and Xe versus O_2 ionized by 25 fs, 795 nm lasers are compared in Figure 3.19 for intensity at $\gamma = 1.3$. The fan structure (white lines) at low energies is very similar for each pair. The difference between the two pairs is due to the difference in ionization potential.

In another example, see Figure 3.20, where O_2, N_2, and CO_2 are compared at the same (or nearly the same) γ, for three values of γ. The nodal structures barely show any

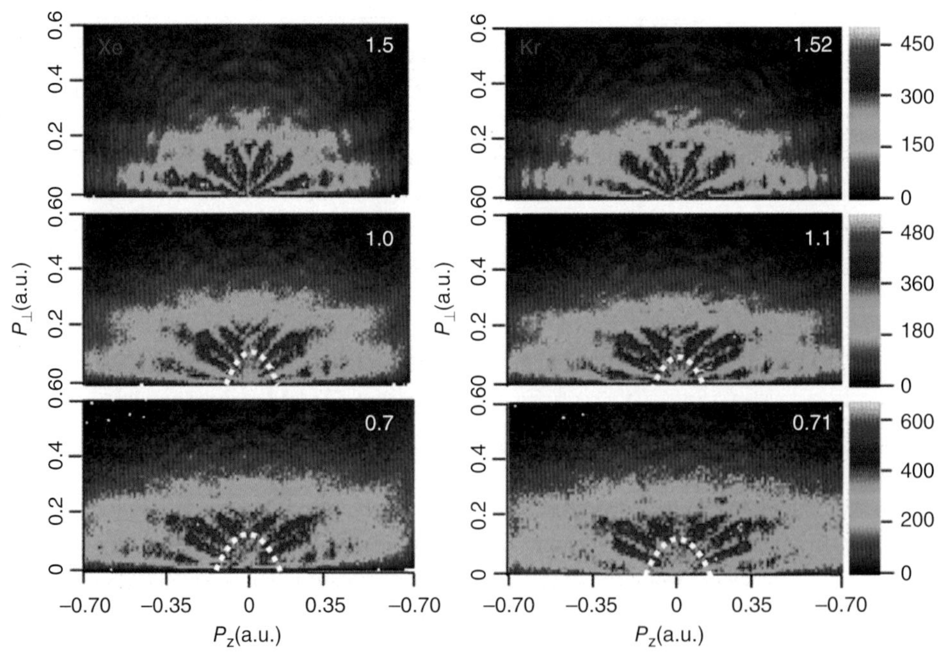

Figure 3.18 2D photoelectron momentum distributions of Xe and Kr from strong-field ionization by 35–40 fs, 1,320 nm lasers. The Keldysh parameters are indicated. With increasing intensity, low-energy electrons are depleted (inside the white semicircle). No differences can be easily identified visually for each pair. (Reprinted from Hong Liu, et al., *Phys. Rev. Lett.*, **109**, 093001 (2012) [35]. Copyrighted by the American Physical Society.)

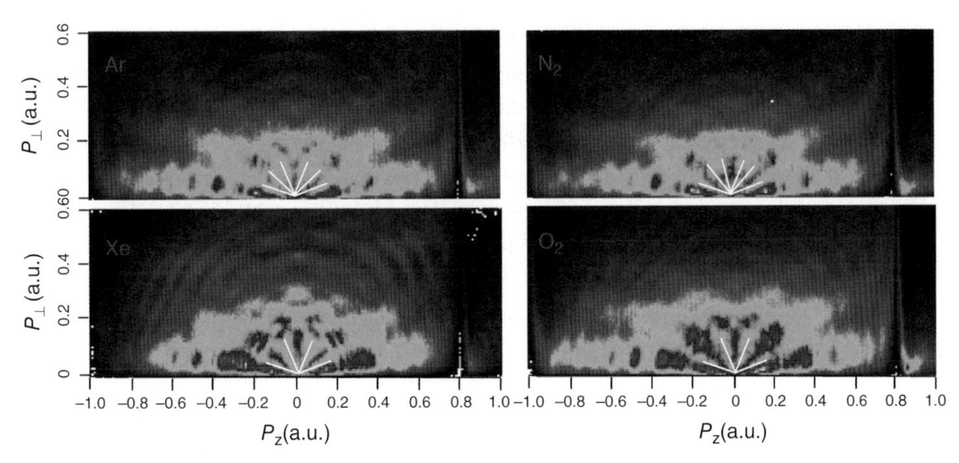

Figure 3.19 2D photoelectron momentum distributions of Ar versus N_2 and Xe versus O_2 ionized by 25 fs, 795 nm laser pulses. The nodal lines and major features for each pair appear to be indistinguishable. (Reprinted from Yongkai Deng et al., *Phys. Rev. A*, **84**, 065405 (2011) [36]. Copyrighted by the American Physical Society.)

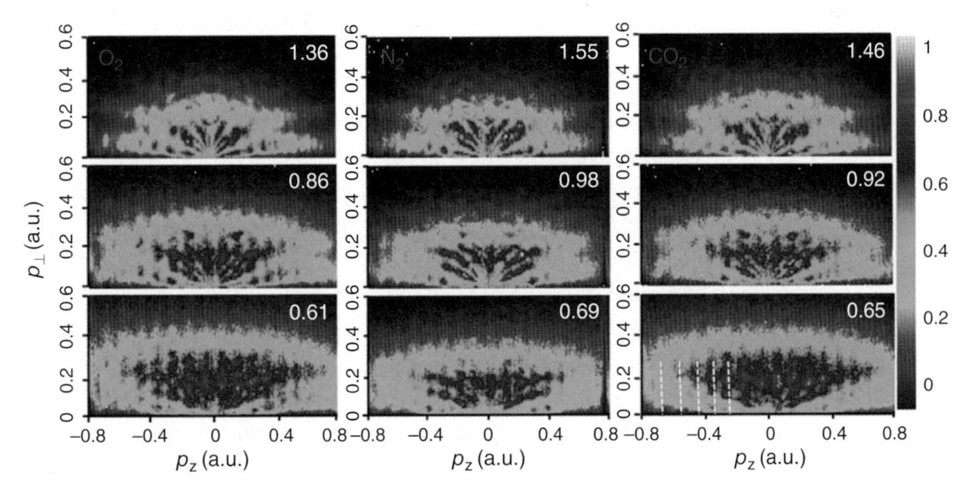

Figure 3.20 2D photoelectron momentum distributions of O_2, N_2 and CO_2 are compared for different targets at about the same Keldysh parameters (γ) (indicated in each frame). The fan structure at larger γ evolves into nearly vertical lines as the γ is decreased toward higher intensities. These results show that low-energy, 2D momentum spectra depend little on the target. (Figures reprinted from Min Li et al., *Sci. Rep.*, **5**, 8519 (2015) [37].)

differences for about the same γ. Note that the typical fan-like nodal structure at larger γ evolves into vertical stripes at small γ, but there are little differences among the targets.

The experimental data shown in Figures 3.19 and 3.20 were carried out with randomly distributed molecules. As will be shown in Section 3.6, molecules can be partially aligned by lasers. Photoelectron energy distributions from preferentially parallel aligned (with respect to the polarization of the probe laser) versus preferentially perpendicularly aligned molecules have been shown [38] to lie on top of each other after they are normalized. The normalization is needed because ionization rates are alignment dependent. In other words, the electron spectra from the threshold to $2U_p$ are determined mostly by the laser field and little influence from the target structure, including the orientation or alignment of the molecules. This same conclusion has been demonstrated in SFA calculations for linear molecules as well as polyatomic molecules [39]. In other words, low-energy, 2D electron-momentum distributions are not sensitive to the target nor to the alignment or orientation of the molecules except for an overall normalization, thus, unlike one-photon ionization, low-energy electron spectra generated from strong-field ionization are not a sensitive probe of the structure of molecules. However, Chapter 4 will show that high-energy photoelectron-momentum distributions (or angular distributions) can be used to extract accurate target-structure information.

3.6 Laser-Induced Rotational and Vibrational Wave Packets of Molecules

3.6.1 General Remarks

On a general basis, molecules in an intense laser field are expected to respond differently depending on their orientations with respect to the laser. Unfortunately, it is quite challenging to orient molecules in the gas phase. Early experiments were mostly done with isotropic molecules, where rich information about the molecular response is lost due to averaging different orientations. Although several alignment techniques based on atom–molecule collisions or strong static electric fields can be used, they are frequently difficult and often not general. The experimentally achievable strength of a static field is still quite limited so alignment can only be performed at very low rotational temperatures on molecules with large dipole moment. Alignment can also be achieved by using resonant laser excitation since the electronic or vibronic transition probability (see Section 1.6) depends on the angle between the laser polarization and the transition dipole. However, the alignment is rather weak since only a small fraction of molecules can typically be excited with weak optical fields.

The situation has changed drastically since the 1990s with the application of intense laser pulses for molecular alignment [40]. The above-mentioned resonant (or near-resonant) laser-excitation technique can be extended to non-perturbative laser intensities with a higher degree of alignment as compared to the weak-field case. In fact, at relatively high intensities, the molecule experiences Rabi oscillations between the ground state and the resonant excited state, exchanging angular momentum and thereby producing a broader rotational wave packet. This book focuses on a more general case of nonresonant laser alignment in which a rotational wave packet is created through sequential Raman transitions (see Section 1.6.1) at even higher laser intensities. In contrast to the resonant case, a wave packet is created and remains in the ground electronic state.

Molecular alignments due to intense lasers can be further classified into two regimes: adiabatic and nonadiabatic (or impulsive). In the adiabatic regime a long laser pulse (relative to the rotational periods of the target), typically of tens to hundreds of picoseconds in duration, is used. This regime was found to be quite similar to the molecular alignment in a strong static-electric field [40]. This book mainly focuses on the nonadiabatic regime in which a short laser pulse, typically of tens to hundreds of femtoseconds, is used to align molecules. The greatest advantage to this regime is that the alignment survives long after the aligning pulse (also called the pump pulse) is over. Indeed, partial alignment occurs periodically (or quasi-periodically) in the form of fractional and full revivals during which experiments with aligned molecules can be performed in the field-free condition.

The nature of the nonadiabatic alignment can be easily understood by looking at a simple, classical picture. Imagine that the pulse is so short that the molecule has no time to rotate during the interaction with laser. Nevertheless, the molecule receives a "kick" toward the laser polarization direction. This kick is proportional to the angle θ between the molecular axis and the laser polarization. Therefore, the molecules at a larger θ get a stronger push so they can catch up with molecules at a smaller θ, which receive a smaller kick. This results in molecular alignment for a brief period of time since the molecules continue to rotate in the field-free condition.

The one-dimensional (1D) laser alignment technique using a single linearly polarized pulse can be extended to three-dimensional (3D) alignment by using elliptically polarized lasers or a combination of laser pulses of different polarization states. Furthermore, the degree of alignment can be improved by using multiple laser pulses with proper time delays and optimal laser parameters. The theoretical basis of these methods will be described in Sections 3.6.2 and 3.6.3. Though molecular alignment is a broad topic that is still under active research, it is relevant, not only to strong-field physics, but also to chemical-reaction dynamics and coherent control as molecular alignment and orientation are known to influence the outcome of chemical reactions.

3.6.2 Laser-Induced Rotational Wave Packets: 1D Alignment

The basics of molecular alignment by a nonresonant intense laser pulse can be understood first for linear molecules.

Most generally, the Hamiltonian of a molecule in an intense laser pulse can be written in the dipole approximation as

$$H = H_0 - \hat{\mu} \cdot \boldsymbol{E}(t), \tag{3.7}$$

where H_0 is the Hamiltonian of the molecule with all nuclear and electronic degrees of freedom, $\hat{\mu}$ is the dipole moment operator, and $\boldsymbol{E}(t)$ is the laser electric vector. Assume that the laser is weak enough such that neither dissociation nor ionization can occur and the molecule remains in the lowest vibronic ground state. Within this subspace the effective interaction is

$$H_{eff}^{int} = -\boldsymbol{\mu_0} \cdot \boldsymbol{E}(t) - \frac{1}{2}\boldsymbol{E}(t)\hat{\alpha}\boldsymbol{E}(t), \tag{3.8}$$

where $\boldsymbol{\mu_0}$ is the permanent dipole moment and $\hat{\alpha}$ is the polarizability tensor of the molecule, which is typically taken at the static limit. The first term exists only for polar molecules. Furthermore, the molecule cannot follow the fast oscillation of the laser so this first term typically averages out to zero over an optical cycle; thus, only the second term remains.

Next by treating the linear molecule as a rigid rotor, the total effective Hamiltonian becomes

$$H_{eff} = B\boldsymbol{J}^2 - \frac{1}{2}\boldsymbol{E}(t)\hat{\alpha}\boldsymbol{E}(t), \tag{3.9}$$

where $B = 1/2I$ is the rotational constant and I is the moment of inertia of the molecule. The first term is the linear rotor model for the molecule while the second term is the effective laser-molecule interaction obtained as the second-order Stark shift.

Since the polarizability of a linear molecule can be decomposed into parallel and perpendicular components with respect to the molecular axes $\alpha_{\parallel}$ and $\alpha_{\perp}$, Equation 3.9 can be rewritten as

$$H_{eff} = B\boldsymbol{J}^2 - \frac{E(t)^2}{2}\left[(\alpha_{\parallel} - \alpha_{\perp})\cos^2\theta + \alpha_{\perp}\right]. \tag{3.10}$$

Here, $E(t)$ is the electric field in the polarization direction, taken to be along the z-axis, and θ is the angle between the molecular axis and the polarization direction. The last term on the right-hand side ($\alpha_{\perp}$) is independent of the angle so it can be dropped from the equation. Thus the Hamiltonian depends only on the polarizability anisotropy $\Delta\alpha = \alpha_{\parallel} - \alpha_{\perp}$.

Assume that the laser pulse is written as

$$E(t) = E_0(t)\cos(\omega t) = \frac{E_0(t)}{2}(e^{i\omega t} + e^{-i\omega t}) \tag{3.11}$$

then the interaction term can be written as

$$\begin{aligned} -\frac{1}{2}\Delta\alpha\cos^2\theta E(t)^2 &= -\frac{1}{8}\Delta\alpha\cos^2\theta E_0(t)^2(e^{i\omega t} + e^{-i\omega t})(e^{i\omega t} + e^{-i\omega t}) \\ &\approx -\frac{1}{4}\Delta\alpha\cos^2\theta E_0(t)^2. \end{aligned} \tag{3.12}$$

In the last equation only the Raman terms resulting from one-photon absorption and one-photon emission are kept. The two-photon excitation terms that oscillate as $\pm 2\omega t$ are neglected as the molecule cannot follow these fast oscillations.

Initially the molecules are assumed to be in thermal equilibrium at temperature T, with the population of rotational states $|JM\rangle$ given by the Boltzmann distribution function

$$\omega_{JM}(T) = \frac{1}{Z(T)}\exp\left[-\frac{J(J+1)}{2Ik_BT}\right], \tag{3.13}$$

where k_B is the Boltzmann constant and $Z(T)$ is the partition function. The above equation is valid only for heteronuclear molecules. For homonuclear molecules, the nuclear spin statistics discussed in Section 1.6.1 must be taken into account. For example, for N_2, the Equation 3.13 has to be multiplied by an additional factor $g_S = 2$ for even Js and $g_S = 1$ for odd Js. Similarly, for O_2 only odd Js are present, while for CO_2 only even Js are present. These results from symmetry constraints were explained in Section 1.6.1.

When a linear molecule is placed in a short intense laser field, the laser excites a coherent superposition of rotational states, also called a *rotational wave packet*, in each molecule. The evolution of the rotational wave packet, starting from the initial state

$\Psi_{JM}(\theta, \phi, t = -\infty) = |JM\rangle$, is described by the TDSE with the effective Hamiltonian given by Equation 3.10 as

$$i\frac{\partial \Psi_{JM}(\theta, \phi, t)}{\partial t} = \left[B\boldsymbol{J}^2 - \frac{E(t)^2}{2}(\alpha_\parallel - \alpha_\perp)\cos^2\theta \right] \Psi_{JM}(\theta, \phi, t). \tag{3.14}$$

Since the optical laser oscillation is much faster than the rotational motion of the molecules, the molecules cannot follow the fast oscillations in $E(t)^2$. Instead, they only "see" the effective intensity $E_0(t)^2/2$, as discussed right after Equation 3.12. In principle, an additional term $-\boldsymbol{\mu}_0 \cdot \boldsymbol{E}(t)$ should be added for a polar molecule where $\boldsymbol{\mu}_0$ is the permanent dipole moment. However, this term vanishes after averaging over the laser cycle.

Equation 3.14 needs to be solved for each initial rotational state $|JM\rangle$. Note that for each Raman transition the selection rule $\Delta J = 0, \pm 2$ applies so the states with initial odd and even Js can be treated separately. Since the quantum number M is conserved for linearly polarized lasers, it is convenient to seek the solution in the form

$$\Psi_{JM}(\theta, \phi, t) = \sum_{J'} C_{J'}(JM, t) Y_{J'M}(\theta, \phi). \tag{3.15}$$

With this expansion, the rotational wave packet after the laser pulse can be calculated by field-free evolution simply as

$$\Psi_{JM}(\theta, \phi, t) = \sum_{J'} C_{J'}(JM, t = t_f) e^{-iE_{J'}t} Y_{J'M}(\theta, \phi) \tag{3.16}$$

with $E_{J'} = BJ'(J' + 1)$.

Once the solutions of the TDSE above are obtained for all initial $|JM\rangle$ that are populated at temperature T, the time-dependent alignment distribution can be obtained as

$$\rho(\theta, t) = \sum_{JM} \omega_{JM} |\Psi_{JM}(\theta, \phi, t)|^2. \tag{3.17}$$

The angular (or alignment) distribution does not depend on the azimuthal angle ϕ in the laboratory frame in which the laser polarization axis is defined as the z-axis. Equations 3.16 and 3.17 allow the determination of the time-dependent alignment distribution of the molecules in the laser field as well as the rotational revivals after the laser has been turned off.

Once the angular distribution is obtained, different observables can be calculated. First, it is conventional to use $\langle\cos^2\theta\rangle$ as a measure of the degree of alignment

$$\langle\cos^2\theta\rangle = 2\pi \int_0^\pi \rho(\theta, t) \cos^2\theta \sin\theta d\theta. \tag{3.18}$$

An example of such an estimate is shown in Figure 3.21(a) for N_2 as a function of time during and after the aligning (pump) pulse. Note the full revival near 8.6 ps as well as a half revival near 4.3 ps. Near these revivals, the molecular alignment changes significantly from a more isotropic distribution (with $\langle\cos^2\theta\rangle \approx 1/3$), which can be seen as the baseline in the figure. For the laser parameters and the temperature used in the simulation, the degree of alignment is rather good with $\langle\cos^2\theta\rangle \approx 0.5$ at full revival near 8.6 ps. This means that

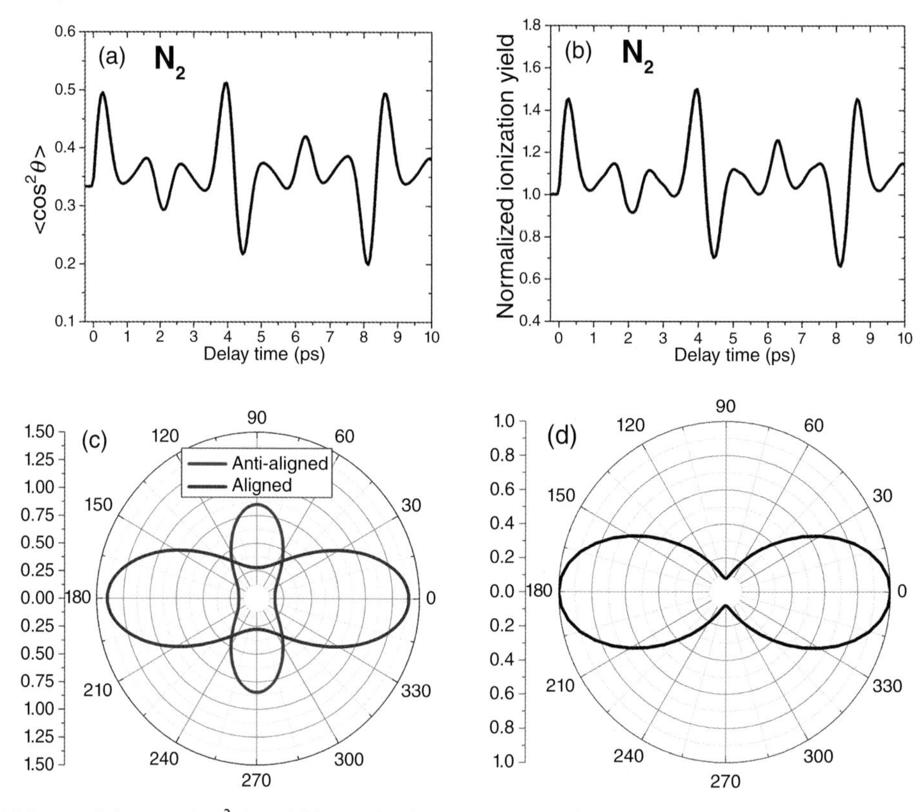

Figure 3.21 (a) Degree of alignment $\langle \cos^2 \theta \rangle$ and (b) normalized ionization yields as functions of pump-probe delay time in N_2. Here, an 800 nm pump pulse with a 60 fs duration (FWHM) and peak intensity of 2×10^{13} W/cm^2 was used and the rotational temperature was taken to be 20 K. (c) Alignment distributions $\rho(\theta, t)$ at delay times of 8.1 ps and 8.6 ps corresponding to anti-aligned and aligned distribution, respectively. (d) Angle dependence of ionization rate for N_2 molecules at the probe laser intensity of 2.5×10^{14} W/cm^2.

the molecules are aligned preferentially along the laser-polarization direction (see the blue curve in Figure 3.21(c)). At the time delay near 8.1 ps, the degree of alignment $\langle \cos^2 \theta \rangle$ is smallest and the molecules are said to be "anti-aligned," i.e., preferentially perpendicular to the laser polarization direction (see the red curve in Figure 3.21(c)).

Now, assume that a second intense pulse (the probe) with polarization parallel to that of the pump pulse is used to ionize the N_2 molecules at different time delays after the pump pulse is over. Clearly, a short pulse, say, of a duration of 30 fs, should be used so molecular rotation during the probe pulse is negligible. With that assumption, the ionization yields for N_2 as a function of time delay between the pump and probe can be calculated as

$$S(t) \propto \int_0^{\pi} P(\theta)\rho(\theta, t) \sin\theta d\theta, \tag{3.19}$$

where $P(\theta)$ is the ionization probability of N_2 aligned at angle θ with respect to the laser-polarization direction. The result by an 800 nm probe laser at intensity of 2.5×10^{14} W/cm^2

is shown in Figure 3.21(b). In this case, the ionization yields are found to follow the shape of $\langle \cos^2 \theta \rangle$ versus time delay shown in Figure 3.21(a). For comparison, Figure 3.21(d) shows the ionization rates versus alignment angle obtained from the molecular ADK theory (MO-ADK) theory at the probe laser intensity.

One can also probe at a fixed delay time, say, at the full revival near 8.6 ps, when the molecules are maximally aligned along the laser-polarization direction by changing the probe polarization with respect to the pump polarization. Both types of measurements – time delays and polarization scans – are useful and give complementary information about the angle-dependent ionization rate of N_2 in the molecular frame.

The treatment above for linear molecules can also be extended to symmetric top molecules. In fact, the formal equations are quite similar [40]. For a prolate symmetric top (with $\alpha_{\parallel} > \alpha_{\perp}$), the molecular axis is aligned along the laser-polarization direction whereas for a oblate symmetric top (with $\alpha_{\parallel} < \alpha_{\perp}$), the molecular axis is aligned perpendicular to it.

3.6.3 Laser-Induced Rotational Wave Packets: 3D Alignment

The next step is to extend the simpler analysis of the previous subsection to 3D alignment of asymmetric top molecules. Generally, the alignment along the laser polarization direction can be created when a broad coherent wave packet in J is formed. A broader range of J typically leads to a better alignment. However, the M quantum number (the projection of $\boldsymbol{J}$ onto the z-axis which is the laser polarization direction) is conserved. So the molecules are free to rotate around the z-axis. Furthermore, the K quantum number (the projection of $\boldsymbol{J}$ onto the molecular z-axis) is conserved or changes by only one or two quanta, see Ref. [40]. Therefore the molecule is essentially free to rotate around the molecular axis. It can be shown that molecules can be excited to a broad range of all J, M, and K by using an elliptically polarized laser pulse such that the rotation in all three Euler angles is restricted, thereby achieving a certain degree of 3D alignment.

The effective time-dependent Hamiltonian Equation 3.9 can be generalized to an asymmetric top molecule as

$$H = H_{rot} - \frac{1}{2}\boldsymbol{E}(t)\hat{\alpha}\boldsymbol{E}(t), \tag{3.20}$$

where $H_{rot} = AJ_a^2 + BJ_b^2 + CJ_c^2$ is the field-free Hamiltonian for an asymmetric rigid rotor with rotational constants A, B, and C, along the three principal axes a, b and c, respectively. The electric field vector for elliptical polarization can be written in the laboratory frame as

$$\boldsymbol{E}(t) = E_0(t)\,(\epsilon_x \mathbf{e}_x \cos\omega t + \epsilon_z \mathbf{e}_z \sin\omega t), \tag{3.21}$$

where $\mathbf{e}_z$ and $\mathbf{e}_x$ are the unit vectors of the major and minor polarization directions and $\epsilon_x^2 + \epsilon_z^2 = 1$ with $0 \leq \epsilon_x < \epsilon_z$. Here, the y-axis is chosen as the laser propagation direction.

The TDSE can be solved more conveniently by expanding the time-dependent wavefunction in the symmetric top basis set $|JKM\rangle$

$$\langle \hat{R}|JKM\rangle = \sqrt{\frac{2J+1}{8\pi^2}} D^{J*}_{MK}(\hat{R}). \tag{3.22}$$

Here, $\hat{R}$ specifies the Euler angles $\{\theta, \phi, \chi\}$ and D^{J*}_{MK} is the Wigner D matrix. After the pulse is over, the wavefunction can be rewritten in the asymmetric top basis so that field-free propagation can be carried out analytically (see Artamonov and Seideman [41]). This basis set is related to the symmetric top basis set $JKM\rangle$ by

$$|J\tau M\rangle = \sum_K a^J_{K\tau} |JKM\rangle. \tag{3.23}$$

As an example, the case of C_2H_4 reported in Rouzée et al. [42] is analyzed. Figure 3.22 shows the angular distributions calculated near the first rotational revival at $t = 8.3$ ps after the aligning laser pulse. For isotropic molecules the distributions are spherical. In case of a linearly polarized pulse, the angular distribution of the molecular Z-axis (along the C–C bond) peaks along the laser polarization direction (chosen as y-axis here), while the two other molecular axes X and Y remain delocalized in the x–z plane (see the right column). This result also reflects on the C–H bonds distribution shown in the left column where it is found to be isotropically distributed around the C–C bond. For the elliptically polarized case with $\epsilon_x^2 = 0.44$ (i.e., the major axis is along y-axis, and the laser propagation is along z-axis), all the three X, Y, and Z molecular axes peak in the three orthogonal x, z and y directions and result in a 3-D alignment of the molecule, see the right column

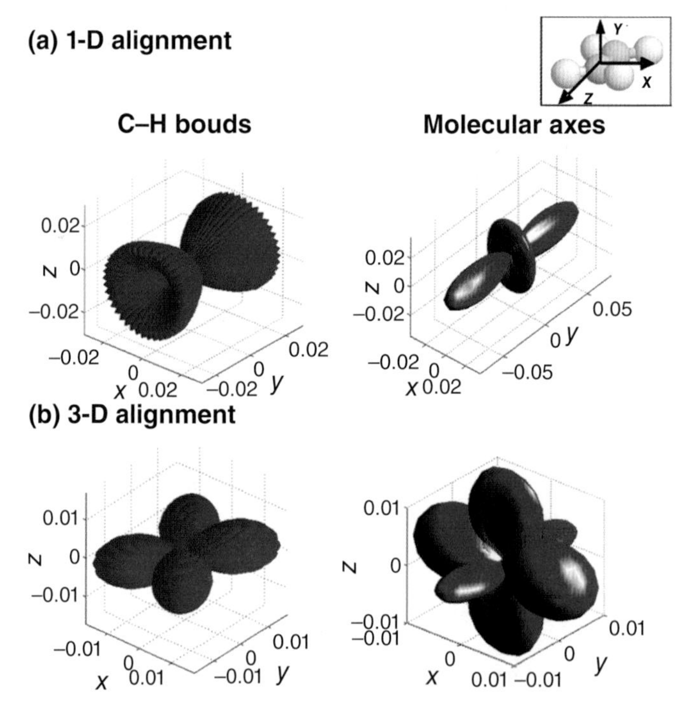

Figure 3.22 Comparison of 1D versus 3D alignment of C_2H_4 in (a) and (b), respectively. Angular distributions of the C–H bonds (left column) and molecular axes (right column) calculated near the first revival at $t = 8.3$ ps after a pump pulse with peak intensity of 20 TW/cm^2 and pulse duration of 100 fs, at the temperature of $T = 0.1$ K. The structure of C_2H_4 and the body-fixed frame are shown in the inset. (Reprinted from A. Rouzee et al., *Phys. Rev. A*, **77**, 043412 (2008) [42]. Copyrighted by the American Physical Society.)

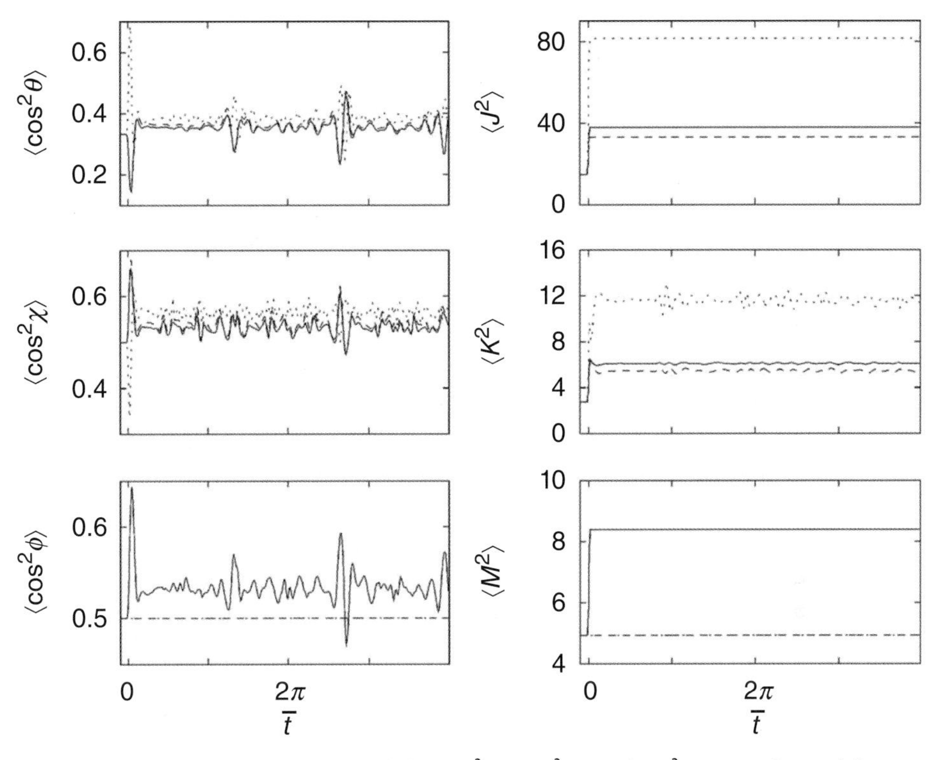

Figure 3.23 Degrees of alignment of a model molecule specified by $\langle \cos^2 \theta \rangle$, $\langle \cos^2 \chi \rangle$, and $\langle \cos^2 \phi \rangle$ versus the rescaled time delay (left column) and the distribution of J, K, and M quantum numbers (right column). Three different aligning laser pulses are used: one with elliptical polarization (solid curve), one with circular polarization (dashed curve), and one with linear polarization (dotted curve). (Reprinted from M. Artamonov and T. Seideman, *J. Chem. Phys.*, **128**, 154313 (2008) [41]. Copyrighted by AIP Publishing LLC.)

of Figure 3.22(b). Furthermore, the distribution of the C–H bonds is localized in the $x-y$ plane and clearly mimics the geometrical shape of the molecule, see the left column of Figure 3.22(b).

To quantify the degree of 3-D alignment, one can use the Euler angles as specified by their expectation values of $\langle \cos^2 \theta \rangle$, $\langle \cos^2 \chi \rangle$ and $\langle \cos^2 \phi \rangle$. These are shown in Figure 3.23 for a nonadiabatic alignment of a model molecule vs rescaled time delay as reported in Artamonov and Seideman [41] for three different laser pulses with elliptical, circular, and linear polarizations. For the elliptical polarized pulse, the revivals are in all three Euler angles. The degree of rotational excitation can be quantified as the expectation values of the angular momentum and its body- and space-fixed z components $\langle J^2 \rangle$, $\langle K^2 \rangle$, and $\langle M^2 \rangle$. These are shown in the right column of Figure 3.23 where the broad range of excitation for all three quantum numbers for the elliptical polarized pulse is clear. Furthermore, from Figure 3.23 it can be seen that $\langle K^2 \rangle$ is not constant after the pulse is turned off. This is expected as K is not a good quantum number for an asymmetric top molecule.

There is no consensus on how to quantify the degree of alignment in three dimensions. Apart from the Euler angles discussed above, different directional cosines have been used.

Quite recently, a single measure for the degree of alignment was proposed [43]. At present, a great challenge is to achieve good degrees of 3D alignment experimentally, especially for the field-free cases (i.e., with nonadiabatic alignment). For example, it was demonstrated in Ren et al. [44] that good 3D alignment can be achieved by using multiple pulses with proper time delays and optimized laser parameters for each pulse. So far, these techniques seem to be quite complicated. Furthermore, they also require very low temperatures which are possible only for gases at a very low density. The practical applications of using field-free 3D aligned molecules remain to be seen. In the meantime, an alternative approach has been proposed quite recently [45]. Relying on a high-quality delay scan for intense laser ionization from a 1D aligned asymmetric top C_2H_4 molecule, it was possible to retrieve the molecular frame ionization rate. It remains to be seen if the method can be extended for other targets and other observables such as high-harmonic generation (HHG) or photoelectron data. If the method proved successful, then it would eliminate the need for 3D alignment.

3.6.4 Laser-Induced Vibrational Wave Packets

Coherent vibrational wave packets can also be created using Raman excitation mechanisms with short intense laser pulses. However, only small-amplitude vibrations are typically created by moderately intense lasers. Nevertheless, such nuclear wave packets can potentially influence the outcomes of experimental results. In certain favorable conditions, these small vibrations can also be observed in intense laser experiments as demonstrated in Wagner et al. and Li et al. [46, 47]. Therefore, this subsection describes the basic idea behind this process.

For simplicity, a diatomic molecule in a moderately intense linearly polarized laser pulse is considered. The field is sufficiently weak such that the molecule remains in the ground electronic state. The Schrödinger equation for the nuclear wavefunction $\chi(R,t)$ is written as

$$i\frac{\partial\chi(R,t)}{\partial t} = \left[-\frac{1}{2\mu}\frac{\partial^2}{\partial R^2} + U(R) - \frac{1}{2}\sum_{ij}\alpha_{ij}(R)E_i(t)E_j(t) \right]\chi(R,t), \tag{3.24}$$

where μ is the reduced mass, $U(R)$ is the field-free potential energy, α_{ij} is the polarizability tensor, and $E_i(t)$ is the component of the electric field of the laser along the i-axis (with $i = x, y, z$). Note that α_{ij} is, in general, R-dependent. This equation resembles Equation 3.9 except that now R is the dynamic variable. In fact, the laser-molecule interaction term is the same in Equation 3.9, but the R dependence is given explicitly. To simplify the simulation, the timescale of the vibrational wave packet is assumed to be much shorter than the molecular rotational period of interest. Thus the rotation of the molecule can be neglected and the molecule is considered to be fixed in space. For a full treatment with a longer timescale, both rotational and vibrational degrees of freedom should be kept.

In practice, instead of using the second-order Stark corrections to the potential, laser-dressed potential energy curves $U(R, E(t))$ at each value of E can also be directly calculated by using standard quantum chemistry packages such as Gaussian or GAMESS. These calculations are only needed near the equilibrium R within the range accessible to the

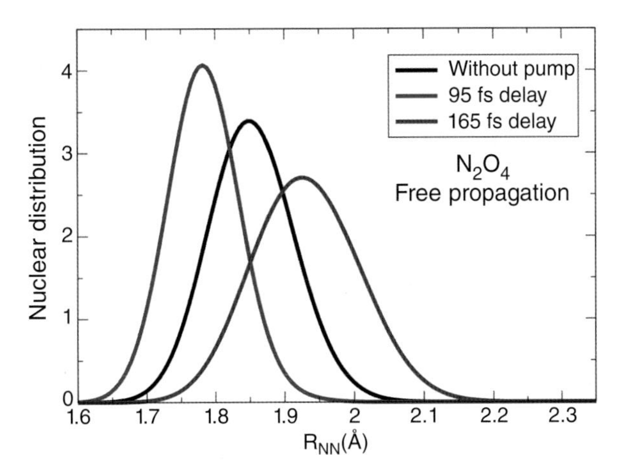

Figure 3.24 N–N distance distribution in N_2O_4 at different time delays of 95 fs and 165 fs after the excitation pulse. For a reference, the distribution of the initial $v = 0$ state (i.e., without the pump) is also shown. Note that these time delays are much shorter than the typical rotational period of N_2O_4, which is larger than 10 ps.

excited nuclear wave packet. This approach aids in going beyond the second-order Stark correction.

As an illustration, Figure 3.24 shows the nuclear wave-packet distribution created in N_2O_4 at different time delays after the 800 nm excitation pulse with an intensity of 2×10^{13} W/cm^2 and a duration of 30 fs. Here, the molecule is treated as an effective linear molecule with the internuclear distance R_{NN} and a range of vibration of about 0.3 Å. Though it is quite small, this range of vibration can be detected experimentally. In fact, in the pump-probe HHG experiments by Li et al. [47], significant changes in HHG spectra were observed as functions of time delay after the excited nuclear wave packet was created. Similarly, vibrational wave packets created in different modes in SF_6 have been observed in time-delay scans of HHG signals [46].

To further illustrate the control over the vibrational wave packets using the Raman excitation mechanism, a simple toy model is considered. This model assumes that there are only two electronic states g and e, for ground and excited states, respectively. The TDSE equation for this two-state model can be written as

$$i \begin{pmatrix} d\chi_e/dt \\ d\chi_g/dt \end{pmatrix} = \left[\begin{pmatrix} T & 0 \\ 0 & T \end{pmatrix} + \begin{pmatrix} U_e & \Omega \\ \Omega & U_g \end{pmatrix} \right] \begin{pmatrix} \chi_e \\ \chi_g \end{pmatrix}. \tag{3.25}$$

Here, χ_g and χ_e are the vibrational wave packets on the ground and excited electronic states with potential energy $U_g(R)$ and $U_e(R)$, respectively, T is the kinetic energy operator, and the electric field $E(t)$ is along the internuclear axis R. The coupling term is $\Omega = -\mu_{ge}(R)E(t) = -\mu_{eg}(R)E(t)$ with the transition dipoles $\mu_{ge}(R) = \langle \psi_g | \hat{\mu} | \psi_e \rangle$ and $\mu_{eg}(R) = \langle \psi_e | \hat{\mu} | \psi_g \rangle$. The dressed-state picture is obtained by diagonalizing the potential energy matrix. This leads to a Stark shift in the ground state given by $-\Delta U + (\Delta U^2 + \Omega^2)^{1/2}$ with $2\Delta U = U_e - U_g$ being the energy gap between the two states. For a relatively weak field such that Ω is smaller than the energy gap ΔU, the Stark shift can be

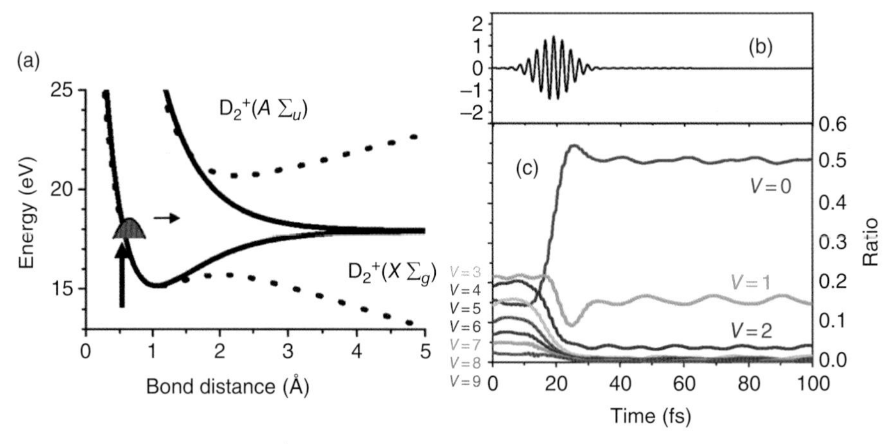

Figure 3.25 (a) The two lowest-energy curves of D_2^+ in the field-free condition and in an electric field (dotted curves). The nuclear wave packet is assumed to be generated by a pump pulse in the ground electronic state. (b) The control pulse. (c) The evolution of the population in the nuclear wave packet when a control pulse is applied at a time delay $\Delta t = 3/4$ of the vibrational period. The control pulse has a wavelength of 800 nm, 18 fs and an intensity of 2×10^{14} W/cm^2. (Reprinted with permission from H. Niikura, D. M. Villeneuve, and P. B. Corkum, *Phys. Rev. Lett.*, **92**, 133002 (2004) [48]. Copyrighted by the American Physical Society.)

approximated by $-\frac{1}{2}\Omega^2/\Delta U$. Since the dipole polarizability for this two-state model is $\alpha = 2\langle\psi_g|\hat{\mu}|\psi_e\rangle\langle\psi_e|\hat{\mu}|\psi_g\rangle/(U_e - U_g)$, the Stark shift can also be written as $-\frac{1}{2}\alpha E(t)^2$, which is the familiar second-order Stark shift used in Equation 3.24. Note that the transition dipole and dipole polarizability are functions of R. This derivation also illustrates the adiabatic-elimination procedure mentioned in Section 3.6.2.

The model in Figure 3.25 was used by Niikura et al. [48] for simulation of nuclear wave-packet dynamics in D_2^+. A nuclear wave packet is assumed to be created by a pump pulse in the ground electronic state of D_2^+ (see Figure 3.25(a)). The second intense laser pulse (control pulse) couples the two electronic states and creates the laser-induced potential energy surfaces shown schematically as the dotted lines in Figure 3.25 for a certain time during the control pulse $E(t)$. By changing the control-pulse parameters, the nuclear wave packet can be manipulated. In particular, with a proper choice of time delay between the pump and control pulses, the population can be transferred quite efficiently to a $v = 0$ state, as shown in Figure 3.25(c).

3.7 Strong-Field Ionization of Molecules

3.7.1 Orientation and Alignment-Dependent Ionization Rates of Molecules

Ionization is the first step for all strong-field phenomena. This section addresses ionization of molecules by generalizing models that were first used for the ionization of atoms.

MO-ADK Theory

In Section 2.4.1, the ionization rate of an atom by a static field was derived. The resulting ADK theory depends on the ionization potential of the atom, the coefficient C_l, and the angular momentum of the asymptotic wavefunction of the ground state. In a diatomic molecule, the orbital angular momentum is not a good quantum number. If the molecular axis lies along the direction of the static field, the two-center wavefunction in the asymptotic region can be expanded in terms of one-center functions as

$$\Psi^m(\mathbf{r}) = \sum_l C_l F_l(r) Y_{lm}(\hat{\mathbf{r}}), \tag{3.26}$$

where m is the magnetic quantum number along the molecular axis and F_l is defined as

$$F_l(r) = r^{Z_c/\kappa - 1} \exp(-\kappa r). \tag{3.27}$$

Here, $\kappa = \sqrt{2I_p}$ with I_p being the ionization potential and Z_c being the asymptotic charge of the molecular ion.

Following exactly the same procedure as in Section 2.4.1, the static-tunneling ionization rate can be calculated as [49]

$$w_{stat}(F, 0) = \frac{B^2(m)}{2^{|m|}|m|!} \frac{1}{\kappa^{2Z_c/\kappa - 1}} \left(\frac{2\kappa^3}{F}\right)^{2Z_c/\kappa - |m| - 1} e^{-2\kappa^3/3F}, \tag{3.28}$$

where

$$B(m) = \sum_l C_l Q(l, m). \tag{3.29}$$

The $Q(l, m)$ was defined in Equation 2.114. If the molecule is oriented at angle $\hat{\mathbf{R}}$ with the polarization axis, then the new $B(m)$ shown in Equation 3.29 is obtained through a rotation

$$B(m') = \sum_l C_l D^l_{m',m}(\hat{\mathbf{R}}) Q(l, m'), \tag{3.30}$$

where D is the rotation matrix and $\hat{\mathbf{R}}$ is the Euler angles between the molecular axis and the laser-polarization axis. The static ionization rate for a molecular oriented by $\hat{\mathbf{R}}$ is then given by

$$w_{stat}(F, \hat{\mathbf{R}}) = \sum_{m'} \frac{B^2(m')}{2^{|m'|}|m'|!} \frac{1}{\kappa^{2Z_c/\kappa - 1}} \left(\frac{2\kappa^3}{F}\right)^{2Z_c/\kappa - |m'| - 1} e^{-2\kappa^3/3F}. \tag{3.31}$$

This equation gives the alignment dependence of the tunnel-ionization rate for molecules, which is a direct extension of the ADK model and is the MO-ADK tunnel-ionization theory for molecules [49]. According to this model, for each linear molecule a set of coefficient C_l's have to be calculated. This set of coefficients is obtained by fitting the molecular wavefunction in the asymptotic region in the form of Equation 3.26. The molecular wavefunction is calculated from well-known quantum chemistry codes such as GAMESS or Gaussian.

The MO-ADK theory can be extended in a straightforward way to nonlinear polyatomic molecules [50]. The asymptotic electronic wavefunction in the molecular frame in the single-center expansion approach can be expressed as

$$\Psi(\mathbf{r}) = \sum_{lm} C_{lm} F_l(r) Y_{lm}(\hat{\mathbf{r}}). \tag{3.32}$$

Let $\hat{\mathbf{R}} = (\phi, \theta, \chi)$ be the three Euler angles of the molecular frame with respect to the laboratory-fixed frame. Following the same derivation for the linear molecule, the static ionization rate is the same as given in Equation 3.31 except that

$$B(m') = \sum_{l,m} C_{lm} D^{l}_{m',m}(\hat{\mathbf{R}}) Q(l, m'), \tag{3.33}$$

where

$$D^{l}_{m',m}(\hat{\mathbf{R}}) = e^{im'\phi} d^{l}_{m',m}(\theta) e^{im\chi}. \tag{3.34}$$

Once the structure parameters are obtained for each molecule, the orientation-dependent tunneling ionization rates based on the simple MO-ADK theory can be calculated analytically.

Molecular PPT (MO-PPT) Theory

The PPT theory is more accurate than the ADK theory since it accounts for the nonadiabatic effect in strong-field ionization (see Section 2.5.7). It is straightforward to generalize the atomic PPT theory for molecular targets [12, 51] by using the same procedure as shown in the "MO-ADK Theory" subsection. The resulting cycle-averaged ionization rate for the MO-PPT theory is given by

$$\begin{aligned} w_{MO-PPT}(F_0, \omega, \mathbf{R}) = & \left(\frac{3F_0}{\pi\kappa^3}\right)^{1/2} \sum_{m'} \frac{B^2(m')}{2^{m'}|m'|!} \\ & \times \frac{A_{m'}(\omega, \gamma)}{\kappa^{2Z_c/\kappa - 1}} (1+\gamma^2)^{|m'|/2+3/4} \\ & \times \left(\frac{2\kappa^3}{F_0}\right)^{2Z_c/\kappa - |m'| - 1} \\ & \times (1 + 2e^{-1}\gamma)^{-2Z_c/\kappa} \\ & \times e^{[-(2\kappa^3/3F_0) g(\gamma)]}, \end{aligned} \tag{3.35}$$

where $B(m')$ is given by Equation 3.30 or 3.33 for linear or polyatomic molecules. All the other parameters are defined in Section 2.5.7 or in the "MO-ADK Theory" subsection.

3.7.2 Molecular Strong-Field Approximation

The KFR theory for the ionization of molecules can be directly generalized from atoms by replacing the atomic ground-state wavefunction by the molecular wavefunction. For molecules fixed in space, where the orientation axis of the molecule makes angle $\hat{R}$ with respect to the axis of the linear laser polarization, the ionization-probability amplitude in the length gauge is given by

$$f(\mathbf{p},\hat{R}) = i \int_{-\infty}^{\infty} \langle \mathbf{p}+\mathbf{A}(t)|\mathbf{r}\cdot\mathbf{E}(t)|\Phi_0(\mathbf{r})\rangle e^{-iS(\mathbf{p},t)}dt, \tag{3.36}$$

where S is the action,

$$S(\mathbf{p},t) = \int_{t}^{\infty} dt' \left\{ \frac{[\mathbf{p}+\mathbf{A}(t')]^2}{2} + I_p \right\} \tag{3.37}$$

and $\mathbf{p}$ is the momentum of the electron at the detector. The total ionization probability is calculated from

$$P(\hat{R}) = \int |f(\mathbf{p},\hat{R})|^2 d^3\mathbf{p}. \tag{3.38}$$

Note that the orientation dependence of the ionization rate calculated using MO-ADK or MO-PPT is the same with the exception of an overall normalization constant for each molecular orbital. For the molecular strong-field approximation (MO-SFA), the calculations are more tedious since integrations have to be carried out in Equations 3.36 and 3.38, but it is very straightforward. In general, the normalized alignment dependence of the ionization rates from the MO-ADK and the MO-SFA agree quite well, but some differences may be observed at higher recision.

The MO-SFA, according to Equation 3.36, can be used to obtain the energy and angular distributions of the photoelectrons (excluding rescattering). From MO-SFA calculations [39] and experimental data [38], it has been found that the electron energy distributions at different orientation angles are nearly identical except for an overall normalization. The normalization factor is due to the orientation dependence of ionization rates that can be calculated from the MO-ADK theory.

3.7.3 Weak-Field Asymptotic Theory for Tunnel Ionization

In recent years, Tolstikhin and coworkers have proposed a weak-field asymptotic theory (WFAT) [52] in which tunneling ionization is treated in parabolic coordinates and the ionization rate is written as an asymptotic expansion in the field strength F. Specifically, for example, the ionization rate for a linear molecule within the single-active electron approximation is given as the sum of partial rates from different channels labeled by parabolic quantum numbers (n_ξ, m) with

$$\Gamma_{n_\xi m}(\beta) = |G_{n_\xi m}(\beta)|^2 W_{n_\xi m}(F), \tag{3.39}$$

where β is the angle between the molecular axis and the field direction. Typically, only the dominant channel, $(0,0)$ or $(0,\pm1)$, is needed. The first factor is called the structure factor which is field independent

$$G_{n_\xi m}(\beta) = \lim_{\eta\to\infty} e^{-\kappa\mu_z}\eta^{1+|m|/2-Z/\kappa}e^{\kappa\eta/2} \int_0^{\infty}\int_0^{2\pi} \phi_{n_\xi m}(\xi)\frac{e^{-im\varphi}}{\sqrt{2\pi}}\psi(\mathbf{r})d\xi d\phi, \tag{3.40}$$

where μ_z is the projection of the dipole moment of the MO on the field direction, ψ is the (field-free) wavefunction of the MO, and $\phi_{n_\xi m}(\xi)$ is a channel function. For degenerate orbitals with $M \neq 0$, a linear combination of the orbitals should be taken. Such real-valued orbitals are readily available from typical quantum chemistry packages such as Gaussian or GAMESS. The second factor (the so-called *field factor*) is given as

$$W_{n_\xi m}(F) = \frac{\kappa}{2}\left(\frac{4\kappa^2}{F}\right)^{2Z/\kappa - 2n_\xi - |m| - 1} \exp\left(-\frac{2\kappa^3}{3F}\right). \tag{3.41}$$

Here, the notations are the same as in Section 3.7.1 for the MO-ADK theory. The field factor only depends on the target molecule through $\kappa = \sqrt{2I_0}$. The WFAT was derived in the laboratory frame in which the field direction is chosen to be along the z-axis. Therefore the angle β in the WFAT is equivalent to angle θ in the MO-ADK theory (see Section 3.7.1). Note that the structure factor as a function of angle β can be expressed in terms of the associated Legendre polynomials. The expansion typically converges very fast so only the first few structure coefficients are needed. These structure coefficients are target specific and can be tabulated for future use.

The main difference between the WFAT and the MO-ADK is the presence of the dipole-correction factor $\exp(-2\kappa\mu_z)$ in the structure factor $|G_{n_\xi m}(\beta)|^2$. Earlier it was suggested that the MO-ADK theory should be modified to account for the permanent dipole [53, 54] using a simple substitution of the ionization potential by a Stark-shifted potential such as $I_p \rightarrow I_p + \mu_z F$. It can be shown that this modification also introduces the dipole-correction factor identical to that of the WFAT. Indeed, for the small Stark correction $\mu_z F$, the substitution

$$\kappa^3 \rightarrow \left[2(I_p + \mu_z F)\right]^{3/2} = \kappa^3\left(1 + \frac{2\mu_z F}{\kappa^2}\right)^{3/2} \approx \kappa^3 + 3\kappa\mu_z F \tag{3.42}$$

modifies the main exponent factor in the MO-ADK theory to

$$\exp\left(\frac{-2\kappa^3}{3F}\right) \rightarrow \exp\left(\frac{-2\kappa^3}{3F} - 2\kappa\mu_z\right). \tag{3.43}$$

As expected, the dipole-correction factor $\exp(-2\kappa\mu_z)$ appears.

For a homonuclear diatomic molecule in which the dipole vanishes, the WFAT and MO-ADK agree with each other if the summation in Equation 3.31 is limited to the lowest m' [52]. In most cases the lowest m' is taken to be zero except near the nodal lines where $m' = 1$ should be taken. This difference arises because in the WFAT the leading order in the asymptotic expansion is kept consistently.

The above analysis shows that, for all practical purposes, the WFAT and MO-ADK results should be quite similar. In the case of polar molecules, the main advantage of the WFAT can be largely reproduced using a Stark-corrected ionization potential in the MO-ADK theory. First-order correction terms to the WFAT have been proposed and the theory has been generalized to many-electron targets.

For applications, the structure parameters needed for the MO-ADK and WFAT theories have been calculated for many molecules. They are tabulated in Zhao et al. and Saito et al. [55, 56].

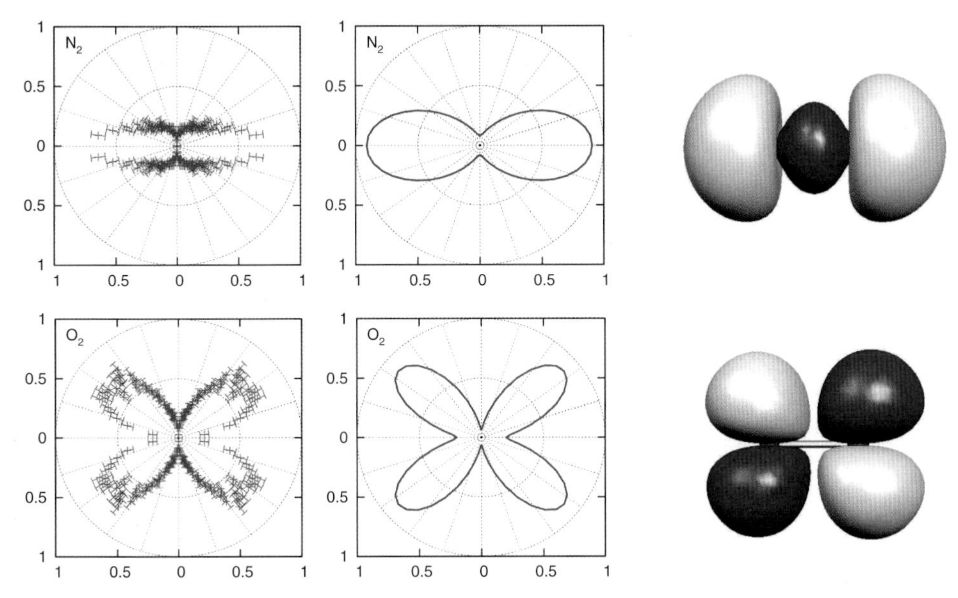

Figure 3.26 Comparison of the measured angular distributions of the fragmented ions from the double ionization of molecules (first column) with the alignment-dependent tunneling ionization rates predicted by the MO-ADK theory for N_2 and O_2 molecules. The angular dependence of the ionization rates is shown to follow the corresponding HOMO electron-density distributions (the last column). The experimental data is taken from [57].

3.7.4 Orientation/Alignment Dependence of Ionization Rate and Symmetry of the Molecular Orbital

According to the tunneling model, the strong-field ionization rate is proportional to the electron density in the direction of the laser field of the highest occupied molecular orbital (HOMO) of a molecule. In turn, the electron density is mostly governed by the orbital symmetry. In Figure 3.26, the calculated alignment-dependent ionization rates are compared to the density of the molecular orbital of N_2 and O_2, respectively. For N_2, the HOMO is a σ_g orbital and for O_2 it is a π_g orbital. The experimentally reported alignment-dependent ionization rates are also shown in Figure 3.26. They are in good agreement with the predictions of the MO-ADK or the MO-SFA models. For a more complex molecule like CCl_4, which has the T_d symmetry, the HOMO is triply degenerate. The calculated orientation dependence of ionization rates of one of them, HOMO-1, is shown in Figure 3.27(a), while the electron density of the orbital is shown in Figure 3.27(b). They are in good agreement with each other.

3.7.5 Experimental Studies of Alignment-Dependent Ionization Rates

Section 3.7.1 discussed how to calculate orientation-dependent ionization rates for molecules using simple MO-ADK, MO-PPT, and MO-SFA models. One can also employ more sophisticated methods like time-dependent density functional theory (TDDFT)

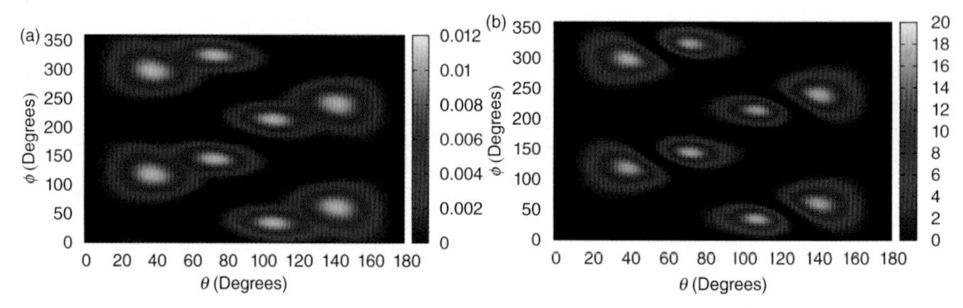

Figure 3.27 (a) Orientation dependence of tunnel ionization rate from HOMO-1 of CCl_4 versus the laser-polarization direction, calculated with the MO-ADK theory for a laser intensity of 0.55×10^{14} W/cm^2. (b) Asymptotic electron-density distribution for HOMO-1. CCl_4 has T_d symmetry and a triply degenerate ($2t_1$) HOMO with ionization potential of 11.47 eV. HOMO-1 is just one of the degenerate HOMO states. The similarity in the two figures demonstrates that tunnel ionization rate is governed by the electron-density distribution, which, in turn, is determined by the symmetry of the molecular orbital. (Figures adopted with permission from Anh-Thu Le, R. R. Lucchese, and C. D. Lin, *Phys. Rev. A*, **87**, 063406 (2013) [58]. Copyrighted by the American Physical Society.)

or other more complex and computationally intensive methods. The accuracy of all of these calculations is best checked against experimental data. However, obtaining accurate experimental orientation-dependent ionization rates for molecules is nontrivial since gas-phase molecules are randomly distributed normally.

For the present discussion, only the alignment dependence will be treated. The alignment of these molecules can be determined by exposing randomly distributed molecules to an intense circularly polarized pulse. The angular distribution of ionization can then be determined by recording the ion yield as a function of the angle between the laser polarization for ionization and the axis of the alignment laser. By a fitting procedure, the angle-dependent ionization rates are retrieved. Since the degree of alignment is not large, this method can suffer from error in the data retrieval, especially in the region of low ionization probability. Alternatively, one can ionize the molecule first and then quickly apply an intense pulse to multiply ionize the molecule. The alignment angle of the molecule is then determined by the Coulomb explosion by recording the fragmented ions in coincidence. In a third method, molecules are doubly ionized by an intense laser pulse in the nonsequential double-ionization regime, where the first electron is emitted by tunnel ionization while the second electron is released by electron-impact excitation, which is followed by ionization from the excited state by the laser. If one assumes that the impact excitation is nearly independent of the alignment angle of the molecules, then the measured angular dependence of the fragmented ions is governed by the first ionization. In the last two methods, short lasers of less than 10 fs are needed to avoid the post-ionization effect [59]. The latter is caused by the additional focusing of molecular ions toward the polarization axis by the ionizing laser. The third method was used to obtain the ionization data in Figure 3.26 [60] and the first method was used in Pavičić et al. [61].

Figure 3.28 depicts the comparison of MO-ADK results and TDSE calculations for H_2^+ and MO-ADK results versus TDDFT for H_2. The larger error in H_2^+ at large angles was traced to the insufficient accuracy of structure parameters obtained from the original MO-ADK model [49]. With the newer, improved parameters, the MO-ADK results are in

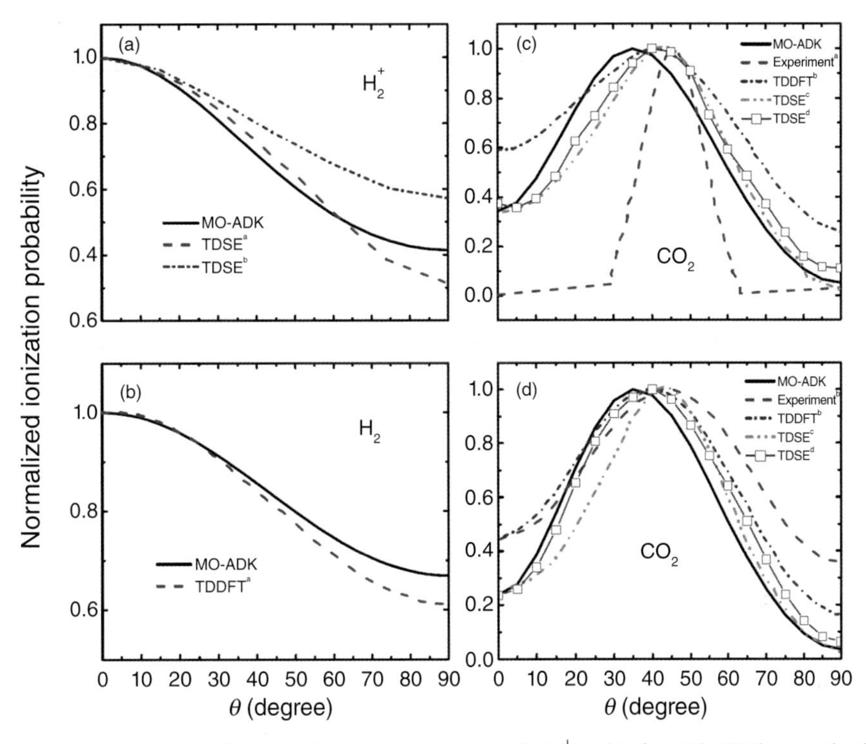

Figure 3.28 (a,b) Comparison of calculated alignment-dependent ionization rates for H_2^+ and H_2 from MO-ADK theory with other theoretical models. (c,d) Alignment-dependent ionization rates of CO_2 at two laser intensities, 1.1×10^{14} W/cm^2 versus 0.3×10^{14} W/cm^2 for (c) and (d), respectively, with the predictions of different theoretical models. Note that the narrow angular distribution obtained in the experiment in (c) is not reproduced by any theories nor in the experiment shown in (d) at the lower laser intensity. A narrow angle-dependent distribution is not expected. (Figures adopted with permission from Song-Feng Zhao et al., *Phys. Rev. A*, **80**, 051402 (2009) [62]. Copyrighted by the American Physical Society.)

good agreement with more elaborate calculations. To illustrate that experimental alignment-dependent rates may also not be accurately derived, the CO_2 results are shown in (c) and (d) of Figure 3.28. In (c) the laser intensity is 1.1×10^{14} W/cm^2 and in (d) the intensity is 0.3×10^{14} W/cm^2. Note that in (c) the "experimental" results are narrowly distributed around 45°, which are in great contradiction with the MO-ADK model and many other elaborate calculations. In (d) the measurement is at a lower intensity, and the reported experimental alignment dependence of ionization rates is much broader and in good agreement with most of the theoretical predictions. The unexpected sharp distribution in the experimental data in (c) is likely due to the problem of extracting the angular ionization dependence from the measured ionization yields of partially aligned molecules.

3.7.6 Ionization from Inner Orbitals of Molecules

Intense laser-field ionization rates depend greatly on the ionization potential of the target. For molecules, the ionization rates also depend on the nature of molecular orbitals.

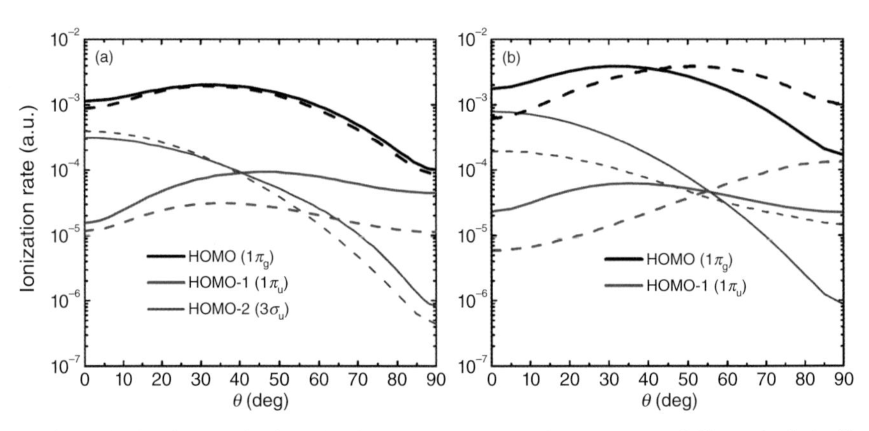

Figure 3.29 Ionization from HOMO and inner orbitals in CO_2. The present MO-PPT results are given in solid lines. The dashed lines in (a) are from [63] and those in (b) are from [64]. (Figures adopted with permission from M. Spanner et al., *Phys. Rev. A*, **80**, 063411 (2009) [63] and from O. Smirnova et al., *Nature* **460**, 972 (2009) [64]. Copyrighted by the American Physical Society and by Nature Publishing Group, respectively.)

Starting with HOMO, it is common to label the subsequent inner orbitals as HOMO-1, HOMO-2, ...etc. As the molecules become more complex, the separation of ionization potentials of successive inner orbitals becomes small. Thus, it is likely that electrons from multiple orbitals will be removed in an intense laser field. Since the alignment dependence is strongly dependent on the symmetry property of the molecular orbitals, ionization from inner orbitals becomes more significant. MO-PPT and MO-SFA can be used to evaluate the relative ionization rates (MO-ADK is known to fail at lower laser intensity; see Section 3.2). To give an example, in CO_2 the ionization potentials of HOMO ($1\pi_g$), HOMO-1 ($1\pi_u$), and HOMO-2 ($3\sigma_g$) are 12.3, 16.7, and 18.2 eV, respectively. The relative ionization rates for these three orbitals calculated using the MO-PPT model are shown in Figure 3.29. The HOMO peaks near 45°, HOMO-1 peaks near 90°, and HOMO-2 peaks near 0°. Their relative rates become closer as the laser intensity increases. The relative MO-PPT ionization rates are compared to those from two other, more elaborate calculations in Figure 3.29. It can be seen that MO-PPT agrees better with one but not the other. When multiple orbitals are contributing to the strong-field ionization, the interpretation of strong-field phenomena can become quite complex. Unfortunately, multiple orbital contributions happen quite often in molecules. Since ionization is the precursor to all strong-field phenomena, multiple orbitals also affect high-order harmonic generation and other nonlinear phenomena. Furthermore, inner-shell holes often lead to dissociation of the molecular ions, which results in multiple fragments.

3.7.7 Ionization Probability of Molecules and Calibration of Laser Intensity

In a typical experiment, accurate calibration of the intensity of a focused laser beam is very challenging. For a loosely focused beam, the intensity variation within the interaction volume may not be too severe. For a tightly focused beam, the intensity variation is large.

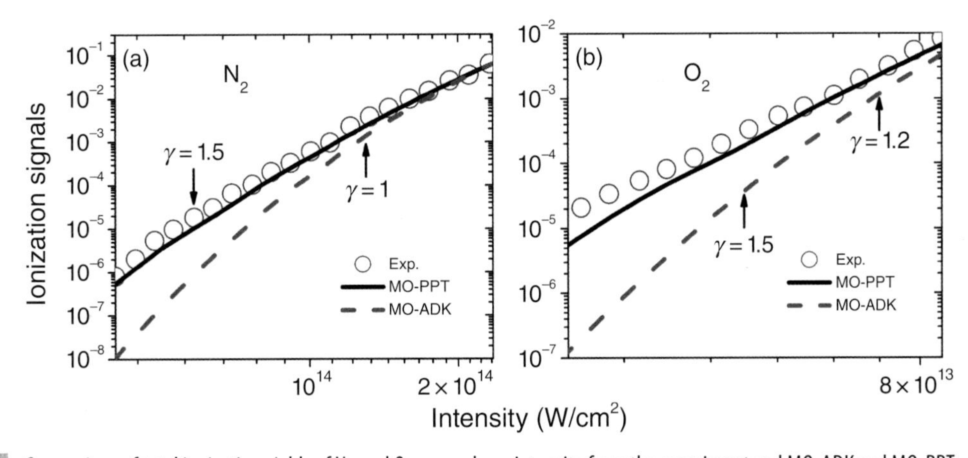

Figure 3.30 Comparison of total ionization yields of N_2 and O_2 versus laser intensity from the experiment and MO-ADK and MO-PPT theories. Volume integration was considered in the theory. Note that MO-ADK theory fails at lower intensities. It is suggested that one can use MO-PPT theory to calibrate absolute experimental peak laser intensity. (Reprinted from Song-Feng Zhao et al., *Phys. Rev. A*, **93**, 023413 (2016) [12]. Copyrighted by the American Physical Society.)

One cannot rely on the so-called $2U_p$ cutoff of photoelectron energy or the $I_p + 3.2U_p$ of the high-order harmonic cutoff to determine the peak laser intensity. These cutoffs are based on a theory of a single laser intensity. With the inclusion of volume integration, such a cutoff is not clearly visible. Section 3.2 shows that the PPT model can predict accurate ionization probability to better than 50% when compared to TDSE calculations. After accounting for volume integration, the calculated ionization yields and the measured ones have been found to be in good agreement after they are normalized. This can be done for molecular targets as well. Figure 3.30 shows that the experimental ionization yields of N_2 and O_2 molecules can be accurately reproduced over a large range of laser intensities using MO-PPT theory, but not MO-ADK theory. Since the calculation based on the MO-PPT theory is relatively easy (including the account of volume integration), it was suggested [12] that experimentalists measure the relative ionization yield over a range of input laser power. For the given target, the MO-PPT theory is used to calculate a range of peak laser intensities, including volume integration. By renormalizing and shifting the two curves until they have the best overall fit, the experimental peak intensity can be derived.

An example of this procedure is illustrated in Figure 3.31. According to the MO-PPT theory, for the ionization of benzene without volume integration, the intensity dependence of ionization yield is shown by the black curve on the left panel of Figure 3.31. In the experiment of Scarborough et al. [65], the measured relative ionization yield is shown by the red curve. In this experiment, ions are collected from a micro volume such that volume integration is not needed. However, the estimated intensities still might not be sufficiently accurate. After shifting the red curve to the left horizontally, the blue curve becomes in good agreement with the black curve. Using this procedure, each of the data points can be assigned a relatively accurate laser intensity. By including the volume integration, the same theory can also calibrate the laser intensity from early measurement [66].

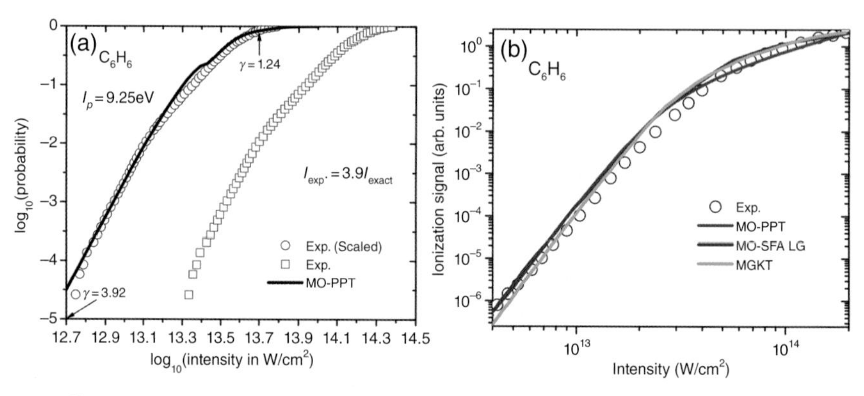

Figure 3.31 Illustration of how MO-PPT theory can be used to calibrate peak laser intensities in experiments. See text for details. (Reprinted from Song-Feng Zhao et al., *Phys. Rev. A*, **93**, 023413 (2016) [12]. Copyrighted by the American Physical Society.)

Notes and Comments

Ionization is the first step of any nonlinear interaction of a molecule with an intense laser pulse. However, the orientation dependence of ionization rates for most molecules is not well determined, either theoretically or experimentally. While MO-ADK, MO-PPT, MO-SFA, or WFAT can be used to calculate the ionization rates, they are carried out mostly only in the one-electron model. *Ab initio* many-electron theories have been carried out for a number of systems, but the accuracy of these calculations in general is difficult to know. Experimentally, the orientation or alignment dependence of ionization rates relies mostly on molecules that are impulsively oriented or aligned. To reach high degrees of orientation or alignment, molecules have to be cooled at the expense of reduced gas density. It is fair to say that optimal alignment or orientation of an ensemble of molecules is an important research topic by itself.

Exercises

3.1 According to Section 3.1.4, excitation probability should occur at the intensity where new channel closing happens. Based on the data in Figure 3.2, confirm that this is indeed correct.

3.2 The ATI peaks and Freeman resonances cannot be separated for short pulses. Mark the position of the possible Freeman resonances in Figure 3.3 for one laser intensity (consider excited states with $n = 4$, 5, and 6). Target is atomic hydrogen.

3.3 Reproduce Figures 3.6(c,e) for ionization probabilities of Ar and Xe using both ADK and PPT models.

3.4 Reproduce Figure 3.8 using the ADK and PPT models, respectively, including volume integration. Assume that the temporal envelope is Gaussian and the focused beam has cylindrical symmetry with Gaussian distributions as well.

3.5 Following the discussion of Section 3.3, show that the model in that section would predict a node at $p_z = 0$ for the first ATI peak in Figure 3.18.

3.6 Starting from the Schrödinger equation for the rotational wave packet Equation 3.14 and the expansion of this wave packet Equation 3.15, derive the coupled equations for the coefficients $C_{J'}(JM, t)$.

3.7 Calculate the alignment-dependent tunneling ionization rates using the MO-ADK theory for N_2 and O_2 molecules. Compare your results with those shown in Figure 3.26. The structure parameters can be found in Zhao et al., *Phys. Rev. A*, **81**, 033423 (2010). You can use an 800 nm laser with peak intensity of 1.0×10^{14} W/cm^2.

3.8 The alignment-dependent ionization rates for the HOMO, HOMO-1, and HOMO-2 of CO_2 molecules have been studied by different groups theoretically. Figure 5 of Zhao et al., *Phys. Rev. A*, **81**, 033423 (2010) examined the comparison of MO-ADK with calculations from other theoretical methods. Use the structure parameters given in that paper to duplicate the MO-ADK results shown in figure 5 of that paper. Since the MO-ADK model is accurate only in the tunnel ionization regime, a better calculation is to use the MO-PPT model. Compare your MO-PPT results with those shown in Figure 3.29.

References

[1] S. Augst, D. D. Meyerhofer, D. Strickland, and S. L. Chin. Laser ionization of noble gases by Coulomb-barrier suppression. *J. Opt. Soc. Am. B*, **8**(4):858–867, Apr. 1991.

[2] X. M. Tong and C. D. Lin. Empirical formula for static field ionization rates of atoms and molecules by lasers in the barrier-suppression regime. *J. Phys. B-At. Mol. Opt.*, **38**(15):2593, 2005.

[3] Q. Li, X.-M. Tong, T. Morishita, C. Jin, H. Wei, and C. D. Lin. Rydberg states in the strong field ionization of hydrogen by 800, 1200 and 1600 nm lasers. *J. Phys. B-At. Mol. Opt.*, **47**(20):204019, 2014.

[4] Q. Li, X.-M. Tong, T. Morishita, H. Wei, and C. D. Lin. Fine structures in the intensity dependence of excitation and ionization probabilities of hydrogen atoms in intense 800-nm laser pulses. *Phys. Rev. A*, **89**:023421, Feb. 2014.

[5] R. R. Freeman, P. H. Bucksbaum, H. Milchberg, S. Darack, D. Schumacher, and M. E. Geusic. Above-threshold ionization with subpicosecond laser pulses. *Phys. Rev. Lett.*, **59**:1092–1095, Sep. 1987.

[6] T. Morishita and C. D. Lin. Photoelectron spectra and high Rydberg states of lithium generated by intense lasers in the over-the-barrier ionization regime. *Phys. Rev. A*, **87**:063405, Jun. 2013.

[7] M. Schuricke, G. Zhu, J. Steinmann, et al. Strong-field ionization of lithium. *Phys. Rev. A*, **83**:023413, Feb. 2011.

[8] R. M. Potvliege and P. H. G. Smith. Adiabatic stabilization of excited states of H in an intense linearly polarized laser field. *Phys. Rev. A*, **48**:R46–R49, Jul. 1993.

[9] T. Nubbemeyer, K. Gorling, A. Saenz, U. Eichmann, and W. Sandner. Strong-field tunneling without ionization. *Phys. Rev. Lett.*, **101**:233001, Dec. 2008.

[10] P. H. Bucksbaum, L. D. Van Woerkom, R. R. Freeman, and D. W. Schumacher. Nonresonant above-threshold ionization by circularly polarized subpicosecond pulses. *Phys. Rev. A*, **41**:4119–4122, Apr. 1990.

[11] A. S. Landsman, A. N. Pfeiffer, C. Hofmann, M. Smolarski, C. Cirelli, and U. Keller. Rydberg state creation by tunnel ionization. *New J. Phys.*, **15**(1):013001, 2013.

[12] S.-F. Zhao, A.-T. Le, C. Jin, X. Wang, and C. D. Lin. Analytical model for calibrating laser intensity in strong-field-ionization experiments. *Phys. Rev. A*, **93**:023413, Feb. 2016.

[13] C. Guo, M. Li, J. P. Nibarger, and G. N. Gibson. Single and double ionization of diatomic molecules in strong laser fields. *Phys. Rev. A*, **58**:R4271–R4274, Dec. 1998.

[14] Z. Chen, T. Morishita, A.-T. Le, M. Wickenhauser, X. M. Tong, and C. D. Lin. Analysis of two-dimensional photoelectron momentum spectra and the effect of the long-range Coulomb potential in single ionization of atoms by intense lasers. *Phys. Rev. A*, **74**:053405, Nov. 2006.

[15] D. G. Arbó, K. I. Dimitriou, E. Persson, and J. Burgdörfer. Sub-Poissonian angular momentum distribution near threshold in atomic ionization by short laser pulses. *Phys. Rev. A*, **78**:013406, Jul. 2008.

[16] C. M. Maharjan, A. S. Alnaser, I. Litvinyuk, P. Ranitovic, and C. L. Cocke. Wavelength dependence of momentum-space images of low-energy electrons generated by short intense laser pulses at high intensities. *J. Phys. B-At. Mol. Opt.*, **39**(8):1955, 2006.

[17] M. Li, Y. Liu, H. Liu, et al. Photoelectron angular distributions of low-order above-threshold ionization of Xe in the multiphoton regime. *Phys. Rev. A*, **85**:013414, Jan. 2012.

[18] T. Marchenko, H. G. Muller, K. J. Schafer, and M. J. J. Vrakking. Electron angular distributions in near-threshold atomic ionization. *J. Phys. B-At. Mol. Opt.*, **43**(9):095601, 2010.

[19] T. Morishita, Z. Chen, S. Watanabe, and C. D. Lin. Two-dimensional electron momentum spectra of argon ionized by short intense lasers: Comparison of theory with experiment. *Phys. Rev. A*, **75**:023407, Feb. 2007.

[20] C. I. Blaga, F. Catoire, P. Colosimo, et al. Strong-field photoionization revisited. *Nat. Phys.*, **5**(5):335–338, May 2009.

[21] W. Quan, Z. Lin, M. Wu, et al. Classical aspects in above-threshold ionization with a midinfrared strong laser field. *Phys. Rev. Lett.*, **103**:093001, Aug. 2009.

[22] C. Y. Wu, Y. D. Yang, Y. Q. Liu, et al. Characteristic spectrum of very low-energy photoelectron from above-threshold ionization in the tunneling regime. *Phys. Rev. Lett.*, **109**:043001, Jul. 2012.

[23] J. Dura, N. Camus, A. Thai, et al. Ionization with low-frequency fields in the tunneling regime. *Sci. Rep.*, **3**:2675, Sep. 2013.

[24] M. G. Pullen, J. Dura, B. Wolter, et al. Kinematically complete measurements of strong field ionization with mid-IR pulses. *J. Phys. B-At. Mol. Opt.*, **47**(20):204010, 2014.

[25] C. Liu and K. Z. Hatsagortsyan. Origin of unexpected low energy structure in photoelectron spectra induced by midinfrared strong laser fields. *Phys. Rev. Lett.*, **105**:113003, Sep. 2010.

[26] A. Kästner, U. Saalmann, and Jan M. Rost. Electron-energy bunching in laser-driven soft recollisions. *Phys. Rev. Lett.*, **108**:033201, Jan. 2012.

[27] B. Wolter, C. Lemell, M. Baudisch, et al. Formation of very-low-energy states crossing the ionization threshold of argon atoms in strong mid-infrared fields. *Phys. Rev. A*, **90**:063424, Dec. 2014.

[28] Q. Z. Xia, D. F. Ye, L. B. Fu, X. Y. Han, and J. Liu. Momentum distribution of near-zero-energy photoelectrons in the strong-field tunneling ionization in the long wavelength limit. *Sci. Rep.*, **5**:11473, Jun. 2015.

[29] Y. Huismans, A. Gijsbertsen, A. S. Smolkowska, et al. Scaling laws for photoelectron holography in the midinfrared wavelength regime. *Phys. Rev. Lett.*, **109**:013002, Jul. 2012.

[30] D. D. Hickstein, P. Ranitovic, S. Witte, et al. Direct visualization of Laser-Driven Electron Multiple Scattering and Tunneling Distance in Strong-Field Ionization. *Phys. Rev. Lett.*, **109**:073004, Aug 2012.

[31] D. Gabor. 1971: For his invention and development of the holographic method. In *Nobel lectures: physics, 1971–1980*, S. Lundqvist, ed., 1–11. World Scientific, Singapore, 1992.

[32] J. J. Barton. Photoelectron holography. *Phys. Rev. Lett.*, **61**:1356–1359, Sep. 1988.

[33] Y. Huismans, A. Rouzée, A. Gijsbertsen, et al. Time-resolved holography with photoelectrons. *Science*, **331**(6013):61–64, 2011.

[34] X. Song, C. Lin, Z. Sheng, et al. Unraveling nonadiabatic ionization and Coulomb potential effect in strong-field photoelectron holography. *Sci. Rep.*, **6**:28392, Jun. 2016.

[35] H. Liu, Y. Liu, L. Fu, et al. Low yield of near-zero-momentum electrons and partial atomic stabilization in strong-field tunneling ionization. *Phys. Rev. Lett.*, **109**:093001, Aug. 2012.

[36] Y. Deng, Y. Liu, X. Liu, et al. Differential study on molecular suppressed ionization in intense linearly and circularly polarized laser fields. *Phys. Rev. A*, **84**:065405, Dec. 2011.

[37] M. Li, X. Sun, X. Xie, et al. Revealing backward rescattering photoelectron interference of molecules in strong infrared laser fields. *Sci. Rep.*, **5**:8519, Feb. 2015.

[38] D. Zeidler, A. B. Bardon, A. Staudte, D. M. Villeneuve, R. Drner, and P. B. Corkum. Alignment independence of the instantaneous ionization rate for nitrogen molecules. *J. Phys. B-At. Mol. Opt.*, **39**(7):L159, 2006.

[39] L. A. A. Nikolopoulos, T. K. Kjeldsen, and L. B. Madsen. Three-dimensional time-dependent Hartree-Fock approach for arbitrarily oriented molecular hydrogen in strong electromagnetic fields. *Phys. Rev. A*, **76**:033402, Sep. 2007.

[40] Henrik Stapelfeldt and Tamar Seideman. Colloquium: Aligning molecules with strong laser pulses. *Rev. Mod. Phys.*, **75**(2):543–557, Apr. 2003.

[41] M. Artamonov and T. Seideman. Theory of three-dimensional alignment by intense laser pulses. *J. Chem. Phys.*, **128**(15):154313, 2008.

[42] A. Rouzée, S. Guérin, O. Faucher, and B. Lavorel. Field-free molecular alignment of asymmetric top molecules using elliptically polarized laser pulses. *Phys. Rev. A*, **77**:043412, Apr. 2008.

[43] V. Makhija, X. Ren, and V. Kumarappan. Metric for three-dimensional alignment of molecules. *Phys. Rev. A*, **85**:033425, Mar. 2012.

[44] X. Ren, V. Makhija, and V. Kumarappan. Multipulse three-dimensional alignment of asymmetric top molecules. *Phys. Rev. Lett.*, **112**:173602, Apr. 2014.

[45] V. Makhija, X. Ren, D. Gockel, A.-T. Le, and V. Kumarappan. Orientation resolution through rotational coherence spectroscopy. arXiv:1611.06476, Nov. 2016.

[46] N. L. Wagner, A. West, I. P. Christov, et al. Monitoring molecular dynamics using coherent electrons from high harmonic generation. *Proc. Natl. Acad. Sci. U.S.A.*, **103**(36):13279–13285, 2006.

[47] W. Li, X. Zhou, R. Lock, et al. Time-resolved dynamics in N2O4 probed using high harmonic generation. *Science*, **322**(5905):1207–1211, 2008.

[48] H. Niikura, D. M. Villeneuve, and P. B. Corkum. Stopping a vibrational wave packet with laser-induced dipole forces. *Phys. Rev. Lett.*, **92**:133002, Mar. 2004.

[49] X. M. Tong, Z. X. Zhao, and C. D. Lin. Theory of molecular tunneling ionization. *Phys. Rev. A*, **66**:033402, Sep. 2002.

[50] S.-F. Zhao, J. Xu, C. Jin, A.-T. Le, and C. D. Lin. Effect of orbital symmetry on the orientation dependence of strong field tunnelling ionization of nonlinear polyatomic molecules. *J. Phys. B-At. Mol. Opt.*, **44**(3):035601, 2011.

[51] A.-T. Le, H. Wei, C. Jin, and C. D. Lin. Strong-field approximation and its extension for high-order harmonic generation with mid-infrared lasers. *J. Phys. B-At. Mol. Opt.*, **49**(5):053001, 2016.

[52] O. I. Tolstikhin, T. Morishita, and L. B. Madsen. Theory of tunneling ionization of molecules: Weak-field asymptotics including dipole effects. *Phys. Rev. A*, **84**:053423, Nov. 2011.

[53] L. Holmegaard, J. L. Hansen, L. Kalhoj, et al. Photoelectron angular distributions from strong-field ionization of oriented molecules. *Nat. Phys.*, **6**(6):428–432, Jun. 2010.

[54] H. Li, D. Ray, S. De, et al. Orientation dependence of the ionization of CO and NO in an intense femtosecond two-color laser field. *Phys. Rev. A*, **84**:043429, Oct. 2011.

[55] S.-F. Zhao, C. Jin, A.-T. Le, T. F. Jiang, and C. D. Lin. Determination of structure parameters in strong-field tunneling ionization theory of molecules. *Phys. Rev. A*, **81**:033423, Mar. 2010.

[56] R. Saito, O. I. Tolstikhin, L. B. Madsen, and T. Morishita. Structure factors for tunneling ionization rates of diatomic molecules. *At. Data Nucl. Data Tables*, **103–104**:4–49, 2015.

[57] A. S. Alnaser, C. M. Maharjan, X. M. Tong, et al. Effects of orbital symmetries in dissociative ionization of molecules by few-cycle laser pulses. *Phys. Rev. A*, **71**:031403, Mar. 2005.

[58] A.-T. Le, R. R. Lucchese, and C. D. Lin. Quantitative rescattering theory of high-order harmonic generation for polyatomic molecules. *Phys. Rev. A*, **87**:063406, Jun. 2013.

[59] X. M. Tong, Z. X. Zhao, A. S. Alnaser, S. Voss, C. L. Cocke, and C. D. Lin. Post ionization alignment of the fragmentation of molecules in an ultrashort intense laser field. *J. Phys. B-At. Mol. Opt.*, **38**(4):333, 2005.

[60] A. S. Alnaser, S. Voss, X. M. Tong, et al. Effects of molecular structure on ion disintegration patterns in ionization of O_2 and N_2 by short laser pulses. *Phys. Rev. Lett.*, **93**:113003, Sep. 2004.

[61] D. Pavičić, K. F. Lee, D. M. Rayner, P. B. Corkum, and D. M. Villeneuve. Direct measurement of the angular dependence of ionization for N_2, O_2, and CO_2 in intense laser fields. *Phys. Rev. Lett.*, **98**:243001, Jun. 2007.

[62] S.-F. Zhao, C. Jin, A.-T. Le, T. F. Jiang, and C. D. Lin. Analysis of angular dependence of strong-field tunneling ionization for CO_2. *Phys. Rev. A*, **80**:051402, Nov. 2009.

[63] M. Spanner and S. Patchkovskii. One-electron ionization of multielectron systems in strong nonresonant laser fields. *Phys. Rev. A*, **80**:063411, Dec. 2009.

[64] O. Smirnova, Y. Mairesse, S. Patchkovskii, et al. High harmonic interferometry of multi-electron dynamics in molecules. *Nature*, **460**(7258):972–977, 2009.

[65] T. D. Scarborough, J. Strohaber, D. B. Foote, C. J. McAcy, and C. J. G. J. Uiterwaal. Ultrafast REMPI in benzene and the monohalobenzenes without the focal volume effect. *Phys. Chem. Chem. Phys.*, **13**:13783–13790, 2011.

[66] A. Talebpour, A. D. Bandrauk, K. Vijayalakshmi, and S. L. Chin. Dissociative ionization of benzene in intense ultra-fast laser pulses. *J. Phys. B-At. Mol. Opt.*, **33**(21): 4615, 2000.

4 Rescattering and Laser-Induced Electron Diffraction

4.1 Introduction

As discussed in Chapter 3, strong-field ionization of atoms by an intense laser generates most of the electrons with energy below $2U_p$ where U_p is the ponderomotive energy. In the ADK model (as well as the PPT model), the target information enters the ionization rate only through the ionization potential and the orbital symmetry of the initial state wavefunction. This is easily understood in the tunneling regime since the electron is emitted at a distance (the tunnel exit) far away from the atomic core, where the electron is mostly governed by the Coulomb potential. After the electron is released, it gains the momentum mainly from the laser field only. The maximum gain is given by the maximum vector potential of the laser pulse $p_{\text{max}} = A_{\text{max}}$. Thus, in Chapter 3, most of the low-energy two-dimensional (2D) momentum spectra look very similar if the laser pulse is about the same. This is especially true if the ionization potentials of the targets are also nearly the same.

It is well known that the above scenario is only valid for direct electrons. In fact, electrons that are released from the target at an earlier time of the laser pulse may be driven back by the oscillating electric field of the laser as it reverses direction. The returning electrons can gain energy from the laser field and may recollide with the target ion. The recollision by the returning electrons is similar to the collision of a beam of laboratory-prepared electrons with an atomic or molecular ion. Thus phenomena like high-order harmonic generation (HHG), high-energy above-threshold ionization (HATI), and nonsequential double ionization (NSDI) can be identified to their respective, equivalent, field-free electron-ion collision processes. Since electrons are powerful tools for interrogating atomic and molecular structure, the rescattering processes offer opportunities for probing the internal structure of the target. More importantly, laser pulses with durations of a few femtoseconds are already available; thus such rescattering processes offer the opportunity for probing the structural changes of atoms or molecules on the femtosecond timescale. In the meantime, rescattering also generates high-order harmonics with photon energy ranging from XUV to X-rays for use as light sources over a broad spectral range. In addition, these harmonics are phase locked and form sources of attosecond pulses. It is fair to say that in the absence of rescattering strong-field and attosecond physics would not exist as we know it today.

The basic concept of rescattering has been known for more than two decades. Rescattering was first used to interpret the generation of high-order harmonics using the so-called *simple-man model*, or the *three-step model*. In this model, an electron is first emitted by tunnel ionization. Then, in the second step, the electron is driven back by the laser field when its electric field changes direction and further accelerated by the laser field toward the parent ion. In the third step the electron recombines with the ion and emits the excessive energy as harmonic photons. Unlike the earlier study of strong-field ionization that was based on the quantum Keldysh–Faisal–Reiss (KFR) or strong-field approximation (SFA) theory, this simple model can be understood classically. The connection with the classical interpretation becomes clear after second-order SFA (SFA2) is examined under the saddle-point approximation. Chapter 5 looks at this historical development in HHG. This chapter considers rescattering in the generation of HATI electrons. Section 2.4.3 illustrated that the classical elastic rescattering of the returning electrons with the ion core can result in a maximal kinetic energy of about 10 U_p, which is much higher than the 2 U_p cutoff for direct electrons. The 10 U_p cutoff energy occurs when the returned electron makes a head-on collision with the ion and is scattered back at 180° from the returning direction.

In conventional elastic or inelastic electron scattering, the strength of the interaction between an electron and a target is measured in terms of differential cross-sections (DCS). In a strong laser field, rescattering occurs in the presence of the laser. To what extent does the laser field modify the electron-ion collision process? This issue was never addressed directly in the three-step model. Thus for years the three-step model was used to interpret the observed experimental data. However, precisely in what way the target structure is revealed in the rescattering process has not been rigorously explored.

This chapter first considers, how SFA2 is used to describe the generation of HATI electron spectra. Then, it introduces the so-called *quantitative rescattering (QRS) theory*. QRS addresses how the measured HATI electron momentum spectra can be used to extract field-free, elastic, electron-ion scattering DCS. According to QRS, two-dimensional (2D) HATI momentum spectra can be expressed as the product of a returning electron wave packet and a field-free electron-ion scattering DCS. Because of this factorization, laser-free electron-ion scattering DCS can be retrieved from the measured 2D HATI momentum spectra. Since the electron pulses only last for a few femtoseconds, if the target structure changes with time, the change will be reflected in the DCS. Thus, through the rescattering process, intense laser pulses can be used to probe dynamic systems. This forms the idea behind the laser-induced electron diffraction method discussed later in this chapter.

4.2 Derivation of the QRS Theory for High-Energy ATI Electrons

4.2.1 High-Energy ATI Spectra from Numerical Simulations

As stated before, it is difficult to judge the validity of strong field physics theory against experiments since experimental conditions are generally not known exactly. Fortunately, for simple atomic targets, accurate solution of the time-dependent Schrödinger equation (TDSE) is possible especially for short pulses. Figure 4.1 presents an example of such calculations. The electric field and the vector potential (both normalized to unit strength) are shown in (a). The central wavelength is 800 nm and the intensity is 1×10^{14} W/cm^2; thus,

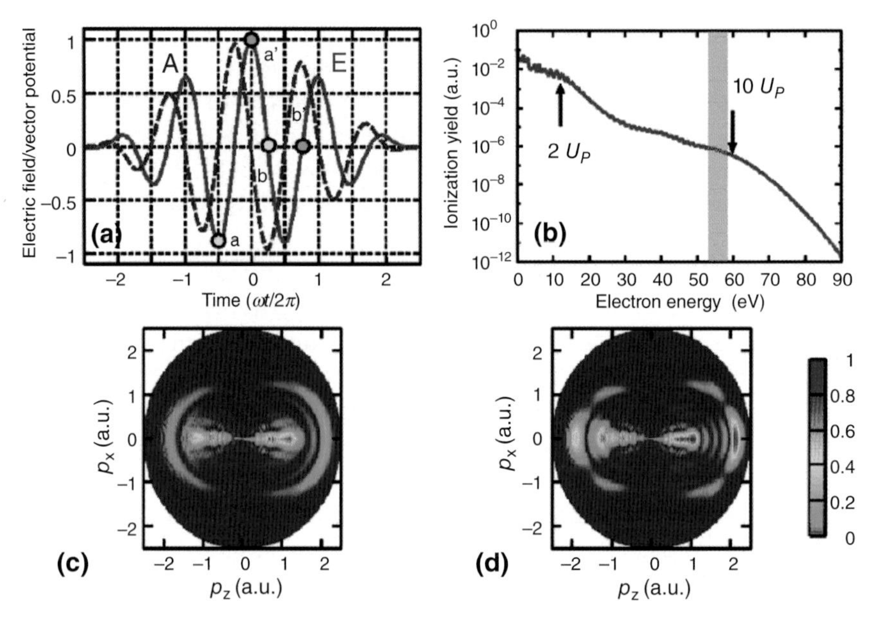

Figure 4.1 Strong field ionization of atomic hydrogen and a model Ar atom by a 5 fs laser pulse with a central wavelength of 800 nm and a peak intensity at 1×10^{14} W/cm^2 calculated by solving the time-dependent Schrödinger equation. (a) Schematic of the electric field E and the vector potential A. (b) Electron energy distribution with the cutoff of direct electrons at 2 U_p and rescattered electrons at 10 U_p. The target is atomic hydrogen. (c) Normalized 2D photoelectron momentum distribution of atomic hydrogen. The images are normalized such that the total electron yield at each given energy is one. The actual differential yield drops by five orders of magnitude from the threshold to about 10 U_p. (d) Normalized 2D photoelectron momentum distribution of a model Ar atom. Note the striking displaced semi-circles at large momenta and the clear difference of intensity distribution along the outermost circle for the two targets. (Reprinted from Toru Morishita, Anh-Thu Le, Zhangjin Chen, and C. D. Lin, *Phys. Rev. Lett.* **100**, 013903 (2008) [1]. Copyrighted by the American Physical Society.)

the ponderomotive energy U_p is 6 eV. The laser has a pulse duration of 5 fs and the carrier-envelope phase is set at zero. In (b), the calculated energy distribution of the photoelectron is shown. Most of the electrons have energies below 2 U_p, or about 12 eV. Beyond 2 U_p, the distribution drops quickly but then flattens out from 4 U_p to 10 U_p. The latter is the rescattering cutoff according to the simple classical model discussed in Section 2.4.3.

In Figures 4.1(c,d), the normalized two-dimensional momentum spectra are shown for atomic hydrogen and a model one-electron Ar atom, respectively. To allow for a closer look at the high energy electrons, the momentum distributions have been normalized individually at each energy. Two important features are clearly seen. First, there are concentric circles with centers shifted from the origin. Along the circumference of each ring, these signals can be identified as elastic scattering events. Since scattering occurs in the laser field, the shift of the center is due to the momentum gained by the scattered electron after it has emerged from the laser field. Neglecting the effect from the ion core, this shift is given by the vector potential of the laser pulse at the time of recollision. Second, along the outermost circle, the intensity distributions for atomic hydrogen and for model

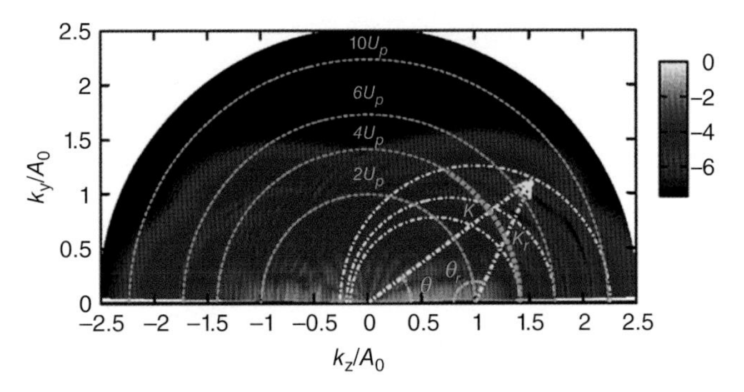

Figure 4.2 2D photoelectron momentum distribution from strong field ionization of atoms. Photoelectrons of constant energy are represented by concentric half circles (green) centered at the origin. Elastic scattering of a returning electron with momentum k_r in the laser field is represented by the half circle (yellow) with its center shifted from the origin by the vector potential. Photoelectrons with momentum k and angle θ are related to the electron-ion elastic scattering momentum k_r and scattering angle θ_r. In the figure, the returning electron enters the target from the right to the left, and the HATI electrons are those that have backscattered. If the scattering is in the forward direction, the electrons are slowed down by the laser field and will emerge from the laser field as low-energy electrons where they will overlap and interfere with direct electrons. According to this figure, one can safely estimate that all electrons with energies higher than 4 U_p are from backscattered returning electrons. (Reprinted from Zhangjin Chen et al., *Phys. Rev. A*, **79**, 033409 (2009) [2]. Copyrighted by the American Physical Society.)

Ar are clearly different. In particular, there is a pronounced minimum for model Ar. This minimum is well-known in field-free DCS in e^-+Ar^+ collisions, see Figure 4.4(b). These TDSE results demonstrate that 2D momentum distributions for HATI electrons contain field-free electron-ion scattering DCS. Note that from Figure 4.1(b), such HATI electrons only account for about 10^{-4}–10^{-5} of the total electrons.

When elaborating on this finding, it is straightforward to extract the field-free electron-ion scattering DCS from the 2D electron momentum spectra using the 2D momentum figure shown in Figure 4.2.

As a further illustration, the 2D electron momentum spectra for H, Ne, Ar, and Xe generated by the same 5 fs, 800 nm and 10^{14} W/cm^2 pulse is shown by solving the TDSE where each atom other than H is represented by a model potential (parameters of the model potential are taken from Tong and Lin [3]). The photoelectron energy spectra are shown in Figure 4.3 for H, Ar, and Xe. From these spectra it is difficult to locate the 2 U_p cutoff or the 10 U_p cutoff. The relative strengths of the low-energy spectra in the order of Xe, H and Ar are easily understood in terms of the ionization potentials (in the order of 12.1 eV, 13.6 eV, 15.6 eV) of the three targets. On the other hand, for energies above 4 U_p, one can clearly see that electron yields of H drop much faster than those of Ar while Xe has the slowest drop as the electron energy increases.

From the calculated 2D momentum spectra following Figure 4.2, the relative DCS for electron–ion collisions can be extracted. Figure 4.4 shows that the DCSs extracted from the HATI spectra are in good agreement with the DCSs calculated from electron–ion scattering over an angular range from about 100°–180°.

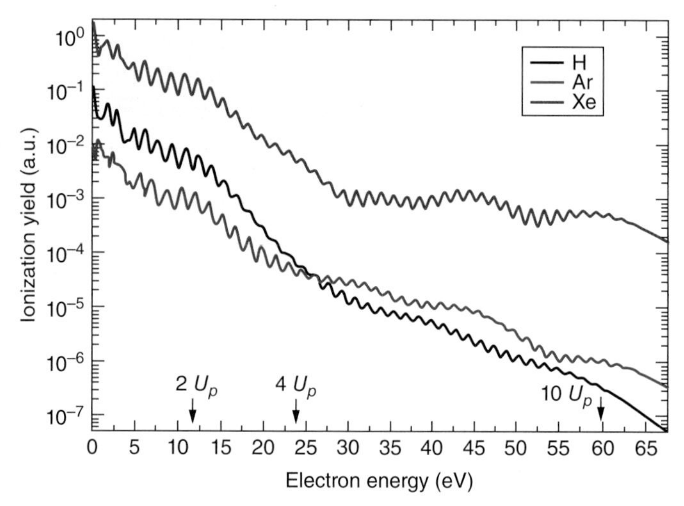

Figure 4.3 Photoelectron energy distributions in logarithmic scale for H, Ar, and Xe obtained by solving TDSE. The laser wavelength is 800 nm, the pulse duration is 5 fs and the peak intensity is 1×10^{14} W/cm^2. (Reprinted from C. D. Lin et al., *J. Phys. B-At. Mol. Opt.*, **43**, 122001, (2010) [4]. Copyrighted by the IOP publishing.)

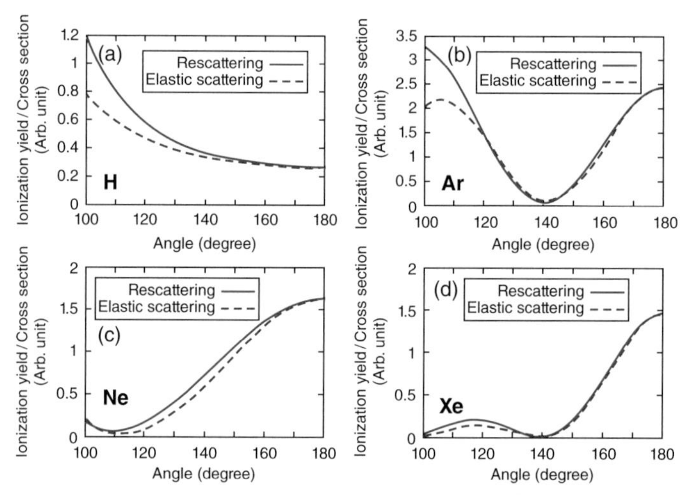

Figure 4.4 Comparing the DCS extracted from the HATI photoelectron momentum spectra for returning electron energy at 3.17 U_p with those directly calculated from electron–ion elastic scattering. Note that the DCSs are expressed on a linear scale. The figures demonstrate that field-free elastic electron–ion DCS can be extracted from photoelectron momentum spectra generated by strong laser fields, i.e., the DCS is not modified by the laser field. (Reprinted from Toru Morishita, Anh-Thu Le, Zhangjin Chen, and C. D. Lin, *Phys. Rev. Lett.*, **100**, 013903 (2008) [1]. Copyrighted by the American Physical Society.)

The TDSE results above prove that HATI spectra are indeed due to the laser-free, large-angle elastic scattering between the returning electron with the parent ion. However, accurate numerical solution of TDSE is only possible for simple, one-electron atomic systems interacting with a short laser pulse, especially for HATI electrons where the total ionization yield is about 10^{-4} of the total continuum electrons. If the simple-man model works well, is it possible to extend the model such that HATI spectra can be accurately obtained at a level close to TDSE results? This goal has been mostly achieved by the QRS model. Before that, one must check how QRS works under the strong-field approximation.

4.2.2 Second-Order Strong-Field Approximation

In Section 2.2.3 the first- and second-order strong-field approximations (or the S-matrix method) were derived using the length gauge. Including these two terms, the ionization probability amplitude can be written as

$$f(\mathbf{k}) = f_1(\mathbf{k}) + f_2(\mathbf{k}), \tag{4.1}$$

where

$$f_1(\mathbf{k}) = -i \int_{-\infty}^{\infty} dt \left\langle \chi_{\mathbf{k}}(t) \left| H_i(t) \right| \Psi_0(t) \right\rangle, \tag{4.2}$$

with the ground-state wavefunction Ψ_0. Here, this term is called SFA1 since SFA has been used in a different context in the literature. The second-order term

$$f_2(\mathbf{k}) = -\int_{-\infty}^{\infty} dt \int_{t}^{\infty} dt' \int d\mathbf{p} \left\langle \chi_{\mathbf{k}}(t') \left| V \right| \chi_{\mathbf{p}}(t') \right\rangle \times \left\langle \chi_{\mathbf{p}}(t) \left| H_i(t) \right| \Psi_0(t) \right\rangle. \tag{4.3}$$

will be called SFA2. SFA2 is the second-order term in the S-matrix expansion but has also been called the *improved strong-field approximation* in the literature. Here, the continuum state is approximated by the Volkov wavefunction. In the length gauge it is given by

$$\left\langle \mathbf{r} | \chi_{\mathbf{k}}(t) \right\rangle = \frac{1}{(2\pi)^{3/2}} e^{i[\mathbf{k}+\mathbf{A}(t)]\cdot\mathbf{r}} e^{-iS(\mathbf{k},t)}, \tag{4.4}$$

where the action S is

$$S(\mathbf{k}, t) = \frac{1}{2} \int_{-\infty}^{t} dt' \left[\mathbf{k} + \mathbf{A}(t')\right]^2. \tag{4.5}$$

For the SFA1 term

$$f_1(\mathbf{k}) = -i \frac{1}{(2\pi)^{3/2}} \int_{-\infty}^{\infty} dt E(t) e^{iS(\mathbf{k},t)} e^{iI_p t} \times \int d\mathbf{r} e^{-i[\mathbf{k}+\mathbf{A}(t)]\cdot\mathbf{r}} r \cos\theta \Psi_0(\mathbf{r}) \tag{4.6}$$

an electron is emitted at time t (the born time) and then propagates in the laser field until it reaches the detector with final momentum $\mathbf{k}$. Initially, the electron is in the ground state. It has an ionization potential I_p. In Equation 4.6, one integrates over all the possible born time that leads to identical $\mathbf{k}$. Using the relation

$$e^{-i\mathbf{q}\cdot\mathbf{r}} = 4\pi \sum_{lm} i^{-l} j_l(qr) Y_{lm}(\hat{\mathbf{r}}) Y^*_{lm}(\hat{\mathbf{q}}), \tag{4.7}$$

where $j_l(qr)$ is the spherical Bessel function, the spatial integration in Equation 4.6 can be simplified to a 1D integral in r

$$\begin{aligned}\Psi_0(\mathbf{q}) &\equiv \int d\mathbf{r} e^{-i\mathbf{q}\cdot\mathbf{r}} r\cos\theta\,\Psi_0(\mathbf{r}) \\ &= 4\pi\sqrt{\frac{4\pi}{3}}\sum_{lm} i^{-l}Y_{lm}(\hat{\mathbf{q}})\int dr r^3 R_{n_0l_0}(r) j_l(qr) \\ &\times \int d\hat{\mathbf{r}} Y^*_{lm}(\hat{\mathbf{r}})Y_{10}(\hat{\mathbf{r}})Y_{l_0m_0}(\hat{\mathbf{r}}),\end{aligned} \tag{4.8}$$

where the initial state wavefunction is $\Psi_0(\mathbf{r}) = R_{n_0l_0}(r)Y_{l_0m_0}(\hat{\mathbf{r}})$. For a linearly polarized laser field, only $m_0 = 0$ is considered, and

$$\int d\hat{\mathbf{r}} Y^*_{lm}(\hat{\mathbf{r}})Y_{10}(\hat{\mathbf{r}})Y_{l_0m_0}(\hat{\mathbf{r}}) = \sqrt{\frac{3(2l_0+1)}{4\pi(2l+1)}}C(1l_0l;000)C(1l_0l;000)\delta_{m0}, \tag{4.9}$$

where the Cs are the Clebsch–Gordan coefficients. The remaining integration over r is done either analytically or numerically. The integration over time is carried out numerically.

For the second-order strong-field approximation, the amplitude for SFA2, Equation 4.3, can be understood by reading the integrand from right to left. First, the electron is released at the born time t to a Volkov state $\mathbf{p}$. The electron then propagates in the laser field until time t' when it recollides with the ion via the potential V. After the elastic collision, the electron's momentum changes from $\mathbf{p}$ to $\mathbf{k}$. The electron then continues to travel in the laser field until it reaches the detector. In SFA2 one has to sum over all the possible intermediate states $\mathbf{p}$ for each final momentum $\mathbf{k}$. Given t and t', for high-energy electrons, the integral over $\mathbf{p}$ is sharply peaked such that one may evaluate the integral using the saddle-point approximation. The resulting equation is then simplified to

$$\begin{aligned}f_2(\mathbf{k}) &= -\int_{-\infty}^{\infty} dt \int_{-\infty}^{t} dt' \left[\frac{2\pi}{\epsilon + i(t-t')}\right]^{3/2} E(t')e^{iI_pt'} \\ &\times e^{-i[S(\mathbf{p}_s,t)-S(\mathbf{k},t)]}e^{iS(\mathbf{p}_s,t')} \\ &\times \frac{1}{(2\pi)^3}\int d\mathbf{r}' e^{i(\mathbf{p}_s-\mathbf{k})\cdot\mathbf{r}'}V(\mathbf{r}') \\ &\times \frac{1}{(2\pi)^{3/2}}\int d\mathbf{r} e^{-i[\mathbf{p}_s+\mathbf{A}(t')]\cdot\mathbf{r}} r\cos\theta\,\Psi_0(\mathbf{r}),\end{aligned} \tag{4.10}$$

where the saddle point is given by

$$\mathbf{p}_s(t,t') = -\frac{1}{t-t'}\int_{t'}^{t} dt''\mathbf{A}(t''). \tag{4.11}$$

A regulation parameter ϵ was introduced in Equation 4.10 to avoid the divergence of this integral. For energies above $1U_p$, the integral is independent of the ϵ used. In principle, when the integral diverges, the saddle-point approximation has to be evaluated up to the next order or by direct numerical integration. In Equation 4.10, the integral involving $V(r)$ would also introduce singularity since the Fourier transform of a Coulomb potential diverges at a small momentum. In actual numerical calculations, a damping term $\exp(-\lambda r)$ is multiplied to $V(r)$ to avoid the divergence. For low-energy electrons, the calculation of

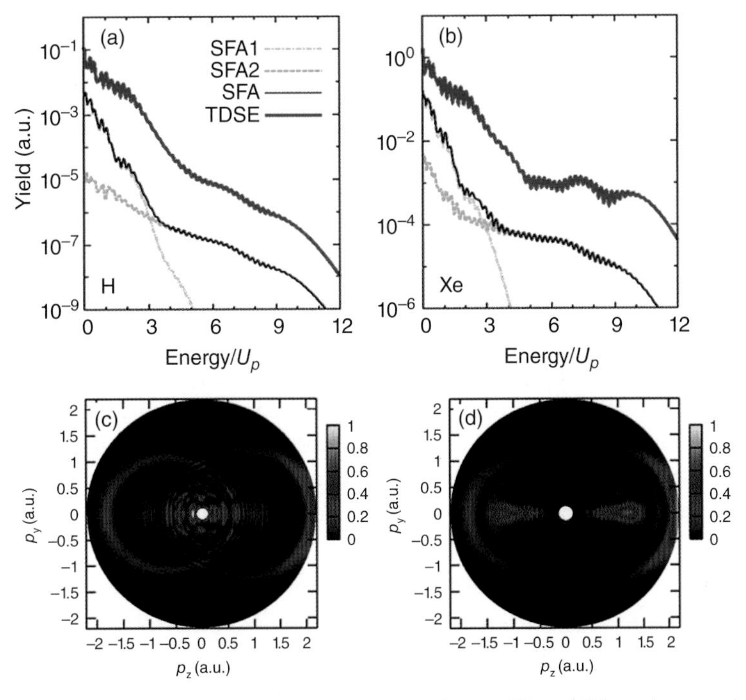

Figure 4.5 Angle-integrated energy spectra (in units of U_p) calculated separately using SFA1 and SFA2, and compared to the TDSE results for H (a) and Xe (b). The laser is 800 nm, 5 fs with an intensity of 1.0×10^{14} W/cm^2. Note that SFA1 dominates at low energy, while SFA2 dominates at high energy. In (c) and (d) the normalized 2D momentum spectra for H calculated using SFA2 and TDSE, respectively, are shown. The ionization yield for each fixed electron energy is normalized to 1.0. The rescattering ring can be seen in both the SFA2 result and the TDSE result. (Reprinted from Zhangjin Chen et al., *Phys. Rev. A*, **79**, 033409 (2009) [2]. Copyrighted by the American Physical Society.)

Equation 4.10 becomes problematic. Thus evaluation of SFA2 for low-energy electrons would have to deal with this difficulty. On the other hand, for low-energy electrons, strong-field approximation is known to be inaccurate since the Coulomb potential is expected to play an important role.

Figure 4.5 compares the angle-integrated, electron-energy spectra calculated using SFA1, SFA2, SFA, and TDSE for (a) H and (b) Xe targets. Figures 4.5(c,d) show the 2D momentum distributions for the hydrogen target calculated from SFA2 and TDSE, respectively. The ionization yields in (c) and (d) at each photoelectron energy are normalized to 1.0 to show a clear rescattering ring in both SFA2 and TDSE.

4.2.3 Extracting the Returning Electron Wave Packet

According to the 1D classical theory of a free electron in a laser field and Figure 2.5, an electron released earlier in the laser pulse may be driven back to the ion core within the same optical cycle. Figure 2.5(b) shows the vector potential $A(t)$ and the electron momentum $k_r(t)$ at time $t = t_r$ when the electron returns to the core. Simple calculations based on the classical equations given in Section 2.4.3 show that if $\omega t_i = 14°$, the electron will return at $\omega t_r = 265°$. If the collision with the ion results in 180° backscattering, a maximal kinetic energy of the electron is $10\,U_p$. For an electron born before $\omega t_i = 14°$, it will return at ωt_r larger than 265°. If it is born after $\omega t_i = 14°$, it will return earlier with ωt_r less than 265°. Thus electrons that were born earlier will return later. From 1D classical equations, there are two (t_i, t_r) pairs for the electron to return to the core with the same k_r; the one with the longer (shorter) excursion time $\tau = t_r - t_i$ belongs to long-(short-) trajectory electrons. Since long-trajectory electrons are emitted earlier in the laser cycle when the laser field is strong, they have higher ionization yields. For short-trajectory electrons, their yields are smaller since they are emitted farther away from the peak field of the laser cycle. Here, sinusoidal waveform pulses are considered. For a different waveform, new values of t_i and t_r should be calculated. The calculation mentioned here neglects the interaction between the electron and the ionic-core potential. They are equivalent to the SFA2 model except that the electron is treated as a classical particle in deriving the time information. One can use quantum orbits theory to calculate the complex born and return time (see Section 5.2.3).

Continuing with the classical model, if the electron is scattered by the ion to emerge at angle θ_r and time t_r, after it exits from the laser field it will gain an additional momentum given by $A(t_r)$ in the laser-polarization direction. In the perpendicular direction there is no change of momentum from the laser field. If the polarization axis is along z and y is perpendicular to it, the relationship between the electron momentum in the laser field right after collision and the momentum detected in the laboratory are related by

$$k_z = k\cos\theta = \pm A_0 \mp k_r \cos\theta_r, \tag{4.12}$$

$$k_y = k\sin\theta = k_r \sin\theta_r. \tag{4.13}$$

The upper signs in Equation 4.12 refer to the right side ($k_z > 0$) while the lower ones refer to the left side ($k_z < 0$). For backscattering, the angle θ_r is greater than 90°. The parameters (k, θ) are in the laboratory frame and (k_r, θ_r) are the same quantities in the laser field. In the equations, A_0 is taken to be the magnitude of the vector potential at the time of recollision $t = t_r$. If only long-trajectory electrons are considered, there is a one-to-one relation between $A_0 = A(t_r)$ and k_r. Figure 2.5 shows that k_r/A_0 ranges from about 1.07 at $\omega t_r = 320°$ to 1.26 at 265°, which corresponds with high-energy electrons from about $4\,U_p$ to $10\,U_p$. Equations 4.12 and 4.13 show that HATI electrons are generated from backscattered electrons. Since large-angle scattering means close collisions, HATI momentum spectra contain important target-structure information.

Since SFA2 is a model for recollision, one may want to take this model more seriously. According to the simple-man picture, the first step is ionization and the second step is electron propagation to generate a beam of returning electrons that aims at the ionic target. The subsequent third step is elastic collision, where the strength is described by a DCS. According to this scattering picture, the electron yield right after elastic scattering for angle θ_r is given by $W(k_r)\sigma(k_r, \theta_r)$, where $W(k_r)$ is the electron intensity integrated over the duration of the pulse. Thus, to connect with the electrons observed in the laboratory, one may write the standard scattering relation

$$N(k, \theta) = W(k_r)\sigma(k_r, \theta_r). \tag{4.14}$$

This equation defines a beam of electrons with flux $W(k_r)$. This "flux" is not calculated nor measured since the relation was written in the presence of the laser field. The (k,θ) and (k_r,θ_r) are related by Equations 4.12 and 4.13 with the k_r/A_0 determined from the classical equations in Section 2.4.3.

Note that Equation 4.14 was not derived from an approximate solution of the TDSE, but rather based on an idea from the rescattering model. While $N(k,\theta)$ is the 2D electron momentum spectrum of an atom measured at the end of a laser field, $\sigma(k_r,\theta_r)$ is the elastic electron–ion elastic DCS in the absence of the laser field. As stated earlier, there is no description of how to measure or calculate the $W(k_r)$. If the scattering picture is correct, then the ratio $N(k,\theta)/\sigma(k_r,\theta_r)$ should be independent of angle θ_r. One can test this with a model one-electron atom, where $N(k,\theta)$ can be obtained accurately from solving the TDSE, and $\sigma(k_r,\theta_r)$ can be calculated as in Section 1.1.3. Furthermore, Equation 4.14 should also be valid for the SFA2 theory except $\sigma(k_r,\theta_r)$ should be calculated by the first Born approximation. Figure 4.6 shows the derived wave packets for θ_r ranging from 155° to 180° for two different peak laser intensities. Indeed, the wave packets $W(k_r)$ derived do not depend on θ_r, and the results from SFA2 and TDSE are identical in shape, i.e., they differ only by an overall normalization that is due to the difference in the ionization rates calculated using TDSE and SFA, respectively. In hindsight, the agreement of $W(k_r)$ between the two methods is surprising. The shape of the wave packet $W(k_r)$ is dominated by the propagation of electrons in the laser field after ionization and before recollision. During this time, the electron is located mostly far away from the ionic core, where it is primarily driven by the laser field. The SFA2 accounts for the effect of laser field correctly, and thus the wave packets from SFA2 and TDSE have nearly the same shape. A similar equation, like Equation 4.14, can be derived for HHG. Further discussion will be given in Section 5.3.

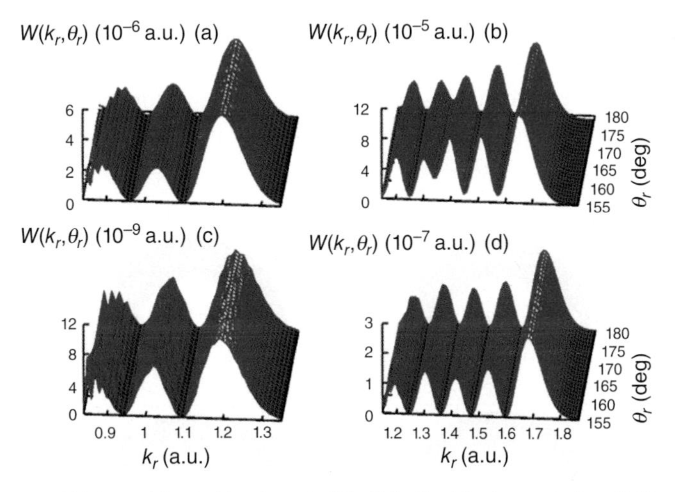

Figure 4.6 Electron wave packet $W(k_r)$ derived from taking the ratio $N(k,\theta)/\sigma(k_r,\theta_r)$ using results from SFA2 and TDSE at intensities of 1.0 and 2.0×10^{14} W/cm^2 for a 5 fs, 800 nm pulse. The figures show that the ratio is independent of angle θ_r. Furthermore, the modulation of the wave packets (or the intensity) is very similar between the SFA2 and TDSE. This suggests that electron wave packets can be calculated from SFA2. (Reprinted from Zhangjin Chen et al., *Phys. Rev. A*, **79**, 033409 (2009) [2]. Copyrighted by the American Physical Society.)

4.2.4 Calculation of HATI Spectra Using the QRS Theory

One of the important consequences of the QRS model is that the wave packet only depends on the laser (except for the ionization probability), and the DCS only depends on the target. To use Equation 4.14 to obtain 2D momentum spectra of HATI electrons, one can use

$$N_{QRS}(k,\theta) = W_{SFA2}(k_r)\sigma_{sw}(k_r,\theta_r), \tag{4.15}$$

where $W(k_r)$ is obtained from

$$W_{SFA2}(k_r) = N_{SFA2}(k,\theta)/\sigma_{B1}(k_r,\theta_r). \tag{4.16}$$

Here $\sigma_{sw}(k_r,\theta_r)$ is obtained from the standard, time-independent atomic and molecular scattering theory using continuum scattering waves (sw). The latter has been studied extensively since the 1960s including many-electron effects. Thus QRS theory can be used to obtain HATI electron spectra without solving the many-body TDSE.

Figure 4.7 compares 2D momentum spectra of HATI electrons calculated from TDSE and from the QRS theory for Ar and Xe, by an identical laser pulse. Under this situation, the wave packet $W(k_r)$ for the two targets is identical (except for an overall normalization). The difference in the momentum spectra between the two targets originates from the large difference in the DCS, $\sigma_{sw}(k_r,\theta_r)$. Integrating over the angles, the energy spectra calculated from QRS (after the absolute values are renormalized) reproduce the HATI spectra (above $4\,U_p$) calculated using TDSE well, as seen in Figure 4.7(e).

When comparing with experimental data, the electron spectra calculated by the theory should be integrated over the volume of the focused laser beam. Using the QRS theory, this only requires repetitive calculations of $W_{SFA2}(k_r)$ for each intensity using the SFA2 model. The DCS part depends only on the electron energies and requires no additional calculations for each laser intensity. Since each SFA2 calculation takes only a few seconds on ordinary desktop computers, volume integration can be easily carried out.

TDSE calculations are especially difficult for long-wavelength lasers. In Figure 4.8, angle-integrated photoelectron spectra of Ar measured using 800, 1,300, 2,000, and

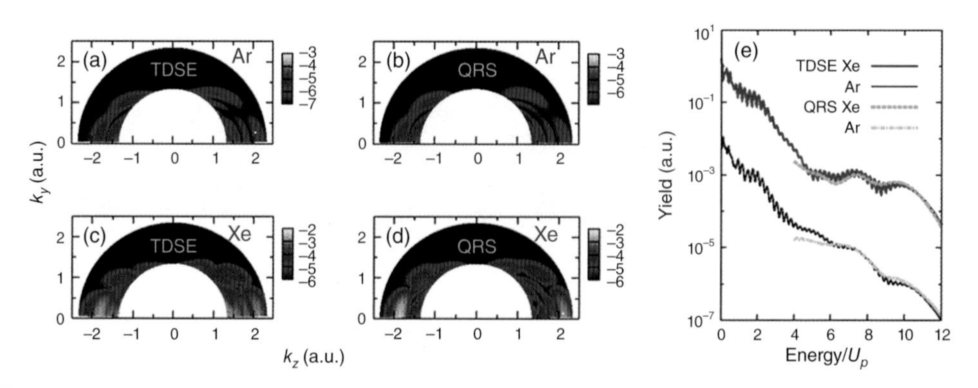

Figure 4.7 Comparison of the 2D HATI momentum spectra and the angle-integrated energy spectra using TDSE and QRS. The QRS results are normalized to the TDSE results at only one point. (Reprinted from Zhangjin Chen et al., *Phys. Rev. A*, **79**, 033409 (2009) [2]. Copyrighted by the American Physical Society.)

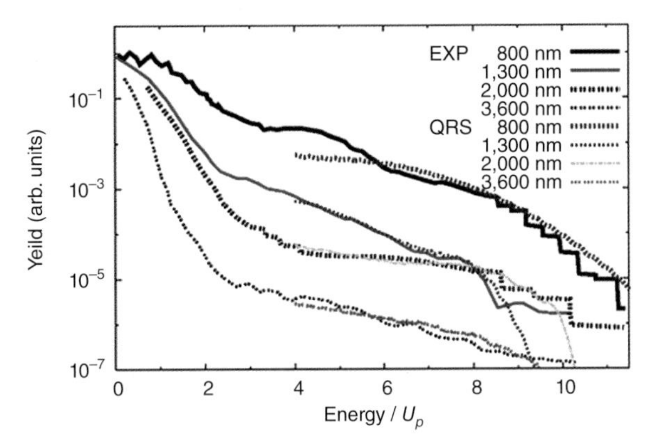

Figure 4.8 Angle-integrated photoelectron spectra in units of U_p for the ionization of Ar by lasers of wavelengths at 800, 1,300, 2,000 and 3,600 nm, respectively, and at a peak intensity of 0.8×10^{14} W/cm^2. The results from the QRS are shown. For each curve, the theory and experiment are normalized at one point to obtain the best overall fit in the high-energy portion above 4 U_p. The experimental data are from [5]. (Reprinted from Zhangjin Chen et al., *Phys. Rev. A*, **79**, 033409 (2009) [2]. Copyrighted by the American Physical Society.)

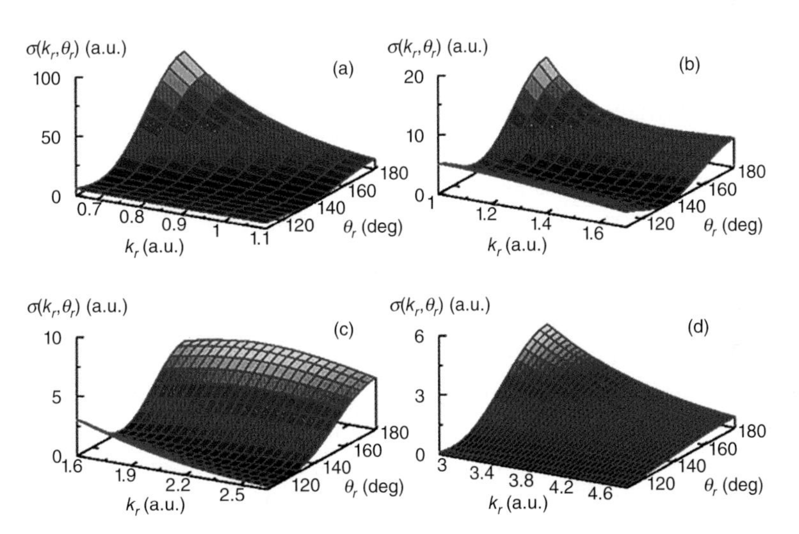

Figure 4.9 The DCS of Ar^+ scattered by electrons in different energy ranges suitable for the HATI spectra of Figure 4.8. In (c), the DCS is large at large angles. This is responsible for the slow drop in the energy spectra for the 2,000 nm laser data seen in Figure 4.8. (Reprinted from Zhangjin Chen et al., *Phys. Rev. A*, **79**, 033409 (2009) [2]. Copyrighted by the American Physical Society.)

3,600 nm lasers at a peak intensity of 0.8×10^{14} W/cm^2 are compared to calculations based on the QRS theory, where the electron energy is expressed in units of U_p's. Note how the shape of the HATI spectra changes with the laser's wavelength. These variations can be traced to the energy dependence of the DCS at different collision energies (see Figure 4.9). By examining the electron–ion elastic scattering DCS, the behavior of the HATI spectra can be readily understood. The differences owe to the strong energy dependence of the field-free elastic DCS.

4.3 Extracting Structure Information from Experimental HATI Spectra

Experimental electron spectra are collected from a focused laser beam, where the intensity distribution is not uniform in space. The need to perform volume integration in theoretical calculations to compare with experimental data was discussed in Section 3.1.1.

$$S(\mathbf{k}, I_0) = \rho \int_0^{I_0} N_I(k,\theta) \left(\frac{\partial V}{\partial I}\right) dI, \tag{4.17}$$

where I is the local peak intensity, I_0 is the peak intensity at the laser focus, and ρ is the density of the gas that was assumed to be uniform.

Since the DCS does not depend on the laser intensity, from Equation 4.14, the DCS term can be pulled out from the integration in Equation 4.17 to obtain

$$S(\mathbf{k}, I_0) = \overline{W}_{I_0}(k_r)\sigma(k_r, \theta_r), \tag{4.18}$$

where $\overline{W}_{I_0}$ is a volume-integrated wave packet when the peak intensity is I_0 at the focus. It is assumed that the gas volume is larger than the focused volume. Equation 4.18 allows one to extract the DCS for a given k_r without knowing the precise laser parameters in the experiment.

HATI electron momentum spectra of rare gas atoms have been measured using 800 nm lasers and Equation 4.18 was used to extract the DCS. In Figure 4.10(a–c), the DCS from experimental HATI spectra for Ne, Ar, and Xe is compared to the accurate DCS from theory [6]. The agreement is considered to be very good. There are discrepancies at large angles close to 180° since the experiment was carried out with long pulses. In Figure 4.10(d), the DCS was extracted from 5 fs laser pulses with different CEPs, i.e., without CEP stabilization. The retrieved DCS indeed agrees well with the experiment in each case. Figure 4.10(e) shows the DCS obtained from electron–Ar^+ experiments and the DCS extracted from the HATI spectra. The former can reach small angles up to about 90° and the HATI method can reach large angles from about 100° to 180°. The black and red curves are from two theoretical calculations. The dashed purple curve is obtained from the Born approximation that is known to be invalid. These result, show that HATI momentum spectra can be used to obtain electron–ion collision cross-sections. Such experiments are very difficult to perform since they require cross-beam or merged-beam measurements with a very limited number of charged particles.

The QRS theory has also been applied to strong-field ionization of rare gas atoms using long-wavelength, mid-infrared laser pulses with a wavelength of 2,000 nm. Longer-wavelength driving lasers allow the generation of higher returning electron energies. Figure 4.11 shows the DCSs derived from such experiments based on the QRS theory. They are compared to DCSs from other experiments and theoretical DCSs. The different curves agree with each other quite well. Thus, at large scattering angles and large scattering energies, there are few differences in the DCS between a neutral atom and its cation. In both cases, scattering occurs close to the atomic core, where the electric fields for neutral and

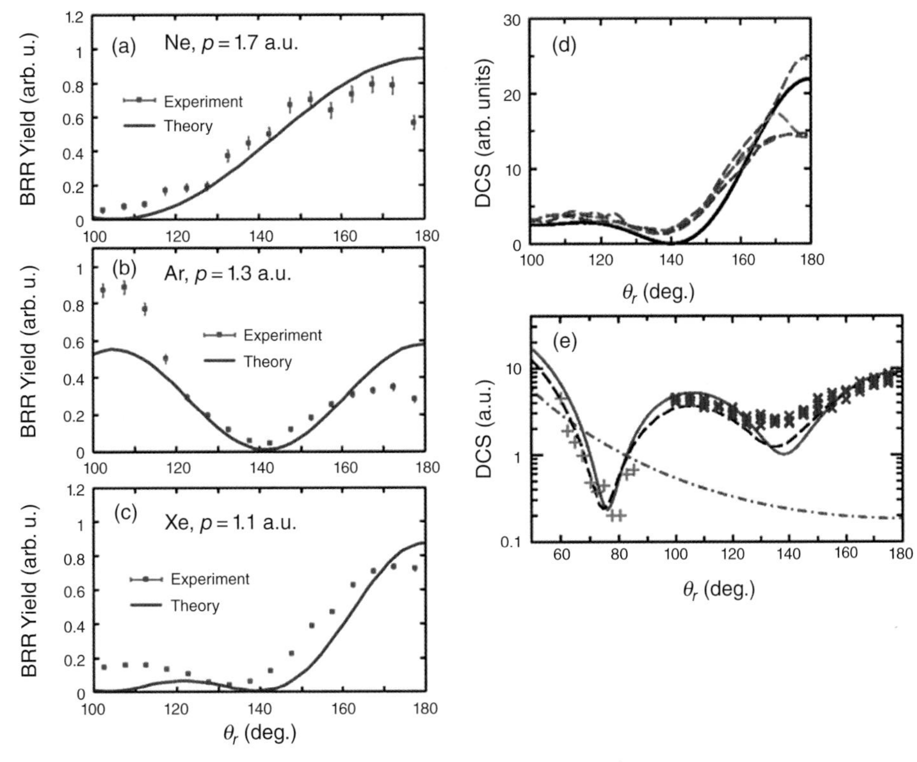

Figure 4.10 (a-c) DCSs extracted from experimental HATI momentum data using 800 nm laser. From [6]. (d) The DCS extracted from HATI spectra using 5 fs, 800 nm laser with different CEPs. (e) The DCS (green data points) for electron-Ar^{+} scattering from electron-ion collision experiment and the DCS (blue data points) extracted from HATI spectra through strong-field ionization by 800 nm laser. The red and black dashed lines are from the scattering theory calculations, while the purple dashed-dotted lines are from Born approximation. (Reprinted from M. Okunishi et al., *Phys. Rev. Lett.*, **100**, 143001 (2008) [6] and Micheau et al., *Phys. Rev. Lett.*, **102**, 073001 (2009) [7]. Copyrighted by the American Physical Society.)

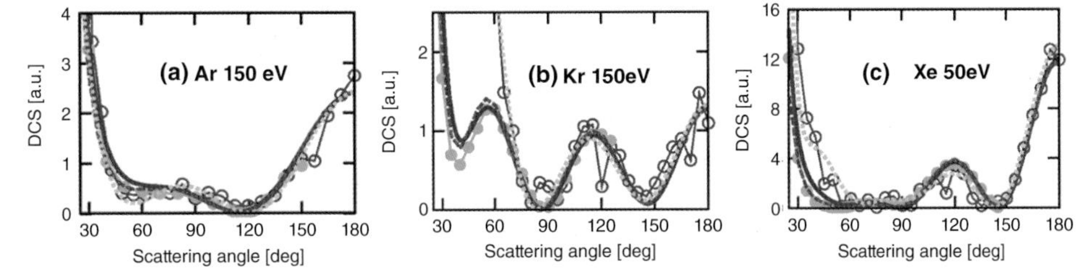

Figure 4.11 Extracted DCS of the Ar and Kr ions at 100 eV and the Xe ion at 50 eV from HATI momentum spectra using mid-infrared lasers. Open symbols are data from different laser intensities and/or wavelengths of 2.0 or 2.3 μm. Green-filled circles are data from electron-neutral atom collisions. The blue full curves are theoretical electron–ion DCSs. These results demonstrate that (1) the DCS extracted at a given energy from the laser experiment is independent of the wavelength and laser intensity. (2) The DCSs for electron-neutral and electron–ion collisions for large-angle scattering are essentially the same. (Figures adopted with permission from Junliang Xu et al., *Phys. Rev. Lett.*, **109**, 233002 (2012) [8]. Copyrighted by the American Physical Society.)

cation are essentially identical. This equality makes HATI electron spectra very powerful tools for probing the structure of a molecule.

4.4 LIED for Ultrafast Self-Imaging of Molecules: Theory

4.4.1 Extracting DCS from Molecules Using HATI Electrons

Consider molecular targets. For each fixed-in-space molecule exposed to a linearly polarized laser pulse, the QRS theory can be extended to obtain the HATI spectra. Let Ω_R be the polar angles of a body-fixed axis of the molecule with respect to the polarization axis of a linearly polarized laser. The HATI photoelectron angular distributions can be expressed as

$$S(\mathbf{k};\Omega_L) = N(\Omega_L)W(k_r)\sigma(\mathbf{k}_r;\Omega_L), \tag{4.19}$$

where $\mathbf{k} = (k,\theta,\phi)$ is the momentum of the photoelectron with the quantization axis, which is chosen to be along the laser polarization, and $\mathbf{k}_r = (k_r,\theta_r,\phi_r)$ is the electron momentum in the laser field at the time of recollision. In Equation 4.19, $N(\Omega_L)$ is the tunnel-ionization rate of the molecule that depends on the orientation angle Ω_L. If molecules are isotropically distributed or partially oriented/aligned with respect to the polarization axis, then the scattering amplitude only depends on $\phi - \phi_r$. Thus the rotation-averaged momentum distribution is cylindrically symmetric with respect to the polarization axis,

$$S(k,\theta) = W(k_r)\bar{\sigma}(k_r,\theta_r), \tag{4.20}$$

where

$$\bar{\sigma}(k_r,\theta_r) = \int N(\Omega_L)\rho(\Omega_L)\sigma(\mathbf{k}_r;\Omega_L)d\Omega_L. \tag{4.21}$$

Here, $\rho(\Omega_L)$ is the angular distribution of the molecules. Note that in this expression the returning electron wave packet is written as $N(\Omega_L)W(k_r)$ to account for the fact that the returning electron flux is proportional to the orientation-dependent ionization rate $N(\Omega_L)$, which can be calculated using the MO-ADK theory or the SFA model as explained in Section 4.2.4. The relations between (k,θ) and (k_r,θ_r) are the same as in the atomic target.

To account for the effect of volume integration, one only has to interpret the left-hand side of Equation 4.20 as the volume-integrated photoelectron spectra and the returning wave packet on the right as the volume-integrated wave packet. Thus, for molecular targets, the retrieved DCS is weighted by the angular distributions of the molecules and the angle-dependent tunnel ionization rate in Equation 4.21.

Large-angle elastic DCS for HATI photoelectrons generated by 800 nm lasers for simple molecules like O_2, CO_2, and CH_4 have been reported since 2011 [6, 9]. In Figure 4.12 the weighted large-angle DCSs for O_2 and CO_2 have been extracted and the results are compared to the DCSs in Equation 4.21 calculated for electron-molecular ion collisions theory. The agreement is quite good.

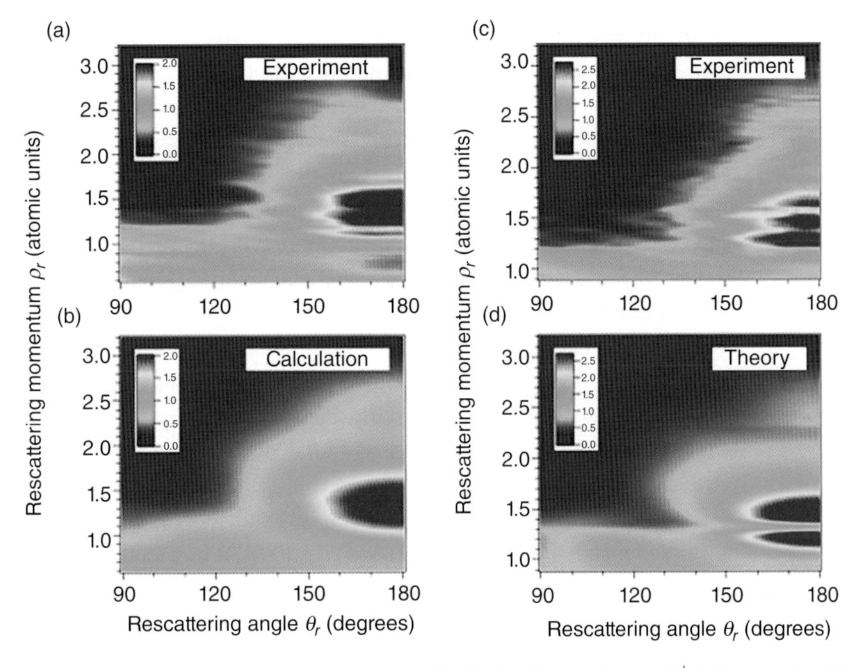

Figure 4.12 Comparison of the experimentally extracted and theoretically calculated elastic electron-O_2^+ (left column) and the electron-CO_2^+ collisions (right column) as a function of electron momentum and scattering angle. In the experiment, 2D photoelectron energy and angular distributions at high energies from the ionization of molecules by intense 800 nm lasers are used. (Reprinted from M. Okunishi et al., *J. Electron Spectrosc. Relat. Phenom.*, **195**, 313 (2014) [9]. Copyrighted by Elsevier.)

The calculation of elastic DCS between an electron and a molecular ion is much more complicated but it is doable with reasonable accuracy for small molecules. The same computer package developed for carrying out electron-molecule collisions has been modified to obtain the DCS defined in Equation 4.21 for electron collisions with molecular ions [9]. The calculations employed the so-called *Schwinger configuration interaction method*, where electron correlation has been included in both the ground state and the ionic final states. In other words, they are full many-body calculations that have been found essential in order to adequately describe low-energy electron collisions. In the calculation, the interatomic positions are set at the equilibrium geometry. While the DCS can be calculated accurately, it is difficult to extract target geometry structure information because the scattering theory is too complicated when the returning energy of the HATI electron is small, as is the case for returning electrons generated by 800 nm lasers.

4.4.2 Conventional Electron Diffraction and the Independent Atom Model

Electron-molecule collisions are well-known conventional tools for studying the structure of molecules. Still, while the elastic DCS depends critically on the molecular structure, exactly what structure information can be retrieved from the DCS, especially for collisions by low-energy electrons? For example, elementary information like the bond length and bond angle between atomic constituents is not easily extracted from the experimental DCS. In fact, extracting target structure from the scattering data is an example of inverse scattering theory. Inverse scattering theory is only possible when the scattering theory is simple. The theory of collisions of high-energy electrons with molecules is simple since the scattering amplitude can be calculated using the first Born approximation. Thus conventional gas-phase electron diffraction (GED) has been used for more than half a century to determine the bond lengths between pairs of atoms in the molecule. Here, a well-collimated electron beam with energies on the order of hundred keVs is impinged on a gas cell or gas jet. The angular distribution of the forwardly scattered electrons – typically up to 10 to 20 degrees – are collected experimentally. In most GED experiments, the molecules are isotropically distributed. From the measured DCS (or the diffraction image), molecular structure information in the form of radial distance between each pair of atoms can be retrieved (see [Hargittai, Hargittai]).

The theoretical tool used for extracting bond lengths in a molecule in GED is based on the independent atom model (IAM). In IAM, a molecule is modeled as a collection of atoms fixed in space. These atoms do not interact and there is no consideration of chemical bonding nor molecular orbital. The potential seen by the incident electron is taken to be the sum of potentials from all the atoms. Let $\boldsymbol{R}_i$ be the position vector of atom i with respect to a fixed coordinate system. The interaction potential of each atom with respect to the incident electron is represented by a short-range potential. The scattering amplitude by the incident electron on atom i is represented by a complex scattering amplitude f_i. According to the IAM, the total scattering amplitude for a molecule fixed in space is given by

$$F(k,\theta,\varphi;\Omega_L) = \sum_i f_i e^{i\boldsymbol{q}\cdot\boldsymbol{R}_i}. \tag{4.22}$$

The IAM model is an extension of the first-order Born (B1) approximation given by Equation 1.49. Here, Ω_L denotes the orientation/alignment angles of the molecule and $\boldsymbol{q} = \boldsymbol{k} - \boldsymbol{k}_0$ is the momentum transfer. The incident electron momentum $\boldsymbol{k}_0$ is taken to be along the z-axis, and $\boldsymbol{k} = (k,\theta,\varphi)$ is the momentum of the scattered electrons. The DCS is given by

$$I_{\text{tot}}(\theta,\varphi;\Omega_L) = I_A + \sum_{i\neq j} f_i f_j^* e^{i\boldsymbol{q}\cdot\boldsymbol{R}_{ij}}, \tag{4.23}$$

where $\boldsymbol{R}_{ij} = \boldsymbol{R}_i - \boldsymbol{R}_j$, and $I_A = \sum_i |f_i|^2$ is the incoherent sum of scattering cross-sections from all the atoms in the molecule. The last term of Equation 4.23 is called the *molecular interference term*. Note that the bond length R_{ij} appears explicitly as a factor in the phase. For a sufficiently large q, the second term introduces modulation into the otherwise smooth atomic term I_A. The modulation in the molecular DCS on top of the smooth atomic background is sensitive to R_{ij}. It is analogous to diffraction in a double-slit experiment, which is sensitive to the distance between the two slits. For a sample of randomly distributed molecules, the rotational average of Equation 4.23 gives

$$\langle I_{\text{tot}}\rangle(\theta) = I_A + \sum_{i\neq j} f_i f_j^* \frac{\sin(qR_{ij})}{qR_{ij}} \tag{4.24}$$

in which q and R_{ij} are the moduli of $\boldsymbol{q}$ and $\boldsymbol{R}_{ij}$, respectively. The modulation is more easily visualized by defining a molecular contrast factor (MCF) γ

$$\gamma = \frac{1}{I_A} \sum_{i \neq j} f_i f_j^* \frac{\sin(qR_{ij})}{qR_{ij}}. \tag{4.25}$$

The magnitude and the number of oscillations in the MCF determine the quality of the retrieved molecular-structure parameters. In fact, from the oscillation in the molecular interference term, the interatomic distance R_{ij} can be obtained using the sine Fourier transform. Clearly, to observe the modulation in the DCS, it is desirable that qR_{ij} for each pair of atoms be at least on the order of π. Since R_{ij} is on the order of one angstrom, the desirable q should be at least on the order of four (1/Å). To use the inverse Fourier transform to obtain R_{ij}, the range of q is taken to be from about five to 20 (1/Å). If one takes the incident electron energy at 100 keV, the DCS will be taken at from about 1.5° to 15°. In this angular range, the electron is scattered off from each atomic center at a distance less than 0.01 au. At such close collisions, the scattering amplitude is proportional to the nuclear charge of the atom. Thus, in GED, only high-Z atomic constituents usually contribute to the diffraction image. In other words, only bond lengths involving high-Z atoms are retrieved with the GED method.

4.4.3 Need of New Tools for Dynamic Chemical Imaging

One of the grand research goals in the twenty-first century is to image the evolution of a chemical reaction or a biological function in real time. To achieve such a goal, new tools that are capable of probing molecules at the few-to-tens femtosecond timescale and spatial resolution of sub-angstroms are needed. While conventional X-ray and electron diffractions are capable of achieving a few picometers of spatial resolution, their temporal resolution is on the order of hundreds of picoseconds. An obvious route to realize these new dynamic imaging tools is to shorten X-ray and electron beams to a few femtoseconds. Indeed, few-femtosecond, intense X-ray free electron lasers are now available [10, 11] and many exciting results have already been obtained. However, these results have been obtained only at costly, large-scale national facilities, and not widely available for general chemistry and physics laboratories. Similar efforts have been made on ultrafast electron diffraction (UED). While electron pulses have been reduced from tens or hundreds of picoseconds to a few hundred femtoseconds, achieving temporal resolution toward tens of femtoseconds is not yet possible for gas-phase diffraction (see Section 4.5.7). Electron diffraction is the preferred tool for imaging gas-phase molecules because its cross-section is typically six orders higher than X-ray diffraction. Since high-energy electrons (above hundreds of keVs) interact weakly with light atoms, UED with high-energy electrons is not able to probe chemical processes that involve light atoms. Thus, it is not suitable for probing important dynamic processes such as proton migration, roaming, isomerization, and ring opening of small molecules.

Section 4.3 demonstrates that angle- and energy-resolved photoelectron momentum spectra generated by intense lasers can be used to extract field-free, electron–ion elastic scattering DCSs. The extracted DCS encodes the structure information. Since laser pulses with duration from a few tens to a few femtoseconds are already available today in many intense-laser laboratories if one can identify the conditions where accurate interatomic distances of a molecule can be extracted from such DCS, then laser-induced electron diffraction (LIED) would offer an alternative powerful table-top tools for dynamic imaging of molecules.

4.4.4 LIED: What Are the Requirements?

What are the conditions of lasers such that molecular bond lengths can be accurately retrieved from the photoelectron momentum distributions? This question was first carefully examined in Xu et al. [12]. The maximal kinetic energy of laser-induced rescattering electrons is given by $3.17U_P$. To increase the returning electron energy without excessive ionization of the molecule, one can increase the wavelength of the laser. For example, with a 6 μm laser at an intensity of 1×10^{14} W/cm^2, the maximal that return energy can reach is about 1 keV. This energy is still much lower than the hundred keV electrons used in GED or UED. As discussed in Chapter 5, the flux of the returning electrons scales very unfavorably with wavelength of the laser as λ^{-5} or λ^{-6}. It is unrealistic of to expect that there will be enough returning rescattering electrons at keV energies for the generation of diffraction images. Thus, LIED does not appear to be promising.

But does one really need to use keV electrons for imaging molecules? High-energy electrons of tens to hundred keVs have the advantage of shorter de Broglie wavelengths for better spatial resolution, and they are easier to guide with electron optics. Furthermore, the IAM is expected to work well at high-collision energies such that the inverse Fourier transform method can be used to retrieve the bond distances from the DCS (or the diffraction image). However, in Equation (4.24) the modulation term is $\sin(qR_{ij})$, where $q = 2k\sin(\theta/2)$. To reach large q, one can choose large k and small θ. This is used in GED or UED. However, in LIED, one can use small k but large θ. Since HATI electrons are those that were backscattered by the target ion, they have large θ and thus k can be smaller. In collision physics, large q means close collisions at small impact parameters. The HATI electrons are from returning electrons that have encountered close collisions with atoms in the molecule. These collisions can reach the same range of momentum transfer as in GED and UED yet at much lower electron energies. Since a 150 eV electron has a de Broglie wavelength of one angstrom, it is speculated that electron energy as low as 100 eV may be possible for LIED imaging if the images are taken at large scattering angles.

To quantify this analysis, one has to first test how well the IAM can describe experimental electron-molecular, ion-scattering cross-sections at large angles around 100 eV; otherwise, a new method of retrieving the bond lengths from the diffraction images has to be developed. Unfortunately, such experimental data is not available in the literature since collisions with molecular ions can only be performed with molecular beams that have severely reduced target densities compared to experiments in the gas cell or jet. However, scattering at large angles (or small impact parameters) is dominated by the strong force near the nucleus. The effect of charge state of the target and the chemical bonding are all expected to be insignificant. Thus, one can use collision data between electrons and neutral

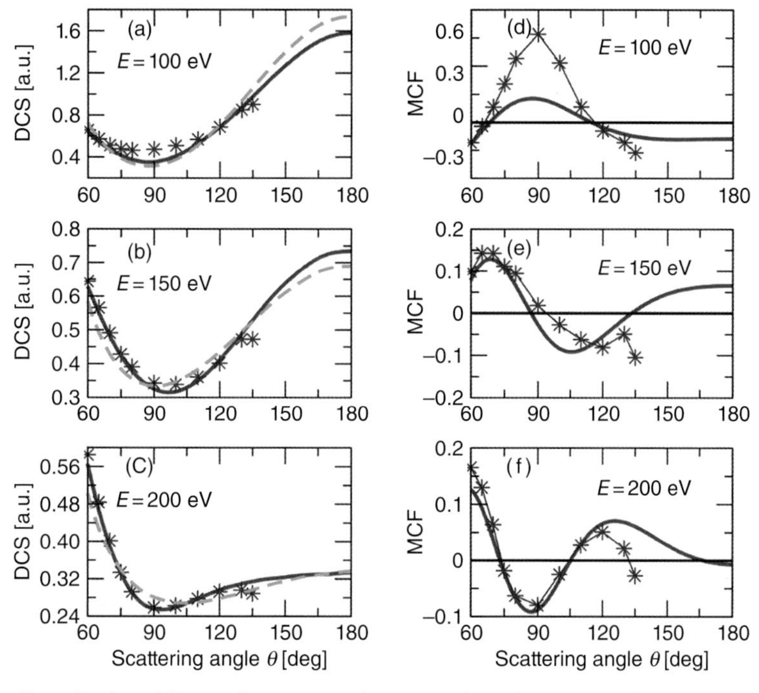

Figure 4.13 Left column: The e–N_2 elastic DCS at incident energies of 100, 150, and 200 eV. Experimental data is marked by symbols. The IAM model is given by the red curve and the atomic term is by the green dashed curve. Note that the molecular DCS modulates about the atomic DCS. Right column: The MCF. The extracted bond lengths from the different energies are shown in Table 4.1. (Reprinted from Junliang Xu et al., *J. Phys.: Conf. Series*, **288**, 012017 (2011) [13].)

molecules around 100 eV to test the validity of the IAM model for electron collisions with molecular cations.

With this theoretical reasoning, it becomes apparent that one can check the validity of the IAM model for the DCS at large angles near 100 eV for electron-molecular ion collision using data from electron-molecule collisions. Such electron-molecule collision data have been reported since the 1970s even though data at large angles is more scarce and errors tend to be larger due to the limited angular range of electrostatic detectors. Figure 4.13 shows the experimental DCSs at 100, 150, and 200 eV for electron–N_2 collisions and the prediction of the IAM model using the known distance between the two nitrogen atoms. The DCS was quite well described by the IAM with the exception of the last two data points at large angles. Indeed, the IAM result shows clear modulation about the atomic term. The modulation is the molecular interference term that contains information on the N–N bond length. By rewriting the molecular interference in terms of the MCF (right column of Figure 4.13) and plotting against the scattering angles, one can see a clear discrepancy between the MCF derived from the experiment and that derived from the IAM. This difference is mostly in the amplitude of the modulation (or modulation depth); the positions of the maximum and minimum agree quite well. To test how accurately the bond length can be extracted in spite of the discrepancies, one can take the N–N distance as an

Table 4.1 Retrieved N–N bond length based on the DCS of e-N_2 collisions using the IAM model

Energy (eV)	N–N bond length (Å)
100	1.161 (+6.5%)
150	1.039 (–4.7%)
200	1.091 (+0.1%)
Experiment	1.09

unknown and use IAM to retrieve the N–N bond distance from the experimental data. The results are displayed in Table 4.1. This table shows that the bond length extracted from the data has about 5% error. While this is not as good as the typical 1% error that can be achieved in GED, it does show that, with an electron energy on the order of about 100 eV, bond length can be retrieved to better than 0.05 Å. For dynamic systems one is looking for large displacement from the equilibrium distance, so a spatial resolution of 0.1 Å is quite adequate. Using DCS extracted from laser-induced HATI electron spectra, one has the immediate advantage of a temporal resolution of a few tens to a few femtoseconds from the femtosecond lasers that are available today.

For rescattering electrons to reach energies of about 100 eV or higher, one would have to use mid-infrared lasers to take advantage of the λ^2 scaling of the ponderomotive energy. This rules out titanium–sapphire lasers for LIED experiments. Note that LIED has been used since the 1990s for somewhat different situations using 800 nm lasers [14, 15]. Here, LIED is placed at the level of GED or UED, where laser-induced electrons are used for generating diffraction images and from which interatomic distances can be directly retrieved using the IAM. More importantly, the goal of LIED is to image a dynamic system with tens of femtoseconds of temporal and sub-angstrom spatial resolution.

4.5 Experimental Demonstration of Dynamic Imaging: LIED and Other Methods

4.5.1 N_2 and O_2

The first LIED experiment that demonstrated sub-angstrom spatial resolution was reported by Blaga et al. [16]. They generated HATI spectra using 1.7 to 2.3 μm lasers on randomly distributed N_2 and O_2 molecules at intensities of a few times of 10^{14} W/cm^2. A typical 2D electron momentum spectrum is shown in Figure 4.14(a). Following the QRS theory, the electron–ion elastic DCS was retrieved (Figure 4.14(b) at $p_r = 2.71$ au). For LIED, since the tunnel-ionization rate depends on the alignment angle of the molecule with respect to the laser-polarization direction, for isotropically distributed molecules the DCS is given by

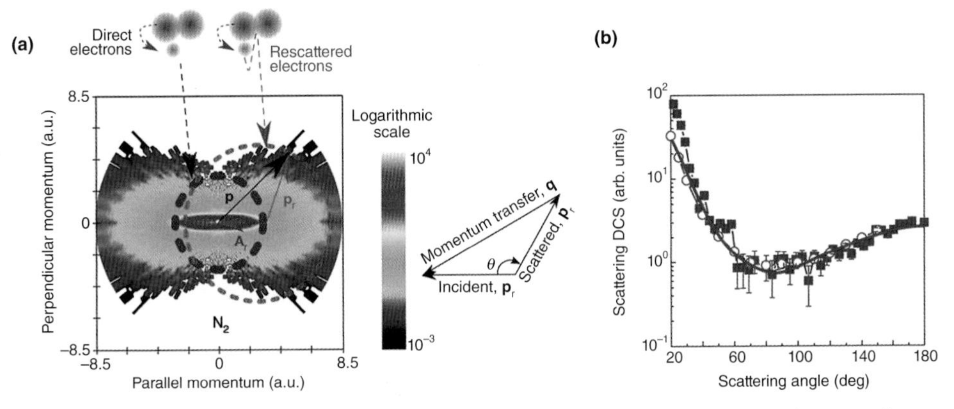

Figure 4.14 (a) The typical 2D electron momentum spectra of N_2 obtained by Blaga et al. [16] showing regions of rescattered electrons and low-energy direct electrons. The DCSs are extracted from the dotted circle (magenta dashed circle) where the electron momentum measured in the laboratory is shifted by the vector potential of the laser at the time of recollision. (b) The extracted DCS from the experimental data in (a) (blue data points), the DCS from experimental e-N_2 collisions (green circles), and the atomic cross-section calculated using the IAM model (red line). At small angles, the DCS is contaminated by direct electrons and is not used in LIED. (Reprinted from C. I. Blaga et al., *Nature*, **483**, 194 (2012). [16])

$$\langle I_{\text{tot}}\rangle(\theta) = \left(\sum_i |f_i|^2\right)\int N(\Omega_L)d\Omega_L + \sum_{i\neq j} f_i f_j^* \int e^{i\boldsymbol{q}\cdot\boldsymbol{R}_{ij}} N(\Omega_L)d\Omega_L, \tag{4.26}$$

where $N(\Omega_L)$ is the tunneling ionization rate calculated using the MO-ADK theory. A corresponding MCF is defined by

$$\gamma = \frac{\sum_{i\neq j} f_i f_j^* \int e^{i\boldsymbol{q}\cdot\boldsymbol{R}_{ij}} N(\Omega_L)d\Omega_L}{(\sum_i |f_i|^2)\int N(\Omega_L)d\Omega_L}. \tag{4.27}$$

Figure 4.15 shows the analysis of the experimental MCFs for N_2 and O_2 versus the momentum transfer, obtained from the photoelectron momentum spectra generated using 2.0 μm and 2.3 μm lasers. Additional data was also taken at 1.7 μm. By fitting the experimental MCF to the IAM theory, the bond length was retrieved (the details of the fitting are explained in the caption of Figure 4.15). According to the rescattering model, it takes close to one optical cycle for the electron to return to the ion core. By changing the wavelength of the driving laser, the return time is varied. Analysis of the experimental data for N_2 showed that the retrieved bond length is 1.15 Å for the 2.0 μm laser. The neutral and the cation bond lengths in N_2 are known to be 1.10 and 1.12 Å, respectively. At the collision energy of about 100 eV, the IAM is expected to have error of about 0.05 Å according to the analysis in Section 4.4.4. Thus the retrieved bond length cannot be used to test whether there is a rearrangement of the N–N bond length between the time of ionization and recollision. Note that in the LIED experiments, the bond length can be retrieved using a range of collision energies to check the consistency of the retrieved result.

For O_2 molecules, the retrieved O–O bond length is 1.10 Å. The neutral's bond length for O_2 is 1.21 Å and that for the cation is 1.10 Å. The retrieved O–O distance of 1.10 Å was interpreted to mean that the O–O distance has shrunk by 0.1 Å. Since it takes about five

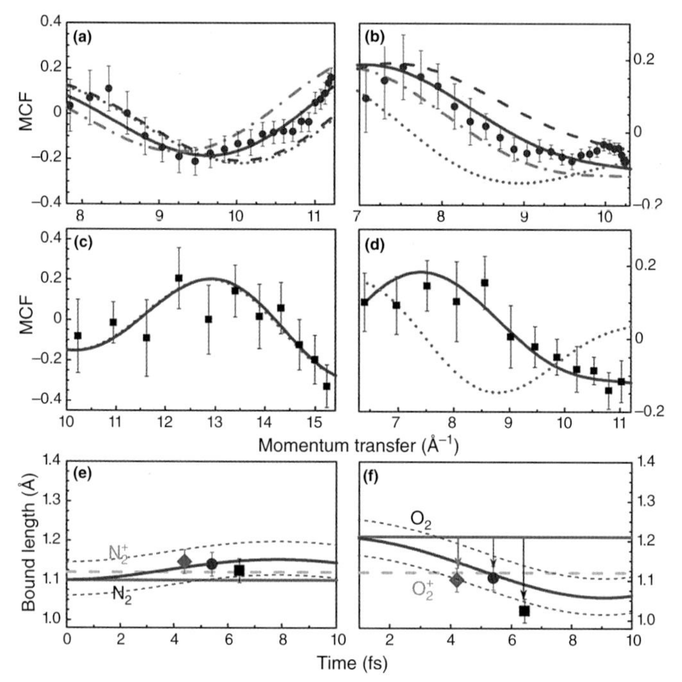

Figure 4.15 LIED for unaligned N_2 (left column) and O_2 (right column) molecules. a-d, MCF extracted from experimental data with error bars, expressed versus the momentum transfer. (a) and (c), for N_2. The data was taken with 2.0 μm and 2.3 μm lasers and with the momentum $p_r = 2.9$ and 4.11 au, respectively. (b) and (d), the same for O_2 with $p_r = 2.91$ and 2.97 au, respectively. The best fit in (a) and (c) are shown as solid lines, the dashed magenta and dash-dotted orange lines are the MCFs using bond lengths that deviate from the best fit by −5pm and +5pm, respectively. In (e) and (f), the bond length data points extracted from the LIED measurements at each wavelength (squares, 2.3 μm; circles, 2.0 μm; diamonds, 1.7 μm) are shown. The known bond length of the neutral (solid) and the ion (dashed) are shown as horizontal lines. The red line gives the mean bond length versus time after tunnel ionization computed under the Franck-Condon approximation. The data points are calculated at the time of recollision after tunnel ionization for each wavelength. The two dotted lines give the range of bond length ±5pm calculated from the FC model. (Reprinted from C. I. Blaga et al., *Nature*, **483**, 194 (2012) [16].)

femtoseconds from the time the electron is removed from the molecule until it returns to recollide with the oxygen cation, this result provided the first direct experimental evidence of sub-angstrom bond length change in about five femtoseconds after ionization.

These conclusions are consistent with experiments using different laser wavelengths and different intensities. Assuming that tunnel ionization is instantaneous, one may use the Franck–Condon principle to estimate the change of bond distance versus time, shown by the red solid lines in Figure 4.15(e,f). The retrieved results from the experiment are consistent with the estimate. In particular, the vibration period of O_2 is about 18 fs, and thus it takes about 5 fs for the O–O distance to change from 1.21 to 1.10 Å. With a longer wavelength, it would take a longer time for the electron to return to probe even smaller bond lengths. The retrieved O–O distance shown in Figure 4.15(f) is consistent with this interpretation when a 2.3 μm laser was used to image the O_2 molecule.

4.5.2 C_2H_2 at Equilibrium

The goal of LIED is to be able to image the structure of more complicated polyatomic molecules. The binding energy of most polyatomic molecules is below 10 eV, less than that of N_2 and O_2. To avoid multiple ionization a lower laser intensity should be used. But reaching returning electron energies of the order of 100 eV would require longer-wavelength lasers. This has been made possible with the unique 3.1 μm laser from the Institute of Photonic Sciences (ICFO) in Barcelona. Their laser system is optical parametric chirped pulse amplification (OPCPA) based with 160 kHz repetition rate. Since the fraction of returning electrons decreases quickly with the wavelength of the driving laser, the two orders of increase in repetition rates is essential. LIED experiments with ICFO's laser system have been carried out on C_2H_2 where the molecules are aligned either parallel or perpendicular to the probing laser's polarization [17]. Figure 4.16 shows the results using returning electrons with different energies for molecules that are parallel aligned or perpendicular aligned. Both C–C and C–H bond lengths have been retrieved. For the C–C bond distance, the retrieved value from the perpendicularly aligned data is 1.24 ± 0.04 Å, and that from the aligned data is 1.26 ± 0.04 Å. For the C–H bond length the extracted value from the perpendicularly aligned data is 1.10 ± 0.03 Å, and that from the parallel aligned data is 1.05 ± 0.03 Å. They are to be compared to the known values of 1.25 Å and 1.10 Å, respectively. Note that the bond lengths in LIED can be retrieved using a range of rescattering energies to help check the consistency of the retrieved data.

It is important to recognize that light atoms such as atomic hydrogen can be imaged using LIED. This is unique since LIED used low-energy electrons of several tens of eVs

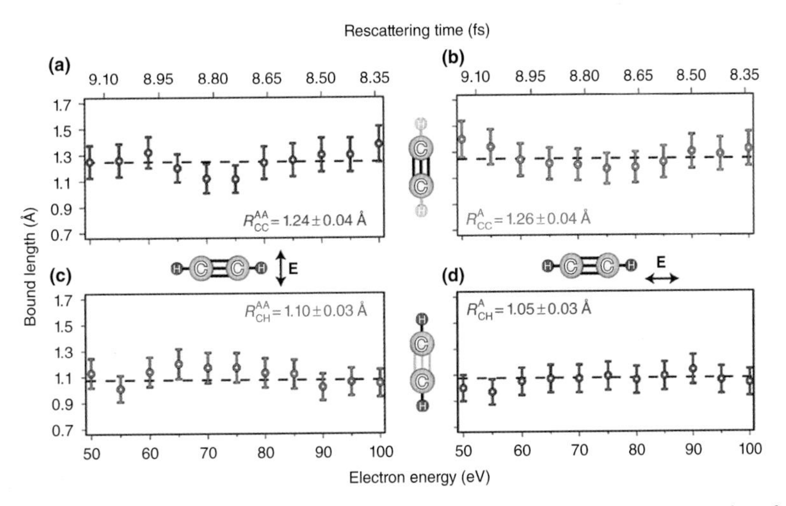

Figure 4.16 Extracted C–C and C–H bond lengths in acetylene using LIED. The bond-length estimates are presented as a function of the scattering electron energy (bottom label) and the rescattering time (top label). The dashed black lines are the expected distance for the cation. The best horizontal fits for each bond are displayed in each panel. Molecules are either perpendicularly (left) or parallel aligned with respect to the polarization of the probing laser pulse. (Reprinted from M. G. Pullen et al., *Nature Commun.*, **6**, 7262 (2015) [17].)

for imaging small molecules. In conventional X-ray or electron diffraction, the small atoms are invisible. Those methods use high-energy X-rays or electrons. Light atoms like atomic hydrogen are hardly scattered, i.e., they do not contribute much to the scattering so it is not possible to retrieve them from the scattering data. In principle, LIED is preferable for imaging dynamics like proton transfer, roaming, or isomerization that involve the motion of light atoms in the molecule.

4.5.3 Bond Breaking in C_2H_2

The aim of LIED and other dynamic imaging methods under development is to probe the conformational change of a molecule. This is ideally carried out using a pump–probe arrangement where the pump creates a dynamic system. A probe pulse is then used at a delayed time to "observe" how the system evolves. Ideally, a probe pulse should be weak such that the dynamic system is not significantly modified by it. Such experiments are clearly difficult to carry out if good temporal resolution is needed, especially for electron-diffraction experiments where the cross-sections are already small even for molecules in the ground state. In spite of this, in a recent experiment Wolter et al. [18] reported the bond breaking of di-ionized acetylene using the setup at ICFO. Here, how the experiment was done and how the data were analyzed are given. Also discussed is the typical challenge of complexity one has to face when probing a polyatomic molecule with an intense laser pulse that results in multiple fragmentation.

The experimental setup is similar to the one used for the LIED experiment in the previous subsection. The C_2H_2 molecules are impulsively aligned with the 1,700 nm laser and probed with the 3,100 nm laser at the time intervals where molecules are either parallel or perpendicularly aligned with a higher-intensity laser of about 65 TW/cm^2. Figure 4.17(a) shows the typical time-of-flight spectrum of the molecular fragments. Besides the dominant $C_2H_2^+$ cations from the removal of one electron studied in Section 4.5.2, the prominent peak of C_2H^+ is clearly seen. These ions are understood to originate from the dissociation process $C_2H_2^{2+} \rightarrow H^+ + C_2H^+$. Figure 4.17(c) shows the ion yields of $C_2H_2^+$, $C_2H_2^{2+}$, and H^+. The knee structure in the nonsequential, double-ionization yield curve is seen clearly near 40 TW/cm^2, beyond which sequential double ionization grows quickly. By expressing the ratios of double versus single ionization, the ratios $C_2H_2^{2+}/C_2H_2^+$ and $H^+/C_2H_2^+$ become nearly equal at 65 TW/cm^2 (see Figure 4.17(d)). Such large ratios can be attributed to mean that fast dissociation of $C_2H_2^{2+} \rightarrow H^+ + C_2H^+$ occurs. At an intensity of 65 TW/cm^2 and above, double ionization occurs sequentially. This can be seen in Figure 4.17(e) where the ion-momentum spectrum shows a transition from the double hump structure in the nonsequential double ionization to the single hump for sequential double ionization.

To understand processes leading to the proton ejection from $C_2H_2^{2+}$, simplified potential curves versus the C H distance are shown in Figure 4.17(f). Two sequential pathways for leading the molecule from the ground state to $C_2H_2^{2+}$ are indicated. These two steps include a first ionization to the highest-energy occupied molecular orbital (HOMO) or HOMO-1 states of the cations. Further absorption of photons from the cation can reach the states highlighted in the potential curves of the dication. In Figure 4.17(f) the two dication potential curves in red can dissociate quickly; they appear in the ion–ion coincidence spectra in

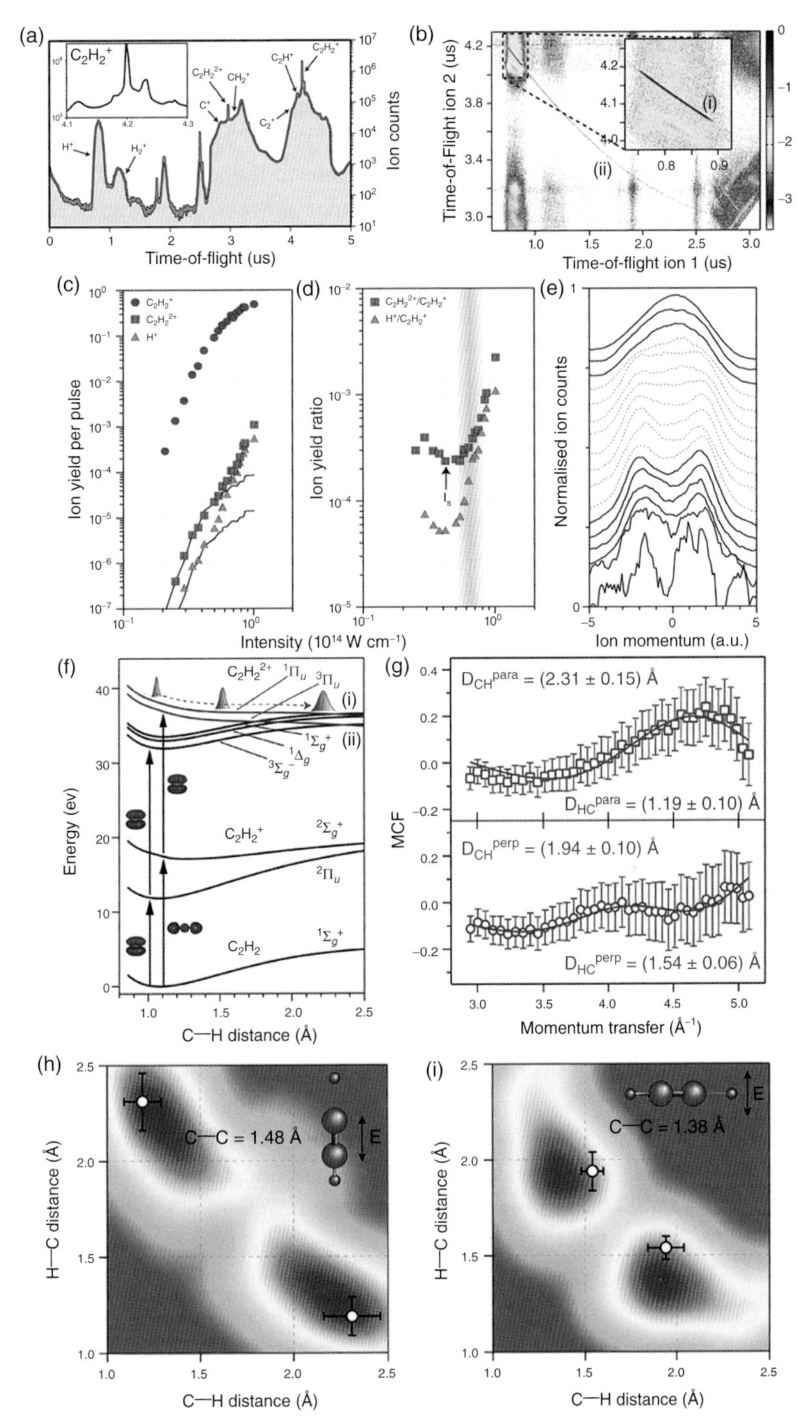

Figure 4.17 Supplementary experimental data and theoretical calculations that lead to the observation and interpretation of the bond breaking of acetylene dication using the LIED method with a 3,100 nm laser. Details are given in the text. (From B. Wolter et al., *Science*, **354**, 308 (2016) [18]. Reprinted with permission from AAAS.)

Figure 4.17(b) (and the inset) as red lines, and are also marked by (i). This is the ultrafast dissociation that will be probed with the LIED method. The other pathway that leads to the lower potential curves of the dication (see the black potential curves in Figure 4.17(f)), results in slow-decaying meta-stable states. Their dissociation is seen as the slow decay curve marked by (ii) in the ion–ion time of flight-coincidence spectra in Figure 4.17(b).

To probe the dissociation-dynamics process involving $C_2H_2^{2+} \rightarrow H^+ + C_2H^+$ that leads to fast proton loss with the LIED method, the momentum spectra of electrons in coincidence with the fast proton channel are measured. This guarantees that only these electrons bear the structure information of the fast dissociation of the dication. Following the same procedure to obtain the MCF from the electron spectra, individual bond lengths are extracted by calculating the MCF patterns for a wide range of possible parameters and comparing with the measured one. The best bond lengths, C–C, C–H, and H–C, are obtained from the minimum of the χ^2 value. The solution space does not assume any relation among the three parameters.

Two different solutions were obtained. For the parallel case, a C–C bond length of 1.48±0.11 Å was retrieved, which is 23% greater than the 1.20 Å equilibrium C–C bond length. The associated C–H and H–C bond lengths are 2.31 ± 0.15 Å and 1.19 ± 0.10 Å (or 118% and 12% elongated), respectively, relative to the equilibrium value of 1.06 Å. The fact that one of the protons in the C–H bond has more than doubled its distance is a clear signature that bond cleavage is happening. In the perpendicular geometry, the C–C bond length retrieved is 1.38 ± 0.06 Å, and the C–H are 1.94 ± 0.10 Å and 1.54 ± 0.06 Å, respectively. The smaller elongation of the C–H bond distance is understandable since, in this perpendicular case, the laser field is pulling the molecule apart almost symmetrically. Figure 4.17(g) compares the quality of the MCF fitting between the experimental values and the MCF with the fitted parameters calculated using the IAM.

To make sense of these extracted, elongated bond lengths from the LIED experiment, it is beneficial to carry out quantum simulations to check whether the retrieved bond lengths are "reasonable." For simplicity, the simulation was carried out for the present breakup channel. Without going into the details of the simulation that were given in Wolter et al. [18], it is interesting to show just the results. Figures 4.17(h,i) show the C–H and H–C bond-distance distributions for a fixed C–C distance (from the retrieved value) for both the parallel and perpendicular alignments obtained from the simulation. Overlaid is the experimentally extracted data (white data points with error bars). The retrieved C–H and H–C lengths lie inside the theoretically calculated distributions well. Since the return time for the electron to rescatter within each cycle of a 3.1 μm laser after tunnel ionization is about 9 fs, this LIED experiment is understood to image the bond distances of a dissociating acetylene dication in about 9 fs; thus providing the first frame of a molecular movie of the dissociation of a molecular ion with sub-angstrom spatial resolution.

4.5.4 Benzene Molecule

The laser used in the experiment at ICFO is not commercially available. On the other hand, for hydrocarbon molecules (as demonstrated in acetylene), bond lengths can be accurately retrieved with electron energy as low as close to 60 eV. In a recent experiment, Ito et al.

[19] used a standard 1.65 μm, 1 kHz, 100 fs laser to extract the C–C and C–H bond lengths of a benzene molecule. The 1.65 μm laser is obtained from OPA. By examining the MCFs for returning electron momenta at 1.9, 2.0, and 2.1 au, from the experimental HATI spectra it was possible to extract the C–C and C–H bond lengths at (1.52, 1.13), (1.47, 1.14), and (1.43, 1.26) Å at the three momenta, respectively. These results are to be compared to the known values (1.39, 1.09) Å. Thus the extracted bond lengths have an error of about 10%. In this example, the MCF fit does not look that good, but the retrieved bond lengths are acceptable.

4.5.5 Complementary Retrieving Method in LIED

In LIED, the electron-diffraction images are generated from a broadband wave packet. In retrieving the molecular structure in Sections 4.5.1–4.5.4, the DCS for a fixed returning electron energy was used. By using different electron energies, the retrieved structure can be compared to check for consistency (see, for example, Figure 4.16). An alternative approach for retrieving the structure is to use the DCS for a fixed scattering angle in the laboratory by varying the electron energy for electrons that are backscattered at 180°. By comparing the electron spectra of N_2 with Ar in Xu et al. [20], the interference signal in N_2 was extracted versus the momentum transfer. Figure 4.18(a) shows the interference term of

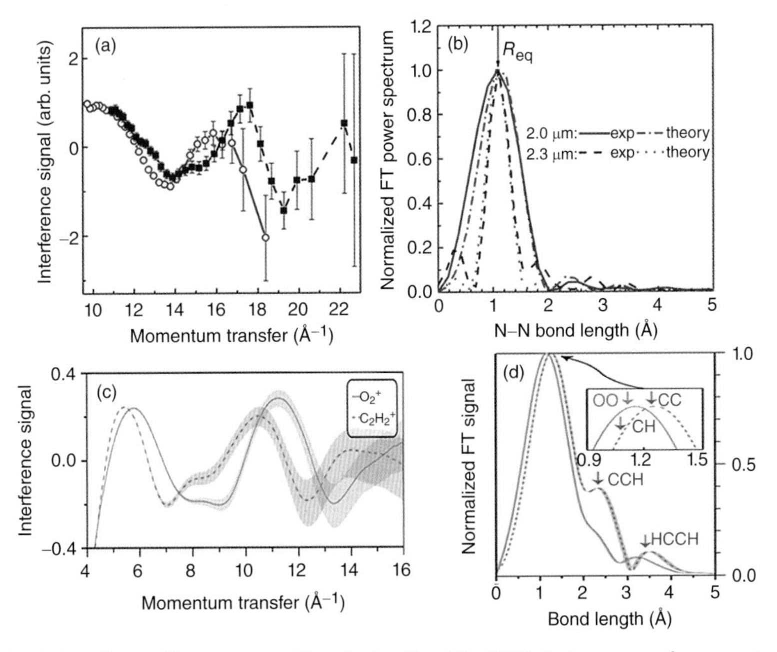

Figure 4.18 (a) The molecular interference fringes extracted from backscattered (by 180°) electrons versus the momentum transfer from nitrogen molecules with a 2 μm laser (blue circles) and a 2.3 μm laser. (b) The bond length of N_2 can be directly retrieved by taking the Fourier transform. From [20]. (c,d) Same as (a,b) but for O_2 and C_2H_2 molecules using a 3.1 μm laser. The electron spectra are obtained in coincidence with the singly charged ions to remove background. (Reprinted from Junliang Xu et al., *Nat. Commun.*, **5**, 4635 (2014) [20], and M. G. Pullen et al., *Nat. Commun.*, **7**, 11922 (2016) [21].)

N_2 versus the momentum transfer for photoelectrons extracted along the laser-polarization axis using 2.0 μm and 2.3 μm lasers, respectively. The momentum transfer covers a much larger range close to the range used in standard gas-phase diffraction experiments, and nearly a full oscillation can be observed. Instead of using the IAM and the fitting method to retrieve the bond length, a direct Fourier transform was able to extract the bond length of N_2 as illustrated in Figure 4.18(b). The peak position thus retrieved accurately matches the known equilibrium N–N bond length R_{eq} within 0.01 Å. The width of the retrieved radial distribution is reduced when a longer-wavelength laser was used because the range of momentum transfer was greater (see Figure 4.18(a)). When using a Fourier transform for inversion in a finite spectral range, the width of the retrieved radial distribution is broad and additional unphysical small subpeak(s) may appear (see the small peak at R near 2.5 Å in Figure 18(b)). In Xu et al. [20], a similar study was carried out for the O_2 and Kr pair, but it was unable to extract the bond length of O_2. It was first speculated that this failure was due to the effect of orbital symmetry since the valence orbital of O_2 has π_g symmetry, which is unfavorable for the initial tunnel ionization in comparison to N_2. In a subsequent publication, Pullen et al. [21] attributed the failure of the Fourier transform method for O_2 in Xu et al. [20] to the inaccuracy of the DCS data. It was found that if the electrons are measured in coincidence with the O_2^+ ions, clear oscillation in the interference spectrum can be observed. Figure 4.18(c) shows the interference signals for O_2^+ and $C_2H_2^+$ where the electrons have been measured in coincidence with the cations for electrons along the polarization axis. Using the Fourier transform method, the retrieved O–O bond length was 1.12 Å, which is consistent with the one reported in Blaga et al. [16]. For $C_2H_2^+$, from Figure 4.18(d), the C–C bond length of 1.25 Å was obtained, but the C–H bond length lay within the first broad peak and was not resolvable. On the other hand, Figure 4.18(d) also reveals two additional peaks: one at 2.33 Å and another at 3.51 Å. The normalized amplitude for the second peak is 0.40 and that for the third peak is 0.11. The second peak may be attributed to C–C–H (and H–C–C) peak (the distance between the two outermost atoms). The third peak is much weaker so it may or may not be related to the H–C–C–H peak even though the distance is quite close. The advantage of the Fourier transform–based retrieval method is that it does not depend on the theoretical model in the analysis. On the other hand, the width of each Fourier-component peak tends to have a larger width. Thus, the peaks may overlap and some unphysical small peaks may appear.

4.5.6 Retrieval of 2D Structure from 1D-Aligned Polyatomic Molecules

The basic retrieval method used in LIED is very similar to the one used in the electron-diffraction community. For simple molecules studied by the LIED so far, the bond distances were retrieved by the fitting procedure. However, it has been shown that the Fourier transform method (Section 4.5.5) can also work, although the width of the radial distribution of the Fourier peak may prevent the distinction of some bond lengths that overlap or nearly overlap, as seen, for example, in C–C and C–H bonds in Figure 4.18. Until recently, electron-diffraction experiments have been carried out for isotropically distributed molecules. Recently, electron-diffraction experiments on aligned molecules have been reported in Hensley et al. [23]. In this experiment, CF_3I molecules are impulsively aligned by a laser. The partially 1D-aligned (along the C–I axis) ensemble of molecules is then exposed to 25 keV electrons to obtain diffraction images, with the electrons imping on the molecules along a direction perpendicular to the C–I axis. An iterative method was used to retrieve the 2D structure information (including bond angle) of the molecule.

The retrieval method used in Hensley et al. [23] has been modified in Yu et al. [22] for LIED experiments from partially 1D-aligned polyatomic molecules to allow the retrieval of 2D structure information of molecules. Two methods were used. One is similar to that used in Hensley et al. [23], where the axis of the aligned molecules is perpendicular to the polarization axis of the laser. The diffraction image to be inverted is from a fixed scattering energy of the electron. Another method is to choose the molecular-alignment axis parallel to the laser-polarization direction. In this second method, due to the cylindrical symmetry, the 2D electron spectrum is taken from the (E, θ) plane, where E is the scattering energy of the returning electron, and θ is the scattering angle. This second method is unique in LIED since it requires the broadband electron pulses from the returning electrons.

For a perfectly 1D-aligned ensemble of molecules, the position of each atom in a molecule is characterized by (z, r) where z is the distance along the axis from a fixed point on the axis, and r is the perpendicular distance. If the 2D diffraction image is expressed as a function of (q_z, q_r), the distribution of (z, r) can be obtained by a Fourier transform. Since molecules cannot be perfectly 1D aligned, an iterative method is first used to extract a diffraction image for perfectly 1D-aligned molecules from the diffraction image of partially 1D-aligned molecules. Clearly, the effectiveness of the method depends on the degree of alignment needed in order to obtain accurate results. Figure 4.19 summarizes how the retrieval method works for partially 1D-aligned $ClCF_3$ molecules. In Figure 4.19(a), the horizontal axis is the direction of laser polarization or the direction of the scattering electrons. Figure 4.19(b) shows three angular distributions for three alignments. Using iterative methods, the 1D, perfectly aligned diffraction images are obtained on the (q_z, q_r) plane (see the left column for the three angular distributions in Figure 4.19). The three pictures look quite close to each other, which shows that the retrieved diffraction images from the three different alignment distributions are essentially the same. Note that part of the 2D momentum-transfer plane is empty because the data was transformed from the (E, θ) plane to the (q_z, q_r) plane. The 2D bond lengths obtained from the Fourier transform are given to the right of each of the figures. The white dots in these figures indicate the input position of the bond (given by its length parallel or perpendicular to the axis). The colors indicate the strength of each peak. The faint ones, FC and FF, are nearly invisible for the experimental setup of Figure 4.19(a). In fact, if the molecules are aligned perpendicularly to the laser polarization, the FF bond will be the most clearly retrieved one. Details of the retrieval method can be found in Yu et al. [22].

4.5.7 Other Tools of Dynamic Chemical Imaging

Section 4.4.3 discussed the need for new tools for imaging the dynamics of molecules using ultrafast X-ray diffraction (UXD) and UED methods. A more complete review of the other tools employed for static imaging as well as dynamic imaging can be found in Xu et al. [24]. This section summarizes methods that are based on electron diffraction,

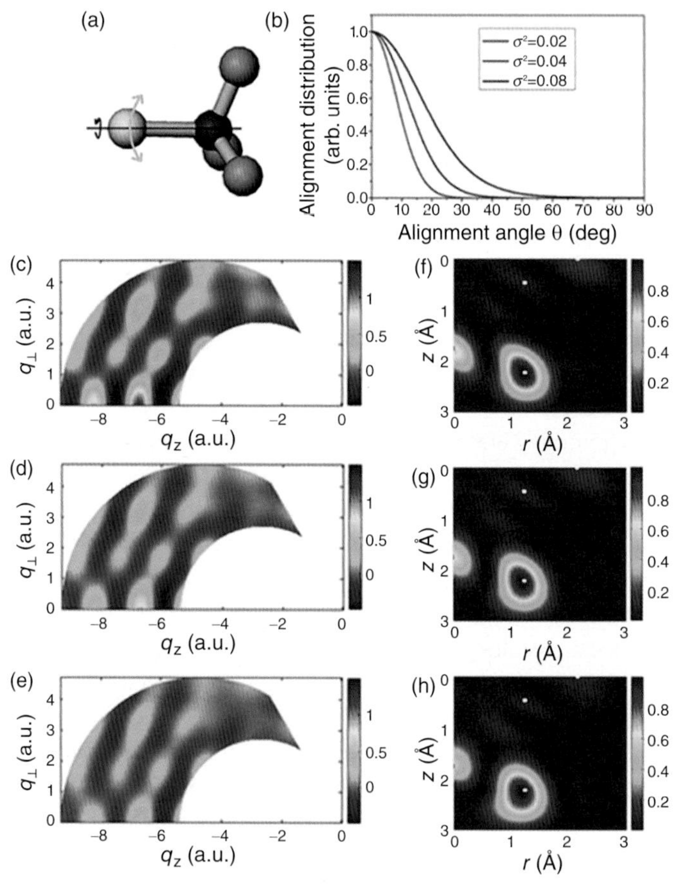

Figure 4.19 Schematic of an experimental setup for retrieving 2D molecular structure from partially 1D-aligned polyatomic molecules $ClCF_3$. (b) The degree of alignment is assumed to be Gaussian in the three examples. (c–e) The retrieved 2D diffraction images for perfectly 1D aligned molecules from images generated by 1D partially aligned molecules for the three angular distributions in (b). (f–h) The 2D molecular structure retrieved. The "experimental" data in the simulation was generated using the IAM. (Reprinted from Chao Yu et al., *Sci. Rep.*, **5**, 15753 (2015) [22].)

which is generally used for imaging gas-phase molecules. Thus the large body of UXD research being carried out with X-ray free-electron lasers at national light sources such as the Linac Coherent Light Source in the United States and the Spring-8 Angstrom Compact Free-Electron Laser in Japan [11] are excluded.

UED

While efforts are being made to develop ultrashort electron pulses down to tens to sub-ten femtoseconds at a number of laboratories, as of 2017 no UED experiments have been performed on gas-phase molecules with pulse durations below 100 fs. Here, the focus will be on recent electron-diffraction experiments from molecules that are impulsively aligned such that UED can be performed under the field-free condition during their full or partial revivals. The degree of alignment typically has $< \cos^2\theta >= 0.6$. To retrieve molecular structure from the diffraction images of such a weakly aligned ensemble, a new iterative algorithm has to be developed. This method allows the extraction of experimental diffraction images from partially 1D-aligned molecules to obtain diffraction images for perfectly 1D-aligned molecules. The latter can then be Fourier transformed to reveal the 2D molecular structure information. Figure 4.20(a) shows the diffraction image for perfectly 1D-aligned CF_3I molecules using the data from experimental partially 1D-aligned molecules [23]. Figure 4.20(b) shows the retrieved bond length r_{CI} and r_{FI} and the bond angle I–C–F to be 2.1, 1.9 Å, and 120°, respectively. In Figure 4.20(b), the white dots show the position of I at $(z, r) = (0, 0)$, of C at $(2.19, 0.0)$ Å and F at $z = (2.92, 1.26)$ Å. Thus,

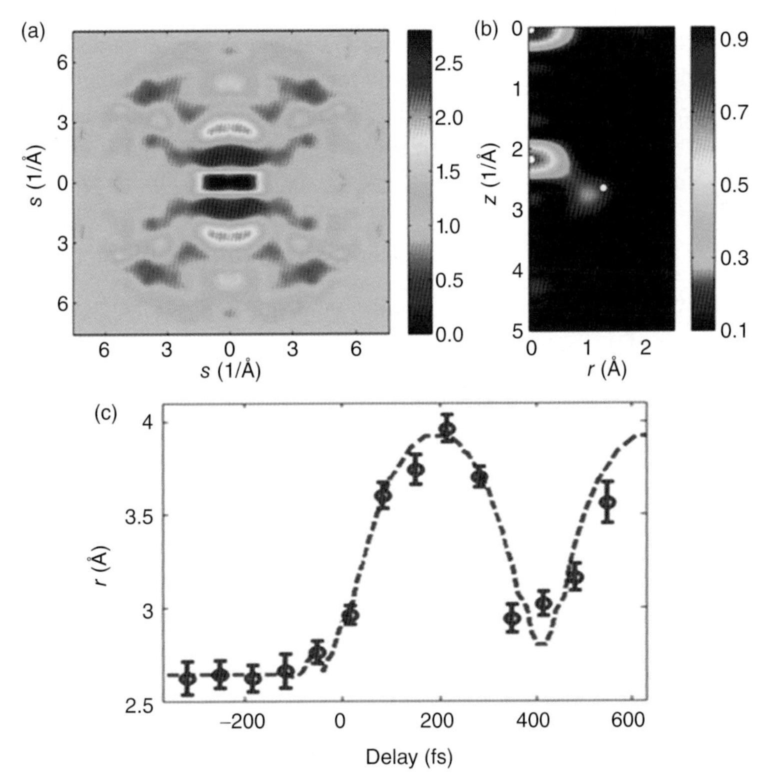

Figure 4.20 (a) Electron-diffraction image obtained for perfectly 1D-aligned CF_3I molecules. The images were extracted from experimental data for partially 1D-aligned molecules through an iterative method. Molecules are impulsively aligned with 800 nm lasers and the diffraction images were taken under the field-free conditions. (b) 2D molecular structure of CF_3I retrieved from (a). From [23]. (c) Retrieved bond length evolution of the vibrational wave packet on the excited, B-state potential curve of I_2. The molecules are first excited by a 530 nm laser from the ground state. The time evolution of the molecules is probed using a 3.7 MeV, 230 fs high-energy electron pulse. By analyzing the difference of diffraction images of laser-on versus laser-off, the mean value of the vibrational wave packet on the excited, B-state potential curve versus the time delay is obtained. (Reprinted from C. J. Hensley, J. Yang, and M. Centurion, *Phys. Rev. Lett.*, **109**, 133202 (2012) [23] and J. Yang et al., *Phys. Rev. Lett.*, **117**, 153002 (2016) [25]. Copyrighted by the American Physical Society.)

from 1D partially aligned molecules, it is possible to obtain 2D structure information. The temporal resolution in this experiment was estimated to be 850 fs.

The main limitation in gas-phase UED has been the velocity mismatch between the laser and electron pulses, and the temporal broadening of the electron pulses due to the repulsive Coulomb force. These factors are mitigated at MeV energies. In Yang et al. [25], 3.7 MeV electron pulses from the SLAC National Accelerator Laboratory in the United States have been used to image the vibrational wave packet of iodine molecules in the gas phase to achieve 0.07 Å spatial resolution and 230 fs temporal resolution. In this experiment, the molecules are first pumped to the excited B state. By taking the difference of the diffraction images without the pump pulse from the one with the pump pulse, the diffraction image due to the vibrational wave packet on the excited B state was obtained. By fitting the difference image, the bond distance at each time delay was retrieved (see Figure 4.20(c)). Since fitting procedure was used, a single averaged value of R is obtained at each time delay. Figure 4.20(c) shows the retrieved oscillation of the mean radius of the vibrational wave packet versus time after the pump pulse. Note that the 230 fs temporal resolution in this experiment is still not short compared to the 400 fs vibrational period of the iodine molecules.

Laser-Assisted Electron Diffraction

Another laser-based electron-diffraction method is called *laser-assisted electron diffraction* (LAED) This is equivalent to carrying out electron diffraction in the presence of a femtosecond laser. In the LAED experiment of Morimoto et al. [26], the electron pulse has 1 keV energy, duration of 15 ps. It was made to collide with CCl_4 in the presence of an 800 nm, 5 kHz, 520 fs titanium–sapphire laser with an intensity of 6×10^{11} W/cm^2. When the electrons and the laser beam overlap in time and space, the electron can scatter elastically or in addition with the absorption or emission of n-photons. Using the Kroll–Watson approximation for the angular distribution of the n-photon peak, the elastic DCS can be extracted. The spatial resolution is determined by the overlap of the laser pulse and the electron pulse. The method has been demonstrated for the static CCl_4 molecule. The diffraction images have been found to agree with the theoretically calculated angular distributions using the known bond-length parameters of this well-studied CCl_4 molecule.

Notes and Comments

Historically, Equation 4.14 was first proposed empirically in Morishita et al. [1] based on the TDSE results as shown in Figure 4.4. The factorization can be derived formally if the interaction is a short-range potential [27–31]. It also can be derived using the quantum orbits theory (see Section 5.2.4 and [32]). The QRS theory for the HATI electron's momentum distributions (Equation 4.14), can be generalized to HHG (Chapters 5 and 6), and also to NSDI. The latter has not been addressed in this book. In this case, the elastic-scattering cross-sections in Equation 4.14 are replaced by the e–2e electron-impact ionization cross-sections (see [33–35]). For NSDI, in general one needs to consider electron-impact excitation followed by laser ionization from the excited states. The separability of Equation 4.14 also allows one to extract the electron wave packet if the atomic scattering cross-sections are taken from accurate theoretical calculations. For short pulses, the electron-momentum distributions on both sides of the polarization axis are different and they vary with the CEP. By analyzing the asymmetry of the electron wave packet on the two sides of the polarization axis, the CEP can be extracted from the experimental data. Since the asymmetry also depends on the peak laser intensity and pulse duration, retrieval of all three parameters has been proposed based on the angular-distribution asymmetry (see Chen et al. and Zhou et al. [36, 37]). Conventional electron diffraction has been discussed in [Hargittai, Hargittai].

Exercises

4.1 Calculate the DCSs between electron and Ar^+ at incident energy of 150 eV and compare with the result shown in Figure 4.11(a). Do the same for the neutral Ar atom. Explain how you want to model the potential between the electron and Ar. Graph the DCS for the two systems in the angular range of $30°$–$180°$.

4.2 Calculate the DCS for e–N_2 collisions at 20, 50, 100, 150, and 200 eV using the IAM. The molecules are isotropically distributed. Describe how you modeled the e–N scattering amplitude. Plot the DCS for e–N_2 scattering using the known N–N bond length (1.10 Å) for scattering angles $\theta > 60°$. Then calculate the MCF for the collision energies asked. Compare your results with figure 2 in Xu et al., *J. Phys.: Conf. Ser.*, **288**, 012017 (2011).

4.3 For the e–N elastic scattering at 50 eV and scattering angles from $60°$ to $180°$, calculate the distances of closest approach if the collision is treated classically. Repeat the calculations if the incident electron energy is 150 keV.

4.4 Repeat Problem 4.2 by comparing the MCF for 150 eV electrons but assuming the N–N distance is two and five times the equilibrium distance. Plot MCF against the momentum transfer. Repeat the calculation assuming the electron energy is 50 eV. Comment on the pros and cons of using 50 eV versus 150 eV electrons for imaging N–N bond distance at five times the equilibrium distance.

References

[1] T. Morishita, A.-T. Le, Z. Chen, and C. D. Lin. Accurate retrieval of structural information from laser-induced photoelectron and high-order harmonic spectra by few-cycle laser pulses. *Phys. Rev. Lett.*, **100**:013903, Jan. 2008.

[2] Z. Chen, A.-T. Le, T. Morishita, and C. D. Lin. Quantitative rescattering theory for laser-induced high-energy plateau photoelectron spectra. *Phys. Rev. A*, **79**:033409, Mar. 2009.

[3] X. M. Tong and C. D. Lin. Empirical formula for static field ionization rates of atoms and molecules by lasers in the barrier-suppression regime. *J. Phys. B-At. Mol. Opt.*, **38**(15):2593, 2005.

[4] C. D. Lin, A.-T. Le, Z. Chen, T. Morishita, and R. Lucchese. Strong-field rescattering physics – self-imaging of a molecule by its own electrons. *J. Phys. B-At. Mol. Opt.*, **43**(12):122001, 2010.

[5] P. Colosimo, G. Doumy, C. I. Blaga, et al. Scaling strong-field interactions towards the classical limit. *Nat. Phys.*, **4**(5):386–389, May 2008.

[6] M. Okunishi, T. Morishita, G. Prümper, et al. Experimental retrieval of target structure information from laser-induced rescattered photoelectron momentum distributions. *Phys. Rev. Lett.*, **100**:143001, Apr. 2008.

[7] S. Micheau, Z. Chen, A. T. Le, J. Rauschenberger, M. F. Kling, and C. D. Lin. Accurate retrieval of target structures and laser parameters of few-cycle pulses from photoelectron momentum spectra. *Phys. Rev. Lett.*, **102**:073001, Feb. 2009.

[8] J. Xu, C. I. Blaga, A. D. DiChiara, et al. Laser-induced electron diffraction for probing rare gas atoms. *Phys. Rev. Lett.*, **109**:233002, Dec. 2012.

[9] M. Okunishi, R. R. Lucchese, T. Morishita, and K. Ueda. Rescattering photoelectron spectroscopy of small molecules. *J. Electron Spectrosc. Relat. Phenom.*, **195**: 313–319, 2014.

[10] SLAC National Accelerator Laboratory, Linac Coherent Light Source, lcls.slac.stanford.edu.

[11] Spring-8 Angstrom Compact Free Electron Laser (SACLA), xfel.riken.jp/eng.

[12] J. Xu, Z. Chen, A.-T. Le, and C. D. Lin. Self-imaging of molecules from diffraction spectra by laser-induced rescattering electrons. *Phys. Rev. A*, **82**:033403, Sep. 2010.

[13] J. Xu, Y. Liang, Z. Chen, and C. D. Lin. Elastic scattering and impact ionization by returning electrons induced in a strong laser field. *J. Phys.: Conf. Series*, **288**(1):012017, 2011.

[14] T. Zuo, A. D. Bandrauk, and P. B. Corkum. Laser-induced electron diffraction: a new tool for probing ultrafast molecular dynamics. *Chem. Phys. Lett.*, **259**(3):313–320, 1996.

[15] M. Meckel, D. Comtois, D. Zeidler, et al. Laser-induced electron tunneling and diffraction. *Science*, **320**(5882):1478–1482, 2008.

[16] C. I. Blaga, J. Xu, A. D. DiChiara, et al. Imaging ultrafast molecular dynamics with laser-induced electron diffraction. *Nature*, **483**(7388):194–197, Mar. 2012.

[17] M. G. Pullen, B. Wolter, A.-T. Le, et al. Imaging an aligned polyatomic molecule with laser-induced electron diffraction. *Nat. Commun.*, **6**:7262, Jun. 2015.

[18] B. Wolter, M. G. Pullen, A.-T. Le, et al. Ultrafast electron diffraction imaging of bond breaking in di-ionized acetylene. *Science*, **354**(6310):308–312, 2016.

[19] Y. Ito, Ch. Wang, A.-T. Le, et al. Extracting conformational structure information of benzene molecules via laser-induced electron diffraction. *Struct. Dyn.*, **3**(3):034303, May 2016.

[20] J. Xu, C. I. Blaga, K. Zhang, et al. Diffraction using laser-driven broadband electron wave packets. *Nat. Commun.*, **5**:4635, Aug. 2014.

[21] M. G. Pullen, B. Wolter, A.-T. Le, et al. Influence of orbital symmetry on diffraction imaging with rescattering electron wave packets. *Nat. Commun.*, **7**:11922, Jun. 2016.

[22] C. Yu, H. Wei, X. Wang, A.-T. Le, Ruifeng Lu, and C. D. Lin. Reconstruction of two-dimensional molecular structure with laser-induced electron diffraction from laser-aligned polyatomic molecules. *Sci. Rep.*, **5**:15753, Oct. 2015.

[23] Ch. J. Hensley, J. Yang, and M. Centurion. Imaging of isolated molecules with ultrafast electron pulses. *Phys. Rev. Lett.*, **109**:133202, Sep. 2012.

[24] J. Xu, C. I Blaga, P. Agostini, and L. F DiMauro. Time-resolved molecular imaging. *J. Phys. B-At. Mol. Opt.*, **49**(11):112001, 2016.

[25] J. Yang, M. Guehr, X. Shen, et al. Diffractive imaging of coherent nuclear motion in isolated molecules. *Phys. Rev. Lett.*, **117**:153002, Oct. 2016.

[26] Y. Morimoto, R. Kanya, and K. Yamanouchi. Laser-assisted electron diffraction for femtosecond molecular imaging. *J. Chem. Phys.*, **140**(6), 2014.

[27] M. V. Frolov, N. L. Manakov, and A. F. Starace. Analytic formulas for above-threshold ionization or detachment plateau spectra. *Phys. Rev. A*, **79**:033406, Mar. 2009.

[28] A. Cerkic, E. Hasovic, D. B. Milosevic, and W. Becker. High-order above-threshold ionization beyond the first-order Born approximation. *Phys. Rev. A*, **79**:033413, Mar. 2009.

[29] O. I. Tolstikhin, T. Morishita, and S. Watanabe. Adiabatic theory of ionization of atoms by intense laser pulses: one-dimensional zero-range-potential model. *Phys. Rev. A*, **81**:033415, Mar. 2010.

[30] M. V. Frolov, D. V. Knyazeva, N. L. Manakov, J.-W. Geng, L.-Y. Peng, and A. F. Starace. Analytic model for the description of above-threshold ionization by an intense short laser pulse. *Phys. Rev. A*, **89**:063419, Jun. 2014.

[31] M. V. Frolov, D. V. Knyazeva, N. L. Manakov, et al. Validity of factorization of the high-energy photoelectron yield in above-threshold ionization of an atom by a short laser pulse. *Phys. Rev. Lett.*, **108**:213002, May 2012.

[32] A.-T. Le, H. Wei, C. Jin, and C. D. Lin. Strong-field approximation and its extension for high-order harmonic generation with mid-infrared lasers. *J. Phys. B-At. Mol. Opt.*, **49**(5):053001, 2016.

[33] Z. Chen, Y. Liang, and C. D. Lin. Quantum theory of recollisional (e, $2e$) process in strong field nonsequential double ionization of helium. *Phys. Rev. Lett.*, **104**:253201, Jun. 2010.

[34] Z. Chen, Y. Liang, and C. D. Lin. Quantitative rescattering theory of correlated two-electron momentum spectra for strong-field nonsequential double ionization of helium. *Phys. Rev. A*, **82**:063417, Dec. 2010.

[35] Z. Chen, Y. Zheng, W. Yang, et al. Numerical simulation of the double-to-single ionization ratio for the helium atom in strong laser fields. *Phys. Rev. A*, **92**:063427, Dec. 2015.

[36] Z. Chen, T. Wittmann, B. Horvath, and C. D. Lin. Complete real-time temporal waveform characterization of single-shot few-cycle laser pulses. *Phys. Rev. A*, **80**:061402, Dec. 2009.

[37] Z. Zhou, X. Wang, and C. D. Lin. Analysis of THz generation through the asymmetry of photoelectron angular distributions. *Phys. Rev. A*, **95**:033418, Mar. 2017.

5 Fundamentals of High-Order Harmonic Generation

5.1 Introduction

The bright, spatially coherent lights emitted by lasers have found numerous applications in spectroscopy, communications, imaging, material processing, and photochemistry. Different applications demand light in specific spectral regions. Thus large synchrotron-radiation, free-electron and other lasers have been built to convert accelerator-based, high-energy electrons into powerful, broadband tunable lights. For tabletop equipment, it has been well known for years that laser light can be converted from one wavelength into another. Familiar examples are second-order and third-order harmonics generated in crystals. The coupling of a laser with an optical waveguide can also generate a broadband supercontinuum, whereas the nonlinear interaction of infrared lasers with gas media can generate high-order harmonics extending to soft and hard X-rays.

The basic setup and principle of high-harmonic generation (HHG) is simple, as illustrated in Figure 5.1. In (a), an intense laser beam is focused into a gas jet or gas cell with typical pressure from a few tens of millibars to a few bars and with laser peak intensity on the order of 10^{14} W/cm^2. When the laser intensity is sufficiently high, HHG can be recorded along the beam direction after a filter is introduced to block the co-propagating, intense-driving infrared laser. After the harmonics are dispersed, the spectral distribution has the typical shape of (b). For the first few harmonics, the yields drop rapidly. The spectrum is then followed by broadband, flat-plateau harmonics until the cutoff energy. The cutoff energy is approximately given by $I_p + 3.17U_p$, where U_p is the ponderomotive energy and I_p is the ionization potential of the target. HHG is easily understood using the three-step model as illustrated in (c). By focusing an intense laser beam into gaseous targets, a bound electron from an atom can be released into the continuum that is accelerated by the driving laser field to gain kinetic energy. As the field direction is reversed, the electron may be driven back to recombine with the parent ion and emit high-energy photons. Since harmonics generated from each atom are coherent with respect to the driving laser (as in every nonlinear process), phase matching plays an important role. In (d), the cartoon illustrates that if the light emitted from each atom is phase matched, the emitted radiation from the gas medium will be built up quadratically with the number of atoms.

HHG is arguably the most important topic in strong-field physics. This chapter covers its generation at the single-atom level and discusses the effects of phase matching when the harmonics are propagated in the macroscopic medium. Harmonics generated in atomic targets will then be presented. In Chapter 6, applications of HHG to molecular targets and methods for enhancing harmonic yields will be covered.

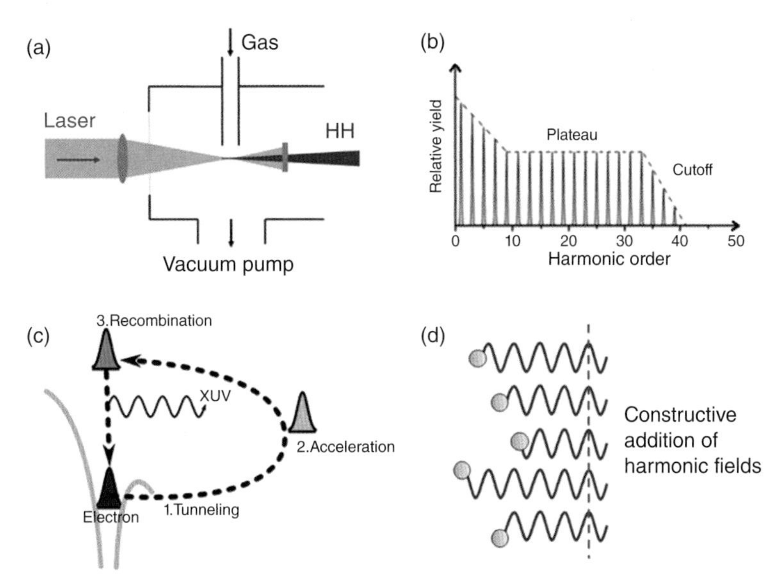

Figure 5.1 Schematic of high-order harmonic generation. (a) HHG is a nonlinear process where high harmonics are generated by focusing an intense light near the infrared-wavelength region on a gaseous medium. (b) Typical spectral shape of high harmonic emission. The yield of the first few harmonics drops quickly and is followed by a broad harmonic plateau before the cutoff limit. (c) The harmonics generated by an individual atom can be described by the three-step model, where an electron first tunnels out of the combined potential from the atomic ion and the laser field. In the second step, the electron is accelerated under the laser's field. In the third step, the energy gained by the electron in the laser field is converted to high-energy radiation when the electron returns to recombine with the parent ion. (d) The harmonics generated by individual atoms are phase locked to the driving laser. Favorable phase matching is needed for the generation of intense high harmonics.

5.2 Theory of High-Order Harmonic Generation by an Atom

5.2.1 Calculation of HHG from Solving the Time-Dependent Schrödinger Equation

When an atom is exposed to an intense laser field, the strong field induces a time-dependent dipole that emits radiation. The time-dependent dipole can be calculated from $\mathbf{D}(t) = \langle \Psi(t)|\mathbf{r}|\Psi(t)\rangle$ once the time-dependent wavefunction has been solved from

$$i\frac{\partial}{\partial t}|\Psi(\mathbf{r},t)\rangle = \left(-\frac{1}{2}\nabla^2 + V(\mathbf{r}) + \mathbf{r}\cdot\mathbf{E}(t)\right)|\Psi(\mathbf{r},t)\rangle. \tag{5.1}$$

The HHG with polarization along a given direction i is taken from the Fourier transform $\mathbf{D}(\omega)$ of $\mathbf{D}(t)$. The power spectrum is given by

$$P(\omega) \propto \omega^4 |D_i(\omega)|^2. \tag{5.2}$$

Here, for simplicity, the total time-dependent Hamiltonian has been written for an one-electron atom in the length gauge

$$H(t) = H_0 + \mathbf{r} \cdot \mathbf{E}(t), \tag{5.3}$$

with the field-free atomic Hamiltonian

$$H_0 = -\frac{1}{2}\nabla^2 + V(\mathbf{r}), \tag{5.4}$$

which has ground state $|g\rangle$, excited states $|e\rangle$, and continuum states $|k\rangle$ with energy eigenvalues $-I_p$, e_i, and $k^2/2$, respectively. The direct numerical solution of the time-dependent Schrödinger equation (TDSE) is much easier for HHG than for calculating ATI spectra. A number of calculations based on solving TDSE or time-dependent density functional theory (TDDFT) (see Sections 2.2.4 and 2.2.6) for many-electron atoms or molecules have been reported. However, such calculations are too time-consuming if one wishes to include phase matching. This chapter focuses on the simple quantitative rescattering (QRS) model or the second-order, strong-field approximation (SFA2) model that can be used readily for studying HHG. Accurate TDSE results for one-electron atoms will mostly be used for the calibration of the models.

5.2.2 Strong-Field Approximation, or Lewenstein Model, for HHG

For more than two decades, the most widely used theory for HHG has been the strong-field approximation (SFA). By the convention used in Section 2.2.3, this model could also be called the SFA2 since HHG is a second-order theory in the strong-field expansion. This theory can be derived from the TDSE with a number of approximations (as first shown by Lewenstein et al. [1]). This is done by assuming that the time-dependent wavefunction can be expanded as

$$|\Psi(t)\rangle = e^{iI_pt}\left\{|g\rangle + \int d^3k\, b(\mathbf{k},t)|\mathbf{k}\rangle\right\}, \tag{5.5}$$

where the excited states have been neglected and the continuum states are approximated by plane waves. By solving the coefficients $b(\mathbf{k},t)$ using the KFR theory, the time-dependent dipole moment $\mathbf{D}(t)$ can be written as

$$\mathbf{D}(t) = \int d^3k\, \langle g|\mathbf{r}|\mathbf{k}\rangle b(\mathbf{k},t) + c.c., \tag{5.6}$$

where $c.c.$ stands for *complex conjugate*. Following the steps in Lewenstein et al. [1] and with the help of a number of approximations, the Lewenstein model for HHG can be derived. However, a more direct way to derive the same expression is to use the integral-equation approach, as discussed in Section 2.2.3.

In this approach, one writes

$$H(t) = H_F(t) + V(\mathbf{r}), \tag{5.7}$$

where

$$H_F(t) = -\frac{1}{2}\nabla^2 + \mathbf{r} \cdot \mathbf{E}(t), \tag{5.8}$$

and the eigenstates of H_F are the Volkov states. In the length gauge it is written as

$$|\chi_{\mathbf{p}}(t)\rangle = |\mathbf{p} + \mathbf{A}(t)\rangle e^{-i\int_{-\infty}^{t} dt'' \frac{1}{2}[\mathbf{p}+\mathbf{A}(t'')]^2}, \tag{5.9}$$

where $\mathbf{A}$ is the vector potential of the laser field. The time-evolution operator of $H_F(t)$ is

$$U_F(t,t') = \int d^3p |\chi_{\mathbf{p}}(t)\rangle\langle\chi_{\mathbf{p}}(t')|. \tag{5.10}$$

With this approach, the first order in H_F, the wavefunction at time t, is given by $-i\int_{-\infty}^{t} dt' U_F(t,t')\mathbf{r}\cdot\mathbf{E}(t')e^{iI_pt'}|g\rangle$, and

$$b(\mathbf{k},t) = -ie^{-iI_pt}\langle\mathbf{k}|\int_{-\infty}^{t} dt' U_F(t,t')\mathbf{r}\cdot\mathbf{E}(t')e^{iI_pt'}|g\rangle. \tag{5.11}$$

Thus,

$$\begin{aligned}\mathbf{D}(t) &= -i\int_{-\infty}^{t} dt' e^{-iI_pt}\langle g|\mathbf{r}U_F(t,t')\mathbf{r}\cdot\mathbf{E}(t')e^{iI_pt'}|g\rangle + c.c.\\ &= -i\int_{-\infty}^{t} dt' \int d^3p e^{-iI_pt}\langle g|\mathbf{r}|\mathbf{p}+\mathbf{A}(t)\rangle\\ &\quad\times \mathbf{E}(t')\cdot\langle\mathbf{p}+\mathbf{A}(t')|\mathbf{r}|g\rangle e^{iI_pt'}\\ &\quad\times e^{-i\int_{t'}^{t} dt''\frac{1}{2}[\mathbf{p}+\mathbf{A}(t'')]^2} + c.c.,\end{aligned} \tag{5.12}$$

or,

$$\begin{aligned}\mathbf{D}(t) = -i\int_{-\infty}^{t} dt' \int d^3p\, \mathbf{d}^*(\mathbf{p}+\mathbf{A}(t))\\ \times \mathbf{E}(t')\cdot\mathbf{d}(\mathbf{p}+\mathbf{A}(t'))e^{-iS(\mathbf{p},t,t')} + c.c.,\end{aligned} \tag{5.13}$$

where

$$\begin{aligned}S(\mathbf{p},t,t') &= \int_{t'}^{t} dt''\frac{1}{2}[\mathbf{p}+\mathbf{A}(t'')]^2 + I_p(t-t')\\ &= \int_{t'}^{t} dt''\left(\frac{1}{2}[\mathbf{p}+\mathbf{A}(t'')]^2 + I_p\right)\end{aligned} \tag{5.14}$$

is the action. Expression Equation 5.13 is the strong-field approximation that corresponds to the quasi-classical, three-step model. Here, $\mathbf{p}$ is the canonical momentum and $\mathbf{p}+\mathbf{A}(t)$ is the instantaneous electron velocity at time t. The equation says that ionization occurs at time t' with the amplitude given by the dipole term at t'. The electron then propagates in the laser field from t' to t, accumulating a phase represented by the action S. It then recombines at time t with the dipole term evaluated at time t.

The SFA expression Equation 5.13 for the laser-induced dipole moment is rarely used. Instead, the integral over the momentum $\mathbf{p}$ is generally carried out under the saddle-point approximation. Assume the laser is polarized along the x-direction, and all the vectors in Equation 5.13 are confined along this direction and then reduced into scalars. Without delving into the mathematical steps (for example, see [2]), we just state that Equation 5.13 is reduced to

$$D_x(t) = -i \int_{-\infty}^{t} dt' \left(\frac{-2\pi i}{t - t' - i\epsilon}\right)^{3/2} d_x^*(p_s + A(t)) \times d_x(p_s + A(t'))E(t')e^{-iS(p_s,t,t')} + c.c., \tag{5.15}$$

where the saddle-point solution p_s is given by

$$p_s = -\frac{1}{t - t'} \int_{t'}^{t} A(t'')dt''. \tag{5.16}$$

In Equation 5.15, ϵ is a small, positive-regularization constant introduced to smooth out the singularity. The $(t - t')^{-3/2}$ factor accounts for the quantum-diffusion effect, i.e., the spreading of the wave packet of the continuum electron. The accuracy of the saddle-point approximation has been checked against actual numerical calculations. In general, the harmonic spectra have the same shape, but the magnitude obtained from the saddle-point approximation has been found to be higher by about 30% at 800 nm and 20% at 1,600 nm lasers.

5.2.3 QO Theory

The HHG power spectrum is calculated by Equation 5.2 in terms of the Fourier transform $D_x(\omega)$ of $D_x(t)$ given by Equation 5.15, where the x-component of the dipole generated by an electric field polarized along the x-direction is considered. Define

$$D_x(\omega) = \int_{-\infty}^{\infty} D_x(t)e^{i\omega t}dt = D_x^{(+)}(\omega) + \left[D_x^{(+)}(-\omega)\right]^*, \tag{5.17}$$

where

$$D_x^{(+)}(\omega) = -i \int_{-\infty}^{\infty} dt \int_{-\infty}^{t} dt' \left(\frac{-2\pi i}{t - t' - i\epsilon}\right)^{3/2} \times d_x^*(p_s + A(t))d_x(p_s + A(t'))E(t')e^{-i\Theta(p_s,t,t')}. \tag{5.18}$$

In Equation 5.18 the phase factor $\Theta(p_s, t, t') = S(p_s, t, t') - \omega t$. The saddle-point momentum p_s and the action $S(p, t, t')$ are defined in Equations 5.16–5.14, respectively.

The basic idea of the quantum orbit (QO) theory is to further apply the saddle-point approximation to the 2D integral over t and t'. The saddle-points of Θ with respect to t' and t lead to two equations

$$\frac{1}{2}[p_s + A(t')]^2 = -I_p, \tag{5.19}$$

$$\frac{1}{2}[p_s + A(t)]^2 = \omega - I_p. \tag{5.20}$$

The first equation implies that the electron is born to the continuum with a "negative kinetic energy" given by $-I_p$. This is partly to account for the quantum-tunneling effect. The second equation gives the energy of the emitted photon (ω) upon recombination of the electron with the ion core. For each ω, the simultaneous solutions of Equations 5.19 and 5.20 give complex saddle-point values t_s' and t_s. Under the saddle-point approximation, Equation 5.18 is reduced to

$$D_x^{(+)}(\omega) = -i\sum_s \sqrt{\frac{(2\pi i)^2}{\det(S'')}}\left(\frac{-2\pi i}{t_s - t_s'}\right)^{3/2} \times d_x^*(p_s + A(t_s))d_x(p_s + A(t_s'))E(t_s')e^{-i\Theta(p_s,t_s,t_s')}. \tag{5.21}$$

In this equation, each pair (t_s', t_s) determines a unique QO labeled by s. In the spirit of Feynman's path integral, $D_x^+(\omega)$ is the superposition of induced dipole from individual QOs. It is reasonable to neglect the contribution from QOs that correspond to negative ω, therefore one can write

$$D_x(\omega) \approx D_x^{(+)}(\omega) = \sum_s D_{xs}(\omega), \tag{5.22}$$

where the induced-dipole moment from each quantum orbit is given by

$$D_{xs}(\omega) = \frac{2\pi}{\sqrt{\det(S'')}}\left(\frac{-2\pi i}{t_s - t_s'}\right)^{3/2} d_x^*(p_s + A(t_s)) \times d_x(p_s + A(t_s'))E(t_s')e^{-i\Theta(p_s,t_s,t_s')}. \tag{5.23}$$

The calculation of the determinant $\det(S'')$ is straightforward (see equations 48–51 of Le et al. [2]). Note that when a singularity occurs in the saddle-point approximation, the expressions above do not hold anymore and the next-order term has to be evaluated.

5.2.4 Factorization of Laser Induced Dipole Moment

From Equations 5.19 and 5.20 one can get

$$d_x(p_s + A(t_s')) = \varepsilon_s' d_x\left(i\sqrt{2I_p}\right), \tag{5.24}$$

$$d_x(p_s + A(t_s)) = \varepsilon_s d_x\left(\sqrt{2(\omega - I_p)}\right), \tag{5.25}$$

where ε_s' and ε_s are either $+1$ or -1 to account for the direction of electron momentum at the moment of ionization or recombination. Consider the harmonics that are above the ionization threshold. The induced dipole in Equation 5.22 can be rewritten as

$$\begin{aligned} D_x(\omega) &= d_x^*\left(\sqrt{2(\omega - I_p)}\right)d_x(i\sqrt{2I_p}) \\ &\quad \times \sum_s \frac{2\pi\varepsilon_s'\varepsilon_s}{\sqrt{\det(S'')}}\left(\frac{-2\pi i}{t_s - t_s'}\right)^{3/2} E(t_s')e^{-i\Theta(p_s,t_s,t_s')} \\ &\equiv d_x^*\left(\sqrt{2(\omega - I_p)}\right)w(\omega). \end{aligned} \tag{5.26}$$

Equation 5.26 shows that the induced dipole can be factored into two parts. One part has to do with the recombination transition dipole-matrix element, while the other part $w(\omega)$ mainly depends on the laser pulse. The latter is called the complex *returning electron wave packet*, combining the ionization at time t' and propagation in the laser field until the recombination time t. Clearly, the contribution from each trajectory to the wave packet can be identified from Equation 5.26. Note that the transition dipole moment is in the field-free form even though recombination occurs in the laser field. The wave packet $w(\omega)$ depends on the target structure only through the ω-independent ionization probability amplitude $d_x(i\sqrt{2I_p})$, which does not affect the shape of the returning electron wave packet. More discussion about the factorization will be given in Section 5.3.

5.2.5 Analysis of HHG by Monochromatic Laser Fields with Analytical QO Theory

Spectrum of Odd Harmonics

In this subsection, the harmonic emission by a monochromatic laser field $E(t) = E_0 \cos(\omega_L t)$ is calculated within the QO theory. In this case, many results can be obtained analytically. The detailed derivations can be found in section 2.2.3 of Le et al. [2]. The main results are summarized here.

Let $T_L = 2\pi/\omega_L$ be the optical period. Clearly,

$$D_x\left(t + \frac{T_L}{2}\right) = -D_x(t). \quad (5.27)$$

Let $\tilde{D}_x(t)$ and $\tilde{D}_x(\omega)$ be the induced dipole within half an optical cycle for $-\frac{T_L}{4} < t \leq \frac{T_L}{4}$. It is straightforward to prove that

$$D_x(\omega) = \sum_{\substack{q=-\infty \\ \text{odd } q}}^{\infty} 2\omega_L \delta(\omega - q\omega_L)\tilde{D}_x(q\omega_L). \quad (5.28)$$

This shows that the spectrum induced by a monochromatic laser field contains odd harmonics only. Thus, one can just focus on the induced-dipole moment generated within half an optical cycle.

Ionization Time and Recombination Time: QO versus Classical Theory

For a monochromatic laser pulse, it is more convenient to use scaled time and energy. Define $\theta = \omega_L t$, $\theta' = \omega_L t'$ and $\tilde{\omega} = (\omega - I_p)/U_p$. Within the QO theory, the saddle point p_s and the angles θ_s and θ_s' can be calculated analytically. The details of such calculations are given in section 2.2.3 of Le et al. [2]. Note that, with scaled quantities, the results are universal and only dependent on the Keldysh parameter $\gamma = \sqrt{\frac{I_p}{2U_p}}$.

In the QO theory, both θ_s and θ_s' are complex numbers. Of course, both quantities are real numbers in the classical theory. Figure 5.2(a) shows the real part of the ionization time and recombination time with the ionization time over a quarter cycle. Electrons that are born before the peak of the laser field ($-90° < \text{Re}\{\theta_s'\} < 0°$) not return to the core. Electrons that are born after the peak of the field ($0° \leq \text{Re}\{\theta_s'\} < 90°$) have a chance to revisit the core with kinetic energy $\tilde{\omega}U_p$. Figure 5.2(a) shows that there are two

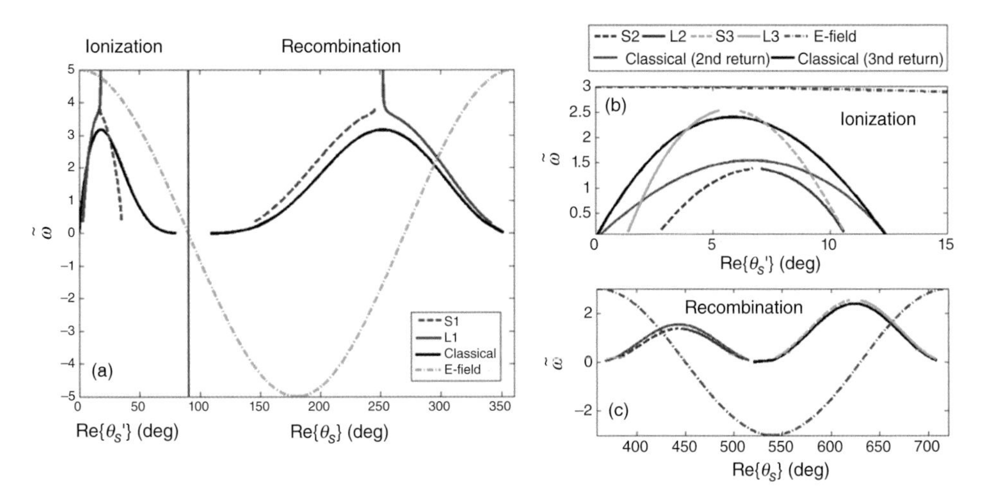

Figure 5.2 Ionization time and recombination time calculated using QO versus the classical theory. In (a), S1 (L1) indicates the first return for the short (long) trajectory. The solid black line is from the classical theory. The red lines are from the real part of the QO theory, where the ionization and recombination times are both complex numbers. Note that short- (long-) trajectory electrons are born later (earlier) and return earlier (later). The electric field over one optical cycle is also shown in green. (b) and (c) are for higher returns. S2 (L2) indicates the second return and S3 (L3) indicates the third return. Target is Ar. Laser is 800 nm with an intensity of 1.5×10^{14} W/cm^2, which corresponds to $\gamma = 0.94$. (Reprinted from Anh-Thu Le et al., *J. Phys. B-At. Mol. Opt.*, **49**, 053001 (2016) [2]. Copyrighted by the IOP publishing.)

pathways below the maximal returning energy that can return with the same kinetic energy. Within the same optical cycle, the one that was born earlier would return later. Classically, this corresponds with a "long"-trajectory electron, while the one that was born later and returns earlier corresponds with a "short"-trajectory electron. The classical ionization and recombination times are also shown in Figure 5.2(a) in solid black lines. Note the small differences between the QO theory and the classical theory.

Electrons that are born within $0^\circ < \mathrm{Re}\{\theta_s'\} < 12^\circ$ may return to the ion core multiple times. The real part of the ionization and recombination time after the second and third returns have been calculated and shown in Figures 5.2(b) and (c), respectively. For the second return, the short-trajectory electron (S2) is born earlier and returns earlier while the long-trajectory electron (L2) is born later and returns later. For the third return, the ionization and return time for the long and short trajectories behave like the first return. According to the classical theory, the maximal scaled kinetic energy for the first return is $\tilde{\omega} = 3.17$ and for the second and third returns they are 1.5 and 2.4, respectively. These cutoff values are changed in the QO theory. For example, when $\gamma = 0.94$, the cutoff for the first return is extended to $\tilde{\omega} = 3.8$.

The imaginary part of the factor $\Theta(p_s, t_s, t_s')$ is related to the tunnel-ionization rate. From Figure 5.3 it has been found that the Im$\{\Theta\}$ are essentially identical for all orbits except for the S1 orbit since emission of electrons corresponding to all other orbits occurs at practically the same small angular range of Re$\{\theta_s'\}$ (see Figure 5.2), while the electron for the S1 orbit is ionized at a larger Re$\{\theta_s'\}$.

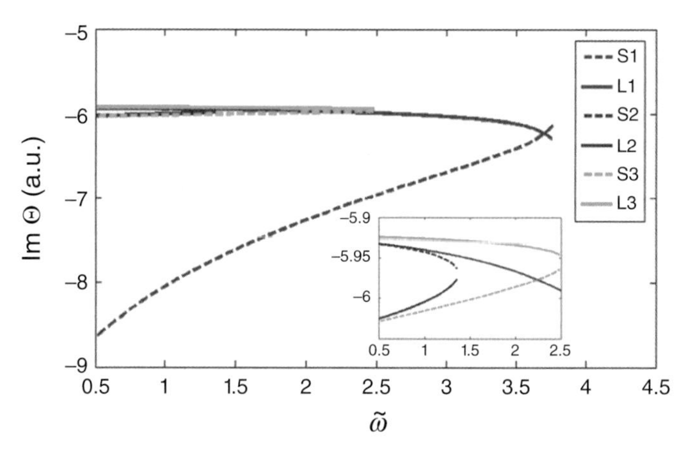

Figure 5.3 Imaginary parts of $\Theta(p_s, t_s, t_s')$ as functions of the scaled energy $\tilde{\omega}$, calculated in the QO theory. The inset shows the zoom-in for the high-order returns. (Reprinted from Anh-Thu Le et al., *J. Phys. B-At. Mol. Opt.*, **49**, 053001 (2016) [2]. Copyrighted by the IOP publishing.)

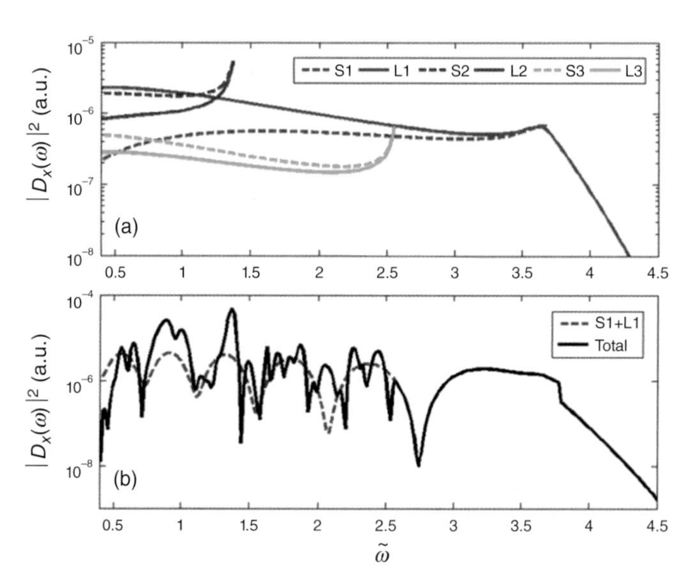

Figure 5.4 (a) HHG spectrum from individual quantum orbits. (b) Summing the contributions coherently shows the complex interference and the contributions from the higher returns, compared with contributions from the first return (red dashed lines) only. Target is Ar. Laser is 800 nm with an intensity of 1.5×10^{14} W/cm^2. (Reprinted from Anh-Thu Le et al., *J. Phys. B-At. Mol. Opt.*, **49**, 053001 (2016) [2]. Copyrighted by the IOP publishing.)

Figure 5.4(a) shows the modulus square of the induced dipole for each individual quantum orbit against the scaled kinetic energy of the returning electron. For $\tilde{\omega}$ greater than 2.5, only first-return electrons contribute to the high harmonics ($\omega = \tilde{\omega}U_p + I_p$ should be used to get the energy of the harmonics). Multiple returns add contributions to the low harmonics. Figure 5.4(b) shows the total harmonic spectra and the contribution from the first-return electrons only, which manifests the noneligible contribution for the higher returns. However, such higher returns become insignificant after phase matching is considered.

Wavelength Scaling of HHG

Since the cutoff energy of the high harmonics is determined by $I_p + 3.2U_p$ and U_p is proportional to λ^2, by using a mid-infrared laser the HHG spectrum may be extended to hundreds of eVs covering the water window or even to the keV regime. It is of utmost importance to determine the wavelength scaling of harmonic yields in order to generate practical HHG light sources in the sub- to few-keV range. The analytical QO theory can be used to derive the scaling law. Such derivation has been given in a tutorial article [2].

In the long-wavelength limit, the recombination time can be taken to be real while the ionization time remains complex. Section 3 of Le et al. [2] works out the derivations to show that, in the long-wavelength limit at a fixed photon energy ω, the HHG scaling law for the L1 orbit is

$$P_{L1}(\omega) \propto \lambda^{-1}\tilde{\omega}^{1.8} \propto \lambda^{-1}U_p^{-1.8} \propto \lambda^{-4.6}. \tag{5.29}$$

For the S1 orbit the HHG scaling law is

$$P_{S1}(\omega) \propto \lambda^{-1}\tilde{\omega}^{4.7} \propto \lambda^{-1}U_p^{-4.7} \propto \lambda^{-10.4}. \tag{5.30}$$

Since L1 predominates, the scaling law will be close to $\lambda^{-4.6}$ in the single-atom theory. Numerical TDSE calculations (including all QOs) show that the scaling is $\lambda^{-4.2}$. This scaling law derived from the QO theory is not different from the solution of TDSE. The discrepancy in this scaling law as compared to the scaling law of $\lambda^{-(5-6)}$ (from [3–5]) is due to the difference in the definition of how HHG yields are compared. In the equations here it was defined as rate per optical cycle while in the other works the rate is defined per unit time. This would account for the $1/\lambda$ difference in the scaling law. As shown in Sections 5.4 and 5.5, harmonics generated by long orbits do not usually phase-match well. Thus the wavelength scaling of HHG is quite unfavorably determined by the scaling law of λ^{-10} under the same conditions.

5.3 QRS Theory

5.3.1 The QRS Theory for HHG

Equation 5.26 shows that the laser-induced dipole $D(\omega)$ can be written as a product of a returning electron wave packet $W(\omega)$ and a field-free transition dipole within the QO theory. This factorization form within the QO theory first appeared in Le et al. [2]. Historically, a similar expression was first postulated by Itatani et al. [6] where the intensity of the emitted HHG from a linear molecule fixed in space was written as

$$S(\omega,\theta) = N^2(\theta)\omega^4 \left|a[k(\omega)]d(\omega,\theta)\right|^2, \tag{5.31}$$

where the molecular axis is aligned by angle θ with respect to the laser-polarization direction. Here, $|d(\omega,\theta)|^2$ is the modulus square of the transition dipole between the ground state and the continuum state approximated by a plane wave, and the remaining factor on the right-hand side can be considered the modulus square of the returning electron wave packet. Between Equations 5.26 and 5.31, the former is a complex induced dipole, while the latter is a positive definite harmonic signal. The former was derived from the QO theory (a simplified form of SFA2), while the latter was from modeling the experimental data. In Itatani et al. [6], the wavefunction of the continuum state was approximated by a plane wave in the calculation of the transition dipole $d(\omega,\theta)$. With many additional assumptions, Equation 5.31 was further used in Itatani et al. [6] to obtain the molecular orbital of the ground state of N_2. This work has generated a great deal of excitement; the procedure is called *tomographic imaging of molecular orbitals by HHG from aligned molecules*. Section 6.1.11 takes a closer look at this topic.

The QRS theory for HHG is readily derived once the QRS theory for HATI spectra has been established. For simplicity, consider the calculation of HHG induced by an intense infrared laser on a linear molecule. Generalization to polyatomic molecules is straightforward and will be discussed in Section 6.2. Within the QRS, the complex-valued induced dipole $D(\omega,\theta)$ for a molecule aligned with a fixed angle θ with respect to the laser polarization is written as

$$D(\omega,\theta) = W(E,\theta)d(\omega,\theta). \tag{5.32}$$

Here, $d(\omega,\theta)$ is the photo-recombination, dipole-transition matrix element, and $W(E,\theta)$ is the complex-valued, returning-electron wave packet with electron energy $E = \omega - I_p$. While factorization can be derived analytically within certain approximate theories, such as the QO theory (see Equation 5.26), the effective-range theory [3, 7], or the adiabatic theory [8], its applicability to real atomic and molecular targets has to be tested. Such tests for HHG have been carried out using numerical solutions of the TDSE [9–11] for one-electron atomic systems and H_2^+ molecular ion [12].

The factorization of Equation 5.32 has many important implications. If it applies to the TDSE result, it should be applicable to the SFA result as well. Thus the wave packet defined from $W(E) = D(\omega)/d(\omega)$ using the TDSE and SFA can differ at most by a constant. Figure 5.5 compares $|W(E)|^2$ obtained from an Ar atom under a four-cycle pulse with a peak intensity of 10^{14} W/cm^2 and wavelength of 1,600 nm. Clearly, the shapes from the two calculations have very similar behaviors. The difference in the magnitude originates from a difference in the ionization rate. This result is actually not surprising. Recall that in the second step of the three-step model, where the electron is far away from the ion core, it is mostly accelerated by the laser field. Both the TDSE and the SFA fully account for this interaction, and thus the spectral distributions (the ω-dependence) of the wave packet from the two theories are very similar. Based on this property, one can obtain the wave packet $W(\omega)$ from the SFA calculation of HHG, i.e.,

$$W(E,\theta) \approx W^{SFA}(E,\theta) = \frac{D^{SFA}(\omega,\theta)}{d^{PWA}(\omega,\theta)}. \tag{5.33}$$

By multiplying this $W(E,\theta)$ to the accurate transition dipole $d(\omega,\theta)$ obtained from the field-free photoionization code, the QRS theory provides a shortcut for calculating

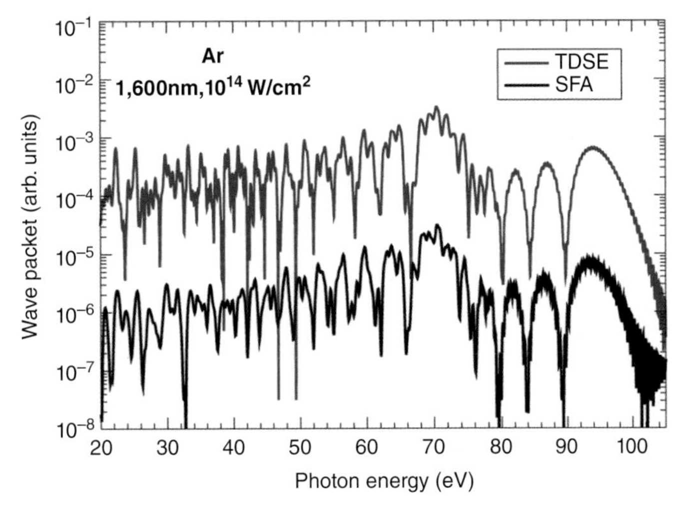

Figure 5.5 Comparison of wave packets defined from $W(E) = D(\omega)/d(\omega)$, where $E = \omega - I_p$ is the photoelectron energy and I_p is the ionization potential, using the TDSE and the SFA model. The modulus squares of the wave packet are shown. The two wave packets differ by mostly a normalization constant. Data has been shifted vertically for clarity. (Reprinted from Anh-Thu Le et al., *J. Phys. B-At. Mol. Opt.*, **49**, 053001 (2016) [2]. Copyrighted by the IOP publishing.)

HHG spectra for molecules. Not only does the QRS theory explicitly identify how the dipole-recombination matrix elements enter the HHG spectra, it also offers an important advantage to using the existing, powerful, quantum chemistry packages for molecular photoionization to calculate HHG spectra, where the many-electron correlation effect can be accounted for without the need to carry out a full, many-electron HHG calculation. Another major advantage is the saving of computer time. The QRS allows a single HHG spectrum to be calculated within seconds to a minute. For another laser intensity, only the wave packet has to be recalculated using the SFA. This simplifies the simulation of macroscopic propagation of the harmonics in the gas medium, where induced dipoles from hundreds of laser intensities in the interaction volume are needed. The wave packet can also be modeled using a reference atom of comparable ionization potential. This method is especially useful when retrieving molecular-transition dipole moments in applications if the latter is accurately known for the reference atom.

5.3.2 Gauge Dependence of HHG Calculation

It is well known that there are different forms for the transition-dipole operator in photoionization theory. For example, for the hydrogen atom, the velocity form and the acceleration form are related to the dipole form by

$$\langle \Psi_0 | \boldsymbol{p} | \Psi_f \rangle = -i\omega \langle \Psi_0 | \boldsymbol{r} | \Psi_f \rangle, \tag{5.34}$$

$$\left\langle \Psi_0 \middle| \frac{\boldsymbol{r}}{r^3} \middle| \Psi_f \right\rangle = \omega^2 \langle \Psi_0 | \boldsymbol{r} | \Psi_f \rangle. \tag{5.35}$$

Similarly, the laser-induced dipole can also be calculated with the velocity gauge and the acceleration gauge as

$$\mathbf{v}(t) = \langle \Psi(t)|\mathbf{p}|\Psi(t)\rangle, \tag{5.36}$$

$$\mathbf{a}(t) = \left\langle \Psi(t) \left| \frac{\partial V}{\partial \mathbf{r}} \right| \Psi(t) \right\rangle, \tag{5.37}$$

where the expressions for the HHG power are given in each gauge, respectively, by

$$P(\omega) \propto \omega^2 |v_x(\omega)|^2, \tag{5.38}$$

$$P(\omega) \propto |a_x(\omega)|^2. \tag{5.39}$$

Here, it is assumed that the laser is polarized along the x-axis and $v(\omega)$, $a(\omega)$ are the Fourier transforms of the corresponding quantities in the time domain. The HHG spectra calculated according to the SFA model are different for length gauge, velocity gauge, and acceleration gauge. However, the returning electron wave packets are nearly independent of the gauge chosen. In other words,

$$W^{SFA}(E) = \frac{D_x^{SFA}(\omega)}{\langle \Psi_0|x|\boldsymbol{p}\rangle} \approx \frac{v_x^{SFA}(\omega)}{\langle \Psi_0|p_x|\boldsymbol{p}\rangle} \approx \frac{a_x^{SFA}(\omega)}{\langle \Psi_0|\partial V/\partial x|\boldsymbol{p}\rangle}. \tag{5.40}$$

Figure 5.6 compares the wave packets from the three different gauges. They are indeed indistinguishable.

It is of interest to emphasize that the returning electron wave packet is a very convenient concept for understanding the three-step model, or equivalently, the rescattering theory. The returning wave packet $W(E)$ (or its modulus square $|W(E)|^2$), within the context of the QRS, is not something that can be directly measured experimentally. In fact, only its expression in the energy domain is defined. The wave packet was cast in the form of standard quantum-collision theory such that scattering cross-sections defined there can be

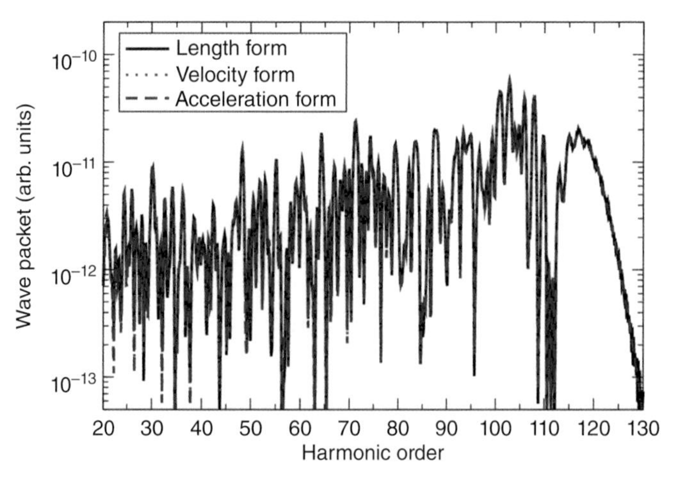

Figure 5.6 Returning electron wave packets extracted from the SFA calculations are shown to be independent of the length, velocity, and acceleration gauges used even though the HHG yields and the transition dipoles are gauge dependent. Shown are the modulus squares of the wave packets. Calculations were performed for atomic hydrogen in a laser field with a wavelength of 1,600 nm, a six-cycle total duration, and a peak intensity of 10^{14} W/cm^2. (Reprinted from Anh-Thu Le et al., *J. Phys. B-At. Mol. Opt.*, **49**, 053001 (2016) [2]. Copyrighted by the IOP publishing.)

used. Unlike simulations using classical theory for the electron, there is no information about the spatial extent of the returning wave packet that can be extracted directly from experiments. Still, the returning electron wave packet is a very powerful concept for rescattering physics. The validity of the factorization of Equation 5.32 renders HHG a powerful spectroscopic tool similar to the conventional photoionization measurements for probing the structure of matter, albeit in a complementary role. While photoionization probes fine details of a molecule in a very narrow energy range, HHG spectroscopy probes a molecule over a broad energy bandwidth, or its short-time behavior. Both approaches rely on the dipole-transition-matrix elements between the initial and final states.

5.4 Phase Matching and Propagation of HHG in the Gas Medium

High-order harmonics are generated by an incident laser beam focused into the nonlinear optical medium to achieve high intensity. Similar to sum or difference frequency generation in crystals, the efficient generation of new frequencies requires good phase matching. This section first discusses the simple phase-matching model that can be solved analytically. It then presents the standard formulation for treating propagation effects of HHG in the gas medium.

5.4.1 General Consideration of Phase Matching with Focused Gaussian Beams

Following Boyd [Boyd] (section 2.10), the electric field $E_q(r,z,t)$ and polarization $P_q(r,z,t)$ of a cylindrical beam can be represented as

$$E_q(r,z,t) = A_q(r,z)e^{i(k_q z-\omega_q t)} + c.c., \tag{5.41}$$

$$P_q(r,z,t) = p_q(r,z)e^{i(k'_q z-\omega_q t)} + c.c. \tag{5.42}$$

It is assumed that the focused laser beam is Gaussian. Such a Gaussian beam can be expressed in a compact form as

$$A_1(r,z) = \frac{\mathcal{A}_1}{1+i\zeta}e^{-r^2/w_0^2(1+i\zeta)}. \tag{5.43}$$

Here,

$$\zeta = 2z/b, \tag{5.44}$$

is a dimensionless longitudinal coordinate defined in terms of the confocal parameter b (see Equation 1.262). Note that Equation 5.43 is equivalent to Equation 1.249 for describing a Gaussian beam and the amplitude of the qth harmonic obeys the equation (using SI units)

$$2ik_q\frac{\partial A_q}{\partial z} + \nabla_\perp^2 A_q = -\mu_0\omega_q^2\chi^{(q)}A_1^q e^{i\Delta k_q z}, \tag{5.45}$$

where $\Delta k_q = qk_1 - k_q$ and $p_q = \chi^{(q)}A_1^q$ is the polarizability for the qth harmonic emission. It turns out that Equation 5.45 can be solved in a closed form

$$A_q(r,z) = \frac{\mathcal{A}_q(z)}{1+i\zeta} e^{-qr^2/w_0^2(1+i\zeta)}, \tag{5.46}$$

where

$$\mathcal{A}_q(z) = i\frac{\mu_0 c q \omega}{2n} \chi^{(q)} A_1^q J_q(\Delta k_q, z_0, z), \tag{5.47}$$

$$J_q(\Delta k_q, z_0, z) = \int_{z_0}^{z} \frac{e^{i\Delta k_q z'} dz'}{(1+2iz'/b)^{q-1}}. \tag{5.48}$$

Here, z_0 is the position of the entrance and z is the position of the exit of the nonlinear medium. From Equation 5.46 the beam waist of the qth harmonic at the far field is $q^{1/2}$ times smaller than that of the incident laser beam. The divergence angle of the harmonic is also $q^{1/2}$ times smaller.

The integral in Equation 5.48 can be evaluated analytically in the plane-wave limit where $b \gg |z_0|, z$. Thus, the integral is reduced to

$$|J_q(\Delta k_q, z_0, z)|^2 = L^2 sinc^2\left(\frac{\Delta k_q L}{2}\right), \tag{5.49}$$

where $sinc(x) = \sin(x)/x$. This equation shows that the harmonic yields are proportional to L^2, when the perfect phase-matching conditions are achieved, i.e., $\Delta k_q = 0$, where $L = z - z_0$ is the length of the interaction region. When Δk is not equal to zero, the harmonic yield decreases and is governed by $sinc^2(\Delta k_q L/2)$. For a tightly focused Gaussian beam, the Gouy phase contributes to the phase mismatch and the analysis of phase matching becomes quite complicated. Note that in this model, Δk_q is assumed to be constant inside the gas medium.

Equation 5.49, seems to imply that harmonic yield can be increased by increasing the length of the gas medium. In reality, this is not so because high harmonics lie in the spectral range from extreme ultraviolet all the way to soft X-rays. They are readily reabsorbed by the gas medium. The absorption length of such ionizing light with frequency ω is given by $L_{abs} = 1/\rho\sigma(\omega)$ where ρ is the gas density and $\sigma(\omega)$ is the ionization cross-section. Assuming that Δk_q is constant, the coherent length is defined as a distance over which the fundamental wave front and the qth harmonic wave front become out of phase and is given by

$$L_{coh} = \pi/\Delta k_q, \tag{5.50}$$

where Δk_q is the phase mismatch between the generated harmonic and the fundamental laser field.

By using a 1D propagation model, it has been found [13] that the condition for optimal harmonic yield is

$$\begin{aligned} L &> 3L_{abs}, \\ L_{coh} &> 5L_{abs}, \end{aligned} \tag{5.51}$$

such as to ensure that the macroscopic response is more than half that of the maximum response. Figure 5.7 illustrates the role of phase mismatch or the coherence length L_{coh} and the length of the medium L in comparison with the absorption length L_{abs} for optimal yield.

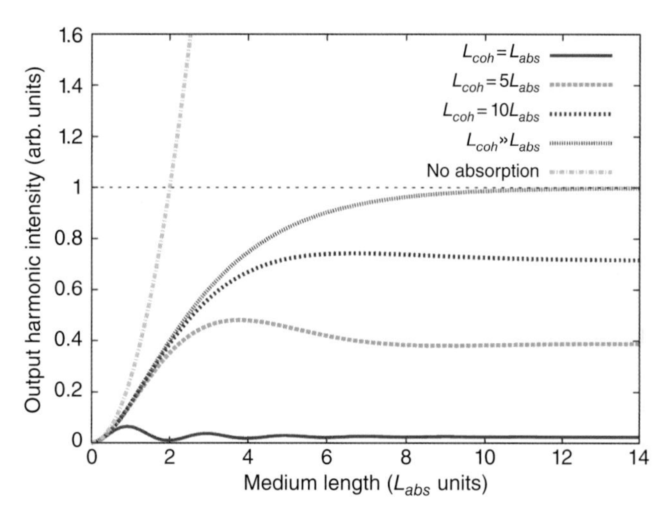

Figure 5.7 The dependence of output photon flux versus the degree of coherence length $L_{coh} = \pi/\Delta k$ expressed in units of absorption length $L_{abs} = 1/\rho\sigma(\omega)$. (Reprinted from E. Constant et al., *Phys. Rev. Lett.*, **82**, 1668 (1999) [13]. Copyrighted by the American Physical Society.)

5.4.2 Factors Contributing to Phase Mismatch in the Optical Medium

For a focused laser beam, the spatiotemporal variation of the laser electric field affects the phase front in position and time. An efficient harmonic-emission process requires that wavefronts of the fundamental laser and generated high harmonics must be in phase. In general, the phase mismatch is the sum of four terms and can be expressed as

$$\begin{aligned} \Delta k_q &= (qk_1 - k_q) + K_{q,dip} \\ &= \Delta k_{q,geo} + \Delta k_{q,el} + \Delta k_{q,at} + K_{q,dip}. \end{aligned} \tag{5.52}$$

Here, k_q and k_1 are wave numbers of the high harmonic and the fundamental laser fields, respectively. In Equation 5.52, phase mismatch is caused by (i) the difference between the diffraction for the individual harmonic and the fundamental beam due to geometrical propagation effects; (ii) the wavelength-dependent index of refraction of the neutral atomic and the ionized medium; and (iii) the dependence of the intrinsic phase of the harmonics on the laser intensity in both longitudinal and radial directions.

According to Equation 5.49, the intensity I_q of the qth harmonic at the end of a nonlinear medium without the absorption effect is given as

$$I_q \propto L^2 sinc^2\left(\frac{\Delta k_q L}{2}\right), \tag{5.53}$$

where L is the medium length. For perfect phase matching with $\Delta k_q = 0$, the harmonic intensity increases quadratically with the propagation distance. Otherwise, the harmonic intensity will oscillate sinusoidally with the propagation distance. The periodicity of this oscillation is twice the coherence length, L_{coh}.

Geometric Dispersion

To initiate a harmonic generation process, a femtosecond laser pulse has to be focused to achieve enough intensity. This introduces a geometric phase for the fundamental laser and the generated harmonics. The phase mismatch is expressed as

$$\Delta k_{q,geo} = qk_{1,geo}(r,z) - k_{q,geo}(r,z). \tag{5.54}$$

There are two main geometries in practice. One is to focus a laser beam in free space, which can be described by a Gaussian beam. The other is to guide a laser beam in a hollow-core waveguide (or fiber) to maintain high intensity over an extended propagation length.

For a focused beam in free space, the geometrical phase shift around the focal point is due to the Gouy phase shift and is given by

$$\phi_{Gouy,1}(z) = -\tan^{-1}\frac{2z}{b}, \tag{5.55}$$

where $b = 2\pi w_0^2/\lambda_1$ is the confocal parameter with beam waist w_0 and laser wavelength λ_1. Assume that the qth harmonic is also Gaussian, its Gouy phase is given in the form of Equation 5.55 as well but its conformal parameter increases by q, thus

$$\phi_{Gouy,q}(z) = -\tan^{-1}\frac{2z}{qb}. \tag{5.56}$$

Thus the Guoy phase from the qth harmonic can be neglected. This gives (at $r = 0$)

$$\begin{aligned}\Delta k_{q,geo}(0,z) &\approx qk_{1,geo}(0,z)\\ &= q\partial\phi_{Gouy,1}(z)/\partial z\\ &= -\frac{2q}{b}\frac{1}{1+(2z/b)^2}.\end{aligned} \tag{5.57}$$

In Equation 5.57, the phase mismatch caused by the Gouy phase varies along the propagation direction z. Thus, it can be adjusted by placing the laser focus at a different position with respect to the nonlinear medium.

Free Electron or Plasma Dispersion

The presence of free electrons in the medium changes the index of refraction. The refractive index of the fundamental laser by a plasma is given by

$$n_p(\omega_1) = \sqrt{1 - \frac{\omega_p^2}{\omega_1^2}}, \tag{5.58}$$

where ω_1 is the laser frequency, and ω_p is the plasma frequency. The plasma frequency is given by

$$\omega_p = e\sqrt{\frac{n_e}{\varepsilon_0 m_e}}, \tag{5.59}$$

where n_e is the density of free electrons, ε_0 is the vacuum permittivity, e is the electron charge and m_e is the electron mass.

From Equation 5.58, the refractive index of a plasma is always less than one and depends on the frequency of the laser beam. Since the laser frequency ω_1 is much larger than the plasma frequency ω_p, i.e., $\omega_p^2 \ll \omega_1^2$, Equation 5.58 can be rewritten as

$$n_p(\omega_1) \approx 1 - \frac{\omega_p^2}{2\omega_1^2}. \tag{5.60}$$

Thus phase mismatch caused by the presence of free electrons (with spatiotemporal dependence expressed explicitly) is

$$\begin{aligned} \Delta k_{q,el} &= qk_{1,el}(r,z,t) - k_{q,el}(r,z,t) \\ &\approx -\frac{e^2 n_e(r,z,t)}{4\pi\varepsilon_0 m_e c^2} q\lambda_1 \\ &= -qr_0 n_e(r,z,t)\lambda_1. \end{aligned} \tag{5.61}$$

This effect is only considered for the fundamental laser and not for high harmonics since the frequencies of high harmonics are much larger than the plasma frequency. In Equation 5.61, the classical electron radius r_0 is defined as

$$r_0 = \frac{1}{4\pi\varepsilon_0}\frac{e}{m_e c^2}. \tag{5.62}$$

Neutral Atom Dispersion

The index of refraction depends on the wavelength. Thus, they are different for the fundamental laser and the high harmonics. The phase mismatch due to neutral atom dispersion is given by

$$\begin{aligned} \Delta k_{q,at} &= qk_{1,at}(r,z,t) - k_{q,at}(r,z,t) \\ &= q\frac{2\pi}{\lambda_1}\delta n(1 - p_{ion}) \\ &\approx \frac{n_0(r,z,t)\pi\alpha_1}{\lambda_q} + n_0(r,z,t) r_0 \lambda_q f_1, \end{aligned} \tag{5.63}$$

where δn is the refractive index difference between the fundamental laser and high harmonic, p_{ion} is the ionization level, $n_0(r,z,t)$ is the neutral atom density, $\lambda_q = \lambda_1/q$, α_1 is the dipole polarizability of the fundamental laser and the atomic scattering factor is $f = f_1 + if_2$. The imaginary part of f is related to absorption length by $L_{abs}^{-1} = 2r_0\lambda_q n_0(r,z,t) f_2$. For the fundamental laser's wavelength far from a resonance, its refractive index in a conversion medium can be obtained using the Sellmeier equations. The refractive index of high harmonics in the XUV or soft X-ray region is generally smaller than one.

Note that when a high intensity laser propagates through a medium, it also causes a change in the refractive index that is proportional to the laser intensity through the third-order susceptibility (usually called the "Kerr effect"). This change is instantaneous and is given by

$$\Delta n = \eta_2 I(t), \tag{5.64}$$

where η_2 is the nonlinear refractive index.

The contribution of neutral gas dispersion to phase mismatch is positive and that of plasma dispersion is negative. The total phase mismatch can be reduced by varying the laser intensity to adjust the fraction of ionization. A critical value of the ionization fraction p_{cr} exists. Above this value, the dispersion from the remaining neutral atoms is not enough to balance the dispersion from the free electrons. From Equations 5.61 and 5.63, this critical value can be expressed as

$$p_{cr}(\lambda_1) = \left[\frac{N_0 r_0 \lambda_1^2}{2\pi \delta n} + 1 \right]^{-1}, \tag{5.65}$$

where N_0 is the initial neutral atom density. It is clear that the critical ionization fraction is independent of the pressure. Values for p_{cr} are on the order of a few percent in the near-infrared region, for instance, $\approx$ 4% (1.5%) for Ar, 1% (0.4%) for Ne and 0.5% (0.2%) for He at 0.8 μm (1.3 μm) driving laser wavelengths [14]. This critical ionization level decreases monotonically as the driving laser wavelength increases.

Induced Dipole Phase

The phase of high harmonics has a strong dependence on laser intensity. Variation of laser intensity in space results in transverse and longitudinal gradients that contribute to the phase mismatch

$$K_{q,dip} = \nabla \varphi_{q,dip}. \tag{5.66}$$

Here, the harmonic phase $\varphi_{q,dip}$ is the action accumulated by an electron during its excursion in the laser field before it recombines with the atomic ion to emit the qth harmonic. The phase has been found to be given approximately by

$$\varphi_{q,dip} = -\alpha_i^q I, \tag{5.67}$$

where I is the instantaneous laser intensity. The proportional constant depends on whether i is a short or long trajectory, with the coefficient $\alpha_{i=S}^q \approx 1\times10^{-14}$ rad cm^2/W and $\alpha_{i=L}^q \approx 24\times10^{-14}$ rad cm^2/W. At the cutoff, $\alpha_{i=S,L}^q \approx 13.7\times10^{-14}$ rad cm^2/W [15, 16]. The intensity dependence of the dipole phase is different for "short" and "long" trajectories; this leads to different good phase-matching conditions for the two trajectories. This dipole phase is also responsible for the spectral broadening of harmonics because the intensity variation $I(t)$ in time causes a frequency chirp $\Delta\omega_q(t) = -\partial\varphi_{q,dip}(t)/\partial t$.

For a Gaussian beam, the contribution of dipole phase to phase mismatch of the on-axis qth harmonic is given by

$$\begin{aligned} \frac{\partial \varphi_{q,dip}(0,z)}{\partial z} &= -\alpha_i^q \frac{\partial I(0,z)}{\partial z} \\ &= \frac{8z}{b} \frac{1}{[1+(2z/b)^2]^2} \alpha_i^q I_0, \end{aligned} \tag{5.68}$$

where I_0 is the peak intensity at the laser focus.

The phase mismatch given in Equation 5.68 also depends on the propagation distance z and its sign is different for positive and negative z. Thus, it is possible to compensate for the geometric phase mismatch in Equation 5.57 by placing the gas medium in the proper position.

5.4.3 Propagation of the Driving Laser and High Harmonics in a Gaseous Medium

As described above, all the phase mismatch factors depend on the spatiotemporal fields of the intense laser and the high harmonics; the evolution of these fields must be calculated numerically. Since the intense laser is much stronger than high harmonics, the propagation of the driving laser field and high-harmonic fields can be treated separately.

Propagation of the Fundamental Laser Field

The 3D Maxwell's wave equation for the fundamental laser is (in SI units) [17, 18]

$$\nabla^2 E_1(r,z,t) - \frac{1}{c^2}\frac{\partial^2 E_1(r,z,t)}{\partial t^2} = \mu_0 \frac{\partial J_{\text{abs}}(r,z,t)}{\partial t} + \frac{\omega_1^2}{c^2}(1-\eta_{\text{eff}}^2)E_1(r,z,t), \quad (5.69)$$

where $E_1(r,z,t)$ is the transverse electric field with the central frequency ω_1. The effective refractive index η_{eff} of the gas medium can be written as

$$\eta_{\text{eff}}(r,z,t) = \eta_0(r,z,t) + \eta_2 I(r,z,t) - \frac{\omega_{\text{p}}^2(r,z,t)}{2\omega_1^2}. \quad (5.70)$$

The first term $\eta_0 = 1 + \delta_1 - i\beta_1$ accounts for refraction (δ_1) and absorption (β_1) by the neutral atoms, the second term accounts for the optical Kerr nonlinearity that is proportional to laser intensity $I(t)$, and the third term accounts for the free electrons expressed in terms of plasma frequency as defined in Equation 5.59. Absorption due to the medium ionization is given by

$$J_{\text{abs}}(t) = \frac{\gamma(t) n_0(t) I_{\text{p}} E_1(t)}{|E_1(t)|^2}, \quad (5.71)$$

where $\gamma(t)$ is the ionization rate, I_{p} is the ionization potential, and $n_0(t)$ is the density of neutral atoms.

For the fundamental infrared laser field, absorption can be neglected. By going to the moving coordinate frame ($z' = z$ and $t' = t - z/c$) and using slow-envelope approximation to neglect the $\partial^2 E_1/\partial z'^2$ term, the equation governing the laser field is given by

$$\nabla_\perp^2 E_1(r,z',t') - \frac{2}{c}\frac{\partial^2 E_1(r,z',t')}{\partial z' \partial t'} = \mu_0 \frac{\partial J_{\text{abs}}(r,z',t')}{\partial t'} + \frac{\omega_{\text{p}}^2}{c^2} E_1(r,z',t') - 2\frac{\omega_1^2}{c^2}[\delta_1 + \eta_2 I(r,z',t')] E_1(r,z',t'). \quad (5.72)$$

This equation can be solved in the frequency domain. By applying the Fourier transform to Equation 5.72, the equation to be solved is

$$\nabla_\perp^2 \tilde{E}_1(r,z',\omega) - \frac{2i\omega}{c}\frac{\partial \tilde{E}_1(r,z',\omega)}{\partial z'} = \tilde{G}(r,z',\omega), \quad (5.73)$$

where $\tilde{E}_1(r,z',\omega)$ and $\tilde{G}(r,z',\omega)$ are the Fourier transform of the electric field and the quantity on the right-hand side of Equation 5.72.

Propagation of High-Harmonic Fields

The starting equation for the propagation of the high-harmonic field is

$$\nabla^2 E_h(r,z,t) - \frac{1}{c^2}\frac{\partial^2 E_h(r,z,t)}{\partial t^2} = \mu_0 \frac{\partial^2 P(r,z,t)}{\partial t^2}. \tag{5.74}$$

By using the slow-envelope approximation and the moving frame, the above equation becomes

$$\nabla_\perp^2 E_\mathrm{h}(r,z',t') - \frac{2}{c}\frac{\partial^2 E_\mathrm{h}(r,z',t')}{\partial z'\partial t'} = \mu_0 \frac{\partial^2 P(r,z',t')}{\partial t'^2}. \tag{5.75}$$

The Fourier transform of the polarization on the right of Equation 5.75 can be written as the sum of the linear and nonlinear terms,

$$\tilde{P}(r,z',\omega) = \chi^{(1)}(\omega)\tilde{E}_\mathrm{h}(r,z',\omega) + \tilde{P}_\mathrm{nl}(r,z',\omega), \tag{5.76}$$

where the linear susceptibility $\chi^{(1)}(\omega)$ includes both linear dispersion and absorption. The nonlinear polarization term $\tilde{P}_\mathrm{nl}(r,z',\omega)$ is given by

$$\tilde{P}_\mathrm{nl}(r,z',\omega) = \hat{F}\left\{[N_0 - n_\mathrm{e}(r,z',t')]D(r,z',t')\right\}, \tag{5.77}$$

where $\hat{F}$ is the Fourier transform operator acting on the temporal coordinate, N_0 is the initial neutral atom density and $D(r,z',t')$ is the single-atom-induced dipole moment caused by the fundamental driving laser field. The refractive index $n(\omega) = \sqrt{1 + \chi^{(1)}(\omega)/\varepsilon_0}$ [Boyd] is related to atomic scattering factors by

$$\begin{aligned} n(\omega) &= 1 - \delta_\mathrm{h}(\omega) - i\beta_\mathrm{h}(\omega) \\ &= 1 - \frac{1}{2\pi}N_0 r_0 \lambda^2 (f_1 + if_2), \end{aligned} \tag{5.78}$$

where r_0 is the classical electron radius, λ is the harmonic wavelength, and f_1 and f_2 are atomic scattering factors. Finally, Equation 5.75 can be written in the frequency domain as

$$\begin{aligned} \nabla_\perp^2 \tilde{E}_\mathrm{h}(r,z',\omega) &- \frac{2i\omega}{c}\frac{\partial \tilde{E}_\mathrm{h}(r,z',\omega)}{\partial z'} \\ &- \frac{2\omega^2}{c^2}(\delta_\mathrm{h} + i\beta_\mathrm{h})\tilde{E}_\mathrm{h}(r,z',\omega) = -\omega^2\mu_0\tilde{P}_\mathrm{nl}(r,z',\omega), \end{aligned} \tag{5.79}$$

where $\tilde{E}_h(r,z',\omega)$ is the Fourier transform of the harmonic field. The nonlinear polarization is the source of the high-harmonic field. The propagation in the medium is carried out to the exit face to obtain the near-field harmonics. To obtain the harmonics at the far field where they are actually measured, a Hankel transformation is further carried out from the near fields. Computer codes for carrying out propagation calculations outlined here have been developed by a small number of research groups.

Harmonic Emission in the Far Field

Once high harmonics are emitted at the exit plane of a nonlinear medium (called near-field harmonics), as shown in Figure 5.8, they are further propagated in the vacuum until they

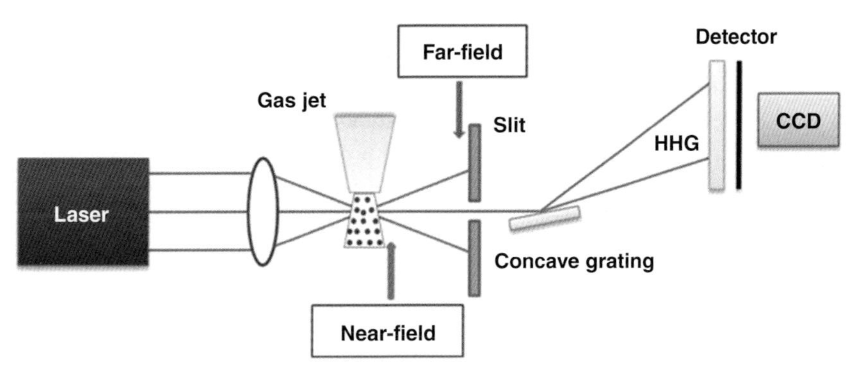

Figure 5.8 Typical configuration for measuring the HHG in the far field.

are detected by the spectrometer. In this process, high harmonics may go through a slit or an iris and be reflected by a mirror or a more complicated optical system before they reach the detector (called far-field harmonics). In an axial-symmetric optical system, the complex electric field on the initial plane (near field) is related to the final plane (far field) by an *ABCD* ray matrix, and $AD - BC = 1$ for a lossless system. Here, only the simplest configuration shown in Figure 5.8 is considered without any additional optics (or within the free-space propagation) between near and far fields. By using Equation 1.268, one obtains $A = 1$, $B = z_\mathrm{f} - z'$, $C = 0$ and $D = 1$ in the *ABCD* matrix, where z_f and z' are the far-field and near-field positions from the laser focus, respectively. According to the diffraction theory in the paraxial approximation, far-field harmonics can be calculated from near-field harmonics using Equation 1.289 [Siegman][1]

$$E_\mathrm{h}^\mathrm{f}(r_\mathrm{f}, z_\mathrm{f}, \omega) = ik \int \frac{\tilde{E}_\mathrm{h}(r, z', \omega)}{z_\mathrm{f} - z'} J_0\left(\frac{krr_\mathrm{f}}{z_\mathrm{f} - z'}\right) \exp\left[-\frac{ik(r^2 + r_\mathrm{f}^2)}{2(z_\mathrm{f} - z')}\right] rdr, \tag{5.80}$$

where J_0 is the zero-order Bessel function, r_f is the transverse coordinate in the far field and the wave vector k is given by $k = \omega/c$ (again, the propagation factor $\exp[-ik(z_\mathrm{f} - z')]$ is not included in the above expression). The conversion process is called a *Hankel transformation*. By integrating the harmonic yield over the far-field plane, the power spectrum is obtained as

$$S_\mathrm{h}(\omega) \propto \int\int |E_\mathrm{h}^\mathrm{f}(x_\mathrm{f}, y_\mathrm{f}, z_\mathrm{f}, \omega)|^2 dx_\mathrm{f} dy_\mathrm{f}, \tag{5.81}$$

where x_f and y_f are Cartesian coordinates on the plane perpendicular to the propagation direction, and $r_\mathrm{f} = \sqrt{x_\mathrm{f}^2 + y_\mathrm{f}^2}$.

Note that, in Equation 5.81, detailed information on the experimental setup is involved. Besides laser parameters such as intensity, duration, wavelength, and spot size, to simulate experimental HHG spectra quantitatively, one needs other information about the experiment: for example, the size and location of a slit if it is used.

[1] $\tilde{E}_\mathrm{h}(r, z', \omega)$ is expressed in the frequency domain and its phase is also involved. Due to the different convention of Fourier transformation, this phase may need to change its sign before entering into the formula.

Propagation of the Driving Laser and High Harmonics in a Hollow Waveguide

High-order harmonics are also often generated in a hollow waveguide. This subsection deals with the solution of propagation equations in such a setup. The propagation equation Equation 5.73 for the fundamental pulse and Equation 5.79 for each harmonic order are to be solved using the operator-splitting method. In this method, the calculation of the advance of the electric field from z' to $z' + \Delta z'$ is separated into two steps as shown in the following:

$$\frac{\partial \tilde{E}_{1,\mathrm{h}}(r,z',\omega)}{\partial z'} = -\frac{ic}{2\omega}\nabla_{\perp}^{2}\tilde{E}_{1,\mathrm{h}}(r,z',\omega), \tag{5.82}$$

$$\frac{\partial \tilde{E}_{1,\mathrm{h}}(r,z',\omega)}{\partial z'} = \frac{ic}{2\omega}R[\tilde{E}_{1,\mathrm{h}}(r,z',\omega)], \tag{5.83}$$

where $\tilde{E}_{1,\mathrm{h}}$ represents the electric field of the fundamental or the harmonics, and $R(\tilde{E}_{1,\mathrm{h}})$ stands for all the linear and nonlinear terms on the right-hand sides of Equations 5.73 and 5.79.

In order to impose the boundary conditions of the hollow waveguide, $\tilde{E}_{1,\mathrm{h}}(r,z',\omega)$ is written as a superposition of eigenmodes as

$$\tilde{E}(r,z',\omega) = F(\omega)\sum_{j} b_j(z')J_0(\mu_j r/a), \tag{5.84}$$

where the μ_j are the roots of the Bessel function of the first kind $J_0(\mu_j)=0$, a is the hollow-core radius, and $F(\omega)$ is the input pulse in the frequency domain. Inserting Equation 5.84 into Equation 5.82, and using the orthogonality relation $\int_0^a J_0(\mu_i r/a)\cdot J_0(\mu_j r/a) r dr = (a^2/2)\delta_{ij}[J_1(\mu_j)]^2$ for each ω, one obtains[2]

$$b_j(z'+\Delta z') = b_j(z')\cdot \exp(-i\kappa_j\Delta z' - \alpha_j\Delta z'), \tag{5.85}$$

where the propagation constant κ_j [19] is

$$\kappa_j = \frac{\lambda}{4\pi}\left(\frac{\mu_j}{a}\right)^2, \tag{5.86}$$

and the mode loss term [19] is

$$\alpha_j = \left(\frac{\mu_j}{2\pi}\right)^2 \frac{\lambda^2}{a^3}\frac{(n_x^2+1)}{2\sqrt{n_x^2-1}}. \tag{5.87}$$

Here, n_x is the refractive index of the cladding and $\lambda = 2\pi c/\omega$.

In the calculation, it is usually assumed that the spatial beam at the entrance of a hollow waveguide is the lowest EH_{11} mode. This can be achieved experimentally by adjusting the ratio between the waist of the incident laser beam and the radius of the waveguide to be about 65%. Once Equations 5.82 and 5.83 are solved inside the gas-filled hollow waveguide to the exit plane, the harmonics emitted at the near field are obtained. To obtain far-field harmonics, a Hankel transformation can be applied.

[2] Note that the sign before $\kappa_j\Delta z'$ may be plus due to the convention of Fourier transform.

5.5 Dependence of HHG Spectra on Macroscopic Conditions

5.5.1 Retrieval of Target Photorecombination Cross-Sections from HHG Generated in a Macroscopic Medium

In recent years, harmonic generation itself has become a tool for obtaining the structural information of atoms and molecules. It was often assumed that the experimental harmonic spectra are perfectly phase matched such that the modulus square of the single-photon atomic-transition dipole can be retrieved from the experimental data. At the single-atom level, the QRS theory has been established to separate the target structural information and laser properties in the harmonic spectrum. Due to the propagation effect, this does not prove that the transition dipole can be directly extracted from the experimental HHG spectra. To show its possibility, it is assumed that, after propagation in the gas medium, HHG spectra can be expressed as

$$S_h(\omega) \propto \omega^4 |W'(\omega)|^2 |d(\omega)|^2, \tag{5.88}$$

where $W'(\omega)$ is called the macroscopic wave packet (MWP) in order to distinguish it from the single-atom response and $d(\omega)$ is the photorecombination transition-dipole moment of a single atom.

To test the validity of Equation 5.88, HHG spectra from two atomic targets of similar ionization potential were investigated in Jin et al. [20], for example, using the MWPs from a hydrogen-like system, where the effective nuclear charge has been adjusted such that its $1s$ binding energy is the same as the $3p$ ground-state energy of Ar. Using the same laser parameters (800 nm and 19.4 fs) and focusing condition and the known transition dipoles for both systems, the two MWPs obtained are shown in Figure 5.9(a). The two MWPs are normalized at the cutoff energy (marked by an arrow) and indeed agree relatively well. The agreement gets better as the laser intensity is decreased. This is shown in Figure 5.9(b) where the laser peak intensity (in the center of the gas jet) is reduced to 1.25×10^{14} W/cm^2. This shows that MWP is mostly determined by the laser parameters and focusing condition and is considered to be independent of the target. The agreement is reflected even when good phase matching is not met (see Figure 5.9(c)), where the laser peak intensity (in the center of the gas jet) is kept at 1.5×10^{14} W/cm^2, but the gas jet is located at 1.5 mm after the focus.

It is mentioned that, in the simulation, the fundamental laser field has been set to be unchanged by the gas medium. This occurs when the laser intensity is low and the gas pressure is also low. The effect of propagation is mainly to phase match short-trajectory harmonics. Modification of the medium at low laser intensity is not serious. Thus, one can neglect the effect of propagation on the MWPs by different media. If Equation 5.88 is valid and the MWP only depends on the laser, then it allows the extraction of the modulus square of the atomic-transition dipole, or equivalently, the photoionization cross-section from the experimental HHG spectra.

Two other examples are given for Xe and Ne targets with each of their hydrogen-like systems, respectively. In Figure 5.9(d), the MWP is obtained from a laser pulse with

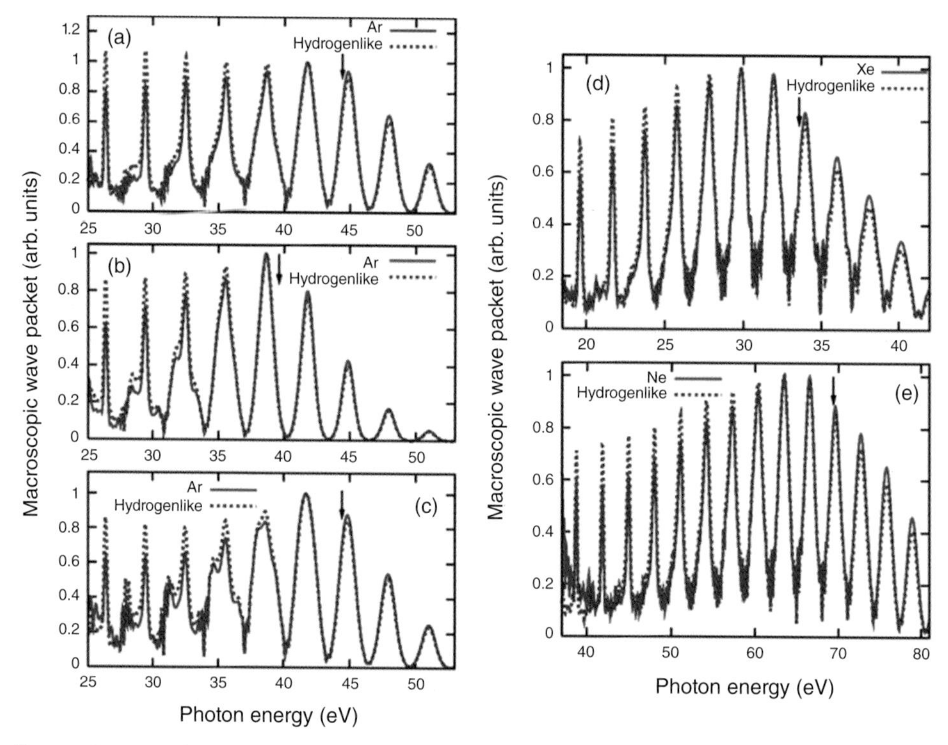

Figure 5.9 Macroscopic wave packet extracted from macroscopic harmonic spectrum based on QRS using a real atom (solid lines) and hydrogenlike atom (dotted lines). (a)–(c) Ar gas jet is 2, 2, and 1.5 mm after the focus, respectively. Laser intensities are 1.5×10^{14}, 1.25×10^{14}, and 1.5×10^{14} W/cm^2, respectively. (d) Xe gas jet is 2 mm after the focus; laser intensity is 5×10^{13} W/cm^2. (e) Ne gas jet is 2.5 mm after the focus; laser intensity is 2.5×10^{14} W/cm^2. The arrows indicate the cutoff energy determined by $I_P + 3.2U_P$; laser intensity given is at the center of the gas jet. (Figure adapted with permission from Cheng Jin, Anh-Thu Le, and C. D. Lin, *Phys. Rev. A*, **79**, 053413 (2009) [20]. Copyrighted by the American Physical Society.)

duration of 21.8 fs, central wavelength of 1,200 nm, and peak intensity of 5×10^{13} W/cm^2 in the center of the Xe gas jet, interacting with the gas jet setting at 2 mm after the focus. In Figure 5.9(e), the MWP is obtained for a laser pulse with duration of 23.3 fs, central wavelength of 800 nm, and peak intensity of 2.5×10^{14} W/cm^2 at the Ne gas jet center, which is 2.5 mm after the focus. These results indeed show that MWPs from different targets with the same I_p agree with each other reasonably well under the same laser condition, even when the two targets have different orbital symmetry in the ground state. These two examples imply that one can extract the transition-dipole of an unknown atom or molecule from one for which the transition dipole moment is known by comparing their measured HHG spectra in the same laser pulse. However, it should be noted that, if the fundamental laser pulse is severely distorted during the propagation, the method is expected to fail since large distortion of the laser field depends on the properties of the gas medium.

5.5.2 Wavelength Scaling of Harmonic Efficiency

To study wavelength scaling of the macroscopic HHG yields, one has to fix all other parameters that may affect the efficiency of harmonic emission. One also has to decide whether it is the total HHG yield or only the HHG yield within a given photon-energy region. For this purpose, we define a parameter that describes the efficiency of harmonic generation. This is the ratio between the output energy (total harmonic energy) with respect to the input energy (fundamental laser energy) for different laser wavelengths.

In Figure 5.10(a) single-atom HHG spectra (only the envelope) calculated for three wavelengths are shown. In the calculation, the laser intensity and duration are kept at 1.6×10^{14} W/cm^2 and 40 fs, respectively. In Figure 5.10(b) the HHG spectra obtained after including macroscopic propagation are shown. In the calculation, the beam waist at the focus is kept at 47.5 μm, a 0.5 mm-long gas jet is placed at 3 mm after the laser focus, and gas pressure is kept at 56 Torr. The yield of each harmonic is obtained by integrating over the whole plane perpendicular to the propagation axis. In Figure 5.10(c), the total harmonic yields are recorded after they have passed a slit (the slit has a width of 100 μm and is placed 24 cm after the gas jet). From Figures 5.10(b) and (c), one can calculate the HHG efficiency per atom versus the wavelength.

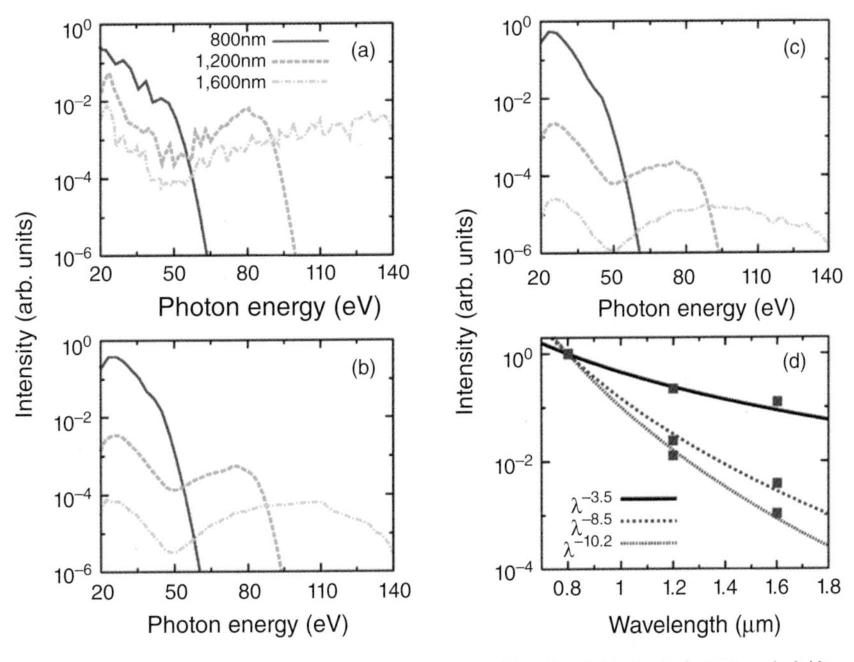

Figure 5.10 (a) Single-atom HHG spectra and macroscopic HHG spectra without (b) and with (c) the slit for 800 nm (solid lines), 1,200 nm (dashed lines), and 1,600 nm (dot-dashed lines) lasers. (d) The wavelength dependence of the integrated HHG yields above 20 eV. The integrated HHG yields in (a), (b), and (c) follow $\lambda^{-3.5\pm0.5}$, $\lambda^{-8.5\pm0.5}$, and $\lambda^{-10.2\pm0.2}$, respectively. (Reprinted from Cheng Jin, Anh-Thu Le, and C. D. Lin, *Phys. Rev. A*, **83**, 023411 (2011) [18]. Copyrighted by the American Physical Society.)

If the HHG yields above 20 eV are integrated as the output energy, the resulting total harmonics energy follows $\lambda^{-3.5\pm0.5}$, $\lambda^{-8.5\pm0.5}$, and $\lambda^{-10.2\pm0.2}$ for the harmonics in Figures 5.10(a–c), respectively, as shown in Figure 5.10(d). It is clear that macroscopic dispersive effects, such as electronic, geometric, dispersion, and the induced-dipole phase result in a more rapid decrease of the HHG scaling with increasing wavelength in comparison to the single-atom response. From these scaling laws, the HHG yields for long-wavelength driving lasers under the same experimental conditions appear quite unfavorable. For practical purposes, experimentally high harmonics with different laser wavelengths are to be generated with different optimized conditions to compensate for the wavelength scaling. In particular, with a long-wavelength driver, the gas pressure can be substantially increased to reach a favorable phase-matching condition. This topic will be further dealt with in Chapter 6.

5.5.3 Macroscopic HHG Spectra of Ar and Xe

Buildup of Harmonics inside the Gas Medium

With the amplitude and phase of single-atom-induced dipole calculated from the TDSE, SFA, or QRS as the source terms for the macroscopic-propagation equations, the macroscopic HHG spectra from these three different models are calculated and compared in this subsection.

Consider an example where Ar atoms in a gas jet are exposed to a 19.4 fs (FWHM) laser pulse with peak intensity of 1.5×10^{14} W/cm^2 and central wavelength of 800 nm, Figure 5.11 compares the HHG spectra after the propagation effect has been included in the simulation. In comparison with the single-atom HHG spectrum, propagation cleans up the spectral features of odd harmonics. In the cutoff region, the SFA gives a correct prediction compared to the TDSE, but fails for the lower-plateau harmonics. After propagation, the QRS model agrees much better with the one obtained from the TDSE over the whole spectral region.

In Figures 5.11(c)–(e), the nineteenth to twenty-third harmonics along the radial distance at the exit after propagation are shown. Comparing to the TDSE, it clearly shows that results from the QRS model are significantly much better than from the SFA model. Next, the phase difference between successive harmonics is shown in Figures 5.12(a) and (b). The results between the two methods (TDSE and QRS) agree very well. In this case, the infrared intensity is relatively weak and the gas jet is placed at 2 mm after the laser focus. It is assumed that the infrared was not modified during its propagation in the gas medium. If the gas jet is placed at the laser focus, Figures 5.12(c) and (d) show that the phase difference calculated from TDSE and QRS show larger differences. The phases of the harmonics are also larger. In the latter case, phase matching is poorly satisfied since the Gouy phase is not well compensated by the dipole phase.

It is important to note that phase-matching conditions are different for each harmonic as it propagates along the beam axis. Thus each harmonic field in space is enhanced or suppressed as it propagates in the medium. Figure 5.13 shows the evolution of four

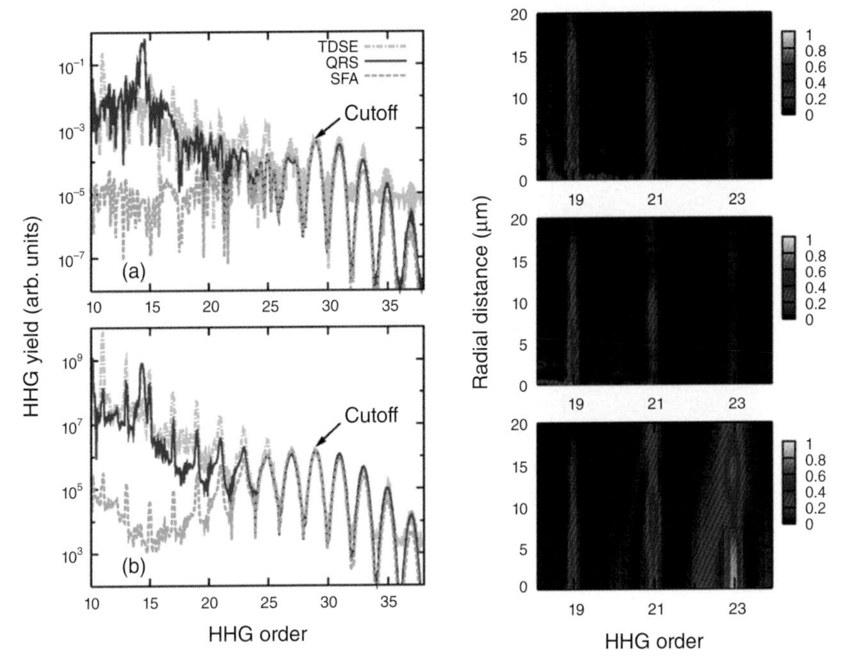

Figure 5.11 (a) Singe-atom HHG of Ar in a 800 nm, 19.4 fs laser with peak intensity of 5×10^{14} W/cm^2, calculated using the TDSE, QRS, and SFA models. (b) The harmonic yields after propagation in the Ar gas medium. The sharp peak in QRS near the fifteenth harmonic occurs because the dipole-matrix element in the SFA model vanishes at that energy such that the retrieved wave packet has a spurious singularity. (c–e) Harmonic yields for H19, H21, and H23 at the exit plane. Clearly, the QRS results compare well with the TDSE, but not the SFA model. (Figure adapted with permission from Cheng Jin, Anh-Thu Le, and C. D. Lin, *Phys. Rev. A*, **79**, 053413 (2009) [20]. Copyrighted by the American Physical Society.)

harmonics inside the gas jet until they reach the exit plane. From these figures, it is clear that reaching the best phase-matching conditions for all the harmonics is difficult.

Comparison of HHG Simulations with Experimental Data

To compare theoretically simulated HHG spectra with experiments, details of the experimental parameters have to be specified. Here, two examples are given in which experimental HHG spectra have been nicely reproduced by the simulations based on the QRS theory. The single-atom response and propagation effects were calculated using parameters provided by the experimentalists.

Figure 5.14 compares the HHG spectra of Ar generated by a 0.5 mm-long gas jet that was located a few mm after the laser focus. Harmonics emitted from the exit plane of the gas jet were further propagated for 24 cm and reached a vertical slit with a width of 100 μm. For a 1,200 (1,360) nm laser in the experiment, the beam waist at the focus is estimated to be 47.5 (52.5) μm, and the pulse duration is ~40 (~50) fs. To reach the best overall fit with experimental data, laser intensity and gas pressure used in the simulation are adjusted. For the 1,200 nm laser, peak intensity for the experiment (theory) is 1.6 (1.5) $\times10^{14}$ W/cm^2

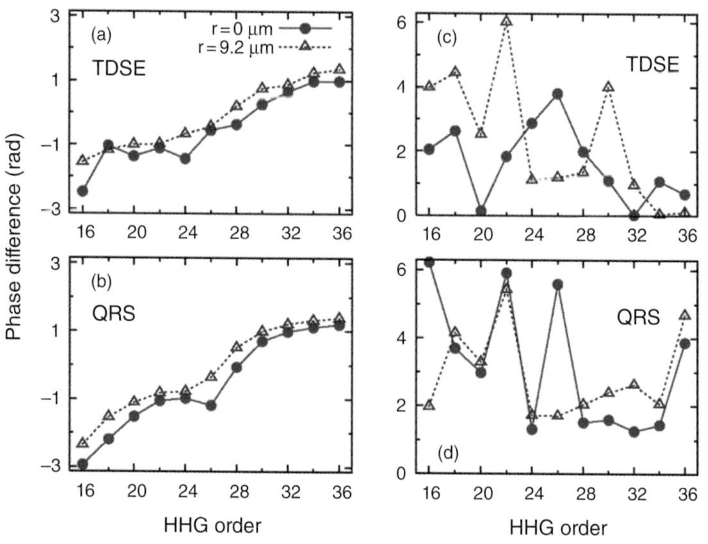

Figure 5.12 Phase difference between successive macroscopic harmonics for Ar from the TDSE (a) and the QRS (b), calculated at the exit plane of the gas jet, at the two radial distances indicated on the graph. The gas jet is located 2 mm after the focus. (c) and (d) are the same as (a) and (b), respectively, but with the gas jet located at the focus. The laser intensity in the center of the gas jet is kept as 1.5×10^{14} W/cm^2. (Figure adapted with permission from Cheng Jin, Anh-Thu Le, and C. D. Lin, *Phys. Rev. A*, **79**, 053413 (2009) [20]. Copyrighted by the American Physical Society.)

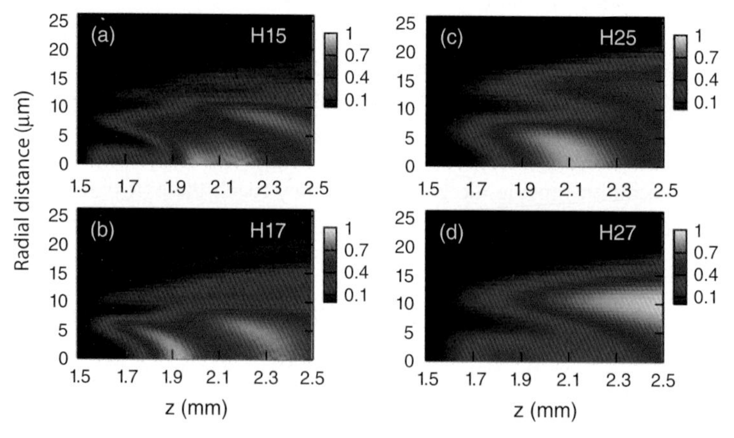

Figure 5.13 Evolution of harmonic intensities on the plane perpendicular to the propagation axis. Because the phase mismatch for each harmonic in space differs, it is not possible to reach best phase matching for all the harmonics on the same plane. (Figure adapted with permission from Cheng Jin, Anh-Thu Le, and C. D. Lin, *Phys. Rev. A*, **79**, 053413 (2009) [20]. Copyrighted by the American Physical Society.)

and gas pressure is 28 (84) Torr. For the 1,360 nm laser, the corresponding intensity and pressure are 1.25 (1.15) $\times 10^{14}$ W/cm^2 and 28 (56) Torr, respectively. In Figure 5.14(a), harmonic distributions along the perpendicular direction have been displayed showing good agreement between the theory and the measured ones for a 1,200 nm laser. The spectra

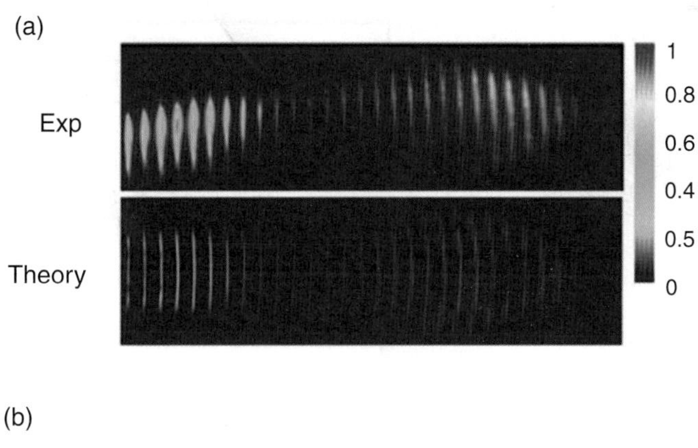

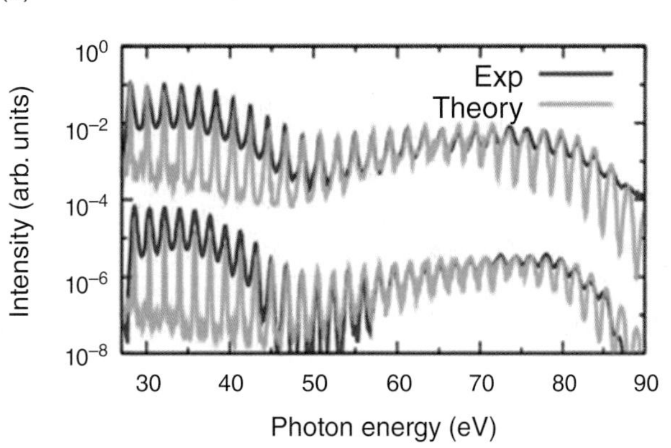

Figure 5.14 HHG spectra of Ar generated by long-wavelength lasers. (a) Spatial distribution of harmonic emission versus photon energy in the far field by a 1,200 nm laser. (b) Comparison of experimental (red lines) and theoretical (green lines) HHG yields integrated over the vertical dimension for 1,200 nm (upper curves) and 1,360 nm (lower curves) lasers. Other laser parameters are given in the text. (Reprinted from Cheng Jin et al., *J. Phys. B-At. Mol. Opt.*, **44** 095601 (2011) [21]. Copyrighted by the IOP publishing.)

integrated over the perpendicular plane are shown in Figure 5.14(b). The agreement between measurement and simulation is very good. The "famous" Cooper minimum near photon energy of about 53 eV is seen clearly in both spectra. Its independence of laser wavelength is consistent with the prediction of the QRS theory when the laser intensity is not too high. Similar extensive measurements by Higuet et al. [23] also confirmed that the Cooper minimum does not change with wavelength and laser intensity except very slightly.

The results above contrast with earlier experiments that look for the Cooper minimum in Ar using 800 nm driving lasers. To see the Cooper minimum, the HHG cutoff energy has to be somewhat higher than 60 eV. With an 800 nm laser, this would require a peak intensity higher than 2.5×10^{14} W/cm^2. At this high intensity, ionization will reach near saturation, and harmonic spectra will be severely modified by the excessive ionization due to the defocusing of the driving 800 nm laser such that the Cooper minimum may not be

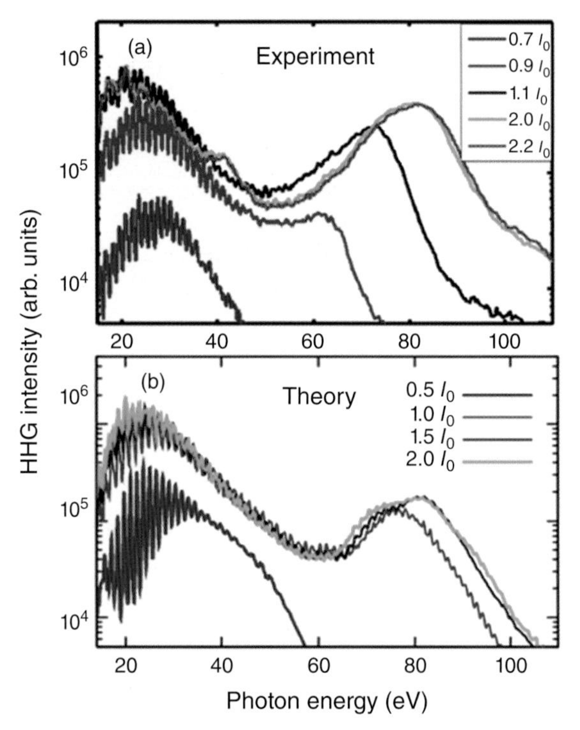

Figure 5.15 (a) Measured and (b) simulated HHG spectra of Xe generated by 1,825 nm lasers for different laser intensities, where $I_0 = 1 \times 10^{14}$ W/cm^2. Laser pulse duration is 14 fs. In the experiment, the carrier-envelope phase (CEP) is not stabilized and theoretical spectra are averaged over random values of the CEP. The broad maximum near 80 eV is due to the intershell-channel coupling in the atomic-transition dipole. Saturation occurs at higher laser intensities, where the harmonic cutoff is truncated due to defocusing in the medium by excessive ionization. (Reprinted from C. Trallero-Herrero et al., *J. Phys. B-At. Mol. Opt.*, **45**, 011001 (2012) [22]. Copyrighted by the IOP publishing.)

observed (see examples in [24, 25]). The deep minimum reported in Wörner et al. [26] for Ar using 800 nm laser at 8 fs is also unexpected and would contradict the prediction of the QRS theory. The origin of the reported deep minimum still remains unsettled. A calculation based on a Bessel beam as the incident wave instead of a Gaussian wave showed that the main features of the HHG spectra are not modified [27].

In another example, the measured and simulated HHG spectra of Xe are compared in Figure 5.15, where the wavelength of the driving laser is 1,825 nm and duration is 14 fs. The most striking features in Figure 5.15(a) are:

- The emergence of a quasi-continuous harmonic spectrum as the laser intensity is increased. These continuous spectra extend over a broad range of photon energy from the cutoff at about 100 eV down to 20–30 eV. Simulations showed that these continuum spectra are indeed capable of producing isolated attosecond pulses if proper spatial and spectral filters are applied [22].

- Saturation occurs at an intensity of about 2×10^{14} W/cm^2.
- A broad enhancement of the spectra near 80 eV.

These features are mostly reproduced using the QRS theory with the inclusion of the propagation effect (see Figure 5.15(b)). Theoretical harmonic spectra for four peak laser intensities after propagation through the gas jet are shown. In the simulation, the experimental parameters including the jet size (1 mm), the slit opening of the spectrometer (190 μm), the distance of the slit from the gas jet (455 mm), and the laser wavelength and pulse duration (1,825 nm and 14 fs) are used. The calculated spectra have been averaged over the carrier-envelope phase (CEP) dependence. From the theory, the emergence of a quasi-continuous spectrum is attributed to the saturation effect at higher laser intensities, where the driving laser is severely reshaped in the medium due to excess ionization. Excess ionization also accounts for the saturation of HHG spectra, where harmonics cannot be further extended to higher energies since the peak laser intensity inside the gas is reduced due to defocusing. Lastly, the enhancement of the HHG spectra is attributed to the well-known 5p partial photoionization cross-section of Xe, which is severely modified by the strong inter-channel coupling with photoionization from the 4d shell [28]. Note that QRS theory was derived using the one-electron model. In the simulation, the broad harmonic enhancement near 80 eV is reproduced by feeding in the theoretical photoionization-transition dipole calculated from many-body perturbation theory [29]. With the QRS theory, one can bypass the complexity of formulating a strong-field theory that also includes the many-electron correlation effect. By taking advantage of the factorization feature of the QRS, the strong-field effect is reflected in the returning electron wave packet, which is mostly a single-electron process, whereas the strong, many-electron correlation effect is reflected in the recombination transition-dipole moment. This "divide-and-conquer" strategy not only makes the calculation much easier, but it also provides a more transparent interpretation of the observed HHG spectra.

Spectral Features of High-Order Harmonics

The examples in Ar and Xe shown in previous subsections demonstrate that global agreement between simulation and experimental data, but the details of the HHG spectra are dependent on the precise experimental conditions. Such dependence is weaker only under the condition that the gas pressure is low and laser intensity is not too high, at the cost of weak harmonic signals. A close examination shows that the harmonic spectral width (or the harmonic chirp) varies as the gas pressure, pulse duration, and laser intensity are modified. In Figure 5.16, the 2D spatial distributions of harmonic emissions are shown for the same peak intensity but different pulse duration and gas pressure. A longer pulse duration and/or a lower pressure tend to generate sharper (narrower width) lower-order harmonics. Careful inspection reveals that the peak position of the harmonic actually blue shifted from one frame to another. This shift is due to the change of the fundamental pulse as it propagates through the nonlinear medium. In addition, the sharpness of the higher harmonics decreases, reflecting the fact that the quality of phase matching deteriorates for higher-order harmonics.

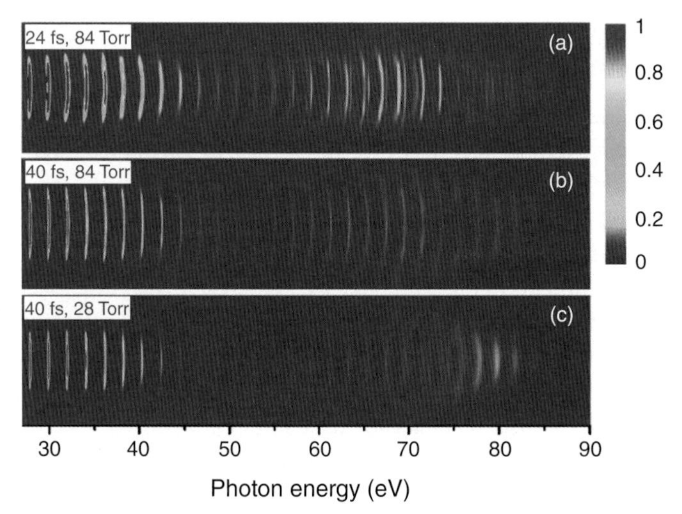

Figure 5.16 Spatial distributions of harmonic emission versus photon energy (normalized using on-axis intensity at 77 eV) in the far field for lasers with different pulse durations and gas jets with different pressures: (a) 24 fs, 84 Torr; (b) 40 fs, 84 Torr; and (c) 40 fs, 28 Torr. The other laser parameters are the same as Figure 5.14(a). (Reprinted from Cheng Jin, Anh-Thu Le, and C. D. Lin, *Phys. Rev. A*, **83**, 023411 (2011) [18]. Copyrighted by the American Physical Society.)

Notes and Comments

This chapter gives the fundamental formulation of high-order harmonic generation with atomic targets. It consists of two parts: first, the calculation of HHG from a single atom; second, the coherent buildup of harmonics in the macroscopic gas medium. For simple atomic targets, the experimental HHG spectra can be accurately reproduced using single-atom harmonics calculated based on the QRS model together with the solution of the Maxwell equations in the propagation medium. The materials presented in this chapter are quite standard. Complications of HHG generated from molecules are addressed in Chapter 6.

Exercises

5.1 From the action given in Equation 5.14, derive the saddle-point equation (Equation 5.16) for p_s.

5.2 From $\Theta(p_s, t, t')$ in Equation 5.18, derive saddle-point (Equations 5.19 and 5.20) for t and t', respectively.

5.3 Using the periodic relation Equation 5.27 for $D_x(t)$, prove that the harmonic spectrum consists of only odd harmonics, as shown in Equation 5.28.

5.4 Let $E(t) = E_0 \cos(\omega_L t)$, rewrite the saddle-point (Equations 5.19 and 5.20) in terms of $\theta = \omega_L t, \theta' = \omega_L t', \tilde{\omega} = (\omega - I_p)/U_p$, and $\gamma = \sqrt{I_p/(2U_p)}$. Reproduce the result from classical theory shown in Figure 5.2 by solving the saddle-point equations with $\gamma = 0$. In the classical theory, both θ_s and θ_s' are real quantities.

5.5 The harmonic buildup in a medium is influenced by the phase mismatch, which can be illustrated using a 1D model. In this model, the harmonic signal is considered a coherent sum over all single-atom emitters in the nonlinear medium of length L and can be written as

$$S_q \propto \left| \int_0^L dz \rho A_q \exp\left[i(\Delta k + i\kappa_q)(L - z)\right] \right|^2, \qquad (5.89)$$

where A_q is the amplitude of the atomic response of harmonic order q, ρ is gas density, Δk is the phase mismatch, and κ_q is the absorption coefficient. Assuming that A_q and ρ are constants and using the relations of $L_{coh} = \pi/\Delta k$ and $L_{abs} = 1/(2\kappa_q)$, derive equation 1 in Constant et al., *Phys. Rev. Lett.*, **82**, 1668 (1999). Then using the derived equation, plot Figure 5.7, and verify the optimal conditions of $L > 3L_{abs}$ and $L_{coh} > 5L_{abs}$ for reaching more than half of maximum harmonic signal.

5.6 Assuming that the refractive index at XUV or soft X-rays is one, use the Sellmeier equation to calculate the refractive index of noble gases at infrared wavelengths under standard conditions. (The coefficients in the Sellmeier equation can be found in *Appl. Opt.* **47**, 4856 (2008).) From the refractive indices obtained, calculate the value of the "critical" ionization fraction using Equation 5.65 for Ar, Ne, and He at the fundamental wavelengths of 0.8 and 1.3 μm.

5.7 Given a Gaussian beam with beam waist w_0 of 25 μm, wavelength of 800 nm, and peak intensity at the focus of 4.0×10^{14} W/cm^2, first calculate the confocal parameter b (twice Rayleigh range). Assuming that this Gaussian beam is interacting with Ne gas at very low pressure such that the phase mismatch caused by the plasma and the neutral atomic dispersion can be neglected, use Equations 5.57 and 5.66 to calculate and plot the total on-axis phase mismatch ($\Delta k = \Delta k_{q,geo} - K_{q,dip}$) for plateau harmonic order $q = 21$ ("short" and "long" trajectories separately) and cutoff harmonic order $q = 63$. Identify the good phase-matching regions from your plots.

References

[1] M. Lewenstein, P. Balcou, M. Y. Ivanov, A. L'Huillier, and P. B. Corkum. Theory of high-harmonic generation by low-frequency laser fields. *Phys. Rev. A*, **49**:2117–2132, Mar. 1994.

[2] A.-T. Le, H. Wei, C. Jin, and C. D. Lin. Strong-field approximation and its extension for high-order harmonic generation with mid-infrared lasers. *J. Phys. B-At. Mol. Opt.*, **49**(5):053001, 2016.

[3] M. V. Frolov, N. L. Manakov, T. S. Sarantseva, M. Yu. Emelin, M. Yu. Ryabikin, and A. F. Starace. Analytic description of the high-energy plateau in harmonic generation

by atoms: can the harmonic power increase with increasing laser wavelengths? *Phys. Rev. Lett.*, **102**:243901, Jun. 2009.

[4] J. Tate, T. Auguste, H. G. Muller, P. Salières, P. Agostini, and L. F. DiMauro. Scaling of wave-packet dynamics in an intense midinfrared field. *Phys. Rev. Lett.*, **98**:013901, Jan. 2007.

[5] K. Schiessl, K. L. Ishikawa, E. Persson, and J. Burgdörfer. Quantum path interference in the wavelength dependence of high-harmonic generation. *Phys. Rev. Lett.*, **99**:253903, Dec. 2007.

[6] J. Itatani, J. Levesque, D. Zeidler, et al. Tomographic imaging of molecular orbitals. *Nature*, **432**:867, 2004.

[7] M. V. Frolov, N. L. Manakov, T. S. Sarantseva, and A. F. Starace. Analytic confirmation that the factorized formula for harmonic generation involves the exact photorecombination cross section. *Phys. Rev. A*, **83**:043416, Apr. 2011.

[8] O. I. Tolstikhin, T. Morishita, and S. Watanabe. Adiabatic theory of ionization of atoms by intense laser pulses: one-dimensional zero-range-potential model. *Phys. Rev. A*, **81**:033415, Mar. 2010.

[9] T. Morishita, A.-T. Le, Z. Chen, and C. D. Lin. Accurate retrieval of structural information from laser-induced photoelectron and high-order harmonic spectra by few-cycle laser pulses. *Phys. Rev. Lett.*, **100**:013903, Jan. 2008.

[10] A.-T. Le, T. Morishita, and C. D. Lin. Extraction of the species-dependent dipole amplitude and phase from high-order harmonic spectra in rare-gas atoms. *Phys. Rev. A*, **78**:023814, Aug. 2008.

[11] A.-T. Le, R. R. Lucchese, S. Tonzani, T. Morishita, and C. D. Lin. Quantitative rescattering theory for high-order harmonic generation from molecules. *Phys. Rev. A*, **80**:013401, Jul. 2009.

[12] A.-T. Le, R. Della Picca, P. D. Fainstein, D. A. Telnov, M. Lein, and C. D. Lin. Theory of high-order harmonic generation from molecules by intense laser pulses. *J. Phys. B-At. Mol. Opt.*, **41**(8):081002, 2008.

[13] E. Constant, D. Garzella, P. Breger, et al. Optimizing high harmonic generation in absorbing gases: model and experiment. *Phys. Rev. Lett.*, **82**:1668–1671, Feb. 1999.

[14] T. Popmintchev, M.-C. Chen, A. Bahabad, et al. Phase matching of high harmonic generation in the soft and hard X-ray regions of the spectrum. *Proc. Natl. Acad. Sci. U.S.A.*, **106**(26):10516–10521, 2009.

[15] M. Lewenstein, P. Salières, and A. L'Huillier. Phase of the atomic polarization in high-order harmonic generation. *Phys. Rev. A*, **52**:4747–4754, Dec. 1995.

[16] M. B. Gaarde and K. J. Schafer. Quantum path distributions for high-order harmonics in rare gas atoms. *Phys. Rev. A*, **65**:031406, Mar. 2002.

[17] M. Geissler, G. Tempea, A. Scrinzi, M. Schnürer, F. Krausz, and T. Brabec. Light propagation in field-ionizing media: extreme nonlinear optics. *Phys. Rev. Lett.*, **83**:2930–2933, Oct. 1999.

[18] C. Jin, A.-T. Le, and C. D. Lin. Medium propagation effects in high-order harmonic generation of Ar and N_2. *Phys. Rev. A*, **83**:023411, Feb. 2011.

[19] E. A. J. Marcatili and R. A. Schmeltzer. Hollow metallic and dielectric waveguides for long distance optical transmission and lasers. *Bell Syst. Tech. J.*, **43**(4):1783–1809, Jul. 1964.

[20] C. Jin, A.-T. Le, and C. D. Lin. Retrieval of target photorecombination cross sections from high-order harmonics generated in a macroscopic medium. *Phys. Rev. A*, **79**:053413, May 2009.

[21] C. Jin, H. J. Wörner, V. Tosa, et al. Separation of target structure and medium propagation effects in high-harmonic generation. *J. Phys. B-At. Mol. Opt.*, **44**(9):095601, 2011.

[22] C. Trallero-Herrero, C. Jin, B. E. Schmidt, et al. Generation of broad XUV continuous high harmonic spectra and isolated attosecond pulses with intense mid-infrared lasers. *J. Phys. B-At. Mol. Opt.*, **45**(1):011001, 2012.

[23] J. Higuet, H. Ruf, N. Thiré, et al. High-order harmonic spectroscopy of the Cooper minimum in argon: experimental and theoretical study. *Phys. Rev. A*, **83**:053401, May 2011.

[24] S. Minemoto, T. Umegaki, Y. Oguchi, et al. Retrieving photorecombination cross sections of atoms from high-order harmonic spectra. *Phys. Rev. A*, **78**:061402, Dec. 2008.

[25] J. P. Farrell, L. S. Spector, B. K. McFarland, et al. Influence of phase matching on the Cooper minimum in Ar high-order harmonic spectra. *Phys. Rev. A*, **83**:023420, Feb. 2011.

[26] H. J. Wörner, H. Niikura, J. B. Bertrand, P. B. Corkum, and D. M. Villeneuve. Observation of electronic structure minima in high-harmonic generation. *Phys. Rev. Lett.*, **102**:103901, Mar. 2009.

[27] C. Jin and C. D. Lin. Comparison of high-order harmonic generation of Ar using truncated Bessel and Gaussian beams. *Phys. Rev. A*, **85**:033423, Mar. 2012.

[28] A. D. Shiner, B. E. Schmidt, C. Trallero-Herrero, et al. Probing collective multi-electron dynamics in xenon with high-harmonic spectroscopy. *Nat. Phys.*, **7**:464–467, Jun. 2011.

[29] M. Kutzner, V. Radojević, and H. P. Kelly. Extended photoionization calculations for xenon. *Phys. Rev. A*, **40**:5052–5057, Nov. 1989.

6 Applications of High-Order Harmonics: HHG Spectroscopy and Optimization of Harmonics

6.1 Studies of High-Order Harmonic Generation from Linear Molecules

6.1.1 High-Order Harmonic Generation and Photoionization

There have been a great deal of high-harmonic generation (HHG) studies on molecular targets in the past decade. HHG has been shown to contain information about molecules that is encoded in the form of amplitude and phase of the emitted harmonics. This section uses a number of examples to illustrate what "structure information" can be extracted from the harmonics measured in the experiment. This topic is often called *HHG spectroscopy* (HHS) in the literature.

According to the quantitative rescattering (QRS) theory, the amplitude and phase of the harmonics are directly related to the photo-recombination transition-dipole matrix element. Such relations were demonstrated for atomic targets in Section 5.3.1. Photo-recombination is an inverse process of photoionization. Photoionization has been a powerful method for studying the properties of molecules, typically carried out at national synchrotron facilities.

As a tool for spectroscopy, HHG has a number of advantages over photoionization. First, HHG is broadband such that a large range of photon energies is covered in a single measurement. Second, harmonics are coherent so the phase of the transition-dipole matrix element can be extracted. This is not possible with narrow-band photoionization experiments. Third, HHG can be measured from aligned (and/or oriented) molecules to obtain molecular-frame photoionization information since the alignment laser and HHG driving laser can be routinely synchronized. This has not yet been possible for synchrotron experiments. Fourth, harmonics are generated by ten or sub-ten femtosecond lasers, thus HHG can be used to probe the structural change of a molecular system with femtosecond temporal resolution.

HHS does have a number of drawbacks when compared to photoionization. First, high-order harmonics are generated coherently in a gas medium. As described in Section 5.4, the phase and amplitude of harmonics are modified by various macroscopic phase-matching effects. To extract a single-molecule, recombination dipole-matrix element from the observed HHG spectra, corrections from the propagation medium should be mitigated. This is usually done under the conditions of low laser intensity and low gas pressure. However, such conditions are not easily met for long-wavelength driving lasers because of the unfavorable wavelength scaling of harmonic generation. Second, due to the short-pulse nature of the driving laser, fine spectral features that require high-resolution are not observed in HHS. Third, HHG is a nonlinear process while photoionization is a linear process. To include many-electron effects in HHG is extremely difficult as compared to photoionization.

On the other hand, the two methods can complement each other. Photoionization can be used to study the fine features of a molecule, while HHG can be used to study a molecule over a broad spectral region quickly. For dynamic systems, HHG has the advantage since it is the global change of a molecule that is of interest, not the small details.

This chapter consists of three main sections. First, HHS for simple linear molecules is examined. Since there have been many theoretical and experimental studies on N_2 molecules, N_2 serves as an example of how various structure information on a molecule can be obtained from the harmonics. Examples of HHG from polyatomic molecules and for dynamic systems are treated in Section 6.2. Experimental data for these systems are more scarce and theoretical modelings are quite limited, but some interesting results have been reported and are presented. Section 6.3 focuses on the various ongoing efforts for enhancing the harmonic yields and for extending harmonics to the soft X-ray region. Success of these efforts will guarantee a bright future for the continuing growth of the field covered in this book.

6.1.2 Historical Background

Photoionization is a basic process that allows direct investigation of molecular structure. Almost all such experiments have been performed from an ensemble of randomly oriented molecules. Thus the rich structure of photoelectron angular distributions for fixed-in-space molecules predicted theoretically since the 1970s has remained largely unexplored. With femtosecond driving lasers, as discussed in Section 3.6, gas-phase molecules can be impulsively aligned. After the pulse is over, molecules will be partially aligned or anti-aligned at the time intervals of full or fractional rotational revivals. During these revivals, which each last for tens to hundreds of femtoseconds, HHG spectra can be generated with a few-ten femtosecond laser. HHG spectra can be recorded experimentally by changing the time delay between the aligning and HHG-generating lasers, with the polarization axes of both lasers being parallel or perpendicular to each other [1]. Alternatively, HHG data can be taken at a fixed time delay but probed by varying the angle between the polarization axis of the HHG-generating laser pulse with respect to the alignment laser [2].

According to the QRS theory, the amplitude and phase of the photoionization dipole-transition matrix element of a fixed-in-space molecule can be extracted from the amplitude and phase of the laser-induced high-order harmonics. Thus studies of HHG generated from aligned or oriented molecules shed light on molecular frame photoelectron angular distributions (MFPAD), as well as the phase of the fixed-in-space dipole-transition matrix element.

Over the years, molecular nitrogen has been a favorite target for many such HHG measurements. Some of these experiments will be discussed to address how and what structure information can be extracted from the HHG spectra, in particular the photoionization-transition dipoles $d(\Omega)$ (Ω is the photon energy in atomic units) in the molecular frame.

As discussed in Section 6.1.7, the phase of the harmonics can also be determined. Thus, the amplitude and phase of $d(\Omega)$ at each harmonic energy can be determined to compare with accurate calculations obtained from state-of-the-art molecular-photoionization codes. However, when experiments are carried out at higher laser intensities, new features appeared and evidence of inner orbitals contributing to the HHG process has been identified. HHG from aligned molecules also gives rise to harmonics with different polarization states. Finally, alignment-dependent HHG spectra has been heralded for providing a means to image the molecular orbital (MO) using the so-called "*tomographic imaging*" method. This claim will be examined in Section 6.1.11.

6.1.3 Minimum in the HHG Spectra and the Two-Center Interference Model

Since the early theoretical work in Lein et al. [3], it has been known that the HHG spectra of some simple molecules may have a minimum at a certain harmonic order. The position of the minimum depends on the molecular alignment angle θ with respect to the laser polarization. This minimum is not unexpected and has the same origin as the Cooper minimum in Ar, which occurs when the transition-dipole matrix element goes through zero. In diatomic molecules, this minimum was interpreted in terms of quantum interference of the emitted harmonic light from the two atomic centers [3]. It has the same origin as the double-slit interference. For molecules such as N_2 or O_2, the ground state is treated as the superposition of two atomic orbitals and the continuum electron is treated as plane waves. According to the simple two-emitter model, the minima satisfy the relation

$$R\cos\theta = (n+1/2)\lambda^{eff}, \quad n = 0, 1, 2, \ldots, \quad \text{symmetric wavefunction,} \tag{6.1}$$

$$= n\lambda^{eff}, \quad n = 1, 2, \ldots, \quad \text{anti-symmetric wavefunction,} \tag{6.2}$$

where R is the internuclear distance, and λ^{eff} is the "effective" wavelength of the continuum electron defined such that the "effective" wave vector is $k^{eff} = \sqrt{2\Omega}$ with Ω being the energy of the emitted photon. This model introduces an energy shifted by I_p with respect to normal relation $k = \sqrt{2(\Omega - I_p)}$.

While this simple model seems to predict results that are in accord with some earlier experimental and/or theoretical harmonic spectra, discrepancies have since been found in many other cases (see Figure 6.1). In addition, the definition of k^{eff} is neither unique nor well justified. Clearly, an improved theory beyond the two-center interference model is needed.

6.1.4 Strong-Field Approximation of HHG for Molecules

The strong-field approximation for HHG for atoms, described in Chapter 5, can be readily generalized to molecules (see [5]). Without loss of generality, assume a linear molecule is lying along the x-axis. It is exposed to a laser field $E(t)$ which is linearly polarized on the x–y plane with an angle θ with respect to the molecular axis. The parallel component of the induced-dipole moment can be written in the form

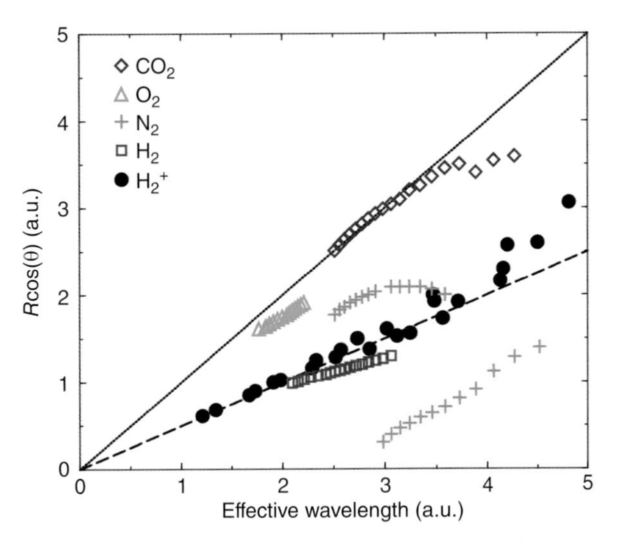

Figure 6.1 Test for the prediction of the position of the minimum in the harmonic spectrum based on the two-emitter model of Equations 6.1 and 6.2, which are depicted by the two straight lines (for different symmetries of the MO). The symbols are from accurate theoretical results for photoionization differential cross-sections. The figure illustrates the limitation of the empirical two-emitter model. (Reprinted from Anh-Thu Le et al., *Phys. Rev. A*, **80**, 013401 (2009) [4]. Copyrighted by the American Physical Society.)

$$D_{\|}(t) = i\int_0^{\infty} d\tau \left(\frac{\pi}{\epsilon + i\tau/2}\right)^{3/2} [\cos\theta d_x^*(t) + \sin\theta d_y^*(t)]$$
$$\times [\cos\theta d_x(t-\tau) + \sin\theta d_y(t-\tau)]E(t-\tau)$$
$$\times \exp[-iS_{st}(t,\tau)]a^*(t)a(t-\tau) + c.c., \qquad (6.3)$$

where $\boldsymbol{d}(t) \equiv \boldsymbol{d}[\boldsymbol{p}_{st}(t,\tau) + \boldsymbol{A}(t)]$, $\boldsymbol{d}(t-\tau) \equiv \boldsymbol{d}[\boldsymbol{p}_{st}(t,\tau) + \boldsymbol{A}(t-\tau)]$ are the transition-dipole moments between the ground state and the continuum state and $\boldsymbol{p}_{st}(t,\tau) = -\int_{t-\tau}^{t} \boldsymbol{A}(t')dt'/\tau$ is the canonical momentum at the stationary points with $\boldsymbol{A}$ being the vector potential. The perpendicular component $D_{\perp}(t)$ is given by a similar formula with $[\cos\theta d_x^*(t) + \sin\theta d_y^*(t)]$ replaced by $[\sin\theta d_x^*(t) - \cos\theta d_y^*(t)]$ in Equation 6.3. The action is independent of the target (with the exception of the term due to the ionization potential) and has the same expression as in Equation 5.16. In Equation 6.3, $a(t)$ was introduced to account for the ground-state depletion.

For molecules, the ground-state wavefunction can be quite accurately obtained from the general quantum chemistry code, such as Gamess or Gaussian. In the spirit of the strong-field approximation (SFA), the transition dipole $\boldsymbol{d}(\boldsymbol{k})$ is given by $\langle \boldsymbol{k}|\boldsymbol{r}|\Psi_0\rangle$, with the continuum wavefunction approximated by a plane wave $|\boldsymbol{k}\rangle$. The ground-state amplitude can be approximated by $a(t) = \exp[-\int_{-\infty}^{t} W(t')/2dt']$, with the ionization rate $W(t')$ obtained from the molecular ADK theory (MO-ADK) theory or from the molecular PPT (MO-PPT) theory.

Based on the study of HHG on atoms, there is no reason to expect that HHG calculated from Equation 6.3 for molecules will be sufficiently accurate. Thus, one needs to extend the QRS theory to HHG generated from molecules.

6.1.5 QRS Theory for HHG from Aligned Molecules

The QRS theory can be easily generalized to molecules. For simplicity, let us consider the parallel component of HHG for a linear molecule that is fixed in space with the molecular axis making an angle θ with respect to the polarization axis of the laser. The induced dipole $D(\omega,\theta)$ can be written as

$$D(\omega,\theta) = W(E,\theta)d(\omega,\theta), \tag{6.4}$$

where $d(\omega,\theta)$ is the "exact" transition dipole that can be calculated from the molecular-photoionization theory and $W(E,\theta)$ is the returning electron wave packet. Here, $D(\omega,\theta)$, $W(E,\theta)$, and $d(\omega,\theta)$ are all complex valued.

According to the QRS theory, the structure information of the molecule in the HHG spectra is all contained in the transition-dipole matrix element $d(\omega,\theta)$. In fact, photoionization and photo-recombination DCSs are related by

$$\frac{d^2\sigma^R}{\omega^2 d\Omega_n d\Omega_k} = \frac{d^2\sigma^I}{c^2k^2 d\Omega_k d\Omega_n}, \tag{6.5}$$

which follows the principle of detailed balancing for the direct and time-reversed processes [4]. The photo-recombination DCS can be written as

$$\frac{d^2\sigma^R}{d\Omega_n d\Omega_k} = \frac{4\pi^2\omega^3}{c^3 k}|\langle\Psi_i|\mathbf{r}\cdot\mathbf{n}|\Psi_{\mathbf{k}}^+\rangle|^2. \tag{6.6}$$

Note that the continuum state in the equation is the scattering wave instead of the plane wave used in the SFA.

Equation 6.4 was written using the one-electron model of the target. However, one can, in principle, generalize it to include the many-electron correlation effect in the transition dipole $d(\omega,\theta)$ by taking it directly from accurate calculations based on different molecular-photoionization codes. For instance, photoionization of N_2 has been extensively studied theoretically since 1982 [6]. Thus, Equation 6.4 can be used to obtain a laser-induced transition dipole for a fixed-in-space molecule. In general, there are two polarization components for each harmonic; one that is parallel to the polarization of the driving laser, and another that is perpendicular to it (and to the laser-propagation direction). The latter component vanishes after integration over the molecular-angular distribution for molecules that are isotropically distributed, but not for molecules that are oriented or aligned. The polarization properties of harmonics for aligned molecules are described in Section 6.1.9.

There are two different ways to obtain the complex-valued wave packet $W(E,\theta)$ using Equation 6.4. The first method is to obtain it from the SFA model, where the wave packet is calculated from

$$W^{SFA}(E,\theta) = \frac{D^{SFA}(\omega,\theta)}{d^{PWA}(\omega,\theta)}. \tag{6.7}$$

Here, $D^{SFA}(\omega,\theta)$ is calculated from the SFA model (see Section 6.1.4), and $d^{PWA}(\omega,\theta)$ is calculated using the plane wave for the continuum state. In another approach, the wave packet is obtained by a numerical solution of the time-dependent Schrödinger equation (TDSE) for a reference atom that has a comparable ionization potential. This is based on the QRS model, where the wave packet is independent of the target except for an overall factor to account for the difference in targets' ionization probabilities. In this approach, the wave packet is obtained from

$$W^{QRS2}(E,\theta) = \left(\frac{N(\theta)}{N^{ref}}\right)^{1/2} W^{ref}(E)e^{i\Delta\eta} = \left(\frac{N(\theta)}{N^{ref}}\right)^{1/2} \frac{D^{ref}(\omega)}{d^{ref}(\omega)}e^{i\Delta\eta}, \tag{6.8}$$

where $N(\theta)$ and N^{ref} are the ionization probabilities for the aligned molecule and the reference atom, respectively. $\Delta\eta$ is introduced to account for the phase difference between the two wave packets, which is nearly independent of the photon energy.

For an ensemble of partially aligned molecules, the laser-induced dipole from each molecule has to be added coherently, weighted by their angular distributions $\rho(\theta,t)$ (see Section 3.6). In particular, if the aligning pulse and the HHG driving pulse are parallel, then the induced dipole is given by

$$\overline{D}(\omega,t) = 2\pi \int_0^\pi D(\omega,\theta)\rho(\theta,t)\sin\theta d\theta. \tag{6.9}$$

Here, it is assumed that the alignment distribution does not change when the HHG driving pulse is on. For the general case where the pump and probe pulses are not parallel, the treatment is a bit more complicated but has been worked out (see [4] or [7]).

6.1.6 HHG Spectra of Aligned N_2 Molecules with 800 nm Lasers: Experiment versus the QRS Theory

This subsection compares the HHG spectra from aligned N_2 molecules with 800 nm laser pulses reported in Ren et al. [8]. N_2 molecules are impulsively aligned with two 800 nm pump lasers and an adjustable delay to obtain the highest possible degree of alignment. The degree of alignment $\langle\cos^2\theta\rangle$ was estimated to be about 0.8. This is high compared to earlier experiments in which the degrees of alignment were typically around 0.6. The harmonic spectra are generated as a function of the pump-probe time delay, as well as by different probe angles (with respect to the pump beams). By varying the focus of the probe pulse relative to the location of the gas jet, phase matching was adjusted to maximize the cutoff harmonics. Under this condition, the probe is focused about 3.5 mm before the jet with the peak intensity estimated to be 2.5×10^{14} W/cm^2. The experiment does not distinguish the polarization of the harmonics; thus, both polarizations are included in the calculation. To highlight the alignment dependence of the harmonics, the data is normalized to the corresponding isotropic signals. Figure 6.2 shows the time delay scan for the normalized harmonics, for orders 19, 25, 27, 31, and 37, and the comparison with results from the QRS calculations. Elaborate simulations have been carried out on laser parameters and

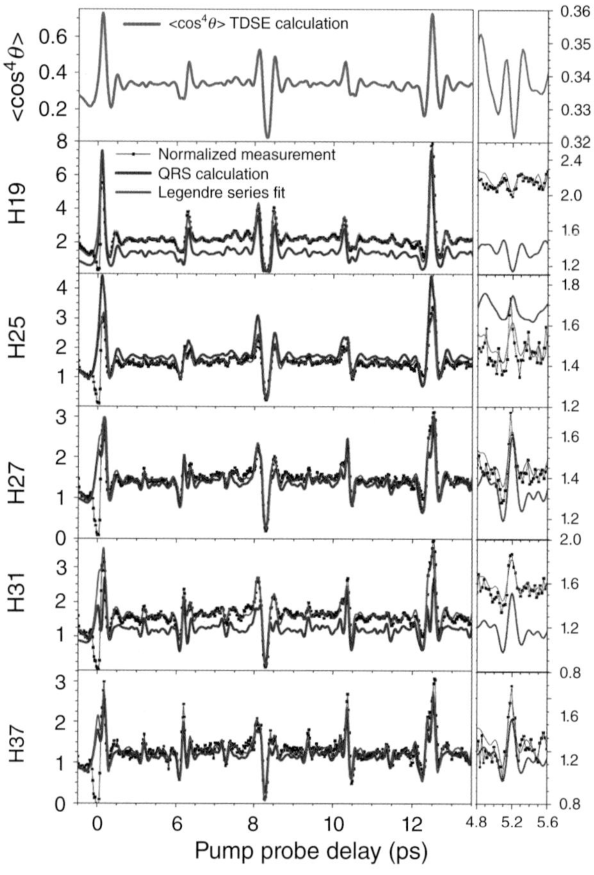

Figure 6.2 Experimental revival scans (solid black curves) of N_2 are shown for harmonic orders 19, 25, 27, 31, and 37, all normalized to the corresponding isotropic signals. Laser parameters obtained from the fit are used in the QRS calculations. (Reprinted with permission from Xiaoming Ren et al., *Phys. Rev. A*, **88**, 043421 (2013) [8]. Copyrighted by the American Physical Society.)

temperatures to optimize the agreement between the data and the simulations. The degree of agreement between the two is quite good. In particular, features around the 1/8th rotational revival (the right-most panel) have not been seen in earlier experiments. The sharp contrast of the observed structures owes much to the significantly improved degree of alignment achieved using two pump pulses adjusted to achieve optimal alignment.

In the experiment of Ren et al. [8], the angle dependence of HHG was also probed directly by rotating the pump polarization at a fixed time delay. The measured normalized harmonic yields versus pump-probe angles are shown in Figure 6.3(a). The yields are compared to results calculated from the QRS theory. The key feature is the strong signal near 30 eV at angles smaller than about 40°. This strong peak is well known and is attributed to a $3\sigma_g \rightarrow k\sigma_g$ shape resonance. What can one learn from the HHG spectra shown on the left of Figure 6.3? Clearly the most important information is the angle dependence, or the molecular frame photoionization differential cross-sections (PIDCS). The latter can be obtained first by deconvoluting the measured HHG data and then by

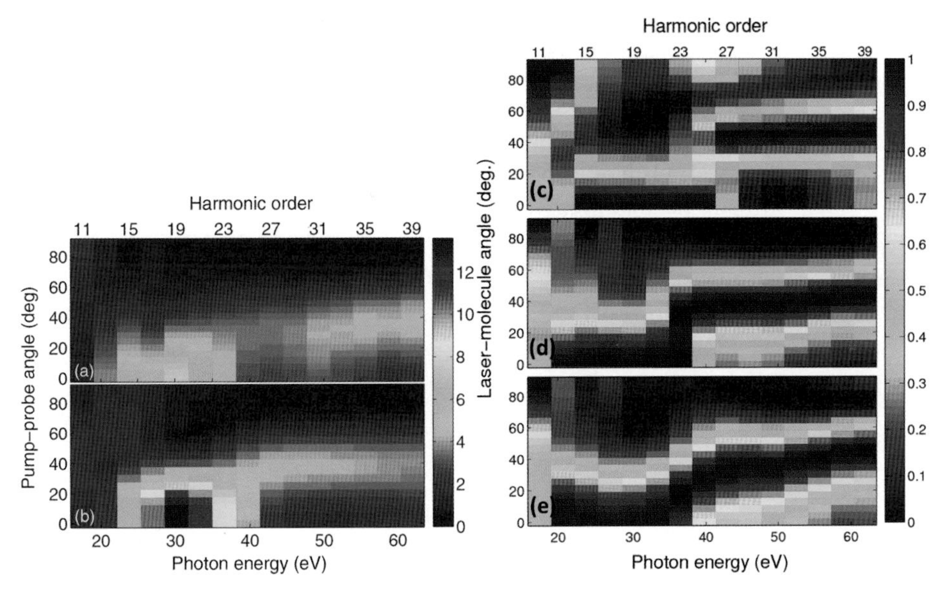

Figure 6.3 Left column: (a) Measured and (b) calculated (based on QRS) HHG intensity as a function of pump-polarization angle and photon energy. The calculations include both polarizations of the harmonics and the harmonic data are normalized to the corresponding values from an isotropic gas. Right column: (c) Theoretical molecular frame differential photoionization cross-sections of the HOMO (both polarization components of the harmonic are included). (d) and (e) Photoionization cross-sections retrieved from the experimental angle and delay scans, respectively. These results show that molecular frame photoelectron angular distributions calculated from the quantum chemistry code are in good agreement with those retrieved from the HHG spectra generated from aligned molecules. (Reprinted with permission from Xiaoming Ren et al., *Phys. Rev. A*, **88**, 043421 (2013) [8]. Copyrighted by the American Physical Society.)

dividing out the angle-dependent tunnel-ionization rate. The results are shown on the right panel of Figure 6.3, where the top frame is obtained from the theoretical calculations while the next two frames are extracted from the experimental HHG spectra, one from the time-delay spectra of Figure 6.2 and the other from the data on the left of this figure. Note that the data for each harmonic has been normalized to its peak value in order to get rid of the energy-dependent factor due to the returning electron wave packet and the detection efficiency of the HHG spectrometer. This step can be avoided, for example, by measuring HHG from Ar gas under the same driving laser. Thus, the results on the right panel provide a direct comparison of molecular frame PIDCS that have not been available in typical photoionization experiments with synchrotron radiation. While such information may not appear as "interesting" or "appealing" as imaging MOs (see Section 6.1.11), or the two-center emitter model, the retrieved molecular frame PIDCSs pertain to the maximal structure information that can be revealed from HHG measurement. The other part of the structure information, photoionization dipole phase, is discussed in Section 6.1.7.

6.1.7 Retrieval of Photoionization Transition-Dipole Phase from HHG Spectra

High harmonics generated by an intense laser field in a gas medium are coherent lights. The intensity of each harmonic can be determined, for example, by ionizing Ar atoms with these harmonics to obtain photoelectron spectra. After correcting for the energy-dependent photoionization cross-section of Ar, the spectral intensity of the harmonics can be obtained. To determine the spectral phase $\phi(\omega)$ of the harmonic, the reconstruction of attosecond harmonic beating by interference in two-photon transition (RABITT) technique can be deployed. (This method is described in more detail in Section 7.2.2). In the RABITT method, photoionization of a target (such as Ar by the harmonics) is carried out in the presence of a time-delayed co-propagating infrared field. With sufficient infrared intensity, the sideband at photon energy of $2N\omega$ can be generated by absorbing one $(2N-1)\omega$ extreme ultraviolet (XUV) harmonic photon and one infrared photon, or by absorbing one $(2N+1)\omega$ XUV harmonic photon and emitting one IR photon. These two processes interfere. By analyzing the peak position of the sidebands versus the time delay, the intensity of the sideband can be shown to modulate following the relation,

$$S_{2N}(\tau) \sim \cos[2\omega\tau + \varphi_{(2N+1)} - \varphi_{(2N-1)} - \Delta\varphi^{at}], \tag{6.10}$$

where $\Delta\varphi^{at}$ is a small-phase correction due to the two-photon transition-dipole matrix element of the target. Figure 6.4 shows the amplitude and phase of the recombination dipoles retrieved from the observed harmonics emitted from aligned molecules between harmonics 17 and 29, as reported in Haessler et al. [9]. To obtain these recombination dipoles, the HHG spectra of Ar were measured at the same time. Since the returning electron wave packet is independent of the target, the complex recombination dipole of the aligned N_2 molecules can be extracted with respect to the transition dipole for Ar. Because the RABITT method can only determine the phase difference between two neighboring harmonics, the phase of the seventeenth harmonic was set to zero in Figure 6.4. The laser intensity used in the experiment was estimated to be about 1.0–1.2 $\times$ 10^{14} W/cm^2. The data in Figure 6.4(a) show the ratios of the transition dipole amplitude of N_2 divided by that for Ar. In Figure 6.4(b), the difference of the phase from N_2 molecules to that of Ar is

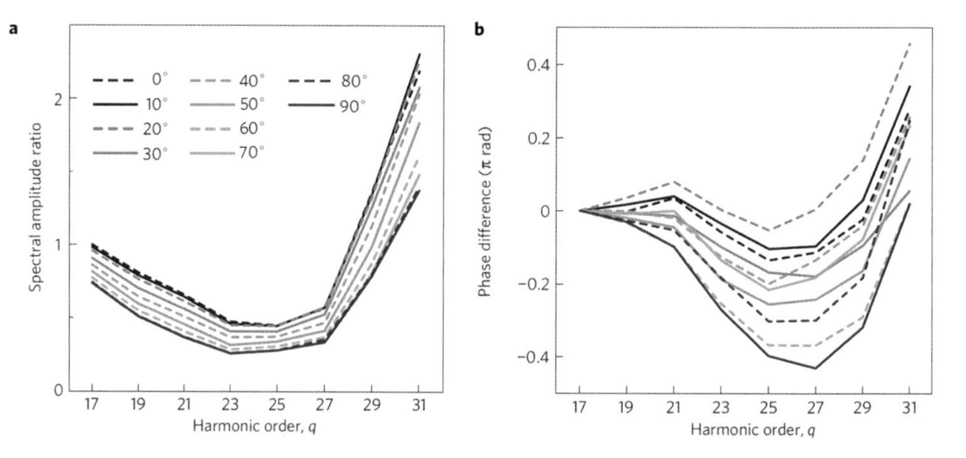

Figure 6.4 Recombination dipole for (a) amplitude and (b) phase, for aligned N_2 molecules at various alignment angles normalized with respect to argon atoms. The phase difference between them was set to zero at the seventeenth harmonic. Assuming that dipole amplitude and phase of Ar are known accurately, these data can be used to obtain molecular frame photoelectron angular distributions. (Reprinted from S. Haessler et al., *Nat. Phys.*, **6**, 200 (2010) [9].)

compared. The strong harmonic-order dependence and alignment dependence are clearly seen. These experimental results have been compared to the predictions of the QRS theory and good agreement was found.

6.1.8 Multiple Orbital Effect in Molecular HHG Spectra

Within the tunnel-ionization theory, the ionization rate decreases exponentially with the ionization potential. For atomic targets, this implies that tunneling mostly removes the electron from only the outermost subshell. For molecules, ionization potentials for the first few inner subshells are relatively close to the ionization potential of the highest occupied molecular orbital (HOMO). The inner orbitals, according to their ionization potentials, are HOMO-1, HOMO-2, etc. Electrons in these MOs have lower ionization rates than the ones from the HOMO in general. However, MOs have directions. At certain alignment angles with respect to the laser polarization axis, some inner orbitals may have higher ionization rates because the orbital has larger electron density along the direction of the polarization of the laser. Similarly, the photo-recombination matrix element also depends on the alignment of the molecule. Take N_2 molecules as an example. The alignment dependence of the ionization rate $N(\theta)$ and PIDCS $\sigma(\omega, \theta)$ are shown in Figure 6.5. According to the QRS theory, the single-molecule harmonic yield should be proportional to $N(\theta)\sigma(\omega, \theta)$. The left panel shows the $N(\theta)$ for HOMO, HOMO-1, and HOMO-2 orbitals at two laser intensities.

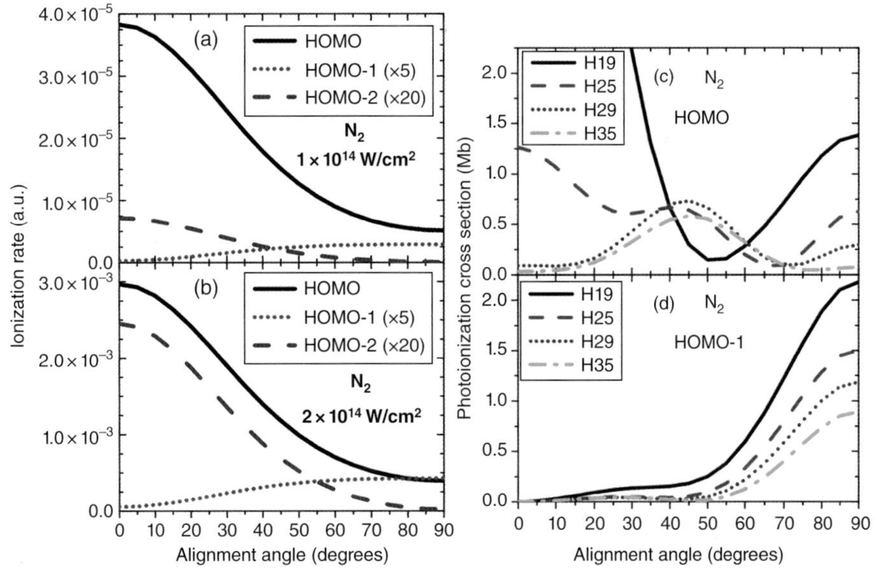

Figure 6.5 The alignment dependence of tunnel-ionization rates $N(\theta)$ (in (a) and (b)), and of photoionization differential cross-section $\sigma(\omega, \theta)$ (in (c) and (d)) for the HOMO (σ_g) and HOMO-1 (π_u) orbitals of N_2 molecules. The alignment dependence of the harmonic signal is roughly proportional to the product of $N(\theta)\sigma(\omega, \theta)$. From these figures, one expects that the HOMO-1 orbital may contribute to HHG spectra at higher laser intensities and large alignment angles where the molecular axis is nearly perpendicular to laser polarization. (Reprinted from Anh-Thu Le, R. R. Lucchese, and C. D. Lin, *J. Phys. B-At. Mol. Opt.*, **42**, 211001 (2009) [11]. Copyrighted by IOP publishing.)

It shows that for small-alignment angles ionization mostly occurs from the HOMO orbital. The contribution from the inner orbitals, the HOMO-1 in particular, would be non-negligible when the alignment angle is near 90°, especially at higher laser intensities. Meanwhile, the angular dependence of the photoionization cross-sections as displayed on the right panels of Figure 6.5 also exhibit rather complicated angular dependence as well as harmonic-energy dependence, for harmonics H19 to H35. Nevertheless, a general trend can be seen from the figure. Namely, at large angles near 90°, the relative contribution from the HOMO-1 to the total photoionization cross-section increases with energy as compared to that from the HOMO. Based on these general observations, the HOMO-1 orbital is expected to make contributions to the HHG spectra when molecules are aligned perpendicularly to the laser-polarization direction, especially at higher intensities. In fact, this was demonstrated experimentally first for N_2 molecules in McFarland et al. [10]. The simulation within the QRS [11] nicely reproduced measured the HHG time-delay scan only when the contribution from the HOMO-1 was included.

This discussion has focused on the importance of inner orbitals contributing to the HHG spectra based on the single-molecule response. The contributions of harmonics from all the MOs should be added up coherently. This may introduce interference depending on the nature of the induced-dipole moment of the orbitals involved. Furthermore, coherent averaging over the molecular-alignment distribution has to be carried out (see, for example, the discussion in connection with Equation 6.9). In practice, high-order harmonics are generated in macroscopic media, so a realistic simulation should involve macroscopic phase matching as described in Section 5.5.

As an illustration, consider HHG from aligned N_2 molecules with a 1,200 nm laser pulse, reported in a joint theory and experimental paper [12]. Experimentally, the molecules are aligned perpendicular to the polarization axis of the probing laser. The laser has a pulse duration of about 40 fs and the degree of alignment is $< \cos^2\Theta >= 0.60 \pm 0.05$. The HHG spectra were reported at three intensities. The experimental laser parameters and the focusing conditions were used in the macroscopic propagation simulation in which the laser-induced dipoles were calculated within the QRS theory. Figure 6.6 shows the measured HHG spectra at intensities of 0.65, 1.1, and 1.3×10^{14} W/cm^2. By adjusting the peak intensity in the simulation, it was found that the intensities 0.75, 0.9, and 1.1×10^{14} W/cm^2, respectively, would give best agreement with the experimental data (see the green curves in Figure 6.6). Note that the HHG spectra are displayed on a linear scale. To get good agreement with the experimental spectra for the two low intensities, only the HOMO orbital was needed in the simulation. In fact, the agreement between the experiment and simulation is about 10%, but at certain harmonics a larger discrepancy of about 30% can be seen. The general shape of the harmonics has been nicely reproduced. Both the experiment and the QRS simulation have shown that there is a minimum near about 40 eV. This minimum has been traced to the photo-recombination dipole for harmonics from the HOMO orbital. The slight shift of the minimum between (a) and (b) of Figure 6.6 is due to the small change of the laser-intensity distribution in the medium when the peak intensity is increased. More importantly, the disappearance of the minimum in (c) of Figure 6.6 and the flattening out of the harmonic signals at higher photon energies cannot be reproduced by the simulation without a contribution from HOMO-1.

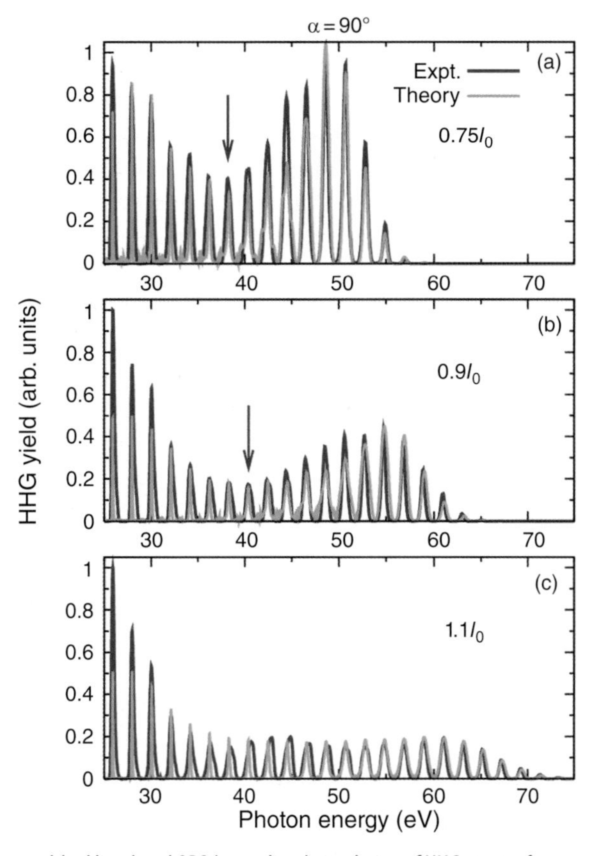

Figure 6.6 Comparison of experimental (red lines) and QRS (green lines) simulation of HHG spectra for perpendicularly aligned N_2 molecules. Laser wavelength is 1,200 nm and the pump-probe angle is 90°. Laser intensities used for the simulation (indicated in the labels) are given in units of $I_0 = 10^{14}$ W/cm^2. They are adjusted to best fit the experimental data. In (a) and (b), only the HOMO contributes and the minimum can be seen in both spectra at about the same location. In (c), HOMO-1 also contributes to the spectra and the minimum disappears. (Reprinted from Cheng Jin et al., *Phys. Rev. A*, **85**, 013405 (2012) [12]. Copyrighted by the American Physical Society.)

Additional HHG spectra from N_2 have been reported in many other experiments. Earlier experiments with 800 nm pulses tended to use laser intensities up to 3×10^{14} W/cm^2 or even higher in order to cover a larger range of photon energies. At such high intensities, good phase matching is hard to achieve. Clearly, it is more difficult to extract an accurate photo-recombination dipole from the HHG spectra when multiple orbitals make contributions to the spectra, even though a good comparison between theory and experiment can still be achieved.

6.1.9 Polarization of High-Order Harmonics Generated from Aligned Molecules

Atomic or randomly distributed molecular gas in an intense linearly polarized laser can only generate harmonics that are linearly polarized in the direction of the driving laser polarization. If the molecules are aligned such that the alignment axis and driving laser's polarization axis form a plane, the emitted harmonics have the perpendicular polarization component in addition to the parallel polarization component. Thus harmonics would generally be elliptically polarized. A few experiments have reported the polarization state of harmonics generated from the aligned N_2 molecules. Assume that the driving laser propagates along the z-axis and is linearly polarized along the x-axis. Within the QRS theory, laser-induced transition dipole in the y-direction can be calculated in a similar manner as in the parallel x-direction simply by replacing the operator x by y in the transition-matrix element. Figure 6.7(a) shows the typical photoionization yields along the x- and y-directions for H17 and H23 versus the alignment angle of the molecule. The elliptical polarization can be described by the ellipse orientation angle ϕ and a relative phase δ between the two electric components. Figures 6.7(b,c) show the experimental measurement [13] and the QRS simulation [14], respectively, for the relative phase δ at the angles $\theta = 40^\circ$, 50°, and 60° between the pump and probe polarizations. The general agreement between theory and experiment is very good. This agreement can also be seen in Figures 6.7(d,e), where the orientation angle ϕ versus the harmonics order and the pump-probe polarization angle from the experiment and the QRS theory are compared. Only the HOMO was included in the calculation.

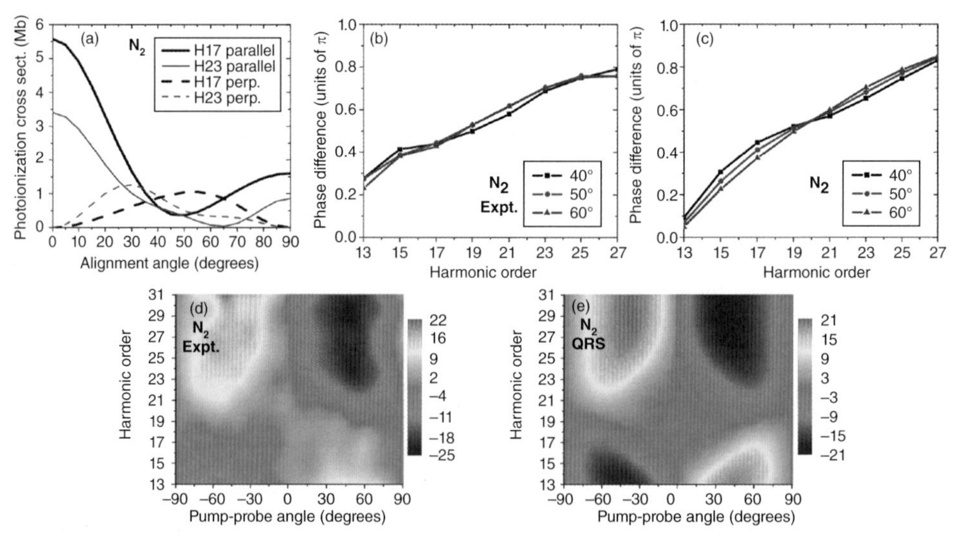

Figure 6.7 Polarization of harmonics generated from aligned N_2 molecules by linear polarized lasers. (a) Comparison of the photoionization cross-sections in the two polarization directions for H17 and H23 versus the alignment angle of the molecule. (b) and (c) Comparison between the experimental data and the QRS simulation for the phase difference (in units of π) between the two polarization components for harmonics from 13 to 27 at the angles $\theta = 40^\circ, 50^\circ$, and 60° between the pump and probe polarizations. (d) and (e) Comparison between the experimental data and the QRS simulation for the orientation angle of the polarization ellipse against the harmonic order and the alignment angle. The experimental data are taken from [13]. (Reprinted from A. T. Le, R. R. Lucchese, and C. D. Lin, *Phys. Rev. A*, **82**, 023814 (2010) [14]. Copyrighted by the American Physical Society.)

6.1.10 How Can HHS Probe the Structure of a Molecule?

HHS has been widely lauded for providing a means of studying the structure of a molecule and, in particular, the structural dynamics of a molecule. Conveniently, unlike photoionization studies that use synchrotron radiation such that a spectrum is built from a sequence of measurement, a broadband spectrum of high harmonics is generated in a single laser measurement. Take N_2 as an example. In Section 6.1.7 it was shown that, by using the QRS theory, the amplitude and phase of the photoionization transition dipole (PITD) can be retrieved from the HHG experiment. The PITDs obtained are independent of the intensity, wavelength, and laser-focusing condition. However, the results shown in Section 6.1.7 are only valid under certain conditions. Since few other theoretical calculations besides the QRS theory have been used to calculate HHG spectra from molecules that can be compared to experiments, it is helpful to recount the steps for applying the QRS theory to obtain HHG spectra:

(i) For each fixed-in-space molecule, the induced dipole (amplitude and phase) generated by a laser with a known laser field is calculated using the QRS theory. The elementary structure information of the molecule used includes the angular dependence of the ionization rate (which can be calculated using MO-ADK or molecular PPT theory), and the PITDs obtained from the molecular photoionization codes. This step has been described in Section 6.1.5.

(ii) If more than one molecular orbital (MO) contributes to the HHG, then the QRS theory should be applied to each MO and the laser-induced dipole from each MO should be added coherently. After the coherent sum, the total laser-induced dipole can no longer be written as the product of a returning electron wave packet multiplied by a PITD matrix element.

(iii) If the molecules are partially aligned/oriented, the induced dipoles from the molecules should also be added and coherently weighted by the angular distributions of the molecules.

(iv) Finally, the induced dipole from (iii) is fed into the Maxwell propagation equation to obtain a harmonic spectrum. Under the loosely focused condition with low laser intensity and low gas pressure, the observed harmonic spectra can be treated as proportional to the HHG from a single molecule obtained in (iii).

It is clear that if only one MO is contributing to the HHG, then the PITD can be directly retrieved from the experimental HHG data. If more than one MO contributes to the HHG, then a separate PITD for each MO cannot be retrieved unless further simplifying assumptions are made.

Compared to traditional photoionization measurements, what "new" structure information can be retrieved from the HHG measurement? First, consider a situation in which only one MO is contributing to the HHG. The harmonic spectra can be used to extract the amplitude of the PITD, and the RABITT method (or similar methods for that purpose) can be used to obtain the phase of the PITD. The latter cannot be obtained in a standard photoionization measurement.

When both HOMO and HOMO-1 are contributing to the HHG spectra, it is difficult to separate the contribution to the HHG from each MO without additional theoretical calculation or modeling. In such cases, retrieving individual PITDs is difficult. The most one can do is compare the harmonic spectra obtained from the experiment and from the theory. For large molecules, it is rare that only one MO contributes to the HHG. Thus retrieving the PITD for each MO is impractical. In this case, extracting the "structure information" from the experiment alone is not possible without involving theoretical simulations.

6.1.11 Tomographic Imaging of MO: Really?

Section 6.1.10 emphasized that the most elementary structure information about the target one can get from HHG measurement is the amplitude and phase of the PITD, for example, in the case of N_2 (See the four graphs in Figure 6.8), where only the HOMO and HOMO-1 orbitals are contributing to the HHG process. In a photoionization experiment the amplitude of the PITD can be obtained but the phase cannot. Since phase is a concept foreign to the general public, one is tempted to present the PITD in an alternative form. One example of such an effort that has already been repeated a few times in the literature is the "experimental measurement" of an MO, or of a hole in a molecule [15]. Since MO is a mathematical construct in quantum mechanics, each time a claim of "experimental measurement" has been made in a "high-impact" journal, it has always been followed by heated debates in the literature. In the context of HHS a *Nature* paper titled, "Tomographic imaging of molecular orbitals," by Itatani et al. [2] generated a great deal of excitement in the HHG

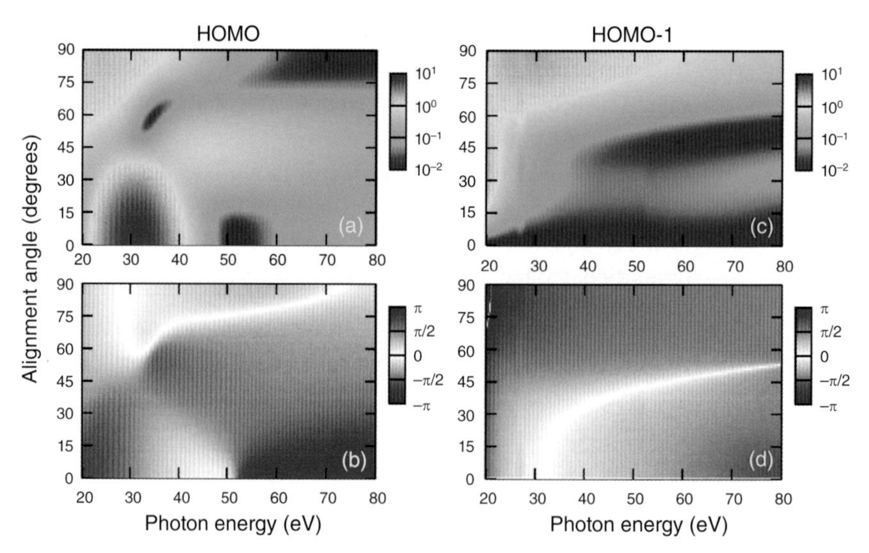

Figure 6.8 Photoionization transition-dipole amplitude and phase for ionization from HOMO and HOMO-1 orbitals of N_2 versus the alignment angle over photon energy from 20 to 80 eV calculated theoretically. These are the complete elementary structure parameters of N_2 that one can try to retrieve from the HHG data. (Reprinted from Cheng Jin et al., *Phys. Rev. A*, **85**, 013405 (2012) [12]. Copyrighted by the American Physical Society.)

community. However, their claims have been questioned and even objected to by the lesser-known theoretical papers (see [16–19]). In spite of these controversies, experimental papers announcing tomographic imaging of MOs continue to appear. A recent example involves using angular-resolved photoemission spectroscopy (ARPES) [20, 21]. These experimental papers all measured the amplitudes of the PITD or, as in the case of Itatani et al. [2], extracted the PITD amplitudes from the measured alignment-dependent HHG spectra.

The debates are centered on two issues. One is technical since the steps leading up to the retrieval of an MO rely on many additional assumptions that are known to be invalid. The other issue is philosophical. It is related to the argument that MO is a mathematical construct in quantum physics and not an "object" that can be measured experimentally.

To understand the debate, consider how MOs are extracted from the PITD [2]. The PITD can also be expressed in terms of $d_x(k,\theta)$ and $d_y(k,\theta)$ on the polarization plane. These are the dipole-transition matrix elements of the many-electron dipole operator involving the initial ground state and the final continuum state. If one assumes that photoionization removes only one electron and that all the other electrons remain unchanged, then the many-electron integral can be reduced to a one-electron integral $\langle\Psi_0|\mathbf{r}|\Psi_k\rangle$. This frozen core model is an approximation that also leads to the more familiar Koopmans' theorem. Furthermore, if one assumes that the continuum wavefunction $|\Psi_k\rangle$ is approximated by a plane wave with momentum k, then, by applying the Fourier slice theorem commonly used in computerized tomographic imaging, one obtains

$$x\Psi(x,y) = \int_0^{\pi} d\theta \int_0^{+\infty} d\omega e^{ik(x\cos\theta+y\sin\theta)} \times [\cos\theta d_x(\omega,\theta) + \sin\theta d_y(\omega,\theta)], \tag{6.11}$$

$$y\Psi(x,y) = \int_0^{\pi} d\theta \int_0^{+\infty} d\omega e^{ik(x\cos\theta+y\sin\theta)} \times [-\sin\theta d_x(\omega,\theta) + \cos\theta d_y(\omega,\theta)]. \tag{6.12}$$

In these equations, the authors [2] adopted the relation $k^2/2 = \omega$ instead of the familiar one, $k^2/2 = \omega - I_p$, from the standard photoelectric effect relation, where I_p is the ionization potential. Experimentally, harmonics for orders from H17 to H51 were measured from N_2 molecules aligned between 0° and 90° with respect to the polarization axis of the 800 nm laser. To obtain $d_x(k,\theta)$ and $d_y(k,\theta)$, the phase of the harmonics should be determined. This was not carried out in the experiment in Itatani et al. [2]. Instead, the phase was modeled based on external theoretical information. One notes that Equations 6.11 and 6.12 are equivalent. In real applications, the retrieved $\Psi(x,y)$ is reported to be the average of the two. Figure 6.8(a) shows the retrieved HOMO of N_2. The retrieved orbital resembles the one obtained from the quantum chemistry calculation shown in Figure 6.8(b). Close inspection shows that the retrieved wavefunction does not go to zero as quickly as the theoretical one since the spectral range of the harmonics in the Fourier transform is too limited.

In a subsequent measurement, Haessler et al. [9] carried out a similar experiment on N_2 at a lower laser intensity. In this experiment, the phase of the PITD was determined. In the analysis, they assumed that the HOMO-1 orbital also contributes to the HHG spectra even though a lower laser intensity was used than in Itatani et al. [2]. To separate the contributions from HOMO and HOMO-1, the authors had to make additional assumptions. Finally, the MOs for HOMO and HOMO-1 are both "measured." Their result for the HOMO orbital is shown in Figure 6.8(c) and it looks quite different from the other two figures in (a) and (b).

It is important to recognize that the MOs obtained using the tomographic-imaging method utilize many additional approximations. The most notable one is to approximate the continuum wavefunction by the plane wave. More than half a century of quantum theory of atoms and molecules has taught us that the plane wave is a very poor approximation for low-energy continuum electrons. However, without making the plane-wave approximation, the whole tomographic-imaging method is no longer available. Since the approximations leading to the retrieval of the MO are not accurate, they would lead to incorrect or inaccurate results. Note there have been no more experimental tomographic imaging studies since Vozzi et al. [22] on CO_2 molecules.

The second point of debate concerns whether measuring the MO is relevant. Recall that any quantum mechanical measurement is a projection. For a many-electron system, atomic or molecular orbitals are mathematical constructs. These orbitals are just basis functions that serve to build up the "exact" many-electron wavefunction. In other words, one cannot uniquely determine a one-electron MO in a many-electron system. Thus, MO wavefunction cannot, in principle, be measured. Any single MO is just one of the many possible representations of an abstract state vector in quantum mechanics. In this way, "measuring MO" can be seen as an offending claim to pure quantum physics followers.

Attempts have been made to argue that the "measured" orbital is not an MO but a Dyson orbital (see [23]). However this does not resolve the conflict. In quantum mechanics, for a many-particle system there is no such operator that will "make" MOs as its eigenstates, thus these orbitals are not measurable.

6.2 High-Harmonics Spectroscopy from Polyatomic Molecules and Dynamically Evolving Targets

6.2.1 Theoretical Treatments of HHG from Polyatomic Molecules

For HHS to become a practical tool, a solid theoretical method needs to be established for polyatomic molecules. Whereas a direct numerical solution of the TDSE can be used for atomic and simple linear molecular targets within the single-active-electron approximation, such an approach is not practical for polyatomic molecules. Significant progress has been achieved in the development of the *ab initio* type of calculation based on the time-dependent density functional theory (TDDFT) and the time-dependent Hartree–Fock (TDHF). However, it is still fair to say that, apart from few exceptions, these theoretical calculations have rarely been compared to experiments. Thus the majority of the current calculations for molecules are still based on simple theories such as the strong-field approximation (SFA) [24], the eikonal–Volkov approximation [25] and the

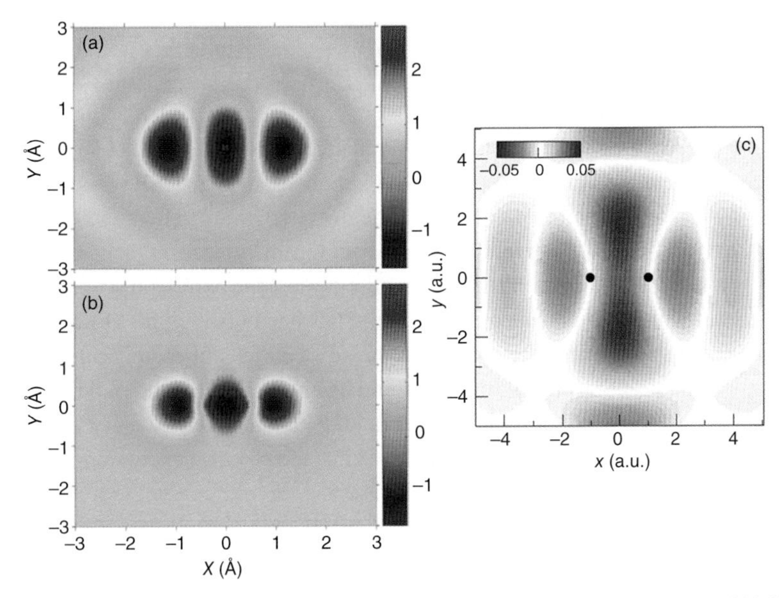

Figure 6.9 Projections of the HOMO wavefunction extracted from HHG spectra from aligned N_2 molecules, (a) and (c), from [2] and [9], respectively, compared to the HOMO wavefunction calculated from the quantum chemistry code (b). (Adopted from J. Itatani et al., *Nature*, **432**, 867 (2004) [2] and S. Haessler et al., *Nat. Phys.*, **6**, 200 (2010) [9].)

QRS theory [4, 26]. Clearly, care must be taken with regard to the choice of a theoretical method because the interpretation of experimental data and the extraction of target structure information rely heavily on the accuracy of those theories. This section describes only two main methods: the SFA (or Lewenstein model) and the QRS. While the former is not accurate, it forms a basis for a qualitative understanding of the HHG mechanism. Presently, the latter is the method of choice due to its simplicity and accuracy as it has been tested for different targets [4, 26].

The SFA for Polyatomic Molecules

The SFA approach can be easily extended to polyatomic molecules under the single-active-electron approximation. In fact, if the active electron is taken to be in a particular MO – typically the HOMO – its wavefunction can be routinely obtained from standard quantum chemistry software such as Gaussian, Gamess, and Molpro. In a majority of these quantum chemistry packages, the wavefunctions are expressed in Gaussian-type orbitals such that all the calculations for transition-dipole matrix elements can be calculated analytically.

The time-dependent induced dipole $\mathbf{D}(t)$ along a particular direction $\mathbf{n}$ can be written as

$$D(t) = i\int_0^t dt' \int d^3\mathbf{p}\mathbf{n}\cdot\mathbf{d}^*(\mathbf{p}+\mathbf{A}(t))\mathbf{E}(t')\cdot\mathbf{d}(\mathbf{p}+\mathbf{A}(t'))\exp\left[-iS(\mathbf{p},t,t')\right] + c.c., \quad (6.13)$$

where the quasiclassical action $S(\mathbf{p},t,t')$ is given, like before, by

$$S(\mathbf{p}, t, t') = \int_{t'}^{t} dt'' \left(\frac{[\mathbf{p} + \mathbf{A}(t')]^2}{2} + I_p \right). \tag{6.14}$$

As mentioned above, the bound-free transition dipole $\mathbf{d}(\mathbf{k})$ can be calculated analytically within the plane-wave approximation used in the SFA once the MO wavefunction has been obtained from quantum chemistry software. Then, the computation of Equation 6.13 can be easily carried out. Furthermore, for typical molecular targets, one can simplify the above equation by using the saddle-point approximation with respect to the integration over momentum $\mathbf{p}$. This approximation has been justified for atomic targets [24, 26] and should be valid unless the molecular size exceed tens of atomic units. In fact, the validity of the saddle-point approximation could be questionable if the molecular size is of the order of the electron excursion in the laser field, which is given by $\alpha_0 = E_0/\omega^2$. For a typical 1,600 nm laser with an intensity of 10^{14} W/cm^2, the electron excursion is 65 au. Equation 6.13 is then reduced to a simpler form

$$\begin{aligned} D(t) &= i \int_0^\infty d\tau \left(\frac{\pi}{\epsilon + i\tau/2} \right)^{3/2} \mathbf{n} \cdot \mathbf{d}^*(\mathbf{p}_{st}(t,\tau) + \mathbf{A}(t)) \\ &\quad \times \mathbf{E}(t-\tau) \cdot \mathbf{d}(\mathbf{p}_{st}(t,\tau) + \mathbf{A}(t-\tau)) \exp\left[-iS_{st}(t,\tau)\right] + c.c., \end{aligned} \tag{6.15}$$

where $\mathbf{p}_{st}(t,\tau) = -\int_{t-\tau}^{t} \mathbf{A}(t')dt'/\tau$ is the canonical momentum at the stationary points and S_{st} is obtained from Equation 6.14 by replacing $\mathbf{p}$ with $\mathbf{p}_{st}$.

More explicitly, the parallel component can be written in the molecular frame as (cf. Equation 6.3)

$$\begin{aligned} D_\parallel(t,\theta,\varphi) &= i \int_0^\infty d\tau \left(\frac{\pi}{\epsilon + i\tau/2} \right)^{3/2} [\sin\theta\cos\varphi d_x^*(t) + \sin\theta\sin\varphi d_y^*(t) + \cos\theta d_z^*(t)] \\ &\quad \times [\sin\theta\cos\varphi d_x(t-\tau) + \sin\theta\sin\varphi d_y(t-\tau) + \cos\theta d_z(t-\tau)]E(t-\tau) \\ &\quad \times \exp[-iS_{st}(t,\tau)] + c.c. \end{aligned} \tag{6.16}$$

Here, the laser-polarization direction is given by spherical polar angle θ and azimuthal angle φ. Shorthand notations $\mathbf{d}(t) \equiv \mathbf{d}[\mathbf{p}_{st}(t,\tau)+\mathbf{A}(t)]$ and $\mathbf{d}(t-\tau) \equiv \mathbf{d}[\mathbf{p}_{st}(t,\tau)+\mathbf{A}(t-\tau)]$ have been used. Note that one can also include additional factors to account for the depletion effect (see Equation 6.3).

The QRS for Polyatomic Molecules

Within the QRS, the laser-induced dipole $D(\omega,\theta,\varphi)$ for a polyatomic molecule in a linearly polarized intense laser pulse can be written as a product of a returning electron wave packet $W(E,\theta,\varphi)$ and a photo-recombination (time inverse of photoionization) transition dipole $d(\omega,\theta,\varphi)$ as

$$D(\omega,\theta,\varphi) = W(E,\theta,\varphi)d(\omega,\theta,\varphi), \tag{6.17}$$

where the electron energy E is related to the emitted photon energy ω by $E = \omega - I_p$, with I_p being the ionization potential of the target. Equation 6.17 is written in the molecular frame with the laser-polarization direction given by spherical angles $\{\theta,\varphi\}$.

PITDs can be calculated using modern molecular photoionization methods that have been developed over the last four decades. In this book, the ePolyScat package [27, 28] is used. Therefore the remainder of this subsection is devoted to discussion of the returning electron wave packet. There are two main methods to obtain this wave packet: the QRS1 and the QRS2 [4]. First, in the QRS1 from Le et al. [4], the wave packet can be conveniently calculated using the SFA

$$W^{QRS1}(E,\theta,\varphi) = \frac{D^{SFA}(\omega,\theta,\varphi)}{d^{PWA}(\omega,\theta,\varphi)}, \tag{6.18}$$

where D^{SFA} is the induced dipole calculated within the SFA and d^{PWA} is the transition dipole in the plane-wave approximation. This method has a serious drawback near a "spurious" Cooper minimum where the transition dipole in the plane-wave approximation vanishes and the wave packet is undetermined. Even though this can be fixed by using a smoothing procedure, the method could be quite tedious for polyatomic molecules since the position of the "spurious" Cooper minimum may depend sensitively on both angles θ and φ.

Therefore a more practical method uses the wave packet from a reference atom with a similar ionization potential. This method was called the QRS2 in Le et al. [4]. More specifically,

$$\begin{aligned} W^{QRS2}(E,\theta,\varphi) &= \left(\frac{N(\theta,\varphi)}{N^{ref}}\right)^{1/2} W^{ref}(E)e^{i\Delta\eta(E,\theta,\varphi)} \\ &= \left(\frac{N(\theta,\varphi)}{N^{ref}}\right)^{1/2} \frac{D^{ref}(\omega)}{d^{ref}(\omega)}e^{i\Delta\eta(E,\theta,\varphi)}. \end{aligned} \tag{6.19}$$

Here, $N(\theta,\varphi)$ and N^{ref} are the strong-field ionization probabilities for electron emission along the laser-polarization direction from the molecule and reference atom, respectively. These can be calculated with different methods as discussed in Chapter 5. The reference atom can be conveniently taken as a scaled "hydrogen-like" atom in the $1s$ ground state with the nuclear charge chosen so the atom has the same ionization potential as the molecular target. One can then calculate D^{ref} by numerically solving the TDSE and using it together with the d^{ref} from the well-known analytical expression for scaled H($1s$). In this case, there are no zeros in the transition dipole. Alternatively, one can also use the SFA and plane-wave approximation for D^{ref} and d^{ref}, respectively, for the reference atom.

In Equation 6.19, $\Delta\eta(\theta,\varphi)$ is introduced to account for the phase difference between the two wave packets. This additional phase can be approximated by the phase of the asymptotic initial wavefunction of the active electron [29]. To illustrate this approximation, Figure 6.10 shows the phase difference between the wave packets at $\{\theta_A = 100°, \varphi = 240°\}$ and $\{\theta_B = 120°, \varphi = 240°\}$ (points A and B in the inset of Figure 6.10) as a function of harmonic order. The calculation was done for one of the degenerate HOMOs of a CCl_4 molecule by using Equation 6.18 with an 1,800 nm wavelength laser pulse and an intensity of 0.55×10^{14} W/cm^2. The phase difference is found to be close to π radians for all harmonics. This is in agreement with the phase difference between point A and point B of the asymptotic wavefunction (see the inset). In contrast, the phase difference between the wave packets at $\{\theta_B = 120°, \varphi = 240°\}$ and $\{\theta_C = 140°, \varphi = 240°\}$ (points B and C in the inset) is quite close to zero radians. This is also in agreement with the phase difference

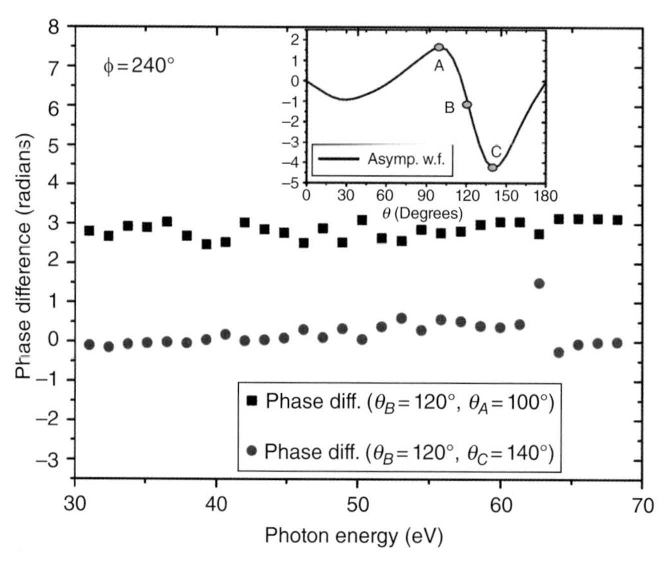

Figure 6.10 Wave-packet phase difference between $\theta_A = 100°$ and $\theta_B = 120°$ (points A and B in the inset) and between $\theta_B = 120°$ and $\theta_C = 140°$ (points B and C in the inset). Here φ is fixed at 240°. The inset shows the asymptotic wavefunction. See text for details. (Reprinted with permission from Anh-Thu Le, R. R. Lucchese, and C. D. Lin, *Phys. Rev. A*, **87**, 063406 (2009) [4]. Copyrighted by the American Physical Society.)

between point B and point C of the asymptotic wavefunction (both have negative values; see the inset).

Within the QRS2, the induced dipole is then expressed as

$$D(\omega, \theta, \varphi) = W^{ref}(\omega) N^{1/2}(\theta, \varphi) e^{i\Delta\eta(\theta,\varphi)} d(\omega, \theta, \varphi), \tag{6.20}$$

where $\Delta\eta(\theta, \varphi)$ is approximated by the phase of the asymptotic initial wavefunction. The appearance of this phase can be understood as due to the missing phase in the ionization-probability amplitude $N^{1/2}(\theta, \varphi)$ that is typically treated as a real number.

6.2.2 Early Studies of HHG from Polyatomic Molecules with 800 nm Lasers

Early investigations of HHG from nonlinear polyatomic molecules were carried out by using an 800 nm laser. Typical polyatomic molecules have quite low ionization potentials, around 10 eV. Thus the HHG process suffers from the strong depletion effect, unless the experiment is done with low laser intensities, which leads to the HHG cutoffs at quite low energies. Theoretically, the early simulations were based on the simple SFA approach. All these circumstances led to a rather qualitative understanding of HHG in polyatomic molecules. In fact, direct comparisons of experimental HHG spectra with theory were rarely presented. Even when they were, the main comparison was for the HHG cutoff, which practically carries no information about the dynamics of the HHG process.

With the application of the impulsive laser-alignment technique to polyatomic molecules, the dependence of HHG yields on molecular alignment can be studied both

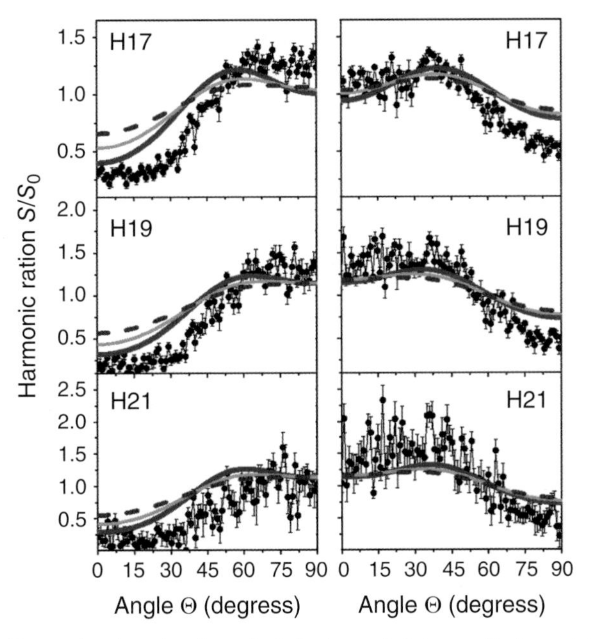

Figure 6.11 Harmonic yields from allene molecules as functions of polarization angle between aligning and driving laser pulses, measured for the aligned (left column) and the anti-aligned distribution (right column). The signals have been normalized to that from the nonaligned molecules. The solid curves show the SFA calculations at different rotational temperatures. Taken from [30]. (Reprinted from R. Torres et al., *Phys. Rev. Lett.*, **98**, 203007 (2007) [30]. Copyrighted by the American Physical Society.)

experimentally and theoretically. In particular, it was revealed that the symmetry of the electronic structure of the HOMO is reflected in the angle dependence of HHG yields. As an example, Figure 6.11 shows HHG yields versus alignment angle for different harmonics from allene [30]. The SFA simulations agree quite well with experimental data. These results indicate that the HHG yields peak near a perpendicular direction with respect to the molecular axis. A close look at the tunneling-ionization rate based on the MO-ADK theory reveals the same pattern. From the MO-ADK theory it is well known that angle-dependent tunneling ionization reflects the shape of the MO. In the case of allene, there is a nodal plane containing the molecular axis along which the ionization is suppressed.

A similar level of agreement of the SFA calculations with experiments has also been seen for linear molecules. This agreement reflects the importance of contributions from the ionization step. In general, the angle dependence of the photo-recombination dipole-matrix element might also influence the angle dependence of HHG yields.

6.2.3 HHG with Mid-Infrared Lasers and the Imprints of Target Structure

With the use of the mid-infrared laser pulses, HHG can be generated to much higher energies without suffering from a depletion effect while maintaining efficient macroscopic propagation [31]. This development also helps with quantitative calibration for different theoretical approaches. As one can expect, the SFA can only provide a qualitative description of HHG. Therefore this subsection uses the QRS theory to compare with experiments. The focus will be on the most pronounced features of HHG spectra that are relatively robust with respect to laser parameters and macroscopic conditions. For this purpose, the Cooper-type minimum in HHG spectra from the CCl_4 molecule is analyzed first. Then the HHG from stereoisomers of 1,2-dichloroethylene is analyzed.

Figure 6.12(a) shows the experimental HHG spectrum from a CCl_4 molecule versus photon energy. The driving laser of 1.8 μm wavelength with an intensity of 9×10^{13} W/cm^2 was used in the experiment. The spectrum has a very pronounced minimum near 40 eV and a cutoff near 65 eV. It is quite interesting to note that the partial photoionization cross (from the $2t_1$ HOMO) has a minimum at nearly the same energy though the minimum is somewhat shallower. Theoretical calculation based on the QRS, shown in Figure 6.12(b), reproduces very well the minimum near 40 eV. The theoretical PICS from the HOMO obtained from the ePolyScat package [27, 28] also agrees quite well with the experimental PICS though the minimum is less pronounced and also shifted to near 48 eV. Note that the theoretical calculations were carried out at a laser intensity of 5.5×10^{13} W/cm^2, which

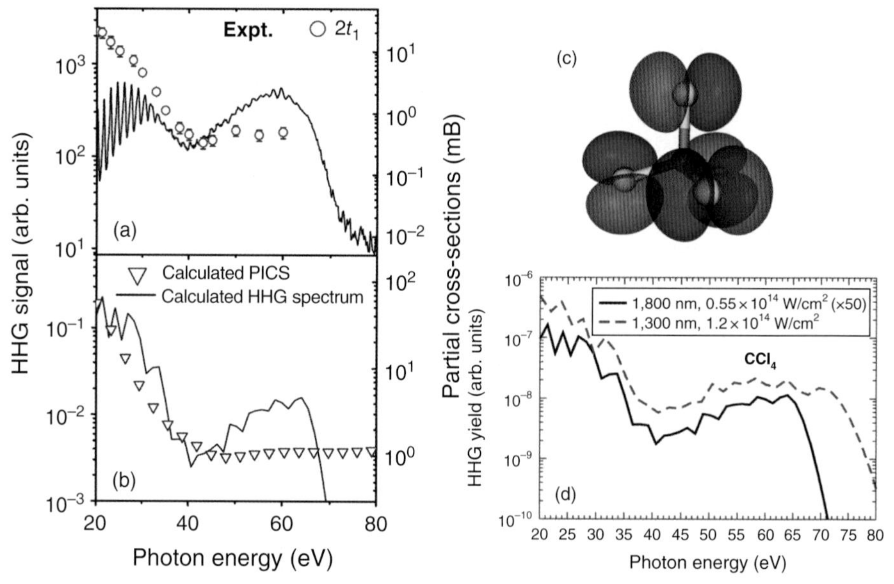

Figure 6.12 (a) Experimental harmonic spectrum from the CCl_4 molecule versus photon energy using a 1.8 μm wavelength laser with intensity of 9×10^{13} W/cm^2. The experimental partial photoionization cross section is also shown (blue circles, right vertical axis). (b) Similar to (a) but from a theoretical calculation based on the QRS. The theoretical PICS (triangle, right vertical axis) was obtained from the ePolyScat package. Figure adopted from [32]. (c) One of the degenerate HOMOs of CCl_4. (d) HHG spectra from CCl_4 with different laser pulses. Laser parameters are given in the text. The yield from an 1,800 nm laser pulse (black line) has been multiplied by a factor of 50. Only envelopes are shown. (Reprinted with permission from M. Wong et al., *Phys. Rev. Lett.*, **110**, 063406 (2013) [32] and from Anh-Thu Le et al., *Phys. Rev. A*, **87**, 063406 (2013) [29]. Copyrighted by the American Physical Society.)

is significantly lower than the experimental estimate. This adjustment was needed in order to get agreement with the cutoff position near 65 eV. The minimum in HHG spectrum near 40 eV is quite stable with respect to laser parameters. As an example, Figure 6.12(d) compares the present HHG spectrum with that of a 1.3 μm wavelength laser pulse with an intensity of 1.2×10^{14} W/cm^2. The stability of the minimum near 40 eV has also been observed in experiment [32].

To understand this behavior, first focus on the PICS. The $2t_1$ HOMO of CCl_4 has a lone-pair (nonbonding) character (see Figure 6.12(c)). Therefore the photoionization from this HOMO should be analogous to that of atomic photoionization from Cl(3p), which exhibits a Cooper minimum near 43 eV. According to the QRS, the minimum in the HHG spectrum can now be interpreted as the consequences in the Cooper minimum in the PICS from CCl_4. In fact, a more involved analysis [29] showed that the angle dependence of the photo-recombination transition dipole is nearly independent of energy in the range of 40–60 eV. Thus the HHG minimum near 40 eV simply reflects the minimum in the PICS in this case. The shift from the theoretical minimum in the PICS at 48 eV to about 41 eV in HHG is partly due to the slope of the returning electron wave packet as a function of energy. Based on the above analysis, the minimum near 41 eV in the CCl_4 HHG spectrum can be classified as a Cooper-type minimum.

Although the calculation is done for the single-molecule response, this pronounced minimum is expected to survive after the macroscopic propagation is carried out since the Cooper-type minimum does not depend on laser parameters. Moreover, the CCl_4 molecule (with T_d symmetry group) has a triply degenerate ($2t_1$) HOMO with an ionization potential $I_p = 11.5$ eV. Induced dipoles from degenerate HOMOs are added up coherently.

The case of CCl_4 is quite simple. In general, the situation could be much more complicated due to the interplay between the angle-dependent ionization rate and the angle-dependent, complex-valued transition dipole, especially when the latter depends sensitively on energy. This scenario can be illustrated by the case of HHG from stereoisomers of 1,2-dichloroethylene (or 1,2-DCE, $C_2H_2Cl_2$), in which HHG measurement has been reported [33]. HHG study of stereoisomers can serve as an interesting test for HHS as the HHG from these targets is expected to be different due to the differences in the geometric arrangement of atoms in the isomers. In fact, it was found in Wong et al. [33] that the spectra from *cis* and *trans* of 1,2-dichloroethylene are quite distinguishable in a broad range of energy, even when the molecules are not aligned (see Figure 6.13(a)). The mechanism behind this was originally attributed in Wong et al. [33] as due to the differences in the strong-field ionization. However, a more detailed analysis based on the QRS, reported in Le et al. [34], rules out that mechanism. The main points of this analysis are summarized below.

As seen from Figure 6.14, the total PICS from the two isomers are practically identical even though the atomic arrangements of the two chlorine atoms are clearly different. The calculation was carried out with the ePolyScat package [27, 28]. Interestingly, there is no clear evidence of a Cooper minimum in this range of energy. This is not entirely surprising as the HOMOs of these isomers have only partial contribution from the Cl(3p) orbital (see the insets of Figure 6.14). Furthermore, the total ionization rates for the isotropic *cis* and *trans*-DCE within the SFA are nearly the same. Note that a recent experimental

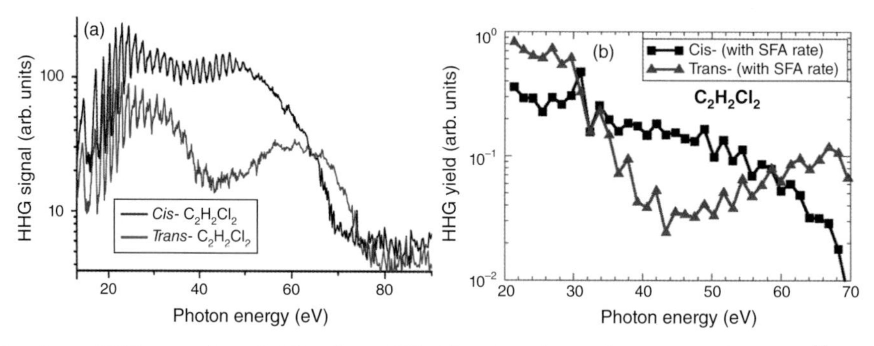

Figure 6.13 (a) Experimental HHG spectra from *cis*-DCE and *trans*-DCE with a 1.8 μm laser and an intensity of 1.1×10^{14} W/cm^2. Figure adopted from [33]. (b) Theoretical HHG spectra (only the envelopes are shown) from *cis*-DCE and *trans*-DCE for a 1.8 μm laser and an intensity of 0.6×10^{14} W/cm^2. The calculations were carried out with the QRS. (Reprinted from M. Wong et al., *Phys. Rev. A*, **84**, 051403 (2011) [33] and from Anh-Thu Le et al., *Phys. Rev. A*, **88**, 021402 (2013) [34]. Copyrighted by the American Physical Society.)

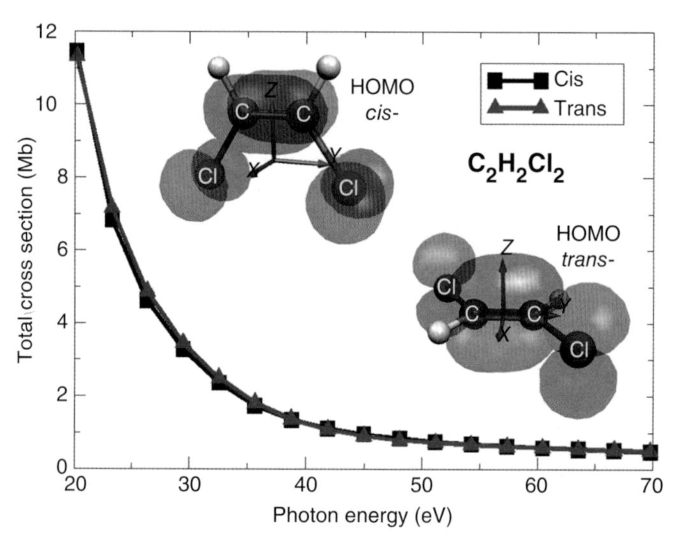

Figure 6.14 Total (integrated) photoionization cross-section from the HOMO of *cis*-DCE and *trans*-DCE. The calculation was carried out with the ePolyScat package. The insets show the HOMOs and molecular frame coordinates used in the calculations. (Reprinted from Anh-Thu Le et al., *Phys. Rev. A*, **88**, 021402 (2013) [34]. Copyrighted by the American Physical Society.)

measurement using an 800 nm laser [35] also showed that the ratio of the ionization yields for the two isomers is close to unity. Nevertheless, both experimental measurement [33] and QRS theory [29] for HHG from *trans*-DCE showed a minimum near 43 eV (see Figure 6.13(a) and (b), respectively), though the minimum is less pronounced as compared to the CCl_4 case. For *cis*-DCE, there is no visible minimum in the QRS simulated spectrum, while experimental spectrum [33] shows a very shallow minimum slightly below 40 eV. A detailed analysis within the QRS theory [29] for the 35–60 eV energy range revealed that the overlap of the angle-dependent ionization rate and transition dipole is quite small for *trans*-DCE, which leads to small HHG yields. Near 45 eV there is a strong destructive interference for contributions from *cis*-DCE with different alignments, which results in a Cooper-type minimum in HHG spectrum. Thus, this minimum is expected to be sensitive to the molecular alignment.

Experimental and theoretical investigations on aligned polyatomic molecules are strongly desirable in the near future as they would provide more detailed information to validate or discriminate different theories. Increased interest in systematic analysis of characteristic features such as the Cooper-type minima and shape resonances in HHG is also anticipated. Regarding the influence of a shape resonance, a first step in this direction has been reported for SF_6 [36, 37].

6.2.4 Probing Dynamically Evolving Molecules with HHG: Nuclear Dynamics

General Scheme

The ultimate goal of HHS is not just to characterize the static molecules, but to follow dynamically changing targets in a chemical reaction, which occurs on a tens to hundreds of femtoseconds timescale. This and the subsequent subsections discuss this capability of HHS with vibrating and dissociating molecules. Before doing so, a simple example is used to illustrate what one could achieve with HHS.

Consider the isomerization of hydrogen cyanide (HCN) to hydrogen isocyanide (HNC) [16]. The left panel of Figure 6.15 gives the schematic of two local potential minima and a

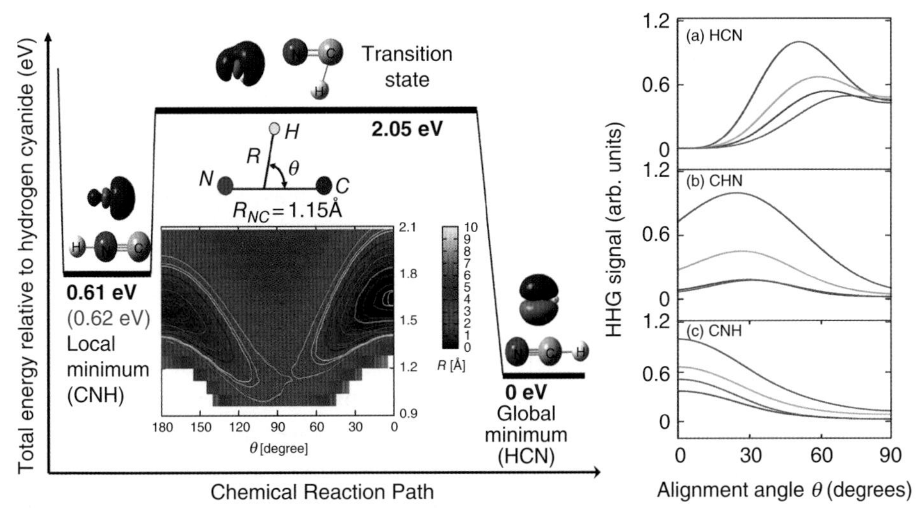

Figure 6.15 Left: Reaction path of hydrogen cyanide (HCN) to hydrogen isocyanide (HCN) calculated by DFT. The contour plot of the PES at the equilibrium distance (R_{NC} = 1.15 Å) of the most stable isomer HCN is shown. Right: Angular dependence of the H21, H25, H29, and H31 harmonics for HCN isomers: HCN (top panel), transition state CHN (middle panel), and HCN (bottom panel). (Reprinted with permission from Van-Hoang Le et al., *Phys. Rev. A*, **76**, 013414 (2007) [16]. Copyrighted by the American Physical Society.)

transition state, the shape of the HOMO, as well as the atomic configurations corresponding to these states. The inset gives the contour plot of the potential energy surface (PES) calculated with Gaussian G03 by using DFT with Becke's three-parameter Lee–Yang–Parr hybrid functional (B3LYP) and within 6-311+G(2*df*, 2*pd*) Gaussian-type basis set. The PES provides information on the reaction path from the linear HCN to the linear HNC, via a transition state. A fairly large energy of 2.05 eV is needed to reach the transition state. The HOMOs for the three molecules are quite different. These differences are reflected on the large differences in the angle dependence of the HHG yields shown in the right panel of Figure 6.15. For simplicity, the simulation was carried out using the SFA model. Thus, imaging the transition from HCN to HNC via the transition state in real time using HHS would appear to be quite feasible. Our goal should be to observe the changes in HHG signals as the chemical reaction progresses in order to extract the changes in molecular structure. As suggested in Le et al. [16], the extraction can be carried out by using an iterative fitting procedure. It would also be interesting to image this reaction using the LIED technique discussed in Chapter 4.

Theoretical Treatment for HHG from Vibrating Molecules

It is generally known that, even within the fixed-nuclei approximation, there are very few reliable theories for HHG from molecules. The situation is even more complicated when one has to include the nuclear degrees of freedom in HHG theory. Different theoretical treatments that are based on simplified models or by direct solution of the TDSE for simple systems in reduced one-dimensional (1D) space have been proposed. Most of the models have not been fully calibrated, and thus their validity is largely unknown. As the QRS is computationally efficient and has been well tested for the fixed-nuclei case, it is desirable to extend this theory to include the nuclear degrees of freedom. In the following, a HHG theory including nuclear degrees of freedom is formulated in such a way that the QRS can be used.

To illustrate the method, consider a homonuclear diatomic molecule in its electronic ground state under a few-cycle intense laser pulse. In Le et al. [38] it was shown that the induced dipole $\bar{D}(t)$ can be calculated as

$$\bar{D}(t) = \int dR|\chi(R,t)|^2 D(t;R), \tag{6.21}$$

where $\chi(R,t)$ is the nuclear wavefunction and $D(t;R)$ is the induced dipole from a molecule with a fixed internuclear distance R. In practice, one can also use the dipole acceleration form in Equation 6.21. As usual, the HHG power spectrum is related to the induced dipole in frequency domain $\bar{D}(\omega)$ by $S(\omega) \sim |\bar{D}(\omega)|^2$ where ω is the emitted photon energy. It is important to realize that $\chi(R,t)$ can generally be significantly modified by the HHG driving laser even when the molecule was prepared in a specific vibrational state, say, for example, the vibrational ground state with $v = 0$. For simplicity, consider the case when laser intensity is well below the target saturation intensity such that the depletion of the ground state is negligible and the molecules remain mostly in the electronic ground state. The main advantage of the present approach is that Equation 6.21 can be readily used in

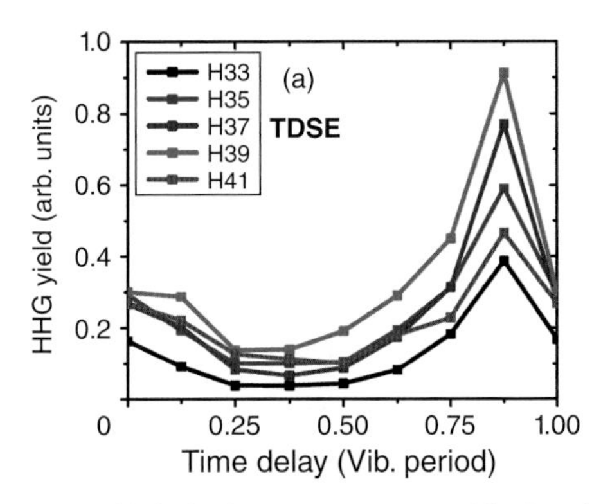

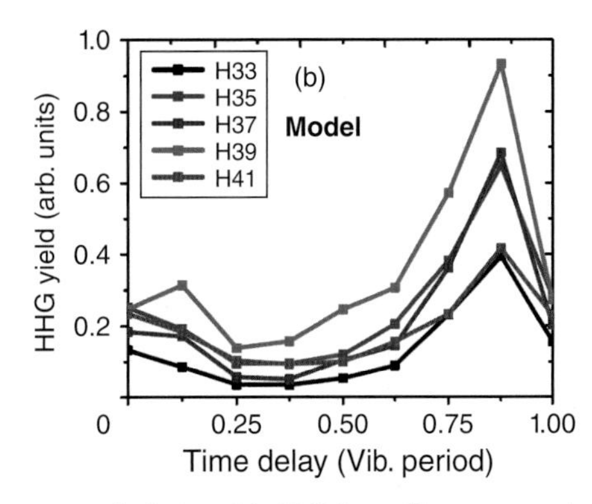

Figure 6.16 (a) HHG yields for few harmonics versus time delay from the numerical solution of the TDSE for a collinear mass-scaled H_2^+ molecule with an initial nuclear wave packet. The "hydrogen" atom mass is 16 M_p. See text for more details. (b) Same as (a) but from the model calculated by using Equation 6.21. (Reprinted from Anh-Thu Le et al., *Phys. Rev. Lett.*, **109**, 203004 (2012) [38]. Copyrighted by the American Physical Society.)

combination with the QRS theory with different fixed internuclear separations. Indeed, the induced dipole $D(\omega; R)$ can be calculated with the QRS for any fixed R and its time-domain function $D(t; R)$ can be inserted into Equation 6.21.

The validity of the model described by Equation 6.21 has been carefully tested against the exact TDSE for a collinear mass-scaled H_2^+ molecule in which the electron and nuclei are restricted to move along the laser-polarization direction (see [38]). In fact, it was found under different laser parameters that the HHG spectra calculated by the two methods are nearly indistinguishable. For illustration, in Figure 6.16 the exact TDSE results are compared with the model calculation based on Equation 6.21 for a few harmonics as a function of pump-probe time delay. Here, the initial vibrational state was assumed to be a nuclear wave packet. It was prepared by a "pump" pulse at time $t_0 = 0$ to be a linear combination of $v = 0, 1$, and 2 vibrational states with the coefficients of 0.8, 0.5, and 0.332, respectively. For simplicity, these coefficients were chosen to be real valued. In case of a real pump pulse, complex-valued coefficients are obtained, in general. An eight-cycle, 800 nm probe pulse, with an intensity of 2.5×10^{14} W/cm^2 was used at different time delays to produce HHG. The mass of "hydrogen" is chosen to be $16M_p$ where M_p is the proton mass. Overall, there is very good agreement between the two methods, as shown in Figure 6.16. The HHG yields were found to modulate as a function of time delay with a period of $T = 64$ fs, equal to the vibrational period of this mass-scaled H_2^+ molecule. This indicates that the HHG signals reflect the motion of the nuclear wave packet as it evolves in time. The peak near the time delay of $0.9T$ in Figure 6.16 is associated with the nuclei distributed at the peak of the HHG driving pulse at large R near the outer turning point where the HHG process is more efficient due to a smaller ionization potential. The precise value of time delay and the sharpness of this peak have no real physical meaning since the initial nuclear wave packet was "created by hand" with an artificial choice of excitation amplitudes without any relative phases, and not by a real pump pulse.

Probing Dynamically Evolving Molecules with HHG: The Case of Vibrating N_2O_4

For another illustration, consider HHG from a large-amplitude vibrating N_2O_4 molecule. This target is chosen since experimental measurements are available. In the experiment by Li et al. [39], a vibrational nuclear wave packet was first initiated by a short laser pulse (the pump) with a relative weak intensity of 2×10^{13} W/cm^2. This process can be understood as an impulsive stimulated Raman excitation, discussed in Section 3.6.4. At different time delays with respect to the pump pulse, high-order harmonics were then generated with a more intense pulse (the probe) at an intensity of 2×10^{14} W/cm^2. Both pulses were of 800 nm wavelength and 30 fs duration (full-width at half maximum (FWHM)). The measured HHG yield was found to modulate as a function of time delay with a period identical to the vibrational period of the symmetric stretch mode ($T \approx 130$ fs) (see Figure 6.17). Some simulation results based on Equation 6.21 in conjunction with the QRS are presented in Figure 6.18. More details can be found in Le et al. [38].

A brief description of the simulation is in order. Since the N–N symmetric stretch is the most dominant mode in this Raman excitation scheme, it is reasonable to approximate the changes in N_2O_4 as due to the change in the N–N distance, R_{NN}, while the N–O bond length and bond angles remain fixed. In the simulation, the nuclear wave packet initiated by the pump pulse can then be calculated by methods discussed in Section 3.6.4, in which N_2O_4 is modeled as an effective diatomic molecule. Subsequently, this wave packet is modified during the HHG driving pulse. This process is also simulated by methods described in Section 3.6.4. PITDs for each fixed geometry are calculated with the molecular photoionization ePolyScat package [27, 28]. For simplicity, only the case of parallel pump

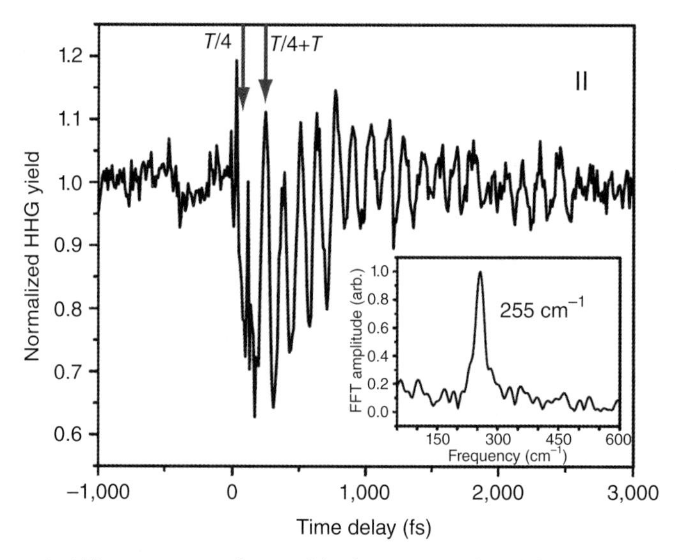

Figure 6.17 Experimental harmonic yield versus pump-probe time delay for H17 in case of parallel pump and probe laser polarizations. The signal has been normalized to the yield at a negative time delay. The oscillation period is $T \approx 130$ fs. (From W. Li et al., *Science*, **322**, 1207 (2008) [39]. Reprinted with permission from AAAS.)

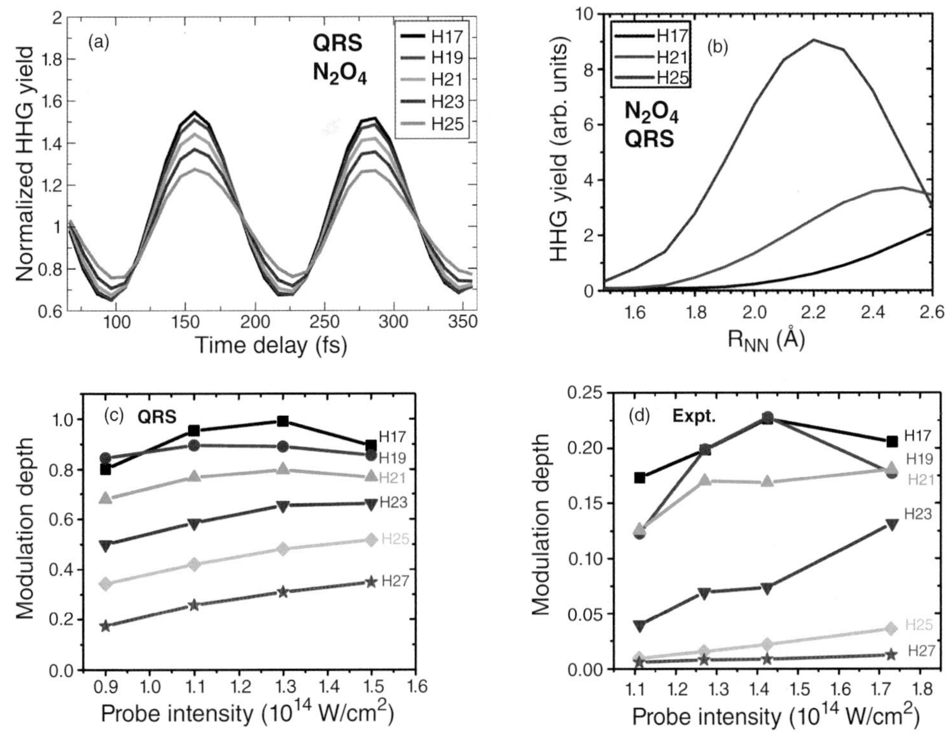

Figure 6.18 (a) Theoretical normalized HHG yields from N_2O_4 versus pump-probe time delay for a few different harmonics. (b) Theoretical HHG yields from the QRS versus R_{NN}. (c) Theoretical modulation depth versus probe laser intensity. (d) Same as (c), but from [39]. In the simulations, the molecules are assumed to be perfectly aligned along the pump laser-polarization direction. This assumption leads to overestimates of the theoretical modulation depths as compared to the experimental ones. (Reprinted from Anh-Thu Le et al., *Phys. Rev. Lett.*, **109**, 203004 (2012) [38]. Copyrighted by the American Physical Society.)

and probe laser polarizations is considered. It was further assumed that all molecules are aligned with the N–N axis along the laser-polarization direction. To minimize the ionization-depletion effect, a probe pulse of a duration of 20 fs and an intensity below about 1.5×10^{14} W/cm^2 is used.

The simulated HHG yields for a few harmonics as functions of the pump-probe delay time are presented in Figure 6.18(a). The results clearly show modulation of HHG signal for all harmonics with a period of about 125 fs, which is the vibrational period of the symmetric-stretch mode in the modeled N_2O_4. This value is quite close to the experimental value of $T \approx 130$ fs. All the observed harmonics are in phase, in good agreement with experiments. If the SFA-induced dipoles are used instead of the QRS for fixed nuclei as inputs to Equation 6.21, some observed harmonics were found to be out of phase. Furthermore, for all these harmonics, the first peak occurs near a time delay of 160 fs in good agreement with the experimental value of 170 ± 10 fs. This can be explained as the HHG yield maximizes near the outer turning point of the N–N stretch at time $T + T/4 \approx 163$ fs. (Note that the earlier peak at delay time of $T/4 \approx 32$ fs is difficult to observe due to the pump-probe overlap for relatively long pulses of 30 fs.) Indeed, the harmonic yields from the QRS as functions of R_{NN}, shown in Figure 6.18(b), are found to increase with R_{NN} in the accessible vibrational region near equilibrium at 1.8 Å. The outer turning point is estimated to be below 2.1 Å for typical pump laser parameters used in experiments (see Figure 3.24). Note that when the influence of the HHG driving laser is taken into account, the vibrational wave packet is further modified [38].

The model calculation also predicted that the modulation depth increases slightly with the probe laser intensity (see Figure 6.18(c)), in good agreement with experiment (see Figure 6.18(d)). This can be understood since, at a higher intensity, the nuclear wave packet extends during the probe pulse to a larger R_{NN} where the HHG process is more efficient. The simulation overestimates the modulation depths by almost a factor of four. This is due to the fact that the molecules were assumed to be perfectly aligned along the laser-polarization direction.

It should be noted that the interpretation presented here is different from the original interpretation given in Li et al. [39] in which it was concluded that both the HOMO and HOMO-1 contribute to the HHG. That simulation was based on the eikonal–Volkov approximation. In the present interpretation, only the HOMO is needed to explain the experimental data. This interpretation is in agreement with a newer *ab initio*–type calculation by Spanner et al. [40] though some discrepancies with experiments still remain in the latter.

The example considered in this section illustrates the complexity of the HHS with dynamically evolving targets: the nuclear wave packet to be retrieved is, in general, modified by the interaction with the intense probe pulse. Therefore an inversion procedure to obtain the "original" nuclear distribution should be more involved than by a perturbative probe.

Although no attempt was made to extract the N–N distance (or nuclear wave packet) versus delay time from experiments, the mapping based on the QRS for the HHG yield versus R_{NN} (see Figure 6.18(b)) together with Equation 6.21 clearly demonstrates the capability of HHS to follow molecular vibration in real time. In fact, if HHG dipoles and phases as functions of R_{NN} are assumed to be known from the QRS, one can retrieve the nuclear wave packet by fitting Equation 6.21 to the measured HHG data at each time delay. A similar idea can also be applied to retrieve electronic wave packet dynamics. This will be discussed in Section 6.2.5.

So far the discussions have been limited to bound nuclear motion. HHS can be extended to unbound nuclear motion as well. In this regard, two pioneering pump-probe experiments performed on Br_2 [41] and NO_2 [42], where HHS is carried out for dissociating molecules initiated by the second harmonic of the titanium-sapphire lasers should be mentioned. The model presented here can be used to provide realistic theoretical simulations for these experiments if it is extended to include multiple electronic surfaces.

6.2.5 Probing Dynamically Evolving Molecules with HHG: Electronic Dynamics

It is quite desirable to extend HHS techniques to study electronic dynamics as well. Since electronic dynamics typically occurs in a much shorter timescale (hundreds of attoseconds to a few femtoseconds) as compared to that of nuclear dynamics (tens of femtoseconds to picoseconds) (see Table 1.1), the probe-pulse duration should be comparable to that of the electronic timescale.

HHG itself can be considered as a special kind of pump-probe process. In fact, within the three-step model, one can think of ionization as the pump while photorecombination is the probe. A similar idea has been utilized in the LIED technique discussed in Chapter 4. This is not a coincidence as both LIED and HHS are based on the rescattering physics. In both cases, the recollision step, either photo-recombination in HHG or rescattering in LIED, serves as the probe. In the classical sense, the time interval between the pump and the probe corresponds to the electron excursion time in the continuum. For a fixed laser wavelength, each harmonic has a specific excursion time, if the short trajectories are selected by a proper choice of the phase-matching condition (see Sections 5.2 and 5.5). This offers the possibility of analyzing target dynamics on the sub-cycle scale, which can also be controlled by changing the laser wavelength.

An early experiment by Baker [43] utilized this feature of HHG to detect the nuclear dynamics [3] during the electron excursion in the continuum. It was also suggested [44] that sub-cycle electron (or hole) dynamics in the target cation is responsible for the observed intensity dependence of the position of the minimum in HHG spectra from aligned CO_2. Such conjecture on sub-cycle dynamics was based on rather involved theoretical modeling. This subsection discusses a recent experiment by Kraus et al. [45], who used HHS techniques to deduce ultrafast "charge migration" in an iodoacetylene (HCCI) molecule. Note that electronic dynamics is one of the main topics of active research in attosecond physics that will be discussed in Chapters 7 and 8.

In the first step of the HHG process from HCCI, an electron can tunnel out most efficiently from the HOMO and HOMO-1, leaving the ion core in a coherent superposition of the ground state (denoted as $\tilde{X}^{+}\,^2\Pi$ in Figure 6.19(d)) and the first excited state (denoted as $\tilde{A}^{+}\,^2\Pi$). The HOMO hole is localized near iodine whereas the HOMO-1 hole is localized near the two-carbon center. The energy gap between these two states is about 2.23 eV, leading to an oscillation of the hole density in space between these two centers with a period of about 1.85 fs. By using a series of HHG measurements together with an elaborate modeling, Kraus et al. [45] managed to obtain the relative populations in $\tilde{X}^{+}\,^2\Pi$ and $\tilde{A}^{+}\,^2\Pi$, and the initial relative phase. This allows them to reconstruct the hole dynamics for the first two femtoseconds after the hole is created. It should be noted that charge migration on the femtosecond timescale caused by electron correlation (or relaxation) was found theoretically even without nuclear dynamics [46]. However, the hole (or electron) dynamics can, in general, initiate a nuclear rearrangement process, which would, in turn, influence and dampen the hole-oscillation dynamics itself.

Kraus et al. [45] measured HHG from aligned HCCI under the field-free condition. The alignment was created impulsively (nonadiabatically) (see Section 3.6). The ratio of aligned and anti-aligned HHG signals are shown in Figure 6.19(a) for the driving laser wavelengths of 800 nm and 1,300 nm. Here, only the 800 nm case is considered. The HHG dipole phase as a function of alignment angle was also measured using the transient grating method and the result is shown in Figure 6.19(c) for a few harmonics. For retrieval, they also measured HHG from oriented HCCI (a polar molecule). To create asymmetry in the orientation, a

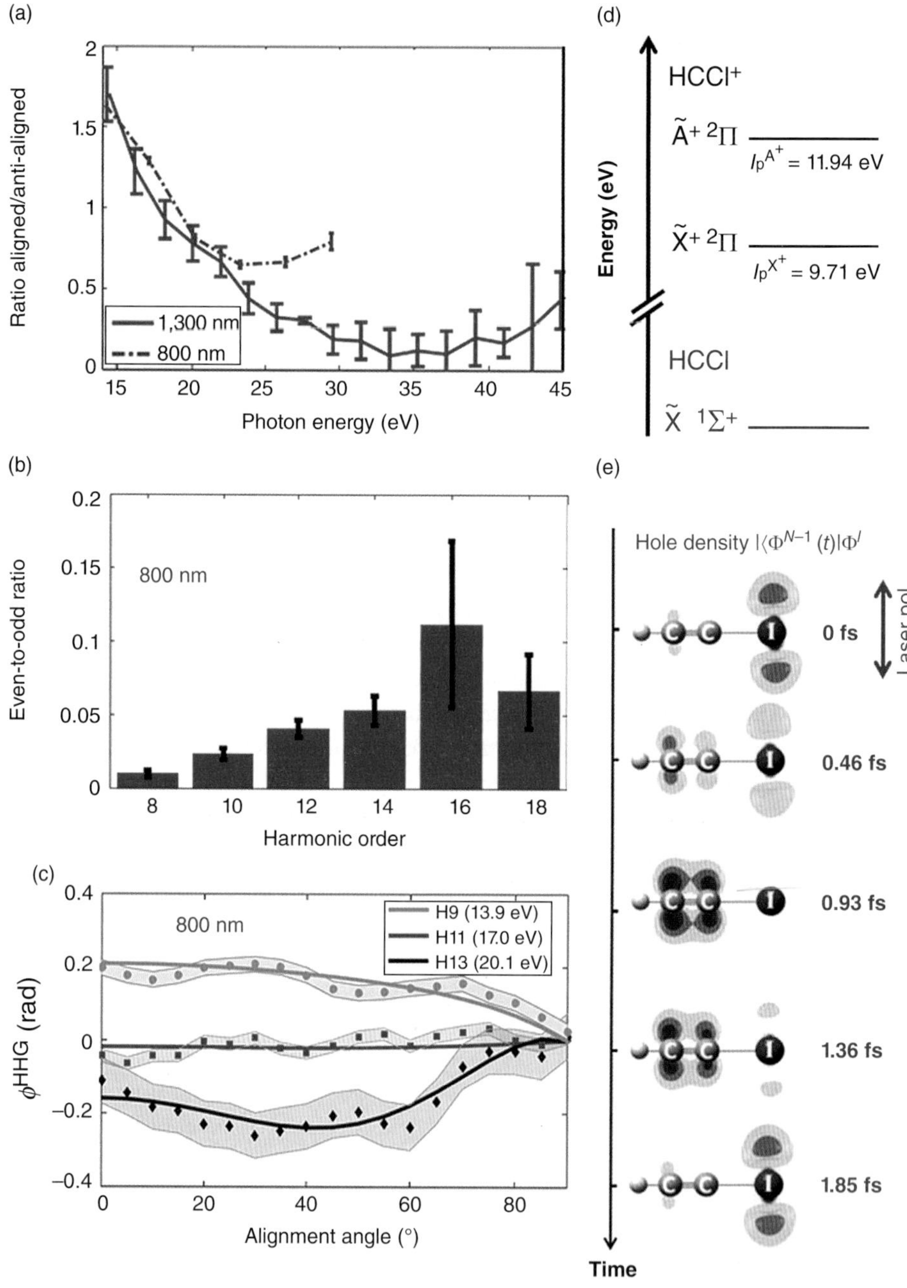

Figure 6.19 (a) Experimental ratio of HHG intensities from aligned and anti-aligned HCCl for 800 nm and 1,300 nm driving lasers. (b) Experimental even-to-odd ratio of HHG yields for 800 nm case. (c) Experimentally measured HHG dipole phase for a few harmonics (indicated in the labels) for 800 nm case with oriented molecules. (d) Schematic of energy levels of the ground state of HCCl and the relevant ionic states. (e) Reconstructed hole density versus time for the case of perpendicular arrangement. (From P. Kraus et al., *Science*, **350**, 790 (2015) [45]. Reprinted with permission from AAAS.)

scheme similar to the impulsive laser alignment was used, but with a short, two-color (800 nm and 400 nm) laser pulse instead of a single-color (800 nm) pulse. The ratio of even-to-odd harmonic yields is shown in Figure 6.19(b).

How was the hole dynamics retrieved from the experimental high-order harmonic spectra in Kraus et al. [45]? Clearly, for retrieval one needs a simple link between the hole population and its relative phase with the observed HHG yields and phases. For that purpose, Kraus et al. used, in essence, a modified version of the QRS. More specifically, the HHG-induced dipole is given as

$$D(\omega,\theta) \propto \sum_{i,f} a_i(\omega) r_i d_{ion,i}(\theta) C_i(\omega) c_{if}(\omega,\theta) d_{rec,f}(\omega,\theta), \tag{6.22}$$

where ω is the emitted photon energy and θ is the angle between the molecular axis and the laser-polarization direction. The first three factors on the right-hand side of the Equation 6.22 give the electron wave packet, initiated from MO i (or leading to ionic state i). This would be the returning electron wave packet used in the standard QRS if both nuclear and hole dynamics were neglected. Here, the first factor plays the role of the "reference atom" wave packet. The strong-field ionization probability in channel i is expressed as a product of angular part $|d_{ion,i}(\theta)|^2$ and relative weights $|r_i|^2$ between the channels. In order to take into account the nuclear dynamics that occurs between the ionization and the recombination steps, the factor $C_i(\omega)$ is introduced in Equation 6.22. This factor was suggested earlier [3, 43]. Similarly, factor $c_{if}(\omega,\theta)$ was used in the above equation to account for the possible transition due to the laser electric field from an initial ionic state i to a final ionic state f, to which the photo-recombination transition occurs. Finally, the last factor describes the photo-recombination dipole that can be calculated using the ePolyScat package [27, 28]. The sum in Equation 6.22 is over all possible ionic states, populated during the HHG process. For retrieval, Kraus et al. [45] included only two states, $\tilde{X}^+\ {}^2\Pi$ and $\tilde{A}^+\ {}^2\Pi$.

To make the retrieval possible, Kraus et al. [45] assumed that the angular-dependent part of ionization $|d_{ion,i}(\theta)|^2$ as well as the photo-recombination dipole were known from theory. To simulate HHG from an ensemble of partially aligned molecules, an average over the molecular distribution for Equation 6.22 should be carried out. Kraus et al. also assumed that the alignment distribution can be accurately calculated using the methods discussed in Section 3.6. The dependence on angle θ in the coefficients $c_{if}(\omega,\theta)$ were also assumed to be simplified as $c_{if,\parallel}(\omega)$ for parallel (aligned ensemble) and $c_{if,\perp}(\omega)$ for perpendicular (anti-aligned ensemble). In the latter case, the laser does not couple the two ionic states due to symmetry reasons (both states are of Π symmetry), so the populations do not change in time. In other words, $|c_{if,\perp}(\omega)| = 1$ for $i = j$ and vanishes otherwise. With the above assumptions, the retrieval was then carried out by a fitting procedure to obtain the initial population and the relative phase of the two ionic states, as well as the coefficients $c_{if,\parallel}(\omega)$. The result is shown in Figure 6.19(e) for the case when the molecule is aligned perpendicularly to the laser polarization. The hole density oscillates in space between iodine and the C–C center as a function of time with a period of 1.85 fs. This oscillation is solely due to the energy difference between the two ionic states. Note that the spatial distribution of the hole density was not directly measured. Instead, it was

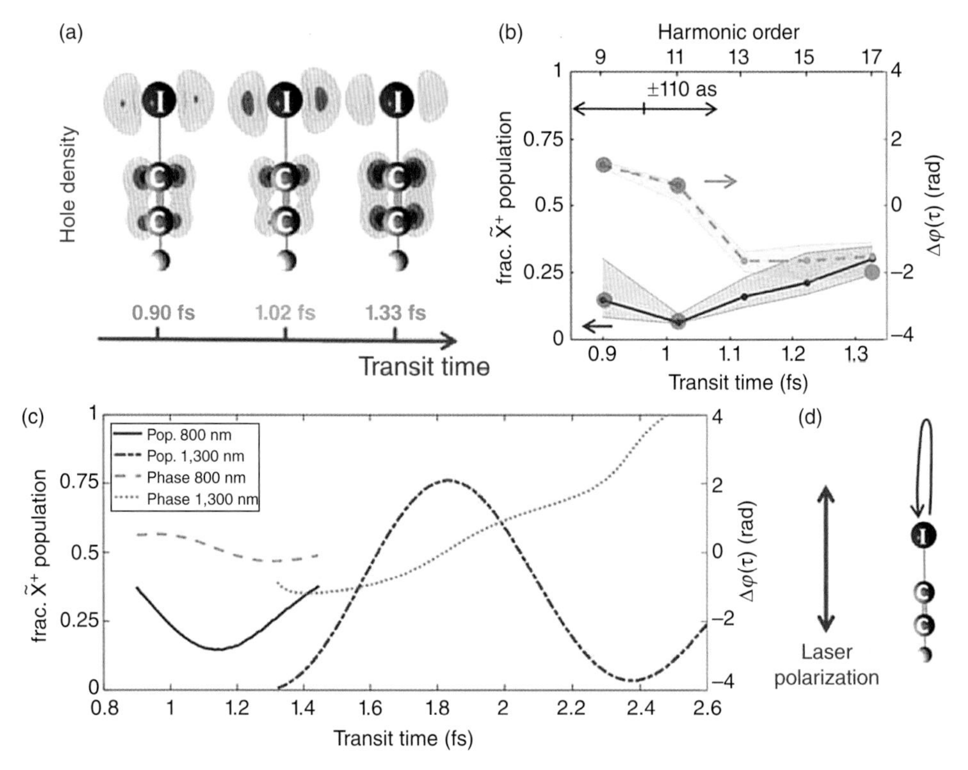

Figure 6.20 (a) Reconstructed hole density versus time for the case of parallel arrangement. (b) Retrieved population and phase of the ground ionic state (relative to the excited state) versus time. (c) Same as (b), but from theoretical simulation. The case of a 1,300 nm laser is also included. (d) Schematic of the parallel arrangement. (From P. Kraus et al., *Science*, **350**, 790 (2015) [45]. Reprinted with permission from AAAS.)

reconstructed using the retrieved populations and relative phase in combination with the theoretically calculated MOs from quantum chemistry software. In Kraus et al. [45], the retrieved amplitudes were found to be 0.9 and 0.43 for $\tilde{X}^{+}\ {}^{2}\Pi$ and $\tilde{A}^{+}\ {}^{2}\Pi$, respectively, to be compared with the theoretical estimates by the weak-field asymptotic theory (WFAT) (see Section 3.7.3) of 0.75 and 0.66, respectively.

The situation is more complicated when the molecule is parallel aligned with respect to the driving laser polarization (see Figure 6.20(d)). This is because the populations of the holes are modified after their initial creations since the two states are strongly coupled by the laser field. Nevertheless, the hole density versus time can be reconstructed (see Figure 6.20(a)), by using the retrieved population and phase of the ionic ground state $\tilde{X}^{+}\ {}^{2}\Pi$ relative to the excited state $\tilde{A}^{+}\ {}^{2}\Pi$, shown in Figure 6.20(b). Note the one-to-one mapping between the excursion time (or transit time – lower horizontal axis) and the harmonic order (upper horizontal axis) is possible because it was assumed that the harmonics are from short-trajectory electrons only. The theoretical simulation for the population and relative phase of the ionic ground state, shown in Figure 6.20(c), compares reasonably well with the retrieved results, although its phase jump near the transit time of 1.1 fs is less pronounced.

This result serves as an independent check for the accuracy of the retrieval procedure. We note that, due to the influence of the laser field for the parallel case, the hole-density oscillation is less pronounced as compared to the perpendicular case.

The reconstructed hole densities versus time, shown in Figures 6.19(e) and 6.20(a), were obtained after elaborate retrieval processes that involve borrowing many parameters from theoretical calculations. These results are interpreted as a measurement of attosecond charge migration. Clearly, the hole densities in these figures are not from direct measurements. They are derived from the measured HHG spectra. But do these derived charge-migration "movies" convey any or better information about the harmonic spectra? It is obvious that harmonic spectra are related to the time-dependent electric dipole of the electron wave packet, not the electron or hole density. In choosing charge migration to characterize electron dynamics the very important phase information is lost. While ultrafast scientists may not agree on how to make an "electron movie" to follow the electron dynamics, charge migration or hole hopping are certainly not the answers. Without tracking the phase there is no electron dynamics.

6.3 Routes to Optimizing Intense HHG

6.3.1 Introduction

One of the most important goals of HHG studies is the realization of table-top high-harmonic light sources for diverse applications to spectroscopy of atoms and molecules, metrology, surface sciences, nanostructures, and many others, to complement expensive large-scale facilities like synchrotrons or free-electron lasers, which restrict user access. HHG light sources offer coherence, broad spectral range, and femto- or attosecond pulse durations, and thus offer additional potential as tools for studying the dynamics of matters at unprecedented timescales. While tremendous progress has been made with high-order harmonics as useful light sources for experiments, HHG has been prevented from wider applications because of its low conversion efficiency, on the order of 10^{-5} for plateau harmonics. The harmonic energy in applications has been restricted mostly to below 100 eV, and photon flux is often too low to use as a probe for typical pump-probe experiments. In recent years, efforts have been made to mitigate these limitations and impressive progress has been made. This section summarizes the various techniques that have been used to generate tunable intense individual harmonics or the generation of attosecond pulse trains. Efforts related to isolated single-attosecond pulses will be addressed in Section 7.4.

Based on the principle of harmonic generation from each atom, higher photon flux can be accomplished by increasing the intensity of the driving laser, while higher-photon energy can be reached by using longer-wavelength driving lasers. However, harmonics are generated from all the atoms in the gas medium. If these harmonics are phase matched, then the harmonic yield will grow quadratically with the gas density, or equivalently, with the pressure of the gas medium. On the other hand, the propagation of the driving laser and the generated harmonics are influenced by the medium, which in turn would subsequently modify the medium. The complicated dynamics in principle can be calculated by solving the Maxwell's wave equations, but it is imperative to understand how to identify the favorable conditions such that harmonics emerging from the gas cell have the desired ingredients for applications.

In Section 6.3.2, phase-matching issues will be addressed. To reach higher-photon energies from 100 eV up to the water-window harmonics between 380 eV and 540 eV, or even keV harmonics, longer-wavelength driving lasers are used. This will be discussed in Section 6.3.3. While most high-order harmonics are generated with single-color driving lasers, it has been found since earlier days that two-color or multicolor laser pulses can enhance harmonic yields by a few orders of magnitude under suitable circumstances. Recent efforts in waveform synthesis for optimizing the cutoff energy or the high-harmonic yields will be discussed in Section 6.3.4. Finally, it is clear that photon fluxes per second can be significantly increased if one can increase the typical kHz repetition rates of the driving laser to MHz. Progress in the development of MHz driving lasers will be briefly addressed in Section 6.3.5.

6.3.2 Optimization by Phase Matching

Phase Matching for Ionization below the Critical Level

To reach the highest conversion efficiency, the macroscopic buildup of the harmonics from all the emitters inside the generation medium should be phase matched. From Section 5.4, this requires that the wave-vector mismatch due to free-electron dispersion, neutral atomic dispersion, the phase of the focusing term, and the atomic dipole-phase term all be added to zero. The phase mismatch for each harmonic order can be written as

$$\begin{aligned}\Delta k_q &= \eta p N_{atm} r_0 \lambda \left(q - \frac{1}{q}\right) - \frac{2\pi q}{\lambda} p \delta n (1-\eta) \\ &\quad + (\text{geometric term}) + (\text{atomic phase}),\end{aligned} \tag{6.23}$$

where λ is the wavelength, p is the gas pressure, η is the ionization fraction, δn is the difference of index of refraction between the laser and the qth harmonic, and N_{atm} is the atomic number density at the pressure of one atmosphere. The first two terms depend on the gas pressure p. The geometric term is negative in a waveguide and for a focused Gaussian beam; this is due to the Gouy phase shift. The last term is related to the atomic-dipole phase of the harmonic, which depends on the laser intensity; thus, it varies in space and time.

In Equation 6.23, phase mismatch due to the free-electron dispersion is positive, and that due to the neutral atom dispersion is negative. Here $\eta = N_e/N_a$ where N_e and N_a are the electron and atomic density, respectively. For small-ionization fraction ($\eta \ll 1$), the neutral-atom phase mismatch is given by

$$\Delta k_q = -\frac{2\pi q}{\lambda} p \delta n (1-\eta). \tag{6.24}$$

Considering only neutral-atom and plasma dispersion, the critical ionization fraction for phase matching to occur can be solved (also see Equation 5.65)

$$\eta_{cr}(\lambda) = \left[\frac{N_{atm} r_0 \lambda^2}{2\pi\,\delta n} + 1\right]^{-1}. \tag{6.25}$$

Based on this equation, for a 800 nm wavelength, the critical ionization fraction is quite low: about 5% for Ar, 1% for Ne, and 0.5% for He. By tuning the gas pressure, these critical values limit the highest harmonic cutoff energies to H35 for Ar, H69 for Ne, and H91 for He. Note that the derivations here are limited to the small-ionization fraction.

For ionization below the critical value, phase matching can be improved by generating harmonics inside a hollow waveguide by using a long focal-length lens in a cell, or by using a Bessel–Gauss beam. By properly adjusting the radius of the waveguide, or the focal position, the gas pressure increase of harmonic yields by one or two orders has been found possible.

Phase Matching at Ionization Level above the Critical Level

If phase matching can be reached, it is clear that the increase of the ionization level and the gas pressure would enhance the harmonic yields. Unfortunately, above the critical ionization level, phase mismatch due to neutral atoms is smaller than that due to free electrons. The negative phase mismatch due to the free electrons would severely defocus the laser beam, resulting in rapid decrease of laser intensity in position and time as it travels inside the gas medium. Since the atomic-dipole phase is proportional to the laser intensity, the phase mismatch due to the induced atomic dipole in a medium may be able to cancel the phase mismatch due to the free electrons. Indeed, such a new phase-matching mechanism has been identified recently in Sun et al. [47]. Using an intense laser in a tight-focusing geometry, where ionization far exceeds the critical ionization level, the excessive free electrons rapidly defocus the laser beam. Phase mismatch due to the copious free electrons is balanced by the large gradient in the high-harmonic phase initiated by the fast drop in the laser intensity over a short distance. This counterintuitive mechanism is called *defocusing-assisted phase matching* (DAPM). It has been shown to enable the extension of harmonic cutoff, with an increase of yields by two orders of magnitude in comparison with the conventional phase matching. To reach the condition for DAPM, an iris is inserted before an intense driving laser before it enters the high-pressure gas cell. The laser intensity can be controlled readily by adjusting the radius of the aperture. With high-intensity incident laser beams, intense higher harmonics beyond the usual cutoff under the low-ionization-level condition have been observed. For example, at a low-ionization level for a typical 800 nm laser with durations of about 25 fs, the highest harmonic generated in the argon gas cell is H35 at about 50 eV. With intensity beyond 2×10^{15} W/cm^2, high-energy harmonics in the 65–70 eV region are clearly observed when the aperture of the iris and the focal point of the laser pulse with respect to the position of the gas jet are adjusted. The observed intensity is 400 times higher than the low-ionization-level approach.

Simulations indeed confirmed that the phase mismatch due to the free-electron dispersion is balanced by the phase mismatch originating from the intrinsic dipole phase of each harmonic. Thus the general statement that harmonics can be generated only when the

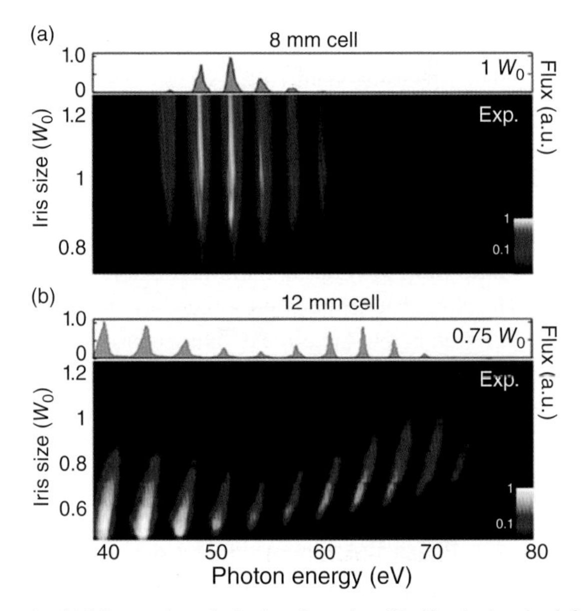

Figure 6.21 Demonstration of generating high harmonics at ionization above the critical level using the defocusing-assisted phase-matching method. Extension of harmonic cutoffs with two orders of increase in yields is accomplished using sharply focused high laser intensity in a short gas cell. The gas cells are (a) 8 mm and (b) 1.2 mm long, respectively. In (b) a wide spectral range of tunability of harmonics is demonstrated from ≈ 35 eV to ≈ 70 eV. The suppression around 51 eV is due to the Cooper minimum of Ar. (Reprinted from H. W. Sun et al., *Optica*, **4** 976 (2017) [47]. Copyrighted by the Optical Society of America.)

ionization level is low is not valid. When a very intense laser enters a gas medium, severe ionization occurs rapidly and thus the laser intensity drops quickly. Under the correct condition, the positive phase mismatch can be balanced by the negative phase mismatch from the intrinsic atomic dipole when the intensity of the laser occurs at the correct value. By fine-tuning the aperture radius of the iris, this can be readily achieved in the laboratory (see Figure 6.21). It will be of great interest to check how this method can be extended to different targets and longer-wavelength lasers.

Quasi-Phase Matching

When normal phase matching cannot be implemented, one of the widely used schemes for improving harmonic-generation efficiency is the quasi-phase-matching (QPM) method, which has been employed in visible nonlinear optics. In QPM, instead of matching the phase velocities of the driving laser to generate harmonics, an additional term $K = 2\pi/\Lambda$, where Λ is the modulation period, is introduced into Equation 6.23 to periodically correct the phase mismatch every two times the coherence length. In other words, in zones that contribute destructively to the harmonic signal, the QPM is used to switch off or phase shift the HHG emission.

Many different ways of implementing QPM have been proposed for enhancing HHG. The first type was achieved with the idea of periodically modulating the laser intensity

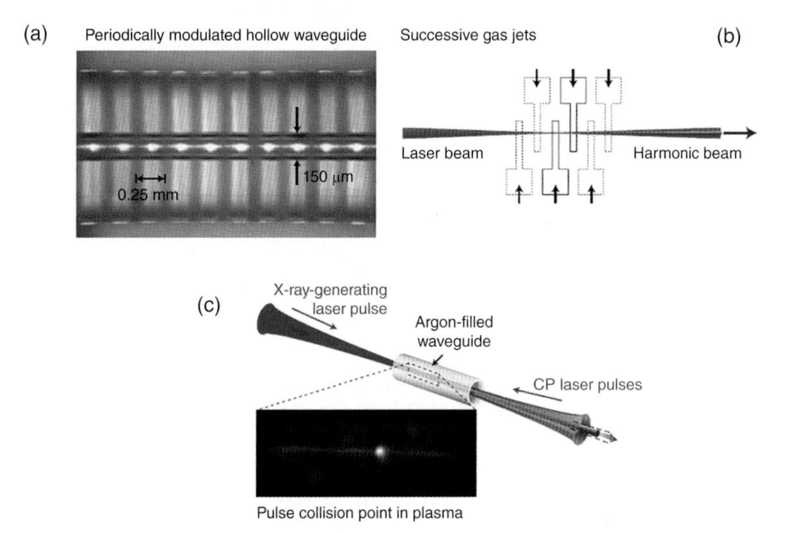

Figure 6.22 Illustration of selected QPM techniques. (a) and (b) Geometrical QPM schemes. A modulated waveguide can periodically modulate the peak intensity of the driving laser beam and hence the HHG phase. A sequence of gas jets can also be used, in which the spacing and density can be varied. (c) All-optical QPM approach. A versatile periodic light structure can adaptively compensate for any laser-propagation dynamics and can be implemented over many coherence zones. A sequence of weak counter-propagating (CP) pulses (of wavelength λ_{CP}, pulse duration τ_{CP}, and electric field strength E_{CP}) can create a light structure inside a waveguide that is matched to the coherence length of a particular harmonic in order to selectively enhance a quasi-monochromatic HHG bandwidth. (From E. A. Gibson et al., *Science*, **302**, 95 (2003) [48], reprinted with permission from AAAS; J. Seres et al., *Nat. Phys.*, **3**, 878 (2007) [49]; and T. Popmintchev et al., *Nat. Photon.*, **4**, 822 (2010) [50], respectively.)

or the medium density. In 2003, experiments by Paul et al. [51] and Gibson et al. [48] demonstrated QPM in a modulated waveguide by periodically modulating its diameter (see Figure 6.22(a)). By expanding the diameter of the waveguide, the laser intensity is reduced in regions where the harmonic light is out of phase by 180° with the driving laser (after one coherence length) so that no new harmonics are generated. This pattern was repeated periodically for achieving the necessary phase modulation on the harmonic field. In 2007, Zepf et al. [52] demonstrated multimode QPM by using a highly tight-focused geometry in a simple hollow waveguide to excite higher-order modes. This was achieved by the beating between the fundamental and higher-order modes to create rapid axial-intensity modulations. In another experiment, Seres et al. [49] suggested a QPM by using successive sources (see Figure 6.22(b)). The sources are arranged along the optical axis of a focused laser beam and pumped by the same laser pulse. The QPM was achieved through a periodic modulation of gas-medium density. However, laser-induced defocusing, laser-mode scrambling, and other geometric limitations together make this QPM scheme difficult to scale to a very long interaction length to achieve increased efficiency.

Other powerful QPM approaches have been suggested. For example, in 2007, Zhang et al. [53] demonstrated experimentally the first use of a train of counter-propagating light pulses to enhance high-harmonic emission (see Figure 6.22(c)). In this all-optical

QPM technique, the correct light patten or sequence of counter-propagating pulses is sent through the medium in the direction opposite of the driving laser, and interferes with the fundamental beam to scramble the quantum phase of the generated harmonics such that the harmonics emission from out-of-phase regions can be suppressed. Other similar schemes have also been proposed by Cohen et al. [54].

These QPM schemes are attractive because they can easily be extended to the driving laser with longer wavelength, and will not be limited by the well-known single-atom-yield, wavelength-scaling law of $\lambda^{-5} - \lambda^{-6}$. However, these QPM schemes are much more complex to implement experimentally than the general phase-matching schemes.

6.3.3 Generation of Harmonics in the Water Window and keV Region

As previously discussed, phase matching is only possible up to a critical ionization level that depends on the gas species and the driving-laser wavelength. The maximum achieved harmonic photon energies under phase-matching conditions are limited to ~150 eV with commonly used titanium-sapphire lasers operating at a wavelength of ~0.8 μm interacting with helium gas. To extend the HHG-cutoff energy to the water window (284–543 eV) or even to the keV region, a longer-wavelength driving laser is preferred since the single-atom cutoff energy is proportional to λ^2. However, single-atom HHG efficiency scales as $\lambda^{-5} - \lambda^{-6}$ [55]. The single-atom wavelength-scaling law can be compensated when phase matching is taken into account. Since the critical ionization level for the phase-matched HHG decreases with the increase of laser wavelength, the laser intensity required to generate the maximum cutoff energy decreases. This defines a phase-matching cutoff photon energy $h\nu_{pm}(\lambda) \propto I(\eta_{cr})\lambda^2$ (neglecting I_p), corresponding to the maximum photon energy that can be generated from a macroscopic medium with near-optimum conversion efficiency (full phase matching). Using three- or eight-cycle laser pulses, Popmintchev et al. [50, 56] theoretically predicted that $h\nu_{pm}(\lambda) \propto \lambda^{1.6-1.7}$ for values of λ up to 10 μm (see region I of Figure 6.23).

Full phase matching with longer-wavelength lasers works due to several factors. The required phase-matching pressure in a waveguide is very high (several tens of atmosphere). At higher-photon energies, the gas medium becomes more transparent (except just above each inner-shell ionization threshold), and thus reabsorption of the generated harmonics by the nonlinear medium lessens, which results in an increased optimum density-length product (around several absorption lengths). Thus water-window harmonics can be generated by a laser wavelength between 1.5 and 2 μm. With laser wavelength approaching 3 μm, the harmonic energy can be extended to 1 keV and even extended to the multi-keV region when even longer mid-infrared-driving lasers are used (see Figure 6.24), by focusing a 3.9 μm pulse into a hollow-core fiber filled with He gas at very high pressure. When driven by few-cycle mid-infrared laser pulses, the extended supercontinuum spans a broadband spectral width, which in principle can support pulses as short as a few attoseconds if a transform-limited condition can be achieved. However, so far such short pulses have not been demonstrated, partly because the harmonic yields in the water-window region are still too weak. Furthermore, such broadband pulses cannot be characterized

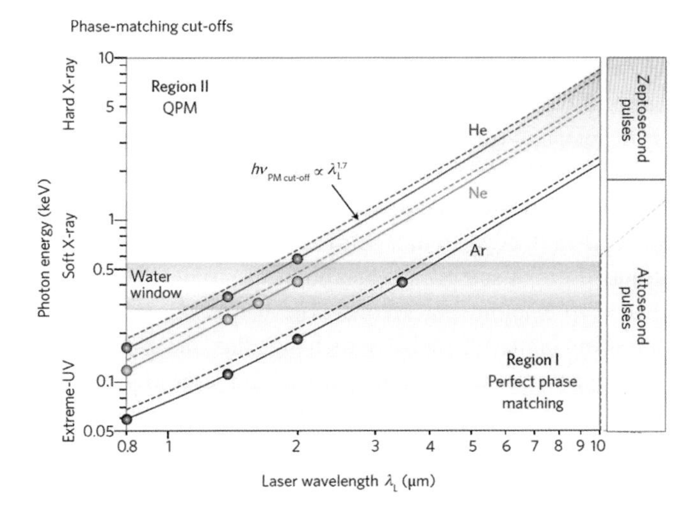

Figure 6.23 Theoretical HHG phase matching cutoff as a function of the driving-laser wavelength for three-cycle (dashed lines) and eight-cycle (solid lines) pulses. (Reprinted from T. Popmintchev et al., *Nat. Photon.*, **4**, 822 (2010) [50].)

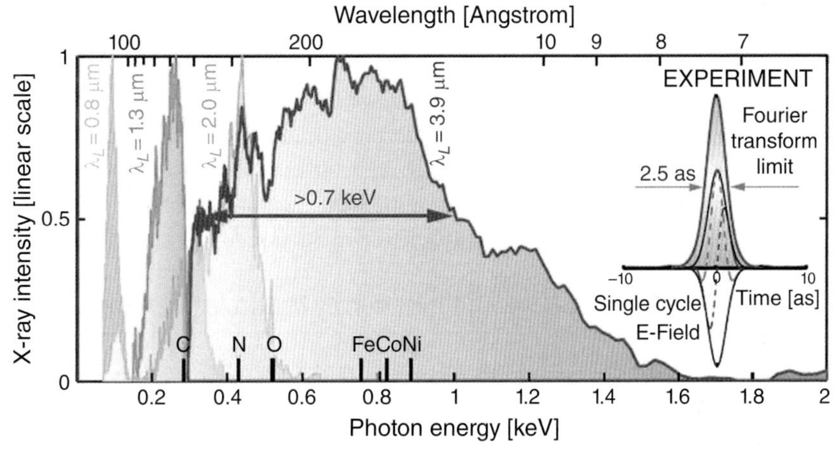

Figure 6.24 Experimental HHG spectra emitted under full phase-matching conditions as a function of driving-laser wavelength (yellow, 0.8 μm; green, 1.3 μm; blue, 2 μm; purple, 3.9 μm). (Inset) Fourier transform-limited pulse duration of 2.5 as that corresponds to the spectral width of about 0.7 keV. (From T. Popmintchev et al., *Science*, **336**, 1287 (2012) [57]. Reprinted with permission from AAAS.)

using the frequency-resolved optical gating for complete reconstruction of attosecond bursts (FROG–CRAB) method. Recently, a new retrieval method has been proposed (see Section 7.5.3).

Water-window harmonics have been reported by a number of experimental groups using mid-infrared lasers generated through the optical parametric amplification (OPA) [57–59] and, most recently, using optical parametric chirped-pulse amplification (OPCPA) technology [60, 61]. In addition, water-window harmonics have been generated using the

0.27 μm third-harmonic of the titanium-sapphire laser [62]. With phase-stabilized one- or two-cycle mid-infrared-driving lasers, supercontinuum harmonics in the water-window region have been demonstrated. By implementing a wavefront rotation technique through spatiotemporal isolation together with theoretical simulation, single-attosecond pulses with a duration of 355 as were deduced [63]. So far, the water-window harmonics are still too weak but they are good enough for soft X-ray absorption, fine-structure measurements at the carbon K-edge from a solid material (see [64]). Most recently, the Barcelona group reported [59] a soft X-ray absorption spectrum from 200 to 543 eV using harmonics with 7.3×10^7 photons per second. The bandwidth would have a duration of 13 as if it is transform limited. To achieve such high flux, the experimentalists used a 1 kHz, the carrier-envelope phase (CEP)-stabilized 1.85 μm laser to generate soft X-ray harmonics in neon or helium, with peak laser intensity of 0.5 PW/cm^2 and pressure up to 12 bars in helium. Figure 6.25 shows a spectrum obtained from helium. The vertical lines indicate the K-shell or the L-shell absorption edges of the elements listed. The two bottom graphs show the absorption measured using a 200 nm carbon foil or titanium foil where strong absorption is clearly visible with the opening of the K-edge of carbon at 284 eV and the L-edge of titanium at 456 eV. Such broadband, water-window-continuum, soft X-rays herald the study of attosecond dynamics of a specific atom or site inside a material with an unprecedented range of applications in chemical processes, structure dynamics of biomolecules, and materials. In Teichmann et al. [59] the mechanism of phase matching in the generation of the soft X-ray continuum harmonics was also studied theoretically. It was found that phase matching at high pressure only occurs in a narrow temporal window that provides the condition for emission of a single attosecond pulse. At high pressure, phase matching was found to be independent of the focusing position if it is within the Rayleigh range. This is consistent with DAPM, discussed in Section 6.3.2.

6.3.4 Optimization by Multicolor Waveform Synthesis

The low conversion efficiency of HHG is the main limitation that prevents HHG from becoming a useful table-top light source so far. In the previous sections, efforts to create favorable phase-matching conditions in the nonlinear gas medium to efficiently build up macroscopic high harmonics have been discussed. Alternatively, modifying the single-atom HHG response is another way to enhance harmonic yield or to extend the cutoff energy. This is to be achieved by altering the driving electric field of the laser pulse at the sub-cycle level such that ionization and propagation of an electron in the laser field can be controlled [65]. This approach becomes attractive because synthesis of the multicolor light fields with complete amplitude and phase control makes it possible to generate arbitrary optical waveform in view of the advance of OPA and OPCPA techniques [66, 67]. However, a large number of free parameters is involved in a multicolor synthesizer such as amplitude, CEP, and relative time delay. It is not possible to scan the whole parameter space routinely in an experiment. For the practical purpose of guiding experiments, several optimal algorithms for fine-tuning the optical waveform have been developed. Here, two examples of optimal waveforms based on genetic algorithms (GA) are discussed. The first example is the so-called *perfect wave* derived by Chipperfield et al. [68] that produces

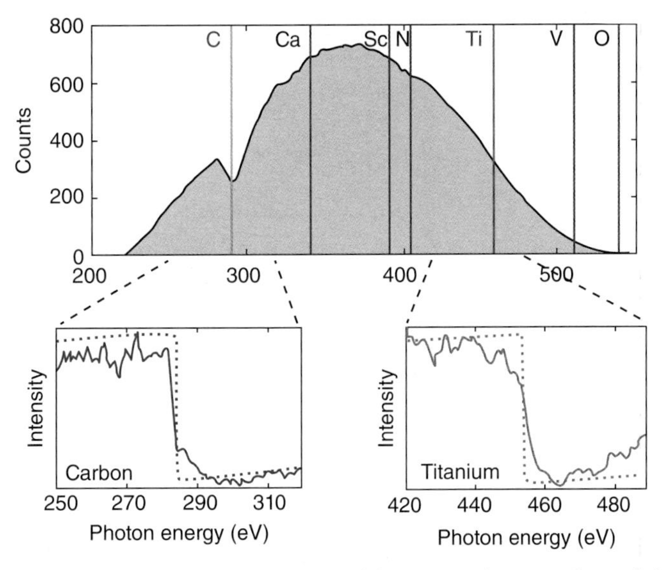

Figure 6.25 HHG spectrum of the soft X-ray generated by the 1 kHz, CEP-stabilized, two-cycle 1.85 μm laser in helium at high gas pressure, covering the energy from about 200 eV to 540 eV. Within the water-window region, the K-edge or L-shell edge of a number of atoms is marked by the vertical lines, making such broadband pulses suitable for probing molecules or materials containing these elements to trace the inner-shell hole hopping at an attoseconds timescale. The two bottom figures show strong absorption occurs when a 200 nm foil of carbon or titanium is inserted into the beam. (Reprinted from S. M. Teichmann et al., *Nat. Commun.*, **7**, 11493 (2016) [59].)

returning electrons with maximum possible energy about three times larger than that of a single-color sinusoidal field for any given oscillation period T and per period fluence F. This ideal waveform is expressed as

$$E(t) = \pm\sqrt{\frac{F}{T}\frac{2}{\epsilon_0 c}}\left(\frac{3t}{T} - 1\right) \tag{6.26}$$

for $0 \leq t \leq T$, which consists of a linear ramp with a dc offset, as illustrated in the red line in Figure 6.26(a). In the same figure, the trajectory of the electron with the maximum returning energy starting at the ionization time $t_i = 0$ and ending at the recombination time $t_r = T$ is shown in green line. However, it is not realistic to create such a perfect wave in the laboratory. Instead, Chipperfield et al. [68] suggest searching for an optimum waveform by coherently combining only a finite number of frequency components. This would generate a returning electron with kinetic energy very close to that given by the perfect wave. The resulting optimum waveform is displayed by the red curve in Figure 6.26(b). In the GA optimization, five colors are used consisting of the first four harmonics of 1/T and a field at half the fundamental frequency 1/2T, limited to half the power of the fundamental. With this five-color optimized waveform, the HHG cutoff energy can be greatly extended. Meanwhile, it can maintain the same harmonic yield as a pure sinusoidal pulse with the same periodicity under typical macroscopic conditions as demonstrated by the calculations [68].

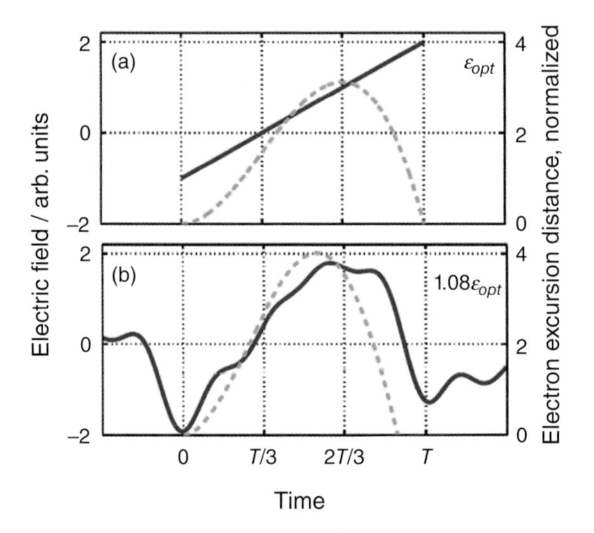

Figure 6.26 The driving electric field (red lines) of (a) a "perfect wave," and (b) an optimized waveform consisting of the first four harmonics of $1/T$ and a field with a period of $2T$. Trajectories of the returning electrons with the maximum kinetic energies are also shown (green lines). $\varepsilon_{opt} = (e^2/m)(FT/4\epsilon_0 c)$ is the maximum returning electron energy generated by the "perfect wave," where e and m are the electron charge and mass, respectively, T is the oscillation period, and F is per period fluence. (Reprinted from L. E. Chipperfield et al., *Phys. Rev. Lett.*, **102**, 063003 (2009) [68]. Copyrighted by the American Physical Society.)

The second example is the two-color waveform for harmonic yield enhancement proposed by Jin et al. [69] that is synthesized by the fundamental laser and its third harmonic field. This waveform is constructed with the idea of generating single-atom harmonics to be much more easily phase matched during the propagation in the macroscopic gas medium. In the optimization, the harmonic yield at the cutoff energy was set as the fitness function. Other constrains were also employed:

(i) the cutoff energy should be maintained at the given value;
(ii) the ionization level should be less than a few percent for favorable phase matching;
(iii) for plateau harmonics, the emission from "short"-trajectory electrons should be stronger than from "long" ones.

In Figure 6.27(a), the optimized waveform is shown (red line) in comparison with the single-color sinusoidal wave of the fundamental (black line) over one optical cycle. This figure clearly demonstrates that: (i) the electric field of the optimized waveform at the ionization time is higher, and thus leads to more electrons that can return to generate harmonics, and (ii) there are more "short"-trajectory electrons than "long" ones. Figures 6.27(b,c) show the temporal HHG emissions with the two waveforms, which were obtained by performing quantum mechanical calculations. For the optimized wave, the strong enhancement of "short"-trajectory electrons are clearly seen. In the simulation, Ne was chosen as the target. For the single-color 1,600 nm laser alone, an intensity of 3×10^{14} W/cm^2

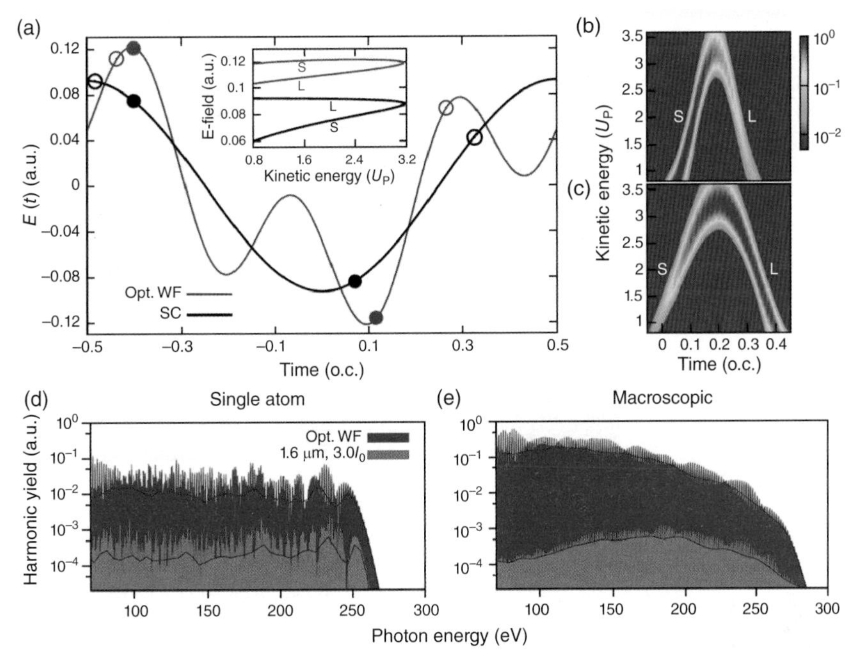

Figure 6.27 (a) Comparison of a single-color (SC) sinusoidal wave and an optimized waveform (Opt. WF) over one optical cycle (o.c.) of the fundamental. Open and filled circles show the tunnel ionization and recombination times for an electron with kinetic energy of 2 U_p following the "long" or "short" trajectories, respectively. The inset depicts the electric field at the ionization time versus the kinetic energy of the returning "short"- and "long"-trajectory electrons, labeled as S and L in the figure, respectively. The yields of harmonic emission versus electron recombination times are shown for (b) the two-color and (c) the single-color waves, respectively. (d) Single-atom harmonic yields for the two waves, showing that the optimized wave is about two orders stronger than the single-color one. (e) The yields of high harmonics after macroscopic propagation. In the simulation, lasers with a beam waist of 40 μm are focused at 2.5 mm before a 1-mm-long gas jet. The gas pressure is 10 Torr. Blue curves in (d) and (e) show the smoothed spectra. (Reprinted from Cheng Jin et al., *Nat. Commun.*, **5**, 4003 (2014) [69].)

was chosen. The CEP of the fundamental was set at zero. The optimization returned the peak intensity for the fundamental at 1.98×10^{14} W/cm^2, and 1.32×10^{14} W/cm^2 for its third harmonic, which had the optimized phase of 1.36π. The single-atom HHG calculations using the QRS model showed that the harmonic yield by the optimized two-color waveform is enhanced by about two orders of magnitude compared to the single-color fundamental laser (see Figure 6.27(d)). By including propagation in the calculation, Figure 6.27(e) shows that the HHG yield near the cutoff (at about 250 eV) from the optimized waveform maintains the enhancement of two orders of magnitude, but at lower photon energies near 100 eV the enhancement is about three orders higher. The additional enhancement reflects that, for the optimized wave, there are more "short"-trajectory electrons contributing to HHG than in the single-color case. The example Figure 6.27(e) demonstrates the success of the waveform-optimization strategy, i.e., optimizing harmonics from "short"-trajectory

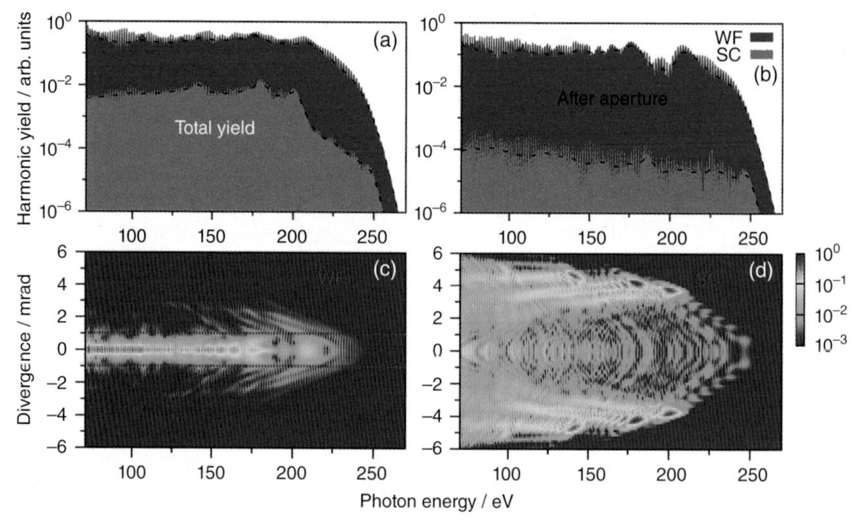

Figure 6.28 (a) Total harmonic yield emitted at the exit of a hollow waveguide and (b) harmonic yield integrated within 1 mrad (indicated by the dot-dashed lines in (c)) using an aperture in the far field for two-color waveform (WF) and single-color (SC) laser. Harmonic divergence in the far field (c) for two-color WF and (d) for SC. Gas pressure is 50 Torr, length and radius of the waveguide are 5 mm and 125 μm, respectively. (Reprinted with permission from Cheng Jin et al., *Phys. Rev. Lett.*, **115**, 043901 (2015) [70]. Copyrighted by the American Physical Society.)

electrons, and not from both the "long"- and "short"-trajectory electrons that coexist in the single-atom harmonics.

The optimized two-color waveform has been further combined with laser guiding in a hollow waveguide filled with gas to efficiently generate low-divergence soft X-ray harmonics in the far field. The harmonic spectra with the two-color waveform are shown in Figure 6.28. To simultaneously achieve the best cutoff energy of about 250 eV (close to the cutoff of the single-atom response) and the best harmonic yield, both the length of the waveguide and the pressure have been varied. The optimal values are found to be 5 mm and 50 Torr, respectively. In Figure 6.28(a), the harmonic spectra generated by the single-color 1.6 μm laser alone under the same condition are also shown. The harmonic emissions in the far field for the two cases are shown in Figures 6.28(c,d). The harmonics generated with the waveform extend over a very broad spectral band from 70 to 250 eV. These harmonics are located close to the axis, and their divergence angles are smaller than 1 mrad, as indicated by the dot-dashed lines in the figure. If an aperture is used to filter out high harmonics with divergence greater than 1 mrad, the resulting harmonic spectra for the single-color and waveform are compared in Figure 6.28(b). The study on the interplay among waveguide mode, atomic dispersion, and plasma effect uncovers how an optimized waveform is maintained through the propagation in the waveguide and how dynamic phase matching is accomplished when optimal waveguide parameters (radius and length) and gas pressure are identified [70].

6.3.5 Optimization by Increase of Repetition Rates of the Driving Laser

The previous sections addressed the issues of extending phase-matching to higher photon energies, increasing the yields of the harmonics in the XUV or soft X-ray region using waveform-optimized lasers. Another course that can be taken to develop more intense harmonics is one that involves increasing the repetition rates of the driving laser. A high-repetition harmonic light source is particularly important for exploiting novel applications in fields such as photoelectron spectroscopy, frequency metrology, coherent diffractive imaging, coincidence detection of ionization fragments, and microscopy when avoiding space-charge effects. Although titanium–sapphire-based based laser systems have been the workhorse for HHG and attosecond science for decades, they are limited by the kHz-type repetition rates and their average power. Nevertheless, employing high-repetition-rate driving lasers for HHG draws stringent requirements on laser technology, but great progress has been made in the last few years.

Several different laser techniques have been placed under development to generate HHG at high repetition rates. The first technique utilizes certain nanostructure materials to create intense, local, nonhomogeneous field strengths directly with multi-ten MHz oscillators. Experimental demonstration of this method was carried out by Kim et al. [71], but debates about whether the emitted radiation was coherent HHG or incoherent atomic line emissions abound. The second approach is the so-called *femtosecond enhancement cavities* (fsEC) operated between 10 MHz and 250 MHz. This approach uses passive-high-finesse external resonators to enhance incoming low-energy, high-repetition-rate ultrashort lasers. Because high harmonics are generated in a focus inside of the evacuated resonator, where a gas jet provides the generating medium, they are called "intra-cavity" harmonics. The challenge with these intra-cavity harmonics is being able to couple them out from the cavity when they co-propagate with the driving fundamental laser. The third technique is to directly use high-average-power lasers to drive the HHG process. This is made possible by the rapid progress in ytterbium-doped femtosecond laser technology where the power level can be raised to kilowatts (as compared to few watts for titanium-sapphire lasers) for InnoSlab [72], thin-disk [73], and fiber lasers [74]. For example, a diode-pumped Kerr-lens, mode-locked ytterbium-doped–yttrium aluminum garnet (Yb:YAG), thin-disk laser combined with extra-cavity pulse compression has been used to generate waveform-stabilized few-cycle pulses (7.7 fs, 2.2 cycles) with a pulse energy of 0.15 mJ that can be scaled to even higher pulse energies.

One of the grand goals of HHG research is to build table-top sources of coherent, extreme ultraviolet to soft x-ray radiations to complement light sources at large-scale facilities like synchrotrons and free-electron X-ray lasers. To this end, ultrashort pulses with high photon flux and high repetition rates at the same time are needed. Recent advances in high-power fiber lasers has made this possible with OPCPA and nonlinear pulse compression. This topic has been recently reviewed in Hädrich et al. [75]. Figure 6.29 shows the photon yields per second per eV for continuum harmonics generated in neon (a) and helium (b) at a repetition rate of 100 kHz. The backing pressures of 11 and 14 bars are shown. For helium,

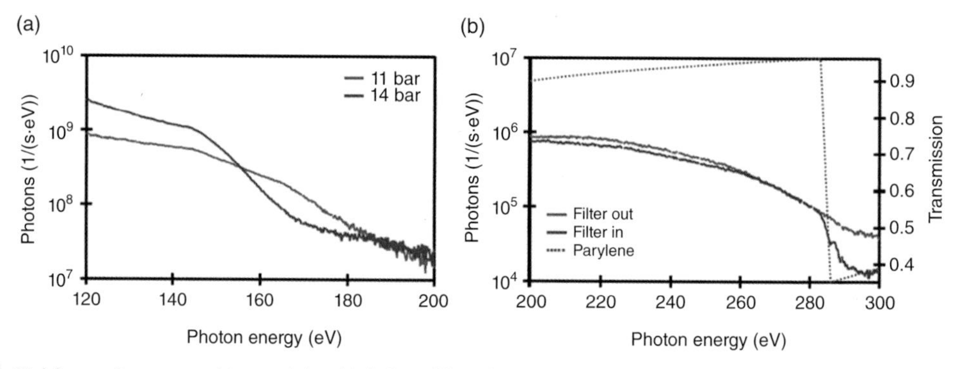

Figure 6.29 High harmonics generated in neon (a) and in helium (b) gas by a 100 kHz laser at different backing pressure of 11 and 14 bar for neon and 45 bar for helium. The vertical scale shows the number of photons per eV per second. In (b) the effect of a parylene filter is used to absorb harmonics beyond the K-edge of carbon at 284 eV. (Figures adopted from S. Hädrich et al., *J. Phys. B-At. Mol. Opt.*, **49**, 172002 (2016) [75], under the terms of the Creative Commons Attribution 3.0 license.)

the backing pressure was 45 bar. In the latter, a parylene filter is used to absorb photons above the carbon edge at 283 eV. After a few stages of amplifications and compressions, sub-8 fs pulses with up to 350 μJ pulse energy at 100 kHz repetition rate were used to generate the continuum harmonics. Clearly, this new laser technology has not reached its end, and further progress to make HHG a useful light source for laboratory experiments appears to be reachable in the next few years. In fact, as described in Rothhardt et al. [76], a novel high-harmonic source generated from a 100 kHz fiber laser system that delivers 10^{11} photons per second in a single 1.3 eV bandwidth harmonic at 68.6 eV has already been built. It has been applied to coincidence measurement to study the dissociation dynamics of inner-shell ionization of CH_3I. Figure 6.30(a) shows the harmonic spectrum where the harmonic at 68.6 eV was optimized with more than 4×10^{10} photons per second per harmonic and 20 hours of stability. Figure 6.30(b) shows the ion–ion coincidence spectrum in the region containing the coincidence between CH_x^+ ($x = 0, \ldots, 3$) and I^+ fragments. To use this facility for pump-probe experiments, it was estimated that the count rate has to increase by two orders of magnitude. This is reachable in the foreseeable future. Such a light source with MHz or sub-MHz repetition rates has time resolution not yet available at today's free-electron-laser facilities.

In summary, recent rapid progress in the development of new driving-laser sources shows that high-harmonic and attosecond sources can be scaled to higher pulse energies or higher repetition rates. Figure 6.31 (adapted from [77]) illustrates the potential of MHz-to-GHz driving lasers for the generation of gas-based HHG sources. Realization of usable table-top coherent light sources from XUV to soft X-rays would allow the monitoring and control of chemical-reaction dynamics at the level of electrons, for ultrafast photonic and electronic signal processing, coherent diffractive imaging and frequency metrology, and many other applications. It is likely that the coming years will reveal whether HHG can, in fact, meet such great expectations.

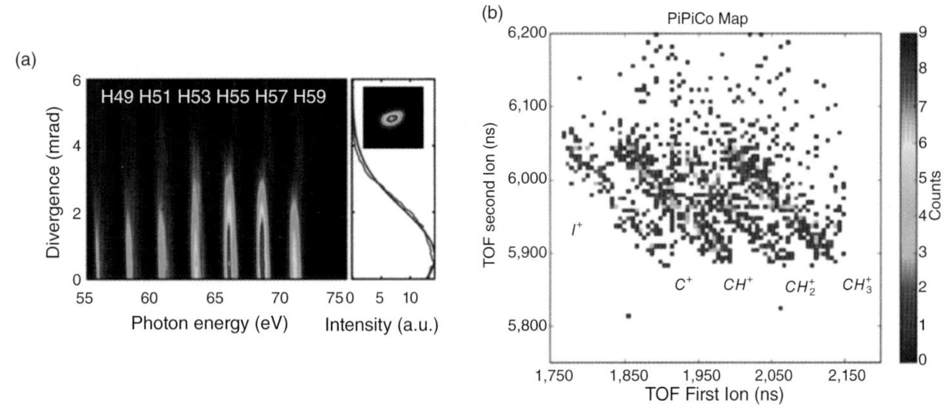

Figure 6.30 (a) Harmonics at 100 kHz generated with MHz lasers. Phase matching was adjusted to optimize the 68.6 eV harmonic, which was then used to ionize the inner shell of iodine in CH_3I molecules. An example of the ion–ion coincidence spectrum between CH_x^+ (x=0, ..., 3) and I^+ fragments are shown in (b). With further improvement, the XUV can be used to perform pump-probe experiments with a repetition rate that is not available at free-electron-laser facilities. (From Jan Rothhardt et al., *Opt. Express*, **24**, 18133 (2016) [76]. Reprinted with permission from Optical Society of America.)

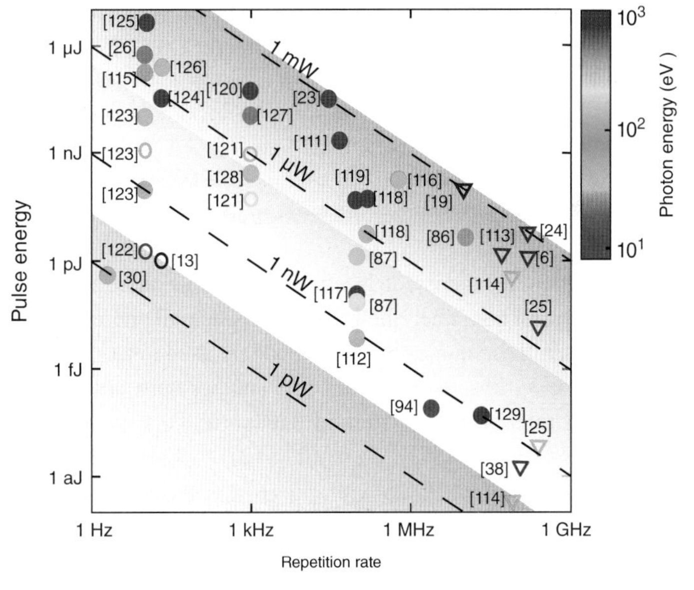

Figure 6.31 Generated pulse energy per single harmonic order versus repetition rate and average power (dashed lines) of state-of-the-art high-harmonic sources. Three types of sources are shown: single-pass HHG in a gas cell or gas jet (filled circles), single-pass in a capillary (open circles), and intra-cavity (triangles). The shaded areas indicate the average power of the reported best values within three different energy ranges ($\geq$ 20 eV, $\geq$ 100 eV, and $\geq$ 1 keV). References in this graph are given in the original article. (Figure adopted from C. M. Heyl et al., *J. Phys. B-At. Mol. Opt.*, **50**, 013001 (2017) [77], under the terms of the Creative Commons Attribution 3.0 license.)

Notes and Comments

High-order harmonics of N_2 have been studied extensively in many experiments. As elaborated in Section 6.1, these experiments can be explained by the QRS theory. When N_2 molecules are well aligned, the experimental HHG data can be used to extract accurate molecular frame PITD matrix elements. The QRS theory has also been applied to polyatomic molecules but experimental data is more limited. Since polyatomic molecules have lower ionization energy, extended harmonic spectra can only be obtained with long-wavelength lasers. Due to the unfavorable wavelength scaling, higher laser intensities are often used in experiments where propagation effects can complicate the interpretation of the data. At the same time, for polyatomic molecules, many inner MOs contribute to the HHG, making the calculations much more tedious. For larger molecules, HHS may not be a viable method to probe the structure of molecules since the calculations of PITDs also become complicated and are, most likely, not very accurate.

The Holy Grail of strong-field physics research is to generate intense coherent light sources over the whole electromagnetic spectrum. While the technological challenge lies in the realm of experimentalists, understanding the underlying phase-matching mechanism is a critical theoretical issue. With the emergence of waveform synthesis, optimization of the best waveform for the most efficient harmonic generation will be of great interest. While the authors of this book are not the right persons to provide guidance on this emerging technology, it is certainly a topic that should not be ignored.

Exercises

6.1 In quantum theory, it is often argued that there is no MO and there are no valence electrons nor inner-shell electrons since all the electrons are indistinguishable, even though these names have been widely used in atomic and molecular physics. Consider two electrons, one spin up and another spin down. The two are coupled to a state with total spin $S = 0$, and $M_S = 0$. This two-electron state can be expressed as $|SM_S >= |00 >= \frac{1}{\sqrt{2}}[\alpha_z(1)\beta_z(2) - \beta_z(1)\alpha_z(2)]$, where α and β can be defined with respect to the z-axis of a chosen frame. Show that the same state can also be expressed as $|00 >= \frac{1}{\sqrt{2}}[\alpha_x(1)\beta_x(2) - \beta_x(1)\alpha_x(2)]$ using spin-up and spin-down states with respect to the x-axis. Thus the spin up with respect to the x-axis is as good a "molecular orbital" as the one with spin up with respect to the z-axis. Therefore the spin-up "orbital" is just a basis state, not a physical "observable" that can be measured. In a similar fashion, the MO of an electron in a many-electron system is just a basis function; it cannot be measured.

6.2 In the tomographic imaging method of retrieving the MO of the HOMO of N_2 molecules from the HHG spectra, list the complete set of approximations that have been made and the validity of these approximations. Note that if these approximations are valid, the HHG spectra can be calculated with the SFA given in Section 6.1.4 and there would be no need to make measurements.

6.3 The example of charge migration discussed in Section 6.2.5 was deduced from measuring the harmonic spectra of oriented molecules. List the actual experimental data reported. To extract charge migration (or hole hopping), discuss the models used in Equation 6.22. Identify what is assumed to be known and what is to be "retrieved" from fitting to the HHG data. In what way would the "retrieved" hole hopping shown in Figure 6.20(a) be reflected in the high-order harmonic spectra? Given the hole density, even with the model of Equation 6.22, the harmonic spectra cannot be recovered. Thus, in a dynamic system, the whole electron wave packet has to be retrieved, not just the hole density, charge migration, or charge oscillation.

References

[1] T. Kanai, S. Minemoto, and H. Sakai. Quantum interference during high-order harmonic generation from aligned molecules. *Nature*, **435**(7041):470–474, May 2005.

[2] J. Itatani, J. Levesque, D. Zeidler, et al. Tomographic imaging of molecular orbitals. *Nature*, **432**(7019):867–871, Dec. 2004.

[3] M. Lein, N. Hay, R. Velotta, J. P. Marangos, and P. L. Knight. Role of the intramolecular phase in high-harmonic generation. *Phys. Rev. Lett.*, **88**(18):183903, Apr. 2002.

[4] A.-T. Le, R. R. Lucchese, S. Tonzani, T. Morishita, and C. D. Lin. Quantitative rescattering theory for high-order harmonic generation from molecules. *Phys. Rev. A*, **80**:013401, Jul. 2009.

[5] X. X. Zhou, X. M. Tong, Z. X. Zhao, and C. D. Lin. Role of molecular orbital symmetry on the alignment dependence of high-order harmonic generation with molecules. *Phys. Rev. A*, **71**:061801, Jun. 2005.

[6] R. R. Lucchese, G. Raseev, and V. McKoy. Studies of differential and total photoionization cross sections of molecular nitrogen. *Phys. Rev. A*, **25**:2572–2587, May 1982.

[7] M. Lein, R. De Nalda, E. Heesel, et al. Signatures of molecular structure in the strong-field response of aligned molecules. *J. Mod. Opt.*, **52**(2-3):465–478, 2005.

[8] X. Ren, V. Makhija, A.-T. Le, et al. Measuring the angle-dependent photoionization cross section of nitrogen using high-harmonic generation. *Phys. Rev. A*, **88**:043421, Oct. 2013.

[9] S. Haessler, J. Caillat, W. Boutu, et al. Attosecond imaging of molecular electronic wavepackets. *Nat. Phys.*, **6**(3):200–206, Mar. 2010.

[10] B. K. McFarland, J. P. Farrell, P. H. Bucksbaum, and M. Gühr. High harmonic generation from multiple orbitals in N2. *Science*, **322**(5905):1232–1235, 2008.

[11] A.-T. Le, R. R. Lucchese, and C. D. Lin. Uncovering multiple orbitals influence in high harmonic generation from aligned N2. *J. Phys. B-At. Mol. Opt.*, **42**(21):211001, 2009.

[12] C. Jin, J. B. Bertrand, R. R. Lucchese, et al. Intensity dependence of multiple orbital contributions and shape resonance in high-order harmonic generation of aligned N_2 molecules. *Phys. Rev. A*, **85**:013405, Jan. 2012.

[13] X. Zhou, R. Lock, N. Wagner, W. Li, H. C. Kapteyn, and M. M. Murnane. Elliptically polarized high-order harmonic emission from molecules in linearly polarized laser fields. *Phys. Rev. Lett.*, **102**:073902, Feb. 2009.

[14] A.-T. Le, R. R. Lucchese, and C. D. Lin. Polarization and ellipticity of high-order harmonics from aligned molecules generated by linearly polarized intense laser pulses. *Phys. Rev. A*, **82**:023814, Aug. 2010.

[15] J. M. Zuo, M. Kim, M. O'Keeffe, and J. C. H. Spence. Direct observation of d-orbital holes and Cu-Cu bonding in Cu2O. *Nature*, **401**(6748):49–52, Sep. 1999.

[16] V.-H. Le, A.-T. Le, R.-H. Xie, and C. D. Lin. Theoretical analysis of dynamic chemical imaging with lasers using high-order harmonic generation. *Phys. Rev. A*, **76**:013414, Jul. 2007.

[17] Z. B. Walters, S. Tonzani, and C. H. Greene. Limits of the plane wave approximation in the measurement of molecular properties. *The J. Phys. Chem. A*, **112**(39):9439–9447, 2008. PMID: 18729430.

[18] W. H. Eugen Schwarz. Measuring orbitals: provocation or reality? *Angew. Chem., Int. Ed. Engl.*, **45**(10):1508–1517, 2006.

[19] J. F. Ogilvie. Is a molecular orbital measurable by means of tomographic imaging? *Found. Chem.*, **13**(2):87, 2011.

[20] M. Wießner, D. Hauschild, C. Sauer, V. Feyer, A. Schöll, and F. Reinert. Complete determination of molecular orbitals by measurement of phase symmetry and electron density. *Nat. Commun.*, **5**:4156, Jun. 2014.

[21] S. Weiß, D. Lüftner, T. Ules, et al. Exploring three-dimensional orbital imaging with energy-dependent photoemission tomography. *Nat. Commun.*, **6**:8287, Oct. 2015.

[22] C. Vozzi, M. Negro, F. Calegari, et al. Generalized molecular orbital tomography. *Nat. Phys.*, **7**(10):822–826, Oct. 2011.

[23] S. Patchkovskii, Z. Zhao, T. Brabec, and D. M. Villeneuve. High harmonic generation and molecular orbital tomography in multielectron systems: beyond the single active electron approximation. *Phys. Rev. Lett.*, **97**:123003, Sep. 2006.

[24] M. Lewenstein, P. Balcou, M. Y. Ivanov, Anne L'Huillier, and P. B. Corkum. Theory of high-harmonic generation by low-frequency laser fields. *Phys. Rev. A*, **49**:2117–2132, Mar. 1994.

[25] O. Smirnova, M. Spanner, and M. Ivanov. Analytical solutions for strong field-driven atomic and molecular one- and two-electron continua and applications to strong-field problems. *Phys. Rev. A*, **77**:033407, Mar. 2008.

[26] A.-T. Le, H. Wei, C. Jin, and C. D. Lin. Strong-field approximation and its extension for high-order harmonic generation with mid-infrared lasers. *J. Phys. B-At. Mol. Opt.*, **49**(5):053001, 2016.

[27] F. A. Gianturco, R. R. Lucchese, and N. Sanna. Calculation of low-energy elastic cross sections for electron-CF_4 scattering. *J. Chem. Phys.*, **100**(9):6464–6471, 1994.

[28] A. P. P. Natalense and R. R. Lucchese. Cross section and asymmetry parameter calculation for sulfur 1s photoionization of SF6. *J. Chem. Phys.*, **111**(12):5344–5348, 1999.

[29] A.-T. Le, R. R. Lucchese, and C. D. Lin. Quantitative rescattering theory of high-order harmonic generation for polyatomic molecules. *Phys. Rev. A*, **87**:063406, Jun. 2013.

[30] R. Torres, N. Kajumba, J. G. Underwood, et al. Probing orbital structure of polyatomic molecules by high-order harmonic generation. *Phys. Rev. Lett.*, **98**:203007, May 2007.

[31] T. Popmintchev, M.-C. Chen, D. Popmintchev, et al. Bright coherent ultrahigh harmonics in the keV X-ray regime from mid-infrared femtosecond lasers. *Science*, **336**(6086):1287–1291, 2012.

[32] M. C. H. Wong, A.-T. Le, A. F. Alharbi, et al. High harmonic spectroscopy of the Cooper minimum in molecules. *Phys. Rev. Lett.*, **110**:033006, Jan. 2013.

[33] M. C. H. Wong, J.-P. Brichta, M. Spanner, S. Patchkovskii, and V. R. Bhardwaj. High-harmonic spectroscopy of molecular isomers. *Phys. Rev. A*, **84**:051403, Nov. 2011.

[34] A.-T. Le, R. R. Lucchese, and C. D. Lin. High-order-harmonic generation from molecular isomers with midinfrared intense laser pulses. *Phys. Rev. A*, **88**:021402, Aug. 2013.

[35] S. Zigo, A.-T. Le, P. Timilsina, and C. A. Trallero-Herrero. Ionization study of isomeric molecules in strong-field laser pulses. *Sci. Rep.*, **7**:42149, Feb. 2017.

[36] A. Ferre, A. E. Boguslavskiy, M. Dagan, et al. Multi-channel electronic and vibrational dynamics in polyatomic resonant high-order harmonic generation. *Nat. Commun.*, **6**, Jan. 2015.

[37] B. P. Wilson, K. D. Fulfer, S. Mondal, et al. High order harmonic generation from SF6: deconvolution of macroscopic effects. *J. Chem. Phys.*, **145**(22):224305, 2016.

[38] A.-T. Le, T. Morishita, R. R. Lucchese, and C. D. Lin. Theory of high harmonic generation for probing time-resolved large-amplitude molecular vibrations with ultrashort intense lasers. *Phys. Rev. Lett.*, **109**:203004, Nov. 2012.

[39] W. Li, X. Zhou, R. Lock, et al. Time-resolved dynamics in N2O4 probed using high harmonic generation. *Science*, **322**(5905):1207–1211, 2008.

[40] M. Spanner, J. Mikosch, A. E. Boguslavskiy, M. M. Murnane, A. Stolow, and S. Patchkovskii. Strong-field ionization and high-order-harmonic generation during polyatomic molecular dynamics of N_2O_4. *Phys. Rev. A*, **85**:033426, Mar. 2012.

[41] H. J. Wörner, J. B. Bertrand, D. V. Kartashov, P. B. Corkum, and D. M. Villeneuve. Following a chemical reaction using high-harmonic interferometry. *Nature*, **466**(7306):604–607, Jul. 2010.

[42] H. J. Wörner, J. B. Bertrand, B. Fabre, et al. Conical intersection dynamics in NO2 probed by homodyne high-harmonic spectroscopy. *Science*, **334**(6053):208–212, 2011.

[43] S. Baker, J. S. Robinson, C. A. Haworth, et al. Probing proton dynamics in molecules on an attosecond time scale. *Science*, **312**(5772):424–427, 2006.

[44] O. Smirnova, Y. Mairesse, S. Patchkovskii, et al. High harmonic interferometry of multi-electron dynamics in molecules. *Nature*, **460**(7258):972–977, Aug. 2009.

[45] P. M. Kraus, B. Mignolet, D. Baykusheva, et al. Measurement and laser control of attosecond charge migration in ionized iodoacetylene. *Science*, **350**(6262):790–795, 2015.

[46] L. S. Cederbaum and J. Zobeley. Ultrafast charge migration by electron correlation. *Chem. Phys. Lett.*, **307**(3):205–210, 1999.

[47] H.-W. Sun, P.-C. Huang, Y.-H. Tzeng, et al. Extended phase matching of high harmonic generation by plasma-induced defocusing. *Optica*, **4**(8):976–981, Aug. 2017.

[48] E. A. Gibson, A. Paul, N. Wagner, et al. Coherent soft X-ray generation in the water window with quasi-phase matching. *Science*, **302**(5642):95–98, 2003.

[49] J. Seres, V. S. Yakovlev, E. Seres, et al. Coherent superposition of laser-driven soft-X-ray harmonics from successive sources. *Nat. Phys.*, **3**:878–883, Dec. 2007.

[50] T. Popmintchev, M.-C. Chen, P. Arpin, M. M. Murnane, and H. C. Kapteyn. The attosecond nonlinear optics of bright coherent X-ray generation. *Nat. Photon.*, **4**:822–832, Dec. 2010.

[51] A. Paul, R. A. Bartels, R. Tobey, et al. Quasi-phase-matched generation of coherent extreme-ultraviolet light. *Nature*, **421**:51–54, Jan. 2003.

[52] M. Zepf, B. Dromey, M. Landreman, P. Foster, and S. M. Hooker. Bright quasi-phase-matched soft-X-Ray harmonic radiation from argon ions. *Phys. Rev. Lett.*, **99**:143901, Oct. 2007.

[53] X. Zhang, A. L. Lytle, T. Popmintchev, et al. Quasi-phase-matching and quantum-path control of high-harmonic generation using counterpropagating light. *Nat. Phys.*, **3**:270–275, Apr. 2007.

[54] O. Cohen, X. Zhang, A. L. Lytle, T. Popmintchev, M. M. Murnane, and H. C. Kapteyn. Grating-assisted phase matching in extreme nonlinear optics. *Phys. Rev. Lett.*, **99**:053902, Jul. 2007.

[55] J. Tate, T. Auguste, H. G. Muller, P. Salières, P. Agostini, and L. F. DiMauro. Scaling of wave-packet dynamics in an intense midinfrared field. *Phys. Rev. Lett.*, **98**:013901, Jan. 2007.

[56] T. Popmintchev, M.-C. Chen, A. Bahabad, et al. Phase matching of high harmonic generation in the soft and hard X-ray regions of the spectrum. *Proc. Natl. Acad. Sci. U.S.A.*, **106**(26):10516–10521, 2009.

[57] T. Popmintchev, M.-C. Chen, D. Popmintchev, et al. Bright coherent ultrahigh harmonics in the keV X-ray regime from mid-infrared femtosecond lasers. *Science*, **336**(6086):1287–1291, 2012.

[58] E. J. Takahashi, T. Kanai, K. L. Ishikawa, Y. Nabekawa, and K. Midorikawa. Coherent water window X ray by phase-matched high-order harmonic generation in neutral media. *Phys. Rev. Lett.*, **101**:253901, Dec. 2008.

[59] S. M. Teichmann, F. Silva, S. L. Cousin, M. Hemmer, and J. Biegert. 0.5-keV soft X-ray attosecond continua. *Nat. Commun.*, **7**:11493, May 2016.

[60] N. Ishii, K. Kaneshima, K. Kitano, T. Kanai, S. Watanabe, and J. Itatani. Carrier-envelope phase-dependent high harmonic generation in the water window using few-cycle infrared pulses. *Nat. Commun.*, **5**:3331, Feb. 2014.

[61] J. Li, X. Ren, Y. Yin, et al. Polarization gating of high harmonic generation in the water window. *Appl. Phys. Lett.*, **108**(23):231102, 2016.

[62] D. Popmintchev, C. Hernández-García, F. Dollar, et al. Ultraviolet surprise: efficient soft x-ray high-harmonic generation in multiply ionized plasmas. *Science*, **350**(6265):1225–1231, 2015.

[63] F. Silva, S. M. Teichmann, S. L. Cousin, M. Hemmer, and J. Biegert. Spatiotemporal isolation of attosecond soft X-ray pulses in the water window. *Nat. Commun.*, **6**:6611, Mar. 2015.

[64] S. L. Cousin, F. Silva, S. Teichmann, M. Hemmer, B. Buades, and J. Biegert. High-flux table-top soft x-ray source driven by sub-2-cycle, CEP stable, 1.85-μm 1-kHz pulses for carbon K-edge spectroscopy. *Opt. Lett.*, **39**(18):5383–5386, Sep. 2014.

[65] C. Jin and C. D. Lin. Optimization of multi-color laser waveform for high-order harmonic generation. *Chin. Phys. B*, **25**:094213, Sep. 2016.

[66] S.-W. Huang, G. Cirmi, J. Moses, et al. High-energy pulse synthesis with sub-cycle waveform control for strong-field physics. *Nat. Photon.*, **5**:475–479, Aug. 2011.

[67] A. Wirth, M. T. Hassan, I. Grguraš, et al. Synthesized light transients. *Science*, **334**(6053):195–200, 2011.

[68] L. E. Chipperfield, J. S. Robinson, J. W. G. Tisch, and J. P. Marangos. Ideal waveform to generate the maximum possible electron recollision energy for any given oscillation period. *Phys. Rev. Lett.*, **102**:063003, Feb. 2009.

[69] C. Jin, G. Wang, H. Wei, A.-T. Le, and C. D. Lin. Waveforms for optimal sub-keV high-order harmonics with synthesized two- or three-colour laser fields. *Nat. Commun.*, **5**:4003, May 2014.

[70] C. Jin, G. J. Stein, K.-H. Hong, and C. D. Lin. Generation of bright, spatially coherent soft X-ray high harmonics in a hollow waveguide using two-color synthesized laser pulses. *Phys. Rev. Lett.*, **115**:043901, Jul. 2015.

[71] S. Kim, J. Jin, Y.-J. Kim, I.-Y. Park, Y. Kim, and S.-W. Kim. High-harmonic generation by resonant plasmon field enhancement. *Nature*, **453**:757–760, Jun. 2008.

[72] P. Russbueldt, T. Mans, J. Weitenberg, H. D. Hoffmann, and R. Poprawe. Compact diode-pumped 1.1 kW Yb:YAG Innoslab femtosecond amplifier. *Opt. Lett.*, **35**(24):4169–4171, Dec. 2010.

[73] J.-P. Negel, A. Voss, M. A. Ahmed, et al. kW average output power from a thin-disk multipass amplifier for ultrashort laser pulses. *Opt. Lett.*, **38**(24):5442–5445, Dec. 2013.

[74] T. Eidam, S. Hanf, E. Seise, et al. Femtosecond fiber CPA system emitting 830 W average output power. *Opt. Lett.*, **35**(2):94–96, Jan. 2010.

[75] S. Hädrich, J. Rothhardt, M. Krebs, et al. Single-pass high harmonic generation at high repetition rate and photon flux. *J. Phys. B-At. Mol. Opt.*, **49**(17):172002, 2016.

[76] J. Rothhardt, S. Hädrich, Y. Shamir, et al. High-repetition-rate and high-photon-flux 70 eV high-harmonic source for coincidence ion imaging of gas-phase molecules. *Opt. Express*, **24**(16):18133–18147, Aug. 2016.

[77] C. M. Heyl, C. L. Arnold, A. Couairon, and A. L'Huillier. Introduction to macroscopic power scaling principles for high-order harmonic generation. *J. Phys. B-At. Mol. Opt.*, **50**(1):013001, 2017.

7 Generation and Characterization of Attosecond Pulses

7.1 Introduction

High-order harmonics were first observed around 1987 and 1988 with a 248 nm excimer laser [1] and then with a 30 ps 1,064 nm neodymium-doped yttrium aluminum garnet (Nd:YAG) laser [2] on Xe. In 1989, the harmonic spectra of the latter were confirmed in theoretical calculations by Kulander and Shore [3]. Since 1993, with the advent of the chirped-pulse amplification technique [4] and titanium–sapphire lasers, high-order harmonic generation (HHG) studies have been in a golden age. However, it took almost 10 years before attosecond pulses were reported, in 2001: first, the attosecond pulse train (APT) by Paul et al. [5] and then the single-attosecond pulse by Hentschel et al. [6].

Since the beginning, high-order harmonics have been known to exhibit a flat plateau. Theory shows that these harmonics have nearly constant phases. It was speculated that these phase-locked harmonics, if superposed, could produce a train of very short attosecond pulses. However, for a long time, the phase of the harmonics was not accessible to experimental determination; thus, the temporal profile of the harmonic emission remains undetermined. The emergence of attosecond pulses is due to credible experimental determination of the phases of the harmonics. By 2001, phase characterization of discrete harmonics had resulted in the first observation of the APT [5] and that of continuum harmonics had resulted in isolated attosecond pulses (IAPs) [6]. Thus attosecond science was born at the dawn of the twenty-first century.

This chapter discusses the APT and IAPs separately since they use different generation and characterization methods. Simple applications of these pulses will also be given. Additional applications of attosecond pulses will be covered in the next chapter.

7.2 APTs

7.2.1 Route to a Transform-Limited APT

Since 2001, it has been observed [5] that HHG is a useful tool for obtaining trains of attosecond bursts when combining a few harmonics typically in the extreme ultraviolet (XUV) regime. The harmonic spectrum consists of a series of narrow peaks separated by twice the frequency of the driving laser field. For the HHG process driven by a long pulse, by symmetry (see Section 5.2.5) only odd harmonics are emitted. However, measurement of the harmonic spectrum only determines the magnitude of these harmonics, while their relative phase remains inaccessible. To determine the pulse in the time domain, the spectral phase of the harmonic has to be determined first. If these harmonics are not in phase, the corresponding temporal profile will be rather erratic and of little interest.

Consider the superposition of a few different harmonics. Since only odd harmonics are generated in a long pulse, one can write the spectral amplitude of each harmonic as U_{2q+1}, and the phase as φ_{2q+1}. The temporal profile $I(t)$ is given by

$$I(t) = \left| \sum_q U_{2q+1} e^{-i(2q+1)\omega t + i\varphi_{2q+1}} \right|^2, \tag{7.1}$$

where ω is the infrared (IR) frequency. In the simplest case, if $\varphi_{2q+1} = (2q+1)\omega t^e$ (where t^e will be called the emission time) and amplitude U_{2q+1} is independent of q, then the summation of Equation 7.1 gives

$$I(t) \propto \frac{\sin^2[N\omega(t - t^e)]}{\sin^2[\omega(t - t^e)]}. \tag{7.2}$$

In the time domain, Equation 7.2 describes a train of bursts separated by $\frac{T}{2}$ where $T = \frac{2\pi}{\omega}$ is the period of the driving laser and N is the number of harmonics included in the summation. Each burst has a half width of $\frac{T}{2N}$. In this case, the temporal profile of each burst is the shortest and the pulse is called Fourier limited or transform limited (TL). TL will be used in this chapter for such a pulse. It is emphasized that a TL pulse does not mean that harmonic components have the same phase, but rather that the phase difference $\Delta\varphi$ between two consecutive harmonics is constant. In other words, it means that the emission time t^e is independent of the harmonic frequency. In fact, the regular pulsed structure of $I(t)$ is not much affected by the amplitude variation between different harmonics. This is the same as a mode-locked laser. If the emission time depends on the harmonic frequency, then the pulse is chirped and the pulse duration will be longer than the TL pulse.

Because of the uncertainty principle, one cannot define an "instant" associated with a given XUV frequency Ω. The emission time $t^e = \frac{\partial}{\partial\Omega}\varphi(\Omega) \approx \frac{\varphi_{2q+1} - \varphi_{2q-1}}{2\omega}$ is always applied to that of harmonics centered about $\Omega = 2q\omega$. By definition, the emission time is the group delay of the pulse. If a synthesized wave consists of a broad range of harmonics where the emission times are different, then the temporal profile of the pulse is not TL. One can employ external phase corrections to make the pulse shorter and closer to the TL pulse.

The experimental steps leading to the generation and characterization of APT used in Paul et al. [5] is illustrated in Figure 7.1. First, harmonics are generated using a 800 nm, 40 fs, 1 kHz laser with Ar as the target. An aperture is used to block out the divergent IR and allow IR and XUV harmonics to co-propagate along the beam axis. The beams are then deflected into another argon gas chamber by a spherical mirror to generate photoelectrons. Such spectra are measured as a function of the time delay between the IR and the XUV. The resulting two-dimensional (2D) electron spectra is called a spectrogram or a streaking trace. The phase difference between any two neighboring harmonics is derived from the spectrogram, as explained in Section 7.2.2.

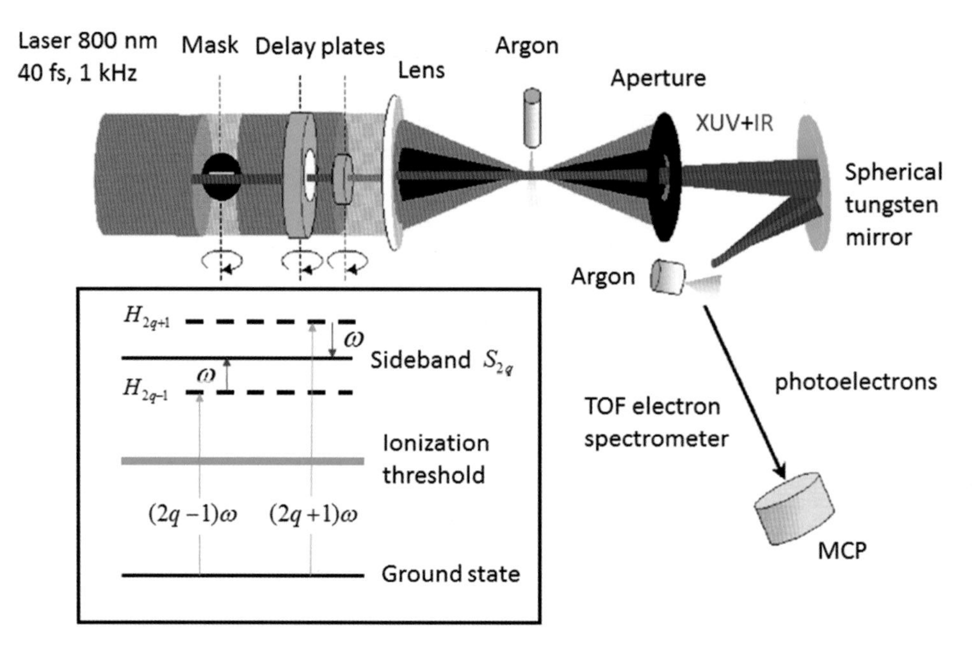

Figure 7.1 The experimental setup for the generation and characterization of an APT. The inset shows the main quantum paths contributing to the sideband photoelectrons generated in the second argon jet. (From P. M. Paul et al., *Science*, **292**, 1689 (2001) [5]. Reprinted with permission from AAAS.)

7.2.2 RABITT Method for Phase Retrieval of an APT

The method to determine the spectral phase of an APT is often referred to as the reconstruction of attosecond beating by interference of two-photon transitions (RABITT) [7]. The idea of this method is closely related to the work of Veniard et al. [8]. One uses the XUV harmonics together with a delayed multi-cycle IR field to ionize target atoms. The intensities of the XUV harmonics are too weak to cause nonlinear effects, and thus only ionize the atom by the single-photon-absorption process. Without the IR field, the photoelectron spectrum will consist of peaks at $E = (2q + 1)\omega - I_p$, through ionization by each odd harmonic. Here, ω is the fundamental frequency used to generate high harmonics and I_p is the ionization potential of the target. The delayed IR field is usually from the one that generates harmonics so it also has frequency ω. In the RABITT scheme, the IR intensity is very low (typically less than 1 TW/cm^2). Thus the electron can only absorb or emit one additional IR photon. In this weak field limit, the generation of the even harmonics, or the sideband, can be treated by the second-order perturbation theory. Figure 7.2 displays the appearance of sidebands at $E = 2q\omega - I_p$ when ionization by the XUV pulse occurs in the presence of the IR field. Moreover, the intensity of the sideband signals varies with the change of the time delay t_d between the XUV harmonic and the IR field. The variation can be explained by the two-path interference model (see the inset of Figure 7.1); the sideband electron S_{2q} is the result of the interference between the electron ionized by the $(2q - 1)\omega$ harmonic followed by absorbing one IR photon and ionized by the $(2q + 1)\omega$ harmonic followed by emitting one IR photon.

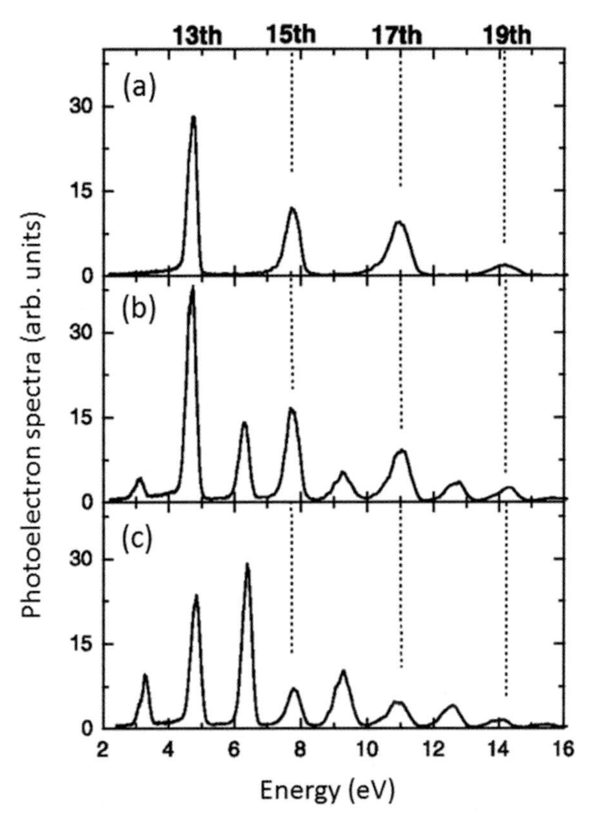

Figure 7.2 Photoelectron spectra of argon ionized by (a) XUV harmonics only. (b) and (c) XUV harmonics plus an IR field. Changing the time delay between IR and XUV harmonics from —1.7 fs in (b) to —2.5 fs in (c) causes a strong amplitude change of the sidebands. (From P. M. Paul et al., *Science*, **292**, 1689 (2001) [5]. Reprinted with permission from AAAS.)

The total sideband signal can be modeled mathematically as

$$S_{2q} = \sum_{f,i} \left| M_{f,i}^{(+)} + M_{f,i}^{(-)} \right|^2, \tag{7.3}$$

where i and f represent the quantum numbers of the initial state and the final continuum state, respectively. The summation includes all possible two-photon ionization channels. The term $M_{f,i}^{(\pm)}$ is the complex two-photon transition amplitude in which $(+)$ denotes the $(2q-1)\omega+\omega$ path (absorption path) and $(-)$ denotes the $(2q+1)\omega-\omega$ path (emission path). To obtain a simple expression of the transition amplitude, the bandwidth of the individual harmonics as well as the bandwidth of the IR field are ignored. Then, one has

$$M_{f,i}^{(+)} = \frac{E_{IR}}{2} U_{2q-1} e^{\varphi_{2q-1}} e^{i\omega t_d} d_{f,i}^{(+)}, \tag{7.4}$$

$$M_{f,i}^{(-)} = \frac{E_{IR}}{2} U_{2q+1} e^{\varphi_{2q+1}} e^{-i\omega t_d} d_{f,i}^{(-)}. \tag{7.5}$$

Here, $U_{2q\pm1}$ and $\varphi_{2q\pm1}$ represent the magnitude and phase of the consecutive odd harmonics $(2q \pm 1)\omega$, respectively. The term $d_{f,i}^{(\pm)}$ denotes the two-photon transition dipole given by

$$d_{f,i}^{(\pm)} = \langle f|z(E_{\mp} - H)^{-1}z|i\rangle = \sum_{\alpha\lambda m} \frac{\langle f|z|\alpha\lambda m\rangle\langle\alpha\lambda m|z|i\rangle}{E_{\mp} - E_{\alpha}}. \tag{7.6}$$

Here, both the XUV and IR fields are polarized along the z axis. $E_{\mp} = (2q \mp 1)\omega - I_p$ are the energies of the virtual intermediate states. H is the Hamiltonian of the unperturbed atom. The summation includes all the eigenstates $|\alpha\lambda m\rangle$ (both bound and continuum) of H in spherical coordinates. Since $E_{\mp}$ lies in the continuum spectrum, additional treatment is needed when E_{α} is close to $E_{\mp}$. The details of evaluating the two-photon transition dipole will be shown in Section 7.3.1.

Plugging Equations 7.4 and 7.5 into Equation 7.3, one can derive the sideband signal as

$$S_{2q} = A_{2q} + B_{2q}\cos[2\omega t_d - (\varphi_{2q+1} - \varphi_{2q-1}) - \Delta\varphi_{2q}^{atomic}]. \tag{7.7}$$

Obviously, the signal S_{2q} oscillates with frequency 2ω as a function of XUV–IR delay t_d. The DC term A_{2q} and the 2ω term B_{2q} are given by

$$A_{2q} = \frac{E_{IR}^2}{4}\left[U_{2q-1}^2\sum_{f,i}|d_{f,i}^{(+)}|^2 + U_{2q+1}^2\sum_{f,i}|d_{f,i}^{(-)}|^2\right], \tag{7.8}$$

$$B_{2q} = \frac{E_{IR}^2}{2}U_{2q-1}U_{2q+1}\left|\sum_{f,i} d_{f,i}^{(+)}d_{f,i}^{(-)*}\right|. \tag{7.9}$$

The atomic phase $\Delta\varphi_{2q}^{atomic}$ can be calculated theoretically as

$$\Delta\varphi_{2q}^{atomic} = \arg\left[\sum_{f,i} d_{f,i}^{(-)}d_{f,i}^{(+)*}\right]. \tag{7.10}$$

The atomic phase is usually small compared to the phase of the harmonics. Once the atomic phase is obtained from the theory (or just ignored), by fitting the oscillation of the measured sideband versus the time delay into the form given by Equation 7.7, the phase difference between two adjacent odd harmonics $\Delta\varphi_{2q}^{XUV} = \varphi_{2q+1} - \varphi_{2q-1}$ can be determined.

In the experiment of Paul et al., four sidebands (S12, S14, S16, S18) were measured as shown in the left side of Figure 7.3. By fitting the sideband signals, the phases of the five odd harmonics (H11 to H19) were found to be −2.6, −1.3, 0, 1.8, and 4.4 radians (the phase of H15 was set to zero). Together with the relative harmonic intensities, the temporal pulse profile of the total electric field was uniquely determined (see the right side of Figure 7.3). This profile shows a sequence of 250 as (in full-width at half maximum (FWHM)) bursts spaced by 1.35 fs, which corresponds with the optical half cycle. This is the first experimental demonstration (or characterization) of APTs obtained from HHG harmonics.

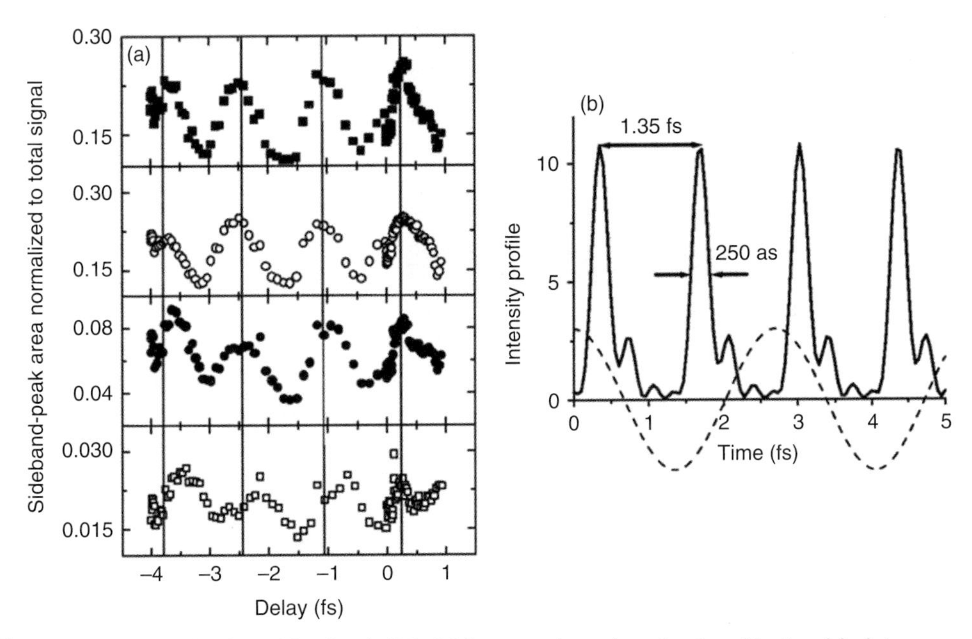

Figure 7.3 (a) Amplitude of the first four sideband peaks (S12–S18 from top to bottom) as a function of the time delay between the IR pulse and the harmonics. The vertical lines are spaced by 1.35 fs, half the cycle time of the driving laser. (b) Temporal intensity profile of a sum of five odd harmonics (H11–H19) as reconstructed from the measured phases and amplitudes of harmonics. (From P. M. Paul et al., *Science*, **292**, 1689 (2001) [5]. Reprinted with permission from AAAS.)

7.2.3 Control of Amplitude and Phase of an APT

The RABITT method described in Section 7.2.2 determines the spectral phase of each frequency component of an APT. To achieve TL attosecond pulses, the spectral phase of each harmonic must vary in such a way that every harmonic in the train has the identical emission time. Systematic studies of the harmonic phase have been carried out by Mairesse et al. [9]. Figure 7.4(a) shows the measured Ar photoelectron spectrum as a function of the XUV–IR delay. Neglecting the atomic phase, Equation 7.7 can be rewritten as

$$S_{2q} \approx A_{2q} + B_{2q} \cos[2\omega(t_d - t_{2q}^e)], \tag{7.11}$$

where $t_{2q}^e = \frac{\Delta\varphi_{2q}^{XUV}}{2\omega}$ is the XUV group delay centered at photon energy $2q\omega$. The measured group delay is shown in Figure 7.4(b) in blue. If the group delay is constant, all the harmonics are synchronized, which results in TL pulses that have the shortest duration allowed by the bandwidth. However, the measured group delay increases linearly with photon energy; that is, the attosecond pulse is positively chirped, which leads to temporal broadening. This group delay can be related to the semiclassical harmonic emission time. When focusing the laser beam ahead of the gas jet, macroscopic selection of the short trajectory can be achieved through phase matching. Semiclassical theory at the single-atom level predicts that the harmonic emission time increases as photon energy increases for the short trajectory, as plotted in Figure 7.4(b) in red. Therefore the positive chirp in the emitted

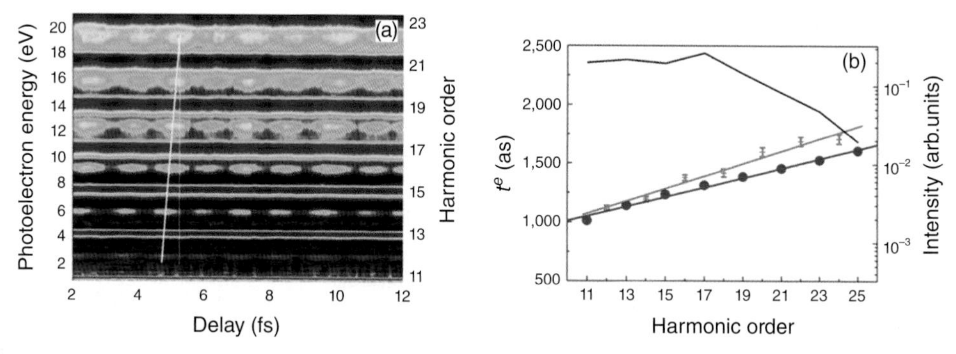

Figure 7.4 Measurement of the XUV group delay in argon using RABITT. (a) Photoelectron spectrum as a function of delay. The thick white line guides the eye to the location of the sideband maxima; the thin line is vertical. (b) (blue crosses with error bars) Measured XUV group delay (or the emission time t^e), which gives a time difference of $t^e_{2q+2} - t^e_{2q} =$ 106 ± 8 as. (red dots) Harmonic emission time predicted by single-atom semiclassical theory, which gives a time difference of 81 ± 3 as. (black line) Measured harmonic intensity. (From Y. Mairesse et al., *Science*, **302**, 1540 (2003) [9]. Reprinted with permission from AAAS.)

attosecond bursts mainly comes from the single-atom response for well-phase-matched harmonics. The latter imposes a limit that can be approached experimentally by optimizing the generating conditions, but never reached. However, one can propagate the harmonics through a negative-chirped medium to compensate for the positive chirp introduced in the HHG process. For example, aluminum films that have negative group delay dispersion (GDD) in the frequency range of the harmonics have been used to compress the original 480 as pulses into near TL 170 as pulses (see Figure 7.5 [10]). The details of how this works are given in the captions.

7.3 Temporal Information Extracted from APT Photoionization Experiments

7.3.1 Extracting Single-Photon Wigner Delay from Two-Photon Atomic Delay

To characterize APTs using RABITT, the atomic phase $\Delta\varphi^{atomic}$ is computed theoretically. Sometimes it is simply ignored because it is usually small. On the other hand, one may wish to extract the atomic phase of the target from the RABITT measurement. Define

$$\tau^{(2)} = \frac{\Delta\varphi^{atomic}}{2\omega} \tag{7.12}$$

as an intrinsic "atomic delay" associated with the two-photon ionization process. Then the sideband signal in Equation 7.7 can be expressed by

$$S_{2q} = A_{2q} + B_{2q}\cos[2\omega(t_d - t^e - \tau^{(2)})]. \tag{7.13}$$

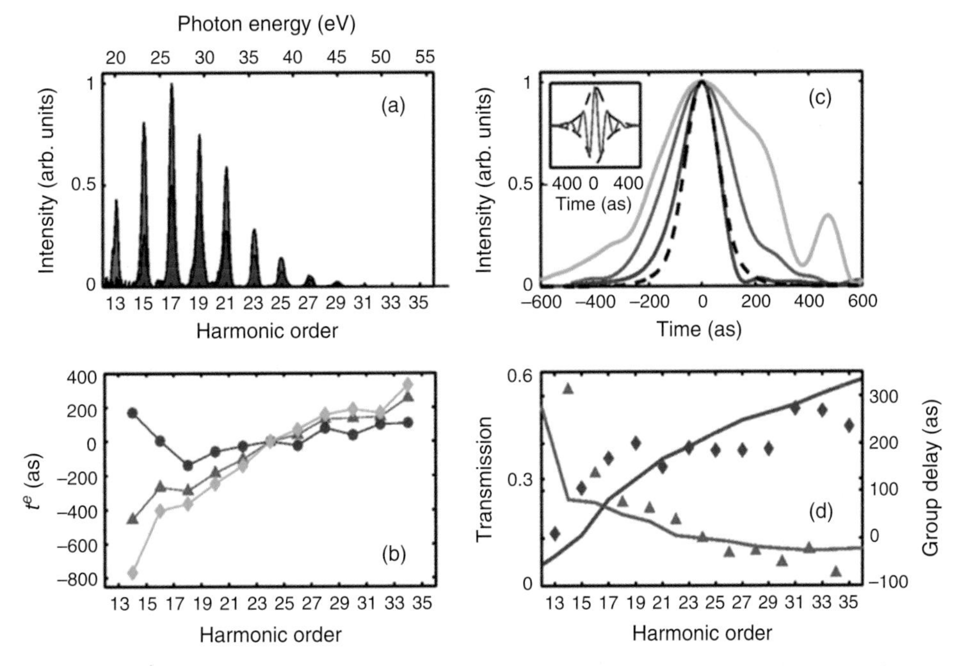

Figure 7.5 Compression of XUV harmonic radiation via amplitude and phase control. The effect of the aluminum films on the intensities and the synchronization of the harmonics is presented in (a) and (b) for one (blue) and three (red) 200 nm-thick films, respectively. These results contain sufficient information to reconstruct the APTs in (c). The green line shows an extrapolation to zero films assuming the same bandwidth as in the one-filter case. The aluminum improves the synchronization of the harmonics resulting in a compression of the pulse duration from 480 as (zero filter, green curve) down to 280 as (one filter, blue curve) and 170 as (three filters, red curve). The latter is very close to the transform limit of 150 as shown by the dashed line. The inset shows the XUV electric field assuming a cosine carrier. The measured transmission (red symbols) and group delay (blue symbols) of the aluminum filter are compared to the tabulated values (solid lines) in (d). (Reprinted from Rodrigo Lopez-Martens et al., *Phys. Rev. Lett.*, **94**, 033001 (2005) [10]. Copyrighted by the American Physical Society.)

In Equation 7.13, t^e is the group delay of the XUV harmonics, while the total time shift $t^e + \tau^{(2)}$ can be directly obtained in a RABITT measurement. To get $\tau^{(2)}$ one first needs to characterize the XUV pulse train using a well-known reference target. One could also use the same XUV+IR pulse to ionize two different targets, or two well-separated ionization channels of the same target. By taking the difference between the measured time shifts from two RABITT traces corresponding to the same photon energy, one thus can obtain the atomic delay difference $\Delta\tau^{(2)}$ between the two targets (channels), at the same photon energy.

The two-photon atomic delay $\tau^{(2)}$ contains the effect of the probing IR field. It is more interesting to obtain the phase information in the single-photon ionization process that is independent of the external field. Fortunately, it is possible to factor out the IR contribution for RABITT. To simplify the derivation, consider sideband electrons measured along the polarization axis ($+z$ direction). In this case, only the $m = 0$ channel exists, and the atomic delay is given by (from Equation 7.10)

$$\tau^{(2)} = \frac{\arg\left[d_{k,i}^{(-)} d_{k,i}^{(+)*}\right]}{2\omega} = \frac{\arg\left[d_{k,i}^{(-)}\right] - \arg\left[d_{k,i}^{(+)}\right]}{2\omega}, \tag{7.14}$$

where $k = \sqrt{2E}$, and $E = 2q\omega - I_p$ is the photoelectron energy of the sideband S_{2q}. The two-photon transition dipole is

$$d_{k,i}^{(\pm)} = \lim_{\epsilon \to 0} \sum_{\alpha\lambda m} \frac{\langle k|z|\alpha\lambda m\rangle\langle\alpha\lambda m|z|i\rangle}{E_{\mp} - E_{\alpha} + i\epsilon}. \tag{7.15}$$

Here, $E_{\mp} = E \mp \omega$, and the final state $|k\rangle$ describes continuum electrons toward the $+z$ direction. Within the single-active-electron approximation, the one-electron model potential is $V(r) = -Z_c/r + V_{sr}(r)$, where $Z_c = 1$ is the asymptotic charge, and V_{sr} is the short-range correction to the Coulomb potential. By expanding $|k\rangle$ in terms of partial waves, Equation 7.15 can be rewritten using spherical coordinates

$$d_{k,i}^{(\pm)} = \sum_{L,\lambda} \sqrt{\frac{2L+1}{4\pi}} e^{i\eta_L(E)} \langle Y_{L0}|\cos\theta|Y_{\lambda 0}\rangle\langle Y_{\lambda 0}|\cos\theta|Y_{l_i 0}\rangle W_{L,\lambda}(E, E_{\mp}). \tag{7.16}$$

Here, the initial state has a well-defined angular momentum l_i, and λ and L are the angular momentum quantum numbers of the intermediate and final partial wave of the photoelectron, respectively. According to the dipole-selection rule, $\lambda = l_i \pm 1$, $L = \lambda \pm 1$. The term $W_{L,\lambda}$ is a two-photon radial matrix element

$$W_{L,\lambda}(E, E_{\mp}) = \lim_{\epsilon \to 0} \int \frac{\langle u_{kL}|r|u_{E_\alpha \lambda}\rangle\langle u_{E_\alpha \lambda}|r|u_i\rangle}{E_{\mp} - E_{\alpha} + i\epsilon} dE_{\alpha} = \langle u_{kL}|r|\rho_{\kappa_{\mp}\lambda}\rangle \tag{7.17}$$

in which $\kappa_{\mp} = \sqrt{2E_{\mp}}$. The energy-normalized radial functions $u_{kL}(r)$ are solutions of the field-free radial Schrödinger equation that have the asymptotic behavior

$$\lim_{r\to\infty} u_{kL}(r) = \sqrt{\frac{2}{\pi k}} \sin\left[kr + \frac{Z_c}{k}\ln(2kr) + \eta_L(E)\right]. \tag{7.18}$$

Here, $\eta_L = -\frac{L\pi}{2} + \sigma_L + \delta_L$ is the phase shift of the partial wave with angular momentum L, including both the Coulomb phase shift σ_L and the phase shift δ_L due to the short-range potential V_{sr} (see Section 1.1.2).

In Equation 7.17 $\rho_{\kappa_{\mp}\lambda}(r)$ was introduced as an intermediate radial-wave function describing the photoelectron after absorbing one XUV photon. This function can be found in the following way:

$$\begin{aligned}
\rho_{\kappa_{\mp}\lambda}(r) &= \left[\lim_{\epsilon\to 0} \frac{1}{E_{\mp} - H_{\lambda} + i\epsilon}\right] r u_i(r) \\
&= \left[\wp \frac{1}{E_{\mp} - H_{\lambda}} - i\pi\delta(E_{\mp} - H_{\lambda})\right] r u_i(r) \\
&= \rho_{\kappa_{\mp}\lambda}^{(R)} - i\pi\langle u_{\kappa_{\mp}\lambda}|r|u_i\rangle u_{\kappa_{\mp}\lambda}(r).
\end{aligned} \tag{7.19}$$

Here, the radial Hamiltonian is introduced as $H_{\lambda} = -\frac{1}{2}\frac{d^2}{dr^2} + V(r) + \frac{\lambda(\lambda+1)}{2r^2}$. The term with $\wp$ prescribes the principal value integration that contributes to the real part of $\rho_{\kappa_{\mp}\lambda}$ (off-shell part). The δ term represents the Dirac delta function that contributes to the imaginary part of $\rho_{\kappa_\mp\lambda}$ (on-shell part). The function $\rho^{(R)}_{\kappa_\mp\lambda}(r)$ is the solution of the Dalgarno–Lewis differential equation

$$(E_\mp - H_\lambda)\rho^{(R)}_{\kappa_\mp\lambda}(r) = ru_i(r) \tag{7.20}$$

with the boundary condition $\rho^{(R)}_{\kappa_\mp\lambda}(r) = 0$ at $r = 0$. Physical solution of $\rho_{\kappa_\mp\lambda}(r)$ requires the asymptotic behavior [11]

$$\lim_{r\to\infty} \rho_{\kappa_\mp\lambda}(r) = -\pi\sqrt{\frac{2}{\pi\kappa_\mp}}e^{i\left(\kappa_\mp r+\frac{Z_c}{\kappa_\mp}\ln(2\kappa_\mp r)+\eta_\lambda(E_\mp)\right)}\langle u_{\kappa_\mp\lambda}|r|u_i\rangle. \tag{7.21}$$

In order to fulfill the asymptotic form Equation 7.21, the physical solution of Equation 7.20 should have the smallest asymptotic amplitude. This is the Dalgarno–Lewis method [12] discussed in detail in Toma and Muller [13].

If $\rho_{\kappa_\mp\lambda}$ is calculated with the Dalgarno–Lewis method, one can obtain the exact two-photon transition dipole. Alternatively, if one uses the asymptotic form of $\rho_{\kappa_\mp\lambda}$ and u_{kL}, Equations 7.18 and 7.21 to evaluate $W_{L,\lambda}$, the approximate result takes the form

$$W_{L,\lambda}(E, E_\mp) \approx ie^{i\{\eta_\lambda(E_\mp)-\eta_L(E)\}}\langle u_{\kappa_\mp\lambda}|r|u_i\rangle T^{cc}(E, E_\mp). \tag{7.22}$$

The term T^{cc} has an analytical expression

$$T^{cc}(E, E_\mp) = -\frac{1}{\sqrt{k\kappa_\mp}}\frac{(2\kappa_\mp)^{iZ_c/\kappa_\mp}}{(2k)^{iZ_c/k}}\left(\frac{i}{\kappa_\mp - k}\right)^{2+i(Z_c/\kappa_\mp - Z_c/k)}\Gamma[2 + i(Z_c/\kappa_\mp - Z_c/k)]. \tag{7.23}$$

Note that this expression does not depend on the target except for the asymptotic charge Z_c, but it does depend on the photoelectron energy. By plugging Equation 7.22 into Equation 7.16 one can obtain the approximate two-photon dipole

$$d^{(\pm)}_{k,i} \approx iT^{cc}(E, E_\mp)\sum_{L,\lambda}\sqrt{\frac{2L+1}{4\pi}}e^{i\eta_\lambda(E_\mp)}\langle u_{\kappa_\mp\lambda}|r|u_i\rangle\langle Y_{L0}|\cos\theta|Y_{\lambda0}\rangle\langle Y_{\lambda0}|\cos\theta|Y_{l_i0}\rangle. \tag{7.24}$$

By working out the angular part for all the possible quantum paths, one can prove that

$$d^{(\pm)}_{k,i} \approx iT^{cc}(E, E_\mp)d^{(1)}(E_\mp), \tag{7.25}$$

where

$$d^{(1)}(E_\mp) = \sum_\lambda\sqrt{\frac{2\lambda+1}{4\pi}}e^{i\eta_\lambda(E_\mp)}\langle u_{\kappa_\mp\lambda}|r|u_i\rangle\langle Y_{\lambda0}|\cos\theta|Y_{l_i0}\rangle \tag{7.26}$$

is the single-photon transition-dipole matrix element to the continuum electron with energy $E_\mp$ along $+z$ axis. Equation 7.25 implies that the two-photon dipole can be approximately separated into two factors: the single-photon transition dipole to the intermediate state and a term T^{cc} accounting for the IR-induced continuum–continuum (CC) transition.

From Equations 7.14 and 7.25 one can deduce that

$$\tau^{(2)} \approx \tau^{(1)} + \tau^{cc} \tag{7.27}$$

in which

$$\tau^{(1)} = \frac{\arg[d^{(1)}(E+\omega)] - \arg[d^{(1)}(E-\omega)]}{2\omega} \tag{7.28}$$

is a finite difference approximation to the Wigner-like delay $\frac{d}{dE}\arg[d^{(1)}(E)]$, which reflects the properties of the electronic wave packet ionized by one-photon absorption. More discussions on this Wigner delay will be given in Section 7.6. The IR-induced CC delay is given by

$$\tau^{cc} = \frac{\arg[T^{cc}(E, E+\omega)] - \arg[T^{cc}(E, E-\omega)]}{2\omega}, \tag{7.29}$$

which is target independent but is a function of the photoelectron energy.

The validity of the delay separation Equation 7.27 is verified in Figure 7.6. In the figure, accurate, two-photon atomic delay and single-photon Wigner delay for each ionization channel in Ne and Ar are calculated based on the many-body perturbation theory (MBPT), from which a CC delay can be determined by $\tau^{cc} = \tau^{(2)} - \tau^{(1)}$. This figure shows that the CC delays extracted from all the targets and channels agree with each other as well as the analytical value. Therefore Equation 7.27 remains a good approximation even after electron correlation has been included. Thus, Equation 7.27 serves as the foundation of extracting the Wigner delay from a RABITT-type measurement unless there is new evidence to prove otherwise.

Note that the above derivation of Equation 7.27 is for photoelectrons measured along the polarization direction. Considering that angular-integrated photoelectron yields are measured, Equation 7.27 is, strictly speaking, valid only when the electron is ionized from the *s* state. In this case, the bound electron can only absorb one XUV photon to

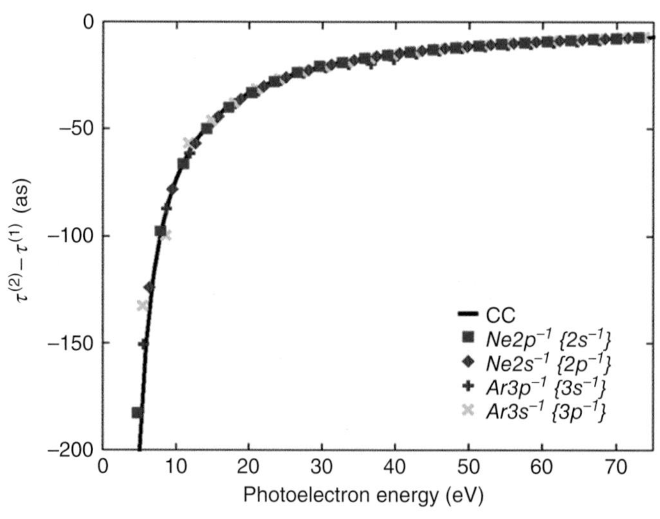

Figure 7.6 The C–C delay $\tau^{cc} = \tau^{(2)} - \tau^{(1)}$ determined for all outermost *n*-shell electrons in neon and argon. The good agreement with the analytical curve (CC) shows that a meaningful separation of the atomic delay $\tau^{(2)}$ can be made for atomic many-electron systems. The wavelength of the probe IR is 800 nm. (Reprinted from J. M. Dahlstrom, T. Carette, and E. Lindroth, *Phys. Rev. A*, **86**, 061402 (2012) [14]. Copyrighted by the American Physical Society.)

the intermediate continuum p partial wave. By contrast, when the electron comes from the p subshell, the intermediate partial wave can be either s or d, and the quantum paths $p \to s \to p$ and $p \to d \to p$ have the same final state so they will interfere. Such interference will make the atomic delay more complicated. However, if the contribution from the $p \to d$ channel is much larger than that from the $p \to s$ channel (a condition that is typically satisfied in noble gas atoms), one may neglect the $p \to s$ channel, then the separation of the atomic delay Equation 7.27 remains a good approximation. Another important point is that the RABITT method is based on second-order perturbation theory; thus, the laser intensity used in the experiment should be kept low to avoid contributions to the sideband harmonics from higher-order terms.

7.3.2 Retrieving Photoionization Time Delay between the 3*s* and 3*p* Subshells of Ar

RABITT has been applied to study photoemission of electrons from the 3s and 3p subshells in argon. Figure 7.7 displays the measured photoelectron spectra as a function of the delay between the XUV and IR pulses. The low-energy spectra (a) show electrons ionized from the 3s subshell and the high-energy spectra (b) for photoelectrons from the 3p subshell. For each ionization channel the sideband signals oscillate. By taking the Fourier transform along the time axis the delay $t^e + \tau^{(2)}$ can be extracted. By taking the difference between the extracted delays from the 3s and 3p channels by harmonics of the same order, the atomic delay difference $\tau_{3s}^{(2)} - \tau_{3p}^{(2)}$ can be obtained. Using Equation 7.27 we have

$$\tau_{3s}^{(2)} - \tau_{3p}^{(2)} = (\tau_{3s}^{(1)} - \tau_{3p}^{(1)}) + (\tau_{3s}^{cc} - \tau_{3p}^{cc}). \tag{7.30}$$

Note that all quantities should be calculated at the photoelectron energies that correspond to the same photon energy. The CC delay τ^{cc} can be calculated using analytical formula Equation 7.29 for the given IR photon energy (1.55 eV). The single-photon Wigner delay $\tau^{(1)}$ can also be calculated theoretically, for example, using the Hartree–Fock (HF) or the

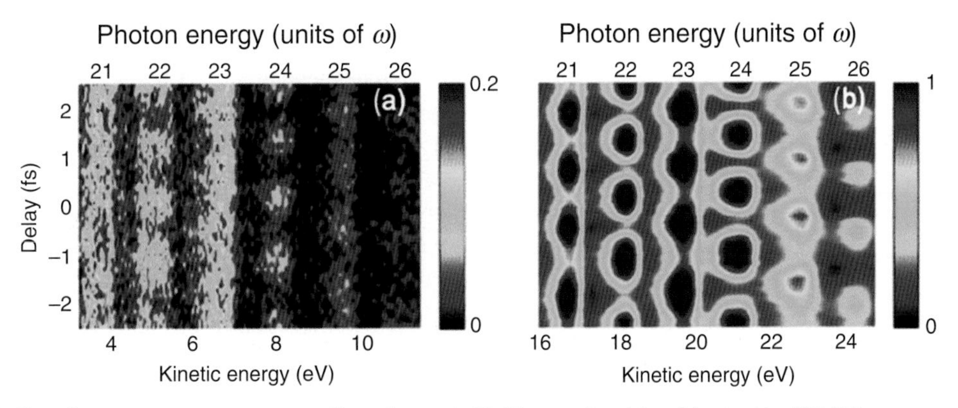

Figure 7.7 Photoelectron energy spectra corresponding to harmonics 21–26 versus time delay of the streaking IR field for electrons liberated from (a) the 3s subshell and (b) the 3p subshell of Ar, respectively. An 800 nm, 30 fs weak IR field ($<10^{12}$ W/cm^2) is used. (Reprinted from K. Klunder et al., *Phys. Rev. Lett.*, **106**, 143002 (2011) [15]. Copyrighted by the American Physical Society.)

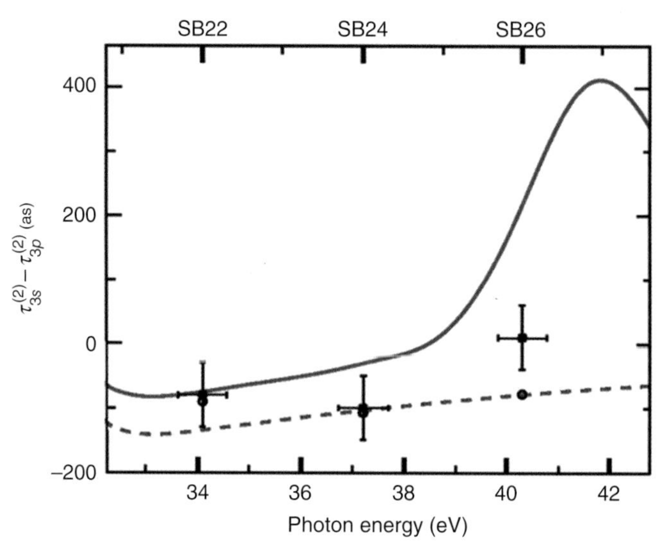

Figure 7.8 Two-photon atomic-delay difference between 3*s* and 3*p* ionization of Ar. Comparison between theoretical calculations (dashed blue line, HF+CC; red solid line, RPAE+CC) and experiments (circles, [15]; crosses, [17]). (Reprinted from D. Guenot et al., *Phys. Rev. A*, **85**, 053424 (2012) [17]. Copyrighted by the American Physical Society.)

random-phase approximation with exchange (RPAE) method. The RPAE theory includes intrashell and intershell correlation effects and thus should yield more accurate results, especially for the 3*s* channel. Comparisons between the measured and calculated atomic delay difference $\tau_{3s}^{(2)} - \tau_{3p}^{(2)}$ are shown in Figure 7.8. The measured delays vary from −110 to +10 as depending on the photon energy. Based on comparisons with experimentally retrieved data for the two lowest energy points, it is difficult to say that the RPAE theory is more accurate than the HF theory. Note that the error bar from the experiment is on the order of 100 as. On the other hand, there is a clear indication that the RPAE+CC delay increases rapidly near and above 40 eV, owing to the strong 3*s*–3*p* intershell correlation. From earlier photoionization studies, it is known that the 3*s* subshell cross-section has a sharp minimum at 40 eV [16] due to the 3*s* and 3*p* interchannel coupling. The dipole phase near this sharp minimum is expected to change rapidly as exhibited by the RPAE calculation. In Figure 7.8, the experimental time delay was obtained from the finite difference of the transition-dipole phase separated by a bandwidth of 3.1 eV (see Equation 7.28). Thus, in the energy region where phase changes rapidly, the experimental value may be significantly reduced as compared to the theoretical one. The averaging effect can be reduced if the streaking IR field is replaced by a mid-IR laser.

Using the RABITT technique, differences in photoionization time delays between outer-shell electrons in helium, neon, and argon have also been reported (see [17, 18]).

7.3.3 Retrieving Phase Information in the Vicinity of a Resonance

According to the formulation of the RABITT method, the atomic phase $\Delta\varphi^{atomic}$ or equivalently, the atomic time delay $\tau^{(2)}$, entering the sideband spectra is determined by the two-photon transition-matrix element. Based on the empirical relation Equation 7.27, where τ^{cc} is "universal," the "Wigner time delay" $\tau^{(1)}$ can be extracted. It is well known in photoionization theory that, in the vicinity of a resonance, the phase shift undergoes rapid change with energy (a change of about π across the resonance width of a few to tens of meVs), thus implying that $\tau^{(1)}$ will be large near a resonance. It may also be expected that $\tau^{(2)}$ has a significant change across a resonance. Recall that in Figure 7.4, the emission time of the XUV increases nearly linearly with the photon energy. Since $\tau^{(2)}$ is usually much smaller than the emission time, the time shift of the sideband peak with photon energy is also linear. This result is expected to change in the vicinity of a resonance and, in fact, has been confirmed in the experiment of Kotur et al. [19] near the $3s^1 3p^6 4p$ resonance of Ar. The excitation energy of this resonance is close to the seventeenth harmonic (H17) of the 800 nm laser. The laser system in Kotur et al. [19] allows the adjustment of the central wavelength of the driving laser such that H17 can be varied from below to above the $3s^1 3p^6 4p$ (or $3s^{-1}4p$) resonance. When the detuning is large, the sideband peak position shows a linear dependence versus photon energy, similar to Figure 7.4(b). When the central energy of H17 is at the resonance position of 26.63 eV, the linear dependence of the sideband position with respect to the harmonic order is no longer observed. Figure 7.9(a) shows the sideband as a function of time delay from S14 to S20. In this figure, the electron peaks from the odd harmonics have been removed for clarity. A long line has been drawn from the peak of S20 to the peak of S14 for reference. To identify the peak positions of S16 and S18, two short lines have been drawn. Both are shifted with respect to the expected positions of S14 and S20 if the resonance were not present, but toward opposite directions. The sideband S16 or S18 involves emission or absorption of an IR photon after being excited by H17. Thus the two-photon atomic phases in sidebands S16 and S18 contain contribution from the H17 Fano resonance. The atomic phases $\Delta\varphi^{atomic}$ of S16 and S18 extracted from the experiment as functions of the photon energy of H17 are shown in Figures 7.9(b) and (c), respectively. They display strong variation across the resonance. However, unlike the simple relation in Equation 7.27, there

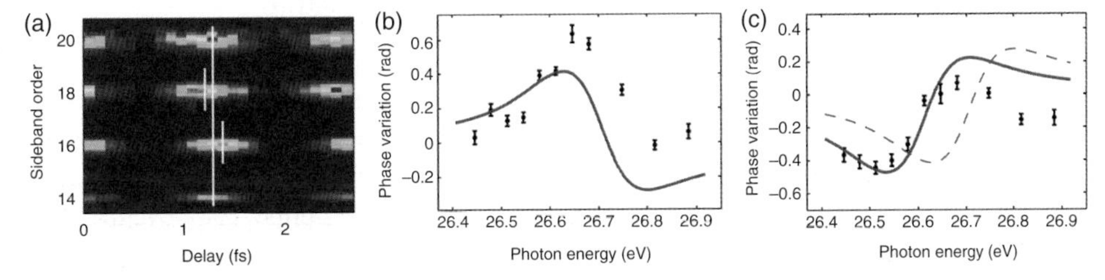

Figure 7.9 (a) Photoelectron signal for sidebands S14–S20 as a function of delay between the XUV and IR pulses. The laser wavelength is chosen such that the central energy of H17 is 26.63 eV, in near resonance with the $3s^{-1}4p$ state of Ar. The results have been corrected for the chirp of the attosecond pulses. (b) and (c) Atomic phase variation of (b) S16 and (c) S18 as a function of the energy of H17. The theoretical results are indicated by the red solid lines whereas the experimental results are shown by the black symbols. The thin dashed red line in (c) is the opposite of the red line in (b), which is close to the solid red line in (c) apart from an energy shift. (Reprinted from M. Kotur et al., *Nat. Commun.*, **7**, 10566 (2016) [19].)

is no simple theory to extract the phase of the single-photon dipole of the $3s^{-1}4p$ resonance from the RABITT data. Employing the RABITT technique, the wave-packet dynamics in the vicinity of the $2s2p$ Fano resonance of helium have also been studied [20].

7.4 Generation of IAPs

Although each burst is of a sub-femtosecond duration in an APT, the whole pulse train is still a few to tens of femtoseconds. For applications in dynamic systems that require sub-femtosecond temporal resolution, one would need IAPs. There are two important problems with IAPs: its generation and its characterization. This section has addressed the generation. The characterization of the IAP will be addressed in Section 7.5.

7.4.1 Methods of Generating IAPs

Amplitude Gating

In HHG, discrete harmonics are the result of the coherent superposition of continuum harmonics generated from each optical cycle, and the cutoff harmonic energy for each cycle is given by $I_p + 3.2U_P$, where U_P is the ponderomotive energy of this cycle within the simple model. For high-energy harmonics that are generated by a single cycle, the harmonic spectrum is continuous. Thus, by filtering out the discrete harmonics, single-attosecond pulses can be obtained. This is how the IAP was first generated [6]. For this method to work, it is desirable to have very short laser pulses and the carrier-envelope phase (CEP) of the pulse to be stabilized. For an 800 nm laser, pulses as short as 3.6 fs have been used for the generation of IAPs. Using this method with Ne as the target, IAP with central energy of 80 eV, and pulse duration of 80 as has been reported [21].

Polarization Gating

Generating a single-cycle laser pulse is difficult, but it is possible to generate an IAP with a long laser pulse, where the polarization of the driving pulse is manipulated such that the rising edge and the trailing edge of the pulse are elliptically polarized while the central cycle is linearly polarized, as originally proposed by Corkum et al. [22]. Harmonic generation is sensitive to the ellipticity, and thus the rising and trailing edges of the pulse do not generate harmonics, and high harmonics are generated only from the central portion of the pulse, which is linearly polarized. IAPs as short as 130 as in the spectral range of 25–50 eV with an energy of 70 pJ were first generated and characterized in 2006 by the Milan group [23]. This method is called *polarization gating* (PG).

A time-dependent ellipticity laser pulse can be formed by combining a left-hand circularly polarized pulse and a right-hand circularly polarized pulse with a time delay T_d [24]. For such a pulse, the time interval wherein the ellipticity $\varepsilon(t)$ is less than a certain threshold ε_{th} is given approximately by

$$\delta t_{PG} = \frac{\varepsilon_{th}}{\ln 2} \frac{\tau^2}{T_d}, \tag{7.31}$$

where τ is the duration of both circularly polarized pulses. Since harmonic yield drops by about a factor of two when the ellipticity changes from 0 to 0.13, and if one chooses T_d to be about equal to τ, then the gate window where ε is less than 0.2 will be given by $\delta t_{PG} = 0.3\tau$. This gate window should be less than the time between two successive emissions. For an 800 nm pulse this would require a pulse duration of about 5 fs and $T_d = 5$ fs. In other words, one would need to start with two 5 fs circularly polarized lights. In addition, by taking the depletion of the ground-state population by the leading edge of the circularly polarized light into account, the method also requires short-driving laser pulses.

To relax the need of starting with very short pulses for the generation of harmonics, one can change the period of HHG from $T_0/2$ to T_0 by adding a second harmonic with proper energy and phase to the driving laser. This method is called double-optical gating (DOG) [25]. For the same pulse energy, this method enhances the generation efficiency because there is less ionization at the leading edge of the driving pulse. One can also use a higher-intensity driving laser to reach a more intense IAP using DOG than with the PG method. To use even longer driving pulses, one can replace each circularly polarized pulse with an elliptically polarized light with ε about 0.5. This method (called *GDOG*) would allow a larger polarization gate such that a 148 as pulse can be generated with a 800 nm, 28 fs driving pulses (see [26]).

At present, DOG appears to be the most commonly used method for IAP generation. Using a 7 fs, 750 nm, 1 kHz titanium-sapphire laser focused to 1×10^{15} W/cm^2 on Ne at high gas pressure, Zhao et al. [27] reported the generation of 67 as pulses with energy centered at 80 eV. The PG method was used to generate continuum harmonics from 50 to 450 eV using two-cycle, 1.7 μm driving mid-IR lasers obtained using OPCPA in Li et al. [28]. The supercontinuum in this high-energy region has not been characterized yet since the standard FROG–CRAB method is not expected to work for such broadband continuum harmonics.

Attosecond Lighthouse

The attosecond lighthouse method was first introduced in Vincenti and Quéré [29]. It is carried out by inserting a pair of glass prisms in the beam path before focusing to generate pulse-front tilt in the driving laser. With such pulse-front tilt, each attosecond burst generated in a train is emitted in a different direction. If the wavefront rotation within one half cycle of the driving laser field is larger than the divergence of the individual attosecond pulses, then each attosecond burst can be separated spatially. This method still requires relatively short pulses and has been demonstrated for a 5 fs CEP-stabilized 800 nm laser [30]. It has also been demonstrated with a 2 cycle 1.8 μm laser [31]. The latter also measured the spatial and temporal profile of each subcycle burst and confirmed that, at the beam center, the near-field pulse duration is 390 as and increases to 420 as in the far field. The pulses were characterized using the *in situ* technique introduced in Dudovich et al. [32] by applying a weak second harmonic of the fundamental to perturb the space and time of the generated harmonics.

Ionization Gating

The ionization gating method relies on generating harmonics using intense lasers beyond the saturation intensity. The neutral medium is fully depleted within the leading edge of the driving pulse. At the trailing half, the medium is severely ionized such that plasma dispersion and the absence of neutral atoms turn off the harmonic emission. The gating obtained by confining the harmonic emission on the leading edge may not produce a very narrow pulse, and thus additional spectral filtering is often used (see [33]).

7.4.2 Enhancing Pulse Energy of IAPs

For applications of IAP, not only the generation of pulses with durations below a few hundred attoseconds is needed, but the pulse energy is also critical. Most IAPs are generated using amplitude-, polarization-, or ionization-gating methods and most IAPs used in applications are generated with titanium-sapphire lasers with wavelengths from about 750 nm–800 nm or with its second harmonic. Clearly, one of the most important technological challenges in ultrafast laboratories is increasing the pulse energy of IAPs.

Consider an APT or IAP with mean photon energy below 50 eV. The typical APT has pulse energy in the microjoule range with conversion efficiency on the order of 10^{-4}. For the IAP, the typical pulse energy is in the nanojoules or even picojoules [34, 35]. Figure 7.10, taken from Takahashi et al. [34], illustrates the progress of the energy of IAPs since the first IAP was reported by Hentschel et al. [6] in 2001 of sub-nanojules

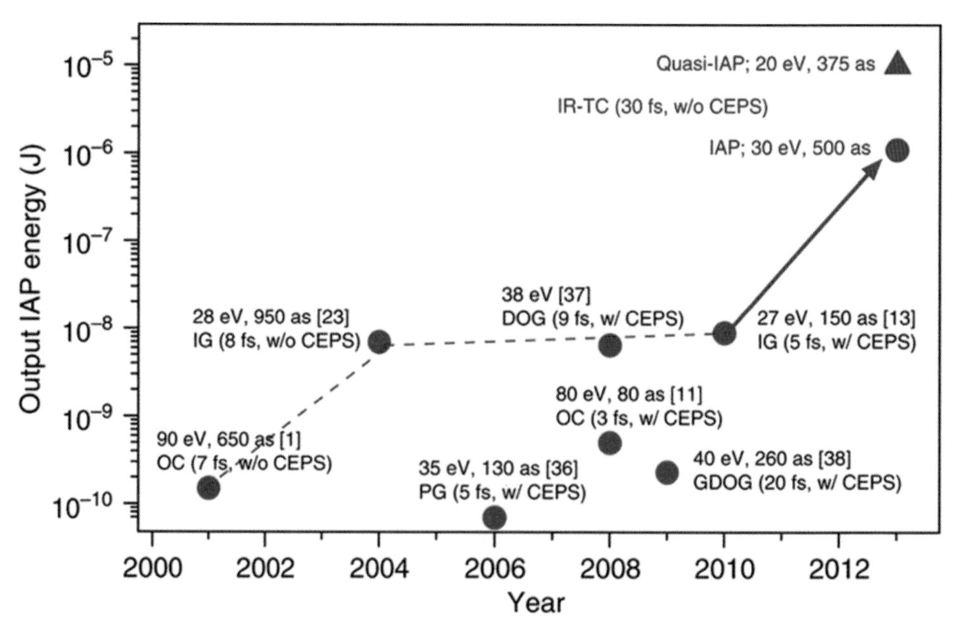

Figure 7.10 This figure shows the progress of the pulse energy of IAPs generated from 2001 to 2012. For the sources of the reported data points, readers should consult the original paper. (Reprinted from E. J. Takahashi et al., *Nat. Commun.*, **4**, 2691 (2013) [34].) Similar documentation can also be found in table 1 of [35].

to 10 nanojules generated with one-color driving lasers. With two-color driving lasers, a 500 as IAP was reported in Takahashi et al. [34] with pulse energy at 1.2 microjoules – an increase of two orders of magnitude from the other IAPs.

Since an IAP is only generated within a single cycle of the driving laser, there are two ways to enhance the pulse energy. The first is to superpose coherently electric fields of multiple-color components, especially when the frequencies of the elementary colors are incommensurate; this has the effect of increasing the optical period to guarantee the generation of continuum harmonics. Second is to increase the repetition rates of the driving laser.

Two-Color and Multicolor Gating for Enhancing Energy of the IAP

From the multiphoton picture, at high laser intensity, sum-frequency and difference-frequency generation of harmonics in a nonlinear medium readily produce a quasi-continuum spectrum of radiations. By correct combination of the amplitude and phase of these multicolor components, it is possible to synthesize a driving field that would enhance the harmonic yield and the cutoff energy of the harmonics. To increase harmonic yield, one would like to have a short-wavelength component, but to increase cutoff energy a long-wavelength laser component is preferred. Recall that cutoff energy scales quadratically with the wavelength, but the HHG yield scales inversely proportional to the fifth or sixth powers of the wavelength.

Two-color and three-color waveform synthesis has been employed in high-harmonic generation for APT and IAP. The challenge experimentally is generating such multicolor elementary waves with well-controlled amplitudes and phases. The latter means that the CEP of each elementary pulse should be well stabilized. In spite of this difficulty, much progress has been made. In Takahashi et al. [34], two synchronized 800 nm and 1,300 nm pulses are combined with parallel polarization, with the intensity of the 1,300 nm at about 15% of the 800 nm one. The intensity of the 800 nm pulse was 5×10^{13} W/cm^2 and the CEP was not stabilized. The repetition rate is 10 Hz and the generation medium is Xe. The IAP generated has central energy of 30 eV. By adjusting the focusing point with respect to the gas cell and varying the gas pressure, the IAP generated becomes intense enough such that the pulse profile can be characterized using the autocorrelation method, by measuring the ionization yield of N_2 molecules versus the time delay between the two split IAPs. In this experiment, even without CEP stabilization, the two-color combination that results in the highest cutoff energy also turns out to be the combination that has the largest harmonic yields. By spectral filtering, it was possible to obtain 500 as IAPs that are intense enough for pump-probe ionization. It is worth mentioning that two-color or three-color gating methods can be used to enhance the generation of harmonics in general, as discussed in Chapter 6.

Synthesized Light Transients for IAP Generation

Generation of arbitrary waveforms of electromagnetic waves in the microwave and radio-frequency regimes is routinely produced using function generators, which are electronic oscillators whose speed is limited to about 100 GHz. If similar methods can be extended to the optical region (10^{15} Hz), then any arbitrary waveform covering from ultraviolet (UV) to mid-IR can be generated. A driving laser made of such a few-octave spanning waveforms would allow experimentalists to design goal-specific pulses for applications. Such synthesized light transients may be much more efficient in generating single-attosecond pulses. This route to a synthesized waveform has been actively pursued at the Max Planck Institute of Quantum Optics in Germany in the past few years. While a more technical description of the method can be found in Keathley et al. [36], here, the setup and recent results reported in Hassan et al. [37] are outlined.

Light waveforms whose spectra extend over more than two octaves (1.1–4.6 eV) are generated by broadening a 22 fs, 1 mJ, 790 nm laser pulse through a hollow-core fiber filled with Ne gas of about 2.3 bars. The generated supercontinuum after the exit has about 550 μJ. Using dichroic beam splitters, the spectrum is divided into four spectral bands: 1.1–1.75 eV, 1.75–2.5 eV, 2.5–3.5 eV, and 3.5–4.6 eV. The pulses in these bands are individually compressed to a few femtoseconds before they are spatially and temporally superimposed to a single beam. The pulse energy of the beam at the exit is 320 μJ. For the synthesis, precise control of the relative phase and intensity of each spectral channel is critical. While how the synthesizer works in principle is known, the devil is in the details. Figure 7.11(a) shows a photograph of the attosecond light-field synthesizer, and an illustration of the four femtosecond broadband pulses. Figure 7.11(b) shows the photoelectron spectrogram generated by an XUV streaked by the synthesized IR waveform. The modulation of the spectrogram mimics the electric field of the synthesized wave (or more accurately, the vector potential), but it can also be characterized from the spectrogram (see Figure 7.11(c)). In the time domain, the synthesized wave has a FWHM of 380 as. Attosecond pulses generated from such fully controllable synthesized waveforms are expected to launch the next level of rapid progress in attosecond sciences.

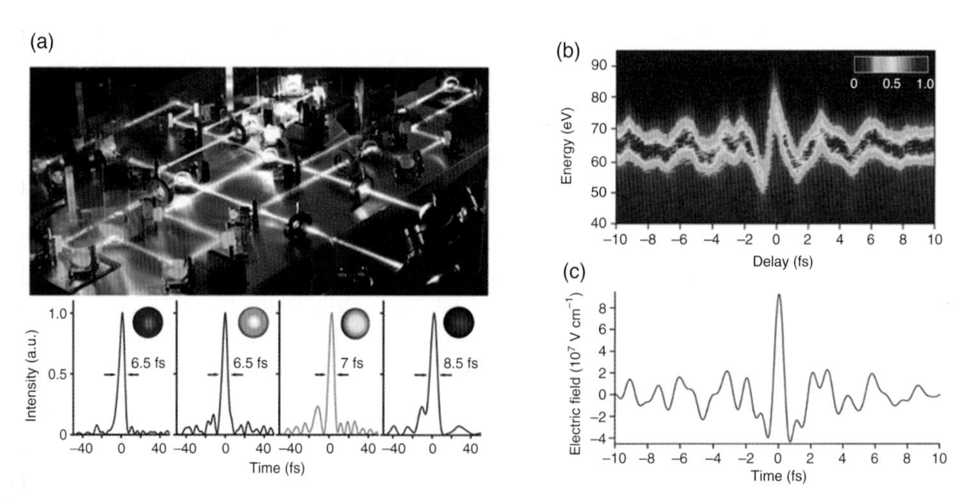

Figure 7.11 (a) Photograph of an attosecond light-field synthesizer. The wide beam (about 1.1–4.6 eV) is divided by dichroic beam splitters into four spectral bands (artificially visualized). The pulse in each band is compressed to a near TL pulse. These pulses are then superimposed in time and space to yield a single pulse. (b) The synthesized broadband pulse is used to generate an attosecond-streaking spectrogram, from which the electric field of the synthesized pulse can be determined. (c) Evaluated electric field of the synthesized pulse. (Reprinted from M. Th. Hassan et al., *Nature*, **530**, 66 (2016) [37].)

7.4.3 Ongoing Developments

Despite tremendous progress in the generation of IAPs in the last decade, the IAPs available today are still quite limited in terms of the spectral region, the pulse energy, and the typical duration of hundreds of attoseconds. Such limitations have restricted the capabilities of these tools for applications. Fortunately, these limitations are well recognized by laser scientists, and thus there is intense research going on worldwide. New methods are constantly emerging. Only a few examples are given here.

Generation of IAPs in the Water-Window Region

Currently, IAPs that have been used for applications are limited to below 150 eV. Clearly, for structure studies, it is desirable to go to the soft or hard X-ray regimes. Due to the $I\lambda^2$ scaling of the harmonic cutoff, higher-energy harmonics can be generated with mid-IR lasers with wavelengths from about 1.5 μm and up. Since the harmonic yield decreases roughly as $\lambda^{(-5\sim-6)}$ at the single-atom level, the efficiency of generating high-energy photons drops rapidly. Fortunately, the attochirp of the harmonics is $1/\lambda$. The efficiency can be improved by using higher gas pressure. Thus, one can use few-cycle mid-IR lasers for attosecond pulse generation. For example, in Teichmann et al. [38] IAPs reaching the carbon K-shell edge (284 eV) are demonstrated. A two-cycle, CEP-stabilized 1,850 nm, 1 kHz, 230 μJ laser system is implemented with wavefront rotation to generate continuum harmonics. After filtering, a broad spectrum covering from 225 to 300 eV was observed. The spectrum moves with the change of the CEP, indicating that IAPs have been generated. There are other reports of IAP generation in the soft X-ray region, but so far there are no actual reports of spectral-phase determination of the IAP since the common FROG–CRAB method for IAP characterization via photoelectron streaking does not work for broadband attosecond pulses.

Generation of Harmonics with Sub-to Few MHz Driving lasers

For IAP or harmonics in general, the average energy per second that can be used for experiments increases with the repetition rates of the driving laser pulse. The development of new, high-repetition-rate HHG sources has advanced high-average-power femtosecond technology, particularly with fiber lasers. In combination with nonlinear compression, efficient HHG generated with hundreds of kHz or even up to 10 MHz lasers have been reported. Such high-repetition harmonics are essential in many applications. For example, multiparticle coincidence momentum imaging techniques such as cold target recoil ion momentum spectroscopy (COLTRIMS) can yield multidimensional data sets that are channel resolved and highly angular resolved in order to reveal details of the reaction products. In such measurements, ionization has to be restricted to one event per pulse. To acquire sufficient statistics, high-repetition-rate pump-probe light sources at MHz or sub-MHz will be required. Fiber lasers with central wavelengths of 1,030 nm are ideal. Already, a coincident experiment for CH_3I with a 68.6 eV, 50–100 kHz repetition-rate source has been reported in Rothhardt et al. [39]. One can expect rapid progress in the similar development but different photon energy range in the coming years. For example, 50 kHz repetition-rate photons at 22.3 eV have been reported in Wang et al. [40] and were generated by the seventh harmonics of the UV laser at 390 nm with the latter frequency doubled in a barium borate (BBO) crystal. Depending on the need of various laboratories, it is increasingly more likely that table-top, tailored, coherent light sources will become possible in the near future.

7.5 Characterization of IAPs

7.5.1 Mathematical Expression of IAPs

In standard pump-probe experiments using IAPs, characterization of the IAPs is the prerequisite to accurately probing dynamical information about the target system. Mathematically, an IAP can be described in the time domain

$$E_{XUV}(t) = \sqrt{I(t)}\cos[\Omega_0 t + \phi(t)], \tag{7.32}$$

or in the frequency domain

$$\tilde{E}_{XUV}(\Omega) = U(\Omega)e^{i\Phi(\Omega)}. \tag{7.33}$$

Equations 7.32 and 7.33 are related by the Fourier transform

$$\tilde{E}_{XUV}(\Omega) = \int_{-\infty}^{\infty} E_{XUV}(t)e^{i\Omega t}dt, \tag{7.34}$$

$$E_{XUV}(t) = \frac{1}{2\pi}\int_{-\infty}^{\infty} \tilde{E}_{XUV}(\Omega)e^{-i\Omega t}d\Omega = \frac{1}{\pi}\int_{0}^{\infty} U(\Omega)\cos[\Omega t - \Phi(\Omega)]d\Omega. \tag{7.35}$$

In Equation 7.32, $I(t)$ is the temporal-intensity profile from which the pulse duration can be deduced, $\phi(t)$ is a temporal phase including attochirps, and Ω_0 is the central frequency of the IAP. In Equation 7.33, $U(\Omega)$ and $\Phi(\Omega)$ are the spectral amplitude and phase, respectively. Since $E_{XUV}(t)$ is real, it is obvious that $U(-\Omega) = U(\Omega)$ and $\Phi(-\Omega) = -\Phi(\Omega)$; thus, in the following discussion, only positive Ω is usually considered.

The temporal profile of the IAP not only depends on the spectral amplitude $U(\Omega)$, but also on the spectral phase $\Phi(\Omega)$. Assume that $U(\Omega)$ takes a simple Gaussian form

$$U(\Omega) = U_0 e^{-2\ln 2\frac{(\Omega-\Omega_0)^2}{(\Delta\Omega)^2}}, \tag{7.36}$$

where its FWHM bandwidth is given by $\Delta\Omega$. Consider the simplest phase $\Phi(\Omega) = \Phi_0$, which is an energy-independent constant. By doing the inverse Fourier transform, one can obtain the pulse in the time domain

$$E_{XUV}(t) = E_0 e^{-2\ln 2\frac{t^2}{(\Delta t)^2}}\cos(\Omega_0 t - \Phi_0). \tag{7.37}$$

Here, Δt is the FWHM duration of the pulse, which satisfies

$$\Delta\Omega\Delta t = 4\ln 2. \tag{7.38}$$

Next, a linear term is added to the spectral phase as $\Phi(\Omega) = \Phi_0 + (\Omega - \Omega_0)\tau$. One can easily find in this case that

$$E_{XUV}(t) = E_0 e^{-2\ln 2\frac{(t-\tau)^2}{(\Delta t)^2}} \cos(\Omega_0 t - \Phi_0). \tag{7.39}$$

Compared to Equation 7.37, the pulse envelope is delayed by an amount of τ with its shape kept the same.

In the next step, again consider an important case that the spectral phase has a quadratic term $\Phi(\Omega) = \Phi_0 + (\Omega - \Omega_0)\tau + \frac{\beta}{2}(\Omega - \Omega_0)^2$. By taking the inverse Fourier transform, in time domain the pulse is

$$E_{XUV}(t) = E_0 e^{-2\ln 2\frac{(t-\tau)^2}{(\Delta t)^2}} \cos\left[\Omega_0 t + \xi\frac{2\ln 2}{(\Delta t)^2}(t-\tau)^2 - \Phi_0 - \frac{1}{2}\arctan\xi\right]. \tag{7.40}$$

Here, the FWHM duration Δt satisfies

$$\Delta\Omega\Delta t = 4\ln 2\sqrt{1+\xi^2}, \tag{7.41}$$

and the parameter ξ is

$$\xi = \frac{\beta(\Delta\Omega)^2}{4\ln 2}. \tag{7.42}$$

From Equation 7.40 we can see that the quadratic term in $\Phi(\Omega)$ leads to a linear chirp in the time domain. β or ξ measures the amount of the attochirp. From Equations 7.38 and 7.41, one can further conclude that, given the same spectral bandwidth $\Delta\Omega$, the TL pulse (corresponding to $\beta = 0$) has the shortest temporal duration, whereas the duration of a chirped pulse will expand by a factor of $\sqrt{1+\xi^2}$ as compared to the TL pulse.

In general, one can approximate the spectral phase $\Phi(\Omega)$ in the vicinity of Ω_0 by Taylor's expansion

$$\Phi(\Omega) \approx \Phi(\Omega_0) + \left.\frac{d\Phi}{d\Omega}\right|_{\Omega_0}(\Omega - \Omega_0) + \frac{1}{2}\left.\frac{d^2\Phi}{d\Omega^2}\right|_{\Omega_0}(\Omega - \Omega_0)^2 + \cdots \tag{7.43}$$

According to the above discussion, much like with femtosecond laser pulses (Section 2.3.1), we can define the group delay of this attosecond pulse

$$\tau_G = \frac{d}{d\Omega}\Phi(\Omega), \tag{7.44}$$

and the GDD

$$\beta = \frac{d\tau_G}{d\Omega} = \frac{d^2}{d\Omega^2}\Phi(\Omega). \tag{7.45}$$

Although the spectral amplitude $U(\Omega)$ can be measured by spectrometers, the pulse duration cannot be determined unless the spectral phase $\Phi(\Omega)$ is retrieved. To measure the phase, certain nonlinear processes are needed so that one can compare the phase at different frequencies. For femtosecond pulses, these nonlinear processes can be autocorrelation, spectral phase interferometry for direct electric field reconstruction (SPIDER), or FROG.

However, these methods cannot be straightforwardly extended to characterize attosecond pulses because of the lack of effective nonlinear materials in the XUV region and the low-photon flux of IAPs that are available today. Instead, photoelectrons play an important role in IAP characterization. The pulse measurement usually relies on the XUV+IR streaking experiment in which the photoelectron is ionized from some target atoms by the XUV pulse in the presence of a delayed IR field. By changing the time delay between XUV and IR, a set of photoelectron spectra (namely, streaking spectrogram or trace) can be obtained. Thus the amplitude and phase of the IAP are transferred to a photoelectron replica and the phase information is embedded in the streaking trace. Extracting the spectral phase from the streaking trace is nontrivial because analytical models that can adequately describe the laser-dressed photoionization process are needed. At low IR intensity, where second-order perturbation theory is valid, one can use the phase retrieval by omega oscillation filtering (PROOF) method. At higher intensity, where many photons are absorbed, the FROG–CRAB method is used. The FROG–CRAB and PROOF methods will be discussed separately in the next two subsections. Both methods have their own limitations. A recent new retrieval method, phase retrieval of broadband pulses (PROBP), for broadband XUV pulses will also be presented. Furthermore, the IAP can be measured in the medium in which it is generated, i.e., *in situ* measurement [32] by applying a weak control field to perturb the trajectory of the re-collision electron. Since XUV pulses may undergo changes in amplitude and/or phase during the transport from the generation point to the gas cell where the experiment is carried out, the *in situ* method is less reliable.

7.5.2 The FROG–CRAB Method

Formulation

Attosecond streaking [41] can be first understood classically. Let $\mathbf{A}(t)$ be the vector potential of the IR field that satisfies $\mathbf{A}(\infty) = 0$. The IR field is strong but it should not directly ionize the atom; therefore the photoelectrons come from single-photon ionization by the IAP (XUV) pulse only. Classically, it is assumed that the photoelectron is released at the time t_r. Once released to the continuum, it is accelerated by the IR field and gains or loses its kinetic energy until the IR field is over. If the electron is released with a kinetic momentum $\mathbf{p}_0$, then the detected momentum after the IR field turns off is given by $\mathbf{p}_0 - \mathbf{A}(t_r)$. The t_r can be changed by sweeping the relative time delay between the XUV and the IR to generate a set of delay-dependent photoelectron spectra, i.e., a streaking trace.

To relate the phase information to the streaking trace, one needs to go further and consider a quantum mechanical model. In this model, the first assumption is that the streaking spectra can be calculated using the strong-field approximation (SFA) [42]

$$S(p, t_d) = \left| \int_{-\infty}^{\infty} E_{XUV}(t - t_d) d(p + A(t)) e^{-i\varphi(p,t)} e^{i\left(\frac{p^2}{2}+I_p\right)t} dt \right|^2. \tag{7.46}$$

Here, the polarization of the XUV, IR, and photoelectrons are all taken along the $+z$ direction so all quantities become scalars. p is the asymptotic momentum of the photoelectron, and the energy of the electron is $E = p^2/2$. t_d is the relative temporal shift between the two fields. A positive t_d means the XUV comes after the peak of the IR field. The function $\varphi(p,t)$ is given by

$$\varphi(p,t) = \int_t^{\infty} \left[pA(t') + \frac{1}{2}A^2(t') \right] dt'. \tag{7.47}$$

Equation 7.46 includes the single-photon transition dipole, $d(p) = \langle p\hat{z}|z|i\rangle$ where $|i\rangle$ is the initial bound state with the ionization potential I_p. In the standard SFA, the continuum state $|p\hat{z}\rangle$ is approximated by a plane wave e^{ipz}. A correct choice is to use the scattering wavefunction that is a continuum eigenstate of the field-free Hamiltonian with asymptotic momentum $p\hat{z}$. When a single-active-electron model potential is used, this single-photon transition dipole $d(E)$ is given in Equation 7.26. Specifically, if the photoionization is from the s states ($l_i = 0$), the transition dipole only involves the continuum p-wave

$$d(E) = \sqrt{\frac{1}{4\pi}} e^{i\eta_1(E)} \langle u_{k1}|r|u_i\rangle. \tag{7.48}$$

However, if the photoionization is from the p states ($l_i = 1$), the transition dipole involves both the continuum s-wave and d-wave

$$d(E) = \sqrt{\frac{1}{12\pi}} \left\{ e^{i\eta_0(E)} \langle u_{k0}|r|u_i\rangle + 2e^{i\eta_2(E)} \langle u_{k2}|r|u_i\rangle \right\}. \tag{7.49}$$

Here, η_L is the phase shift corresponding to the partial wave with angular quantum number L.

Consider Equation 7.46. If the exponential term $e^{-i\varphi(p,t)}$ oscillates as a function of t with a period much shorter than the optical cycle of the laser field, then according to the analysis in Yakovlev et al. [43], the streaking trace can be approximated by

$$S(E,t_d) \approx \left| \int_{-\infty}^{\infty} \chi(t-t_d) e^{-i\varphi(p,t)} e^{iEt} dt \right|^2. \tag{7.50}$$

The function $\chi(t)$ is called the *temporal electron wave packet*. It describes the XUV photoionization process and is related to the energy-domain wave packet $\tilde{\chi}(E)$ by an inverse Fourier transform:

$$\chi(t) = \frac{1}{2\pi} \int_0^{\infty} \tilde{\chi}(E) e^{-iEt} dE. \tag{7.51}$$

The first-order perturbation theory for XUV photoionization predicts that

$$\tilde{\chi}(E) = \tilde{E}_{XUV}(\Omega) d(E), \tag{7.52}$$

in which $\Omega = E + I_p$ is the XUV photon energy. Furthermore, if one assumes that $\varphi(p,t)$ depends weakly on p such that the momentum p in $\varphi(p,t)$ can be replaced by p_0 with p_0 being the center of the momentum of photoelectrons, then Equation 7.50 takes the form

$$S(E,t_d) \approx \left| \int_{-\infty}^{\infty} \chi(t-t_d)G(t)e^{iEt}dt \right|^2, \tag{7.53}$$

with the "gate" function $G(t) = e^{-i\varphi(p_0,t)}$ depending on t only. After taking such a "central momentum approximation," Equation 7.53 fits the mathematical form of the equation for FROG. Therefore iterative algorithms developed for FROG, such as the principal component generalized projection algorithm (PCGPA) [44] or the least squares generalized projection algorithm (LSGPA) [45], can be used to simultaneously extract $\chi(t)$ and $G(t)$ from $S(E,\tau)$. From $G(t)$, the vector potential $A(t)$ of the IR field can be calculated from Equation 7.47 provided that p_0 is given. Then the IR field can be retrieved. If the amplitude and phase of the atomic dipole $d(E)$ is considered known from the theory, then the XUV pulse $E_{XUV}(t)$ can be deduced from the extracted wave packet $\tilde{\chi}(E)$ according to Equation 7.52. These methods are usually called FROG–CRAB [46].

The Applicability and Limitations of the FROG–CRAB Method

FROG–CRAB is the standard method used extensively to retrieve IAPs. Unlike the RABITT method that was used to retrieve APTs "exactly," so long as the second-order perturbation theory is applicable, the FROG–CRAB method relies on (1) the SFA, (2) the central momentum approximation, and (3) the convergence of the iterative solution. Thus the reliability of the FROG–CRAB method has to be examined carefully.

Like FROG, the FROG–CRAB method cannot determine the absolute time t. In other words, the output of the FROG–CRAB could be $\chi(t-t_0)$ and $G(t-t_0)$ where t_0 is arbitrary. Equivalently, such uncertainty would add a linear term Ωt_0 to the spectral phase $\arg\tilde{\chi}(\Omega)$.

Calibration of the FROG–CRAB Method for Retrieving XUV Phase

Figure 7.12 gives two examples of characterizing IAPs from the Ne spectrograms that come from SFA simulation using Equation 7.46. The two IAPs share the same $U(\Omega)$ with $\Omega_0 = 60$ eV and $\Delta\Omega = 23$ eV, but have different $\Phi(\Omega)$. The first IAP is TL and has a duration of 80 as while the second IAP has an attochirp such that its duration is increased to 130 as. The IR field in these simulations is 800 nm in wavelength, 8.8 fs in FWHM duration, and 10^{13} W/cm^2 in peak intensity. Figures 7.12(a,b) clearly demonstrate that the streaking trace is sensitive to the XUV spectral phase. By using the LSGPA algorithm, the spectral amplitude and phase as well as the temporal profile of the two input IAPs can be retrieved successfully. The comparison between the input and the retrieved pulses is given in Figures 7.12(c–e). In the IAP characterization, the output $\tilde{\chi}(E)$ from the FROG–CRAB has been divided by the known atomic dipole $d(E)$ of Ne. According to Equation 7.52, $\tilde{E}_{XUV}(\Omega) = \tilde{\chi}(E)/d(E)$. Additionally, to get rid of the uncertainty of the absolute time in the FROG output, the peak of all IAPs has been moved to $t = 0$, and their $\Phi(\Omega)$ have been readjusted by adding a linear term consistently.

In Figure 7.12, the electron spectrogram was obtained using the SFA theory, and thus the retrieved results show that the FROG–CRAB method works accurately in spite of

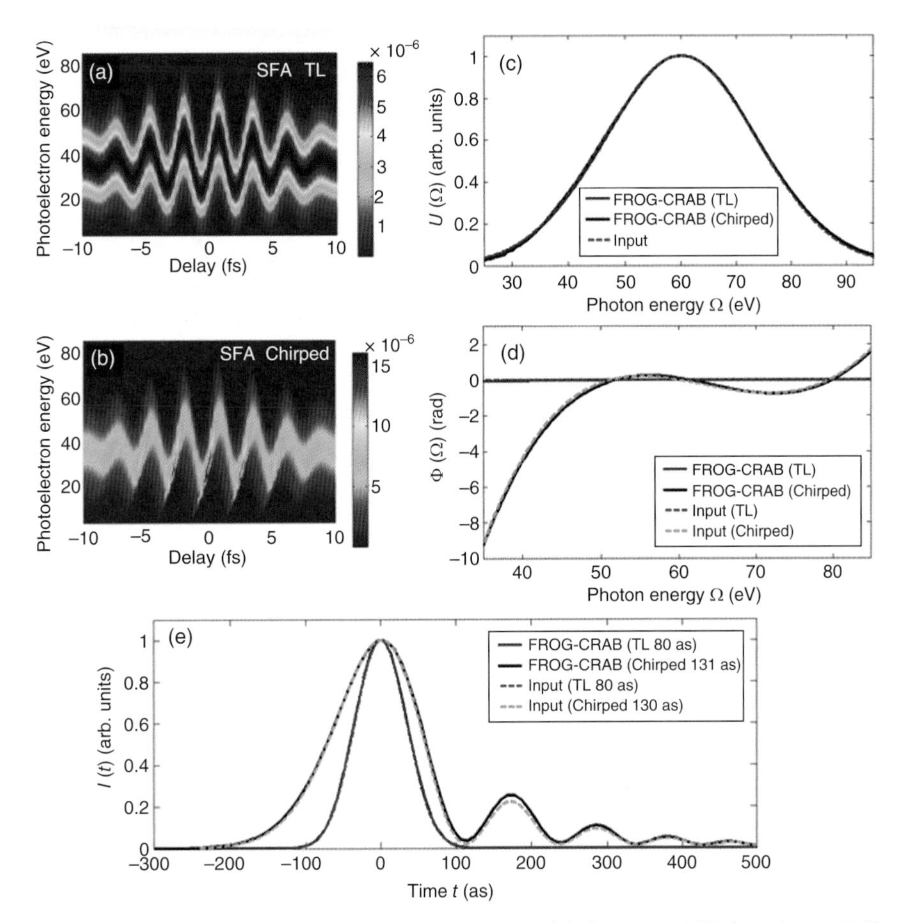

Figure 7.12 Characterizing IAPs from Ne spectrograms generated using the SFA model. The two input IAPs have $\Omega_0 = 60$ eV, $\Delta\Omega = 23$ eV, 80 as duration for the TL pulse and 130 as duration for the chirped pulse. The IR field is 800 nm in wavelength, 8.8 fs in FWHM duration and 10^{13} W/cm^2 in peak intensity. (a) SFA trace for the TL pulse. (b) SFA trace for the chirped pulse. Comparison of input XUV pulses with the retrieved ones: (c) Spectral amplitude, (d) spectral phase, (e) temporal profile of the input IAPs (dashed lines), and FROG–CRAB-retrieved IAPs (solid lines). (Reprinted from Hui Wei, "Characterization and application of isolated attosecond pulses," PhD diss., Kansas State University, 2017.)

the central momentum approximation and the iterative method. The SFA is expected to work better for high-energy photoelectrons, which was the case for Figure 7.12. At lower photoelectron energies (or photon energies) the SFA model is known to be inaccurate. As a test, the FROG–CRAB is used to retrieve IAPs from streaking traces obtained by solving the single-active-electron time-dependent Schrödinger equation (TDSE).

In Figure 7.13, the IAPs have $\Omega_0 = 40$ eV and $\Delta\Omega = 11.5$ eV, and the target is Ne. In Figure 7.14, the IAPs have even lower photon energy $\Omega_0 = 22$ eV, $\Delta\Omega = 5.9$ eV, and the target is Kr. Figures 7.13(a) and (b) compare the traces obtained from TDSE and SFA. Clearly, the traces are visibly different and show the effects of electron–ion interaction for photoelectron spectra below 30 eV. However, Figure 7.13(c) shows that $U(\Omega)$ can still be accurately retrieved by LSGPA. However, Figure 7.13(d) shows that the retrieved spectral

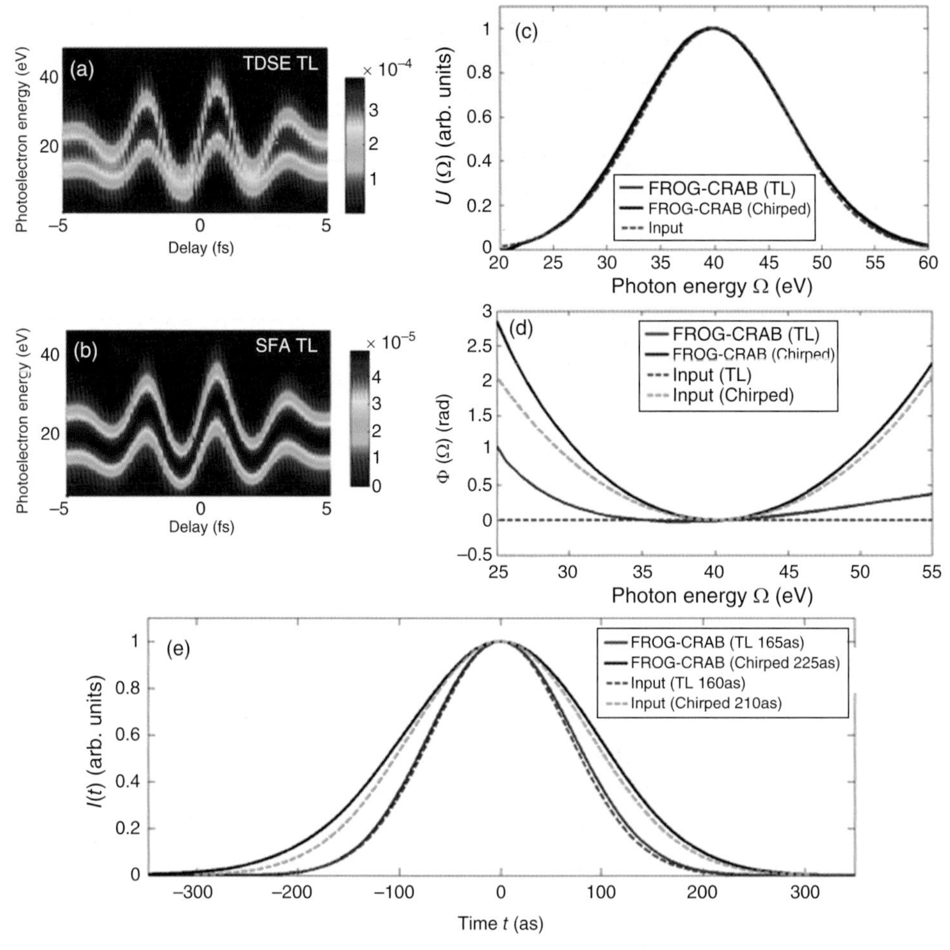

Figure 7.13 Characterizing IAPs from the Ne spectrogram obtained by solving the TDSE. The two input IAPs have $\Omega_0 = 40$ eV, $\Delta\Omega = 11.5$ eV, 160 as duration for the TL pulse and 210 as duration for the chirped pulse. The IR field is 800 nm in wavelength, 4.4 fs in FWHM duration and 10^{13} W/cm^2 in peak intensity. (a) TDSE trace for the TL pulse; (b) SFA trace for the TL pulse; (c) spectral amplitude; (d) spectral phase; (e) temporal profile of the input IAPs (dashed lines) and FROG–CRAB retrieved IAPs (solid lines). The results show that the XUV amplitude and phase can be accurately extracted using the FROG–CRAB method despite the fact that the trace calculated from the SFA does not reproduce the trace calculated using TDSE. (Reprinted from Hui Wei, "Characterization and application of isolated attosecond pulses," PhD diss., Kansas State University, 2017.)

phase $\Phi(\Omega)$ has a larger chirp than the input one for both the TL and the chirped IAPs. Because of the overestimation of the attochirp, the retrieved pulse duration of 165 as is larger than the 160 as of the input TL pulse, and of 225 as compared to the input 210 as for the chirped pulse. Similarly, for the IAP with $\Omega_0 = 22$ eV, the FROG–CRAB retrieved pulses have longer durations than the input ones. Figure 7.14(b) shows the comparison between the input and retrieved spectral phase. The results for the IAP temporal profile are given in Figure 7.14(c). In conclusion, due to the inaccuracy of the SFA model in the

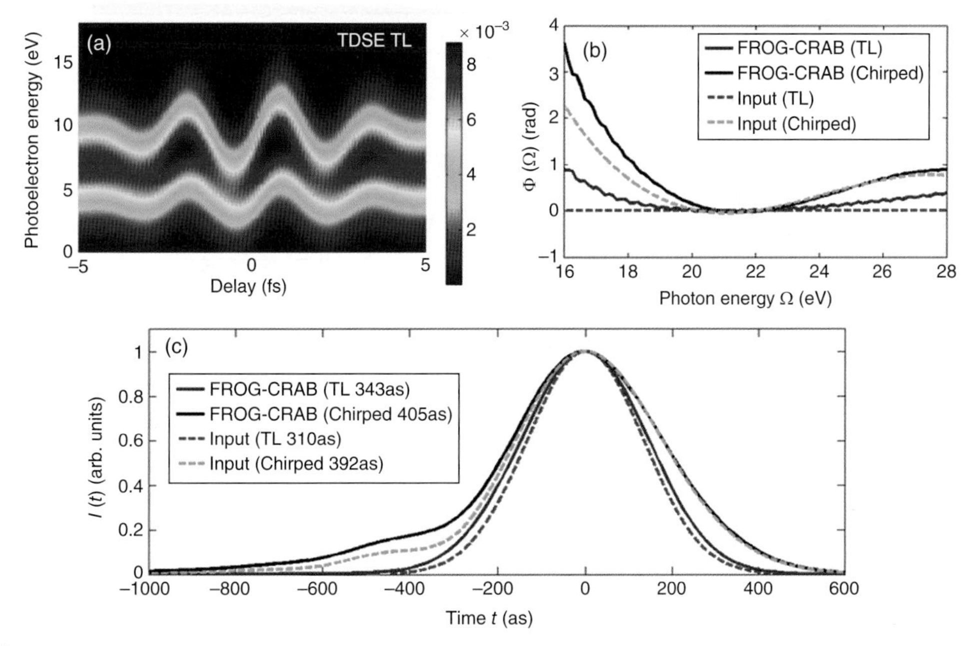

Figure 7.14 Characterizing IAPs from the Kr spectrogram obtained by solving the TDSE. The two input IAPs have $\Omega_0 = 22$ eV, $\Delta\Omega = 5.9$ eV, 310 as duration for the TL pulse and 392 as duration for the chirped pulse. The IR field is 800 nm in wavelength, 4.4 fs in FWHM duration, and 10^{12} W/cm^2 in peak intensity. (a) TDSE trace for the TL pulse. (b) Spectral phase. (c) Temporal profile of the input IAPs (dashed lines) and the FROG–CRAB retrieved IAPs (solid lines). The present result shows that errors become larger when FROG–CRAB is applied to low-energy electrons. About 10% error in the pulse duration was found in the present example. (Reprinted from Hui Wei, "Characterization and application of isolated attosecond pulses," PhD diss., Kansas State University, 2017.)

low-energy region, up to 10% errors in pulse duration will be introduced if one uses FROG–CRAB to characterize IAPs with photon energies below 40 eV. Since most of the errors occur at the wings of the pulse, which have weaker intensity, the error may be considered to be not too severe. On the other hand, the XUV interacts linearly with the medium, and thus photoelectrons generated from the wings of the pulse can still contribute to the photoelectron spectra.

Retrieval of IR Pulse from FROG–CRAB

As previously mentioned, the FROG–CRAB can also extract the IR field. For the TL trace shown in Figure 7.12(a), the comparison between the input $E_{IR}(t)$ and the one obtained from the FROG–CRAB output is shown in Figure 7.15. Although the FROG–CRAB result appears to be in good agreement with the input IR in Figure 7.15(a), the agreement on the attosecond time scale shows its deficiency. According to the zoom-in plot in Figure 7.15(b), the IR peak position was off by more than 100 as. Here, IR-peak positions can be compared because $t = 0$ has been determined by the IAP. To improve the accuracy of the retrieved IR field, a second-fitting approach based on the known $d(E)$ from theory and the known

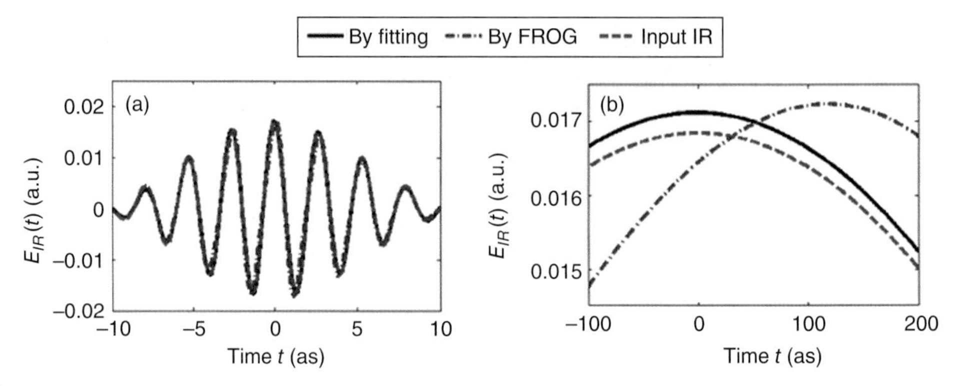

Figure 7.15 Retrieved IR field from the Ne spectrogram Figure 7.12(a) with a TL IAP. (a) (solid black line): The retrieved IR through GA fitting. (dot-dashed blue line): The output IR field from the FROG–CRAB by setting $p_0 = 1.68$. (dashed red line): The input IR field. On this timescale, the IR field appears to be nicely retrieved. On an attoseconds timescale, the error from FROG–CRAB becomes quite apparent, as seen in (b). (b) zoom-in plot of (a) near $t = 0$. (Reprinted from Hui Wei, Toru Morishita, and C. D. Lin, *Phys. Rev. A*, **93**, 053412 (2016) [47]. Copyrighted by the American Physical Society.)

$E_{XUV}(t)$ extracted from the FROG–CRAB was applied. In the second-fitting method, the IR field was modeled by

$$E_{IR}(t) = f(t)\cos[\omega_L(t - \Delta)]. \tag{7.54}$$

The envelope $f(t)$ was constructed by a set of samples (t_i, f_i) through cubic-spline interpolation. The horizontal coordinates t_i were fixed while the vertical coordinates f_i as well as Δ were set as fitting parameters. Then, Equation 7.46 was used to generate trial traces in the genetic algorithm (GA), which was applied to find the optimal parameters by minimizing the error between the input and the trial traces. In Figure 7.15(b), the peak of the IR field extracted via the second fitting is off by only about 2 as. Note that the second fitting is based on the SFA equation without the central momentum approximation that could have limited the performance of the FROG–CRAB.

7.5.3 Characterization of Broadband Attosecond Pulses

Introduction

While FROG–CRAB is the most widely used method for characterizing IAPs, as discussed in Section 7.5.2, it has two limitations. First, the central momentum approximation limits it validity for application to broadband IAPs. The other limitation is the iterative solution used in the FROG algorithm, which may not converge well, especially when mid-IR pulses are used for the streaking field. In this subsection, a newly developed phase-retrieval method for broadband light pulses will be discussed. When applying this method to narrow-band IAPs, it improves the accuracy of the FROG–CRAB. Another method, PROOF, which was originally proposed for characterizing broadband attosecond pulses [48], will also be briefly discussed. The PROOF method is similar to RABITT; it is based on the second-order perturbation theory. Thus the IR intensity has to be limited to well below 10^{12} W/cm^2. With such low IR intensity, oscillation in the spectrogram (streaking) becomes nearly invisible.

The PROBP Method

The starting equation for the streaking spectrogram for the PROBP [49] method is the same SFA equation as in FROG–CRAB,

$$S(p, t_d) = \left| \int_{-\infty}^{\infty} E_{XUV}(t - t_d) d(p + A(t)) e^{-i\Phi(p,t)} e^{i\left(\frac{p^2}{2} + I_P\right)t} dt \right|^2, \tag{7.55}$$

where the phase $\Phi(p, t)$ is given by

$$\Phi(p, t) = \int_t^{\infty} \left[pA(t') + \frac{A^2(t')}{2} \right] dt'. \tag{7.56}$$

In FROG–CRAB, the p in $\Phi(p, t)$ is replaced by p_0. Such an approximation is not made in PROBP, where the goal is to solve Equation 7.55 directly without additional approximations. To speed up the convergency of the iterative procedure, the number of unknown parameters should be limited. In general, these unknowns include the amplitude and phase of the XUV, IR, and atomic-dipole transition-matrix elements. For the characterization of the broadband XUV pulses, it is assumed that the amplitude and phase of the dipole-transition matrix elements can be accurately calculated from atomic-structure calculations. The amplitude of the XUV pulse in the energy domain can be obtained from the ionization of atoms by the XUV alone. Thus the remaining unknowns are the spectra phase of the XUV and the amplitude and phase of the IR. While the XUV is used to represent an IAP, the method also covers IAPs in the soft X-ray region.

In PROBP, the unknown functions are all expanded in the so-called *B-spline functions*. Depending on the nature of the XUV and the IR pulses, these unknown functions are parameterized in the energy domain or in the time domain. For example, the XUV pulse in the energy domain is expressed as

$$E_{XUV}(\Omega) = U(\Omega) e^{-i\phi_{XUV}(\Omega)}. \tag{7.57}$$

A single-color, multi-cycle IR field in the time domain is expressed as

$$E_{IR}(t) = f(t) \cos\left(\omega_L t + \varphi_{IR}(t)\right), \tag{7.58}$$

while a broadband IR pulse consisting of multiple colors is better represented in the frequency domain

$$E_{IR}(\omega) = A(\omega) e^{-i\phi_{IR}(\omega)}. \tag{7.59}$$

To demonstrate how the PROBP method works, it is assumed that the target is a one-electron model Ne atom where the transition dipole $d(p)$ can be calculated accurately. In the simulation, the IR intensity is taken at 10^{13} W/cm^2 and the XUV at 10^{12} W/cm^2. Each unknown function is expanded in terms of B-spline basis

$$f(x) = \sum_{i=1}^{n} g_i B_i^k(x), \tag{7.60}$$

where g_i are the expansion coefficients and i is the index of the B-spline function. The k_{th} order B-spline functions $B_i^k(x)$ are defined through

$$B_i^1(x) = \begin{cases} 1 & x_i \le x \le x_{i+1} \\ 0 & \text{otherwise} \end{cases} \tag{7.61}$$

$$B_i^k(x) = \frac{x - x_i}{x_{i+k-1} - x_i} B_i^{k-1}(x) + \frac{x_{i+k} - x}{x_{i+k} - x_{i+1}} B_{i+1}^{k-1}(x). \tag{7.62}$$

Here, $\{x_i\}$ are the knot points. If there are n B-spline basis functions of order k, the total number of knot points is $n + k$.

As an example, consider three XUV pulses, each with a central photon energy of 40 eV and spectral width of 11 eV. The IR pulse has a central wavelength of 800 nm and duration of 4.4 fs. In this example, each spectrogram is calculated by solving the TDSE instead of using the SFA Equation 7.55. The first XUV pulse is TL, with a duration of 160 as. The second one is chirped with a pulse duration of 200 as. The third one is also chirped with a pulse duration of 250 as. Figure 7.16 documents the results of the simulation. Using the FROG–CRAB and PROBP methods, the XUV phase for each pulse is retrieved and compared to the input values. Both methods accurately retrieved the XUV phases. Most of the discrepancy with respect to the input phase is likely due to the difference in the

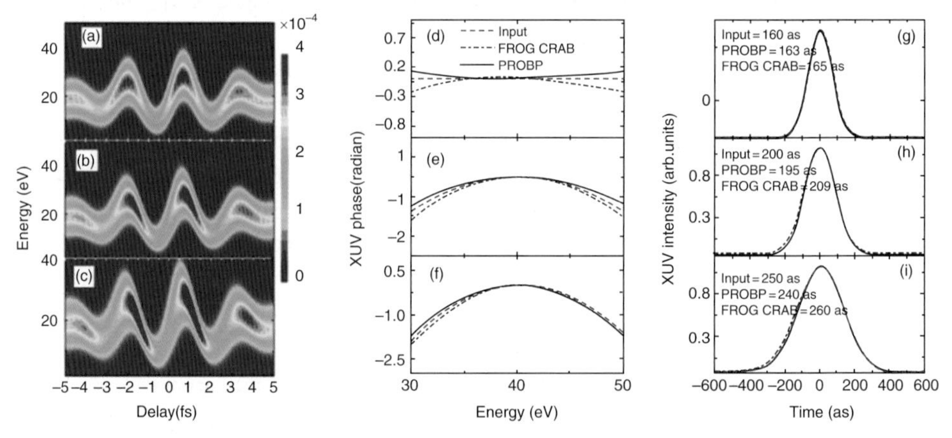

Figure 7.16 (Color online) Characterization of three input XUV pulses centered at 40 eV with a bandwidth of 11 eV using the FROG–CRAB and PROBP methods. The FWHM durations of the pulses are 160 as, 200 as, and 250 as, respectively. In the spectral domain, the pulses have the same amplitude but different phases. The IR is known to have a central wavelength of 800 nm but the amplitude and phase are not known. The spectrograms shown in the first column are generated by solving the TDSE for each input pulse. The spectral phases retrieved using FROG–CRAB and those from PROBP are compared to the input ones, as shown in the middle column. Comparison of the reconstructed XUV intensity in the time domain is shown in the right column. No significant differences in the retrieved results from FROG–CRAB compared to PROBP. (Reprinted from Xi Zhao, Hui Wei, Yan Wu, and C. D. Lin, *Phys. Rev. A*, **95**, 043407 (2017) [49]. Copyrighted by the American Physical Society.)

spectrograms generated by the TDSE and SFA. With the known XUV amplitude and the retrieved phase, one can see that the XUV intensity profile in the time domain for each pulse is accurately retrieved in the right column of Figure 7.16. In this example, Equations 7.57 and 7.58 are used where $\phi_{XUV}(\Omega)$, $f(t)$, and $\varphi_{IR}(t)$ are the unknown functions. These functions are characterized by three sets of unknown B-spline parameters $\{a_i, b_i, c_i\}$. From these parameters, the guessed XUV and IR fields are constructed. For each discrete set of points $\{E_k, t_{d,l}\}$, the corresponding spectrogram is obtained by using Equation 7.55. The error function is defined as

$$E[a_i, b_i, c_i] = \sum_{k,l} [S_0(E_k, t_{d,l}) - S_1(E_k, t_{d,l})]^2, \tag{7.63}$$

where S_0 and S_1 are the input and reconstructed spectrograms, respectively. The GA is used to run a large number of generations (typically 50,000–100,000 generations) until convergence is achieved. The optimal parameters are then used to reconstruct the XUV and IR pulses.

In the next example, a chirped XUV pulse with a duration of 52 as, central photon energy of 80 eV, and bandwidth of 90 eV is to be retrieved. Starting with the spectrogram calculated with the SFA equation, with the same 800 nm IR laser as in the previous example, the retrieved XUV phases and temporal-intensity profiles are shown in Figure 7.17(a) and (b), respectively, using both the FROG–CRAB and PROBP methods. Clearly, FROG–CRAB failed completely, but the PROBP method faithfully retrieved the input broadband XUV phase as well as the XUV intensity in the time domain. In this example, the broad bandwidth of the XUV pulse makes it hard to justify the use of central momentum approximation in the FROG–CRAB method.

The PROBP method can be used to retrieve broadband single-attosecond pulses in the water-window region. In this example, three input attosecond pulses with central energy of 300 eV and bandwidth of 46 eV are considered. In the transform-limited case, this spectrum will support a pulse with a duration of 46 as. The three input XUV pulses differ by their spectral phases (see the left column of Figure 7.18). It is assumed that soft X-ray harmonics are generated by a four-cycle, 2,000 nm mid-IR laser, which is also used as the streaking field. In this example, the FROG–CRAB method was unable to reach converged results. On the other hand, as shown in Figure 7.18, the spectral phases of the soft X-ray pulses and their intensities in the time domain are all accurately retrieved using the PROBP method.

From these examples, it is clear that the PROBP method can not only retrieve narrow-band XUV pulses more accurately than the FROG–CRAB method, but it can also be used to retrieve single-attosecond pulses in the water-window region with mid-IR laser pulses as the streaking field.

The PROOF Method

The PROOF method is a generalization of the RABITT method for IAPs when the IR intensity is weak such that second-order perturbation theory can be applied to model the streaking trace. Consider photoelectrons measured along the polarization direction of the collinear XUV and the IR pulses (chosen to be $+z$ direction) and assume that the IR is

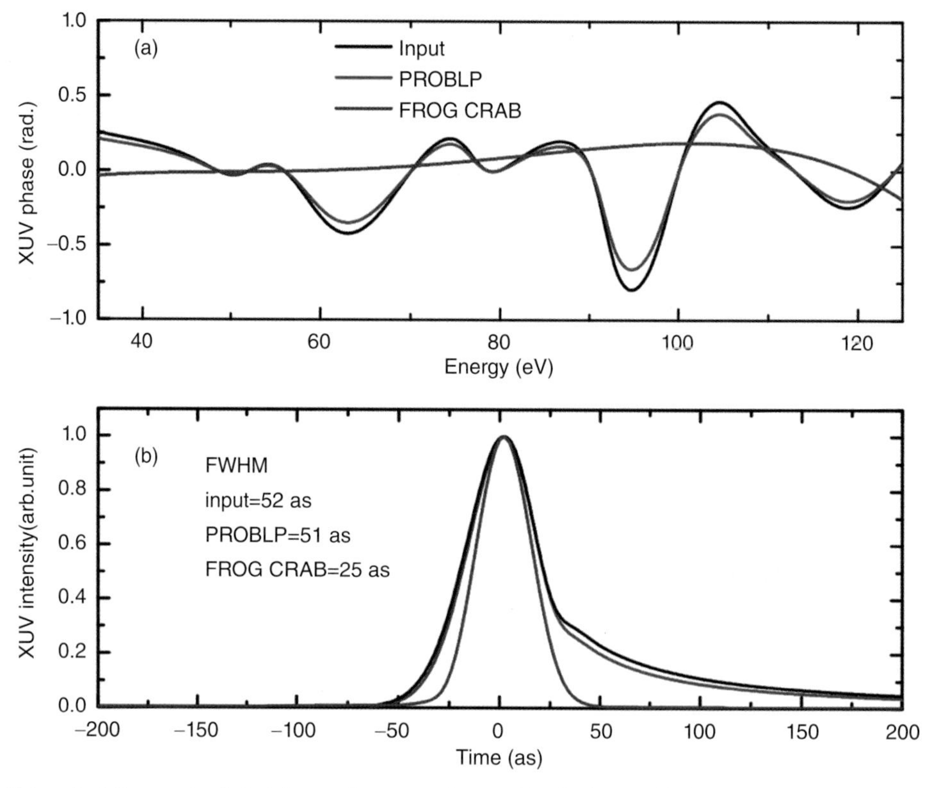

Figure 7.17 (Color online) Phase retrieval of soft X-ray pulses centered at 300 eV with a bandwidth of 46 eV using PROBP. Left column, spectral phases of three different soft X-ray attosecond pulses. The spectral amplitudes for the three pulses are identical. Right column, temporal intensity of the three pulses. The black lines are the input data and the red lines are the retrieved ones from the PROBP method. The FROG–CRAB method does not converge for the present example. The wavelength of the dressing mid-infrared field is 2,000 nm, the duration is four cycles. (Reprinted from Xi Zhao, Hui Wei, Yan Wu, and C. D. Lin, *Phys. Rev. A*, **95**, 043407 (2017) [49]. Copyrighted by the American Physical Society.)

monochromatic. Within the second-order perturbation theory the streaking spectrogram can be calculated from

$$S(E,t_d) = \left|\tilde{E}_{XUV}(\Omega)d^{(1)}(E) + \tilde{E}_{XUV}(\Omega-\omega)\frac{E_{IR}}{2}e^{-i\omega t_d}d^{(+)}(E) + \tilde{E}_{XUV}(\Omega+\omega)\frac{E_{IR}}{2}e^{i\omega t_d}d^{(-)}(E)\right|^2. \tag{7.64}$$

Here, any terms quadratic or higher in the IR field strength are omitted. $E = k^2/2$ is the photoelectron energy, $\Omega = E + I_p$ is the XUV photon energy, ω is the IR frequency, E_{IR} is the IR field strength, $d^{(1)}(E)$ is the single-photon transition dipole given in Equation 7.26, and $d^{(\pm)}(E)$ is the XUV+IR two-photon transition-dipole matrix element given by Equation 7.16, with $(+)$ corresponding to the path that absorbs one IR photon and $(-)$ to the path that emits one IR photon. We can expand Equation 7.64 by taking just the two lowest orders in E_{IR}:

$$S(E,t_d) \approx S_{XUV}(E) + S_{FSI}(E,t_d). \tag{7.65}$$

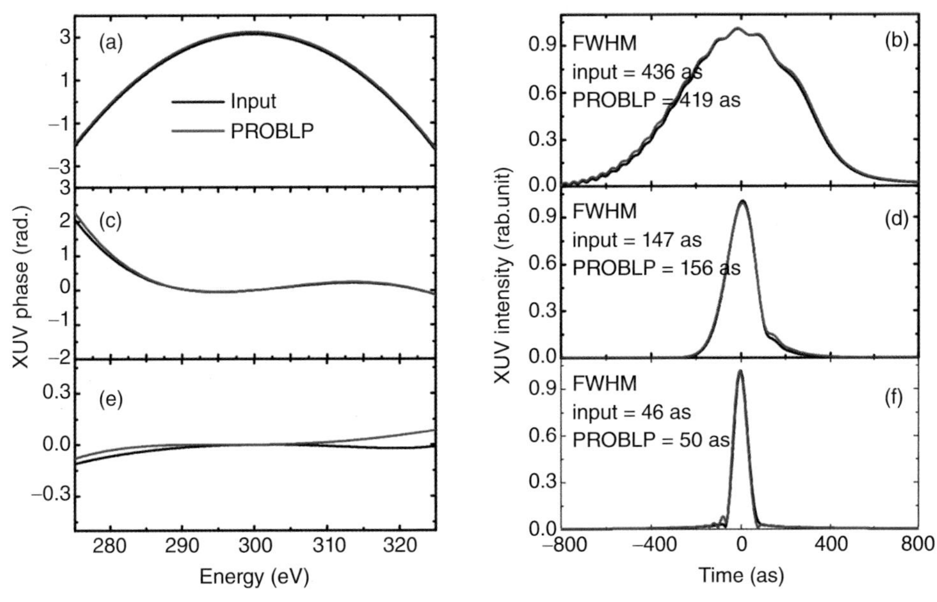

Figure 7.18 (Color online) Comparison between FROG–CRAB and PROBP for characterizing an XUV pulse centered at 80 eV with a broad bandwidth of 90 eV. (a) The XUV spectral phase. (b) The XUV intensity profile in the time domain. The black lines are the input data, and the red lines are retrieved from the PROBP method, while the blue lines are retrieved from the FROG–CRAB. (Reprinted from Xi Zhao, Hui Wei, Yan Wu, and C. D. Lin, *Phys. Rev. A*, **95**, 043407 (2017) [49]. Copyrighted by the American Physical Society.)

Here, the first term is from the XUV only. The second term is first-second order interference (FSI) which is given by

$$S_{FSI}(E, t_d) = E_{IR}\mathrm{Re}\left\{\tilde{E}_{XUV}(\Omega)\tilde{E}^*_{XUV}(\Omega-\omega)e^{i\omega t_d}d^{(1)}d^{(+)*} + \tilde{E}_{XUV}(\Omega)\tilde{E}^*_{XUV}(\Omega+\omega)e^{-i\omega t_d}d^{(1)}d^{(-)*}\right\}. \tag{7.66}$$

For a given electron energy E, the FSI term oscillates as a function of t_d at the IR frequency ω. Equation 7.66 can be rewritten as

$$S_{FSI}(E, t_d) = A(E)\cos[\omega t_d + \Psi(E)], \tag{7.67}$$

where the spectral phase of the XUV pulse is embedded in $A(E)$ and $\Psi(E)$.

From the experimental spectrogram, the FSI term can be obtained by applying a filter on the measured data and then selecting its ω-component. Thus this method is called PROOF (phase retrieval by omega oscillation filtering). The original PROOF method employed a number of additional approximations. These approximations have been removed in a latter publication [50] and the method was called swPROOF. In numerical simulations, it has been found that accurate IAPs can be retrieved, where the details of the retrieving procedure were discussed in Wei et al. [50]. So far PROOF has not been widely used in IAP phase retrieval. As discussed in Section 7.4, currently, the pulse energy of an IAP is about two orders weaker than a typical APT pulse. Since the PROOF method relies on second-order perturbation theory, which requires a weak IR field with intensity less than 10^{12} W/cm^2, the small streaking signal to noise ratio of the photoelectrons may introduce substantial errors on the retrieved phase of the IAP.

7.6 Probing Time Delay in Atomic Photoionization Using an IAP

7.6.1 Wigner Time Delay in a Short-Range Potential

Time is a classical parameter; it is neither a dynamic variable nor an operator in quantum mechanics. For a wave packet, time normally appears in the phase factor e^{-iEt}. This implies that time can be related to the phase, or more accurately, to the derivative of phase with respect to energy. For short-range potential scattering, Eisenbud, Wigner, and Smith introduced a time delay (we call it the Wigner delay) [51] for a given partial wave with angular momentum l:

$$\tau^W(E) = 2\frac{d}{dE}\delta_l(E), \tag{7.68}$$

in which δ_l is the phase shift of the partial wave l due to the short-range potential. The interpretation of this Wigner delay is classical; it can be viewed as the time delay of this particle after moving through the potential compared to the classical free motion when the potential is absent. This concept is quite abstract. In particular, this time delay is defined for a time-independent system. Strictly speaking, one cannot measure the Wigner delay directly.

Now, consider photoionization in which a photoelectron is released and moves in the potential of the atomic core. Usually the core is charged so the potential has an asymptotic Coulomb component. For simplicity, first assume a neutral core, i.e., photo-detachment from negative ions, so that one can only consider short-range potential. This process will still be called *ionization*. This photoionization is equivalent to a "half-scattering" process. In the entrance channel the wavefunction is a bound state instead of a continuum wave. Therefore the Wigner delay for photoionization becomes $\tau^W(E) = \frac{d}{dE}\delta_l(E)$. To see this more clearly, look at the l-component of the continuum photoelectron. The asymptotic form of the outgoing wave packet is

$$\Psi(\mathbf{r},t) \propto \int_0^\infty A(E)Y_{lm}(\hat{r})\frac{e^{i[kr+\delta_l(E)-Et]}}{r}dE. \tag{7.69}$$

Assuming that a short XUV pulse is applied to ionize this target at $t = 0$, according to first-order perturbation theory, $A(E) \propto d_{li}(E)\tilde{E}_{XUV}(\Omega)$, where $\Omega = E + I_p$ is the XUV photon energy and d_{li} is the transition-dipole matrix element between the initial bound state and the final l partial continuum wave. Here, we choose the XUV, a TL pulse, and the dipole d_{li}, as a real quantity so that $A(E)$ is also a real quantity. The relation between classical and quantum descriptions can be established by introducing the "stationary phase condition." The major contribution to the integral Equation 7.69 comes from the "trajectory" that satisfies

$$\frac{d}{dE}[kr + \delta_l(E) - Et] = 0. \tag{7.70}$$

Because $k = \sqrt{2E} = v$, the Equation 7.70 leads to

$$r = v\left[t - \frac{d}{dE}\delta_l(E)\right] = v(t - \tau^W). \tag{7.71}$$

Therefore, from the classical point of view, the photoionization is delayed by an amount of τ^W after the pump pulse. When the photoelectron emission is measured in a particular direction $\hat{k}$ relative to the light polarization (set to $\hat{z}$ direction), the definition of Wigner delay can be generalized to

$$\tau^W(E, \hat{k}) = \frac{d}{dE}\arg d(E, \hat{k}), \tag{7.72}$$

where $d(E, \hat{k}) = \langle E\hat{k}|z|i\rangle$ is the single-photon, dipole-transition matrix element. In general, $d(E, \hat{k})$ contains the contributions from both the $l = l_i - 1$ and $l = l_i + 1$ continuum waves, where l_i is the angular momentum of the initial state. Only when the initial bound state has s-symmetry ($l_i = 0$) is there a single p-component in the continuum wave; then the Wigner delay returns to its original definition $\tau^W(E) = \frac{d}{dE}\delta_1(E)$. In the following discussion, only the forward photoelectrons are studied; thus $\hat{k} = \hat{z}$ and the notation $\hat{k}$ can be dropped.

7.6.2 Time Delay Extracted from IR-Dressed XUV Photo-Detachment for Targets with Short-Range Potentials

Consider the XUV-photo-detachment in the presence of a synchronized IR field in which both fields have the same polarization. The vector potential of the IR is $A(t)$. Classically, the photoelectron is released with kinetic momentum p_0 by the XUV-photodetachment. If the XUV pulse is shifted by a time delay t_d compared to the IR peak field, the detected momentum will be $p(t_d) = p_0 - A(t_d)$ if the XUV-photodetachment happens instantaneously. However, because of the Wigner delay discussed above, the photoelectron appears as if it were released at a time delay τ^W after the XUV pulse, therefore the detected momentum becomes $p(t_d) = p_0 - A(t_d + \tau^W)$. Figure 7.19 gives a TDSE simulation of attosecond streaking using a 200 as TL IAP. The one-electron potential takes a Yukawa-type

$$V(r) = -\frac{Z}{r}e^{-r/a}, \tag{7.73}$$

in which a is the screening length. The initial state is chosen to be an s state. In Figure 7.19(a) the first moment $\bar{p}(t_d)$ of the streaking trace (white dashed line) is compared to the IR vector potential $-A(t_d)$ (orange solid line). The inset clearly shows that the peak of $\bar{p}(t_d)$ is shifted by several attoseconds from the peak of $-A(t_d)$. This shift is treated as the streaking time delay at $p = p_0$. Figure 7.19(b) confirms that the delay extracted from the simulated streaking trace agrees with the Wigner delay calculated from the scattering phase shift by $\tau^W(E) = \frac{d}{dE}\delta_1(E)$, so long as the core potential is short ranged.

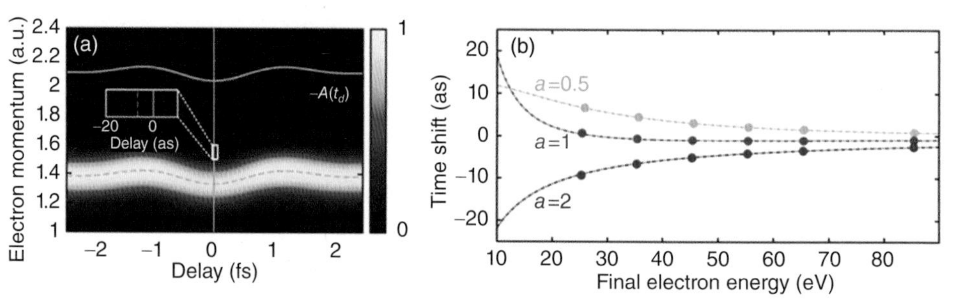

Figure 7.19 (a) TDSE-simulated streaking spectrogram for ionization of a model negative ion with Yukawa-like short-ranged potential. The IR laser field $\lambda = 800$ nm, a sine-squared envelope, total duration of 6 fs, and intensity of 4×10^{11} W/cm^2. The TL XUV pulse has a Gaussian envelope, a central energy of 80 eV, a FWHM duration of 200 as, and an intensity of 10^{13} W/cm^2. The first moment of the spectrogram $\bar{p}(t_d)$ is plotted in the white dashed line while $-A(t_d)$ is plotted in the orange solid line. (b) Comparison between the time delay extracted from the first moment of the quantum mechanical simulated streaking trace (dots) and the Wigner delay calculated from the scattering phase shift (lines) for three screening lengths. Note that the IR dressing field is very weak in this simulation. (Figure adopted with permission from R. Pazourek, S. Nagele, and J. Burgdörfer, *Rev. Mod. Phys.*, **8**, 765 (2015) [52]. Copyrighted by the American Physical Society.)

7.6.3 Time Delay Extracted from IR-Dressed XUV Photoionization for Atoms including Long-Range Coulomb Potential

Next, take the long-range Coulomb potential into account, like in the case of photoionization of neutral atoms dressed by IR fields. In a classical view, the long-range Coulomb interaction between the continuum electron and the ionic core will modify the electron trajectory and the final momentum. The asymptotic momentum can be written as $p(t_d) = p_0 - A(t_d + \tau^S)$ with τ^S being different from the Wigner delay τ^W. The difference is often referred to as the Coulomb-laser-coupling (CLC) delay

$$\tau^{CLC}(E) = \tau^S(E) - \tau^W(E). \tag{7.74}$$

Classically, one can derive an approximate formula for the CLC delay [52]

$$\tau^{CLC}(E) \approx \frac{Z_c}{(2E)^{3/2}}\left[2 - \ln\left(\frac{2\pi E}{\omega_{IR}}\right)\right]. \tag{7.75}$$

Here, ω_{IR} is the frequency of the IR field, and Z_c is the asymptotic charge seen by the photoelectron. From Equation 7.75, one can see that $\tau^{CLC}(E)$ is independent of target or IR intensity. In practice, the IR intensity should be weak enough to prevent field ionization and depletion of the system by the IR field but strong enough to cause easily detectable energy modulations of the emitted electron.

Take the simplest example of photoionization of an H atom from its ground state. The core potential does not have any short-range part, so the Wigner delay only depends on the Coulomb phase shift $\sigma_l(E) = \arg\Gamma[l + 1 - iZ_c/k]$, that is $\tau^W(E) = \frac{d}{dE}\sigma_l(E)$. Both TDSE and classical-trajectory Monte Carlo (CTMC) methods have been used to simulate the streaking trace from which the time delay τ^S is extracted. Figure 7.20(a) shows that τ^S

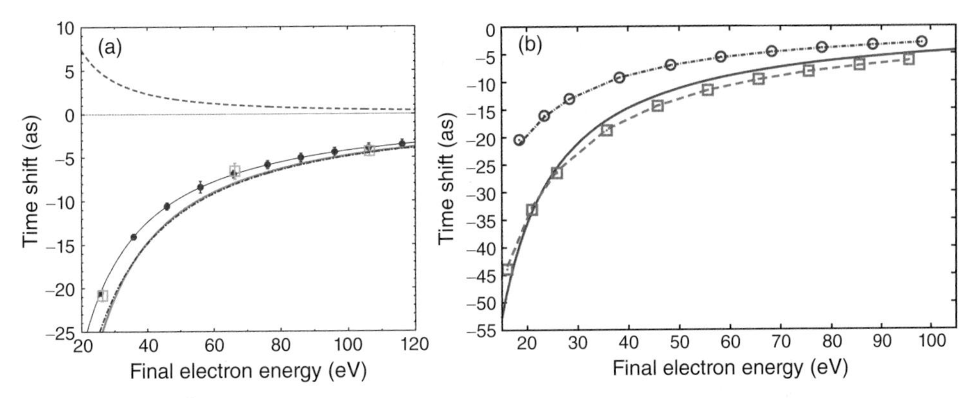

Figure 7.20 (a) Streaking delay τ^S as a function of final electron energy extracted from TDSE simulation (dots) and CTMC simulation (squares) for photoionization of atomic hydrogen. The red dashed line is the Wigner delay τ^W. The red solid line is the extracted CLC delay by Equation 7.74, while the dot-dashed purple line is the CLC delay calculated by Equation 7.75. From [52]. (b) τ^S extracted from single-active-electron TDSE simulations for 2s (green squares) and 2p_0 (blue circles) initial states of Ne. The results are obtained from spectrograms taken in the forward direction with respect to the laser polarization axis. The dashed green and blue lines show the results of adding the corresponding Wigner delay τ^W to the CLC delay (red solid line) using Equation 7.76. From [53]. The XUV pulses used in these simulations are TL with a FWHM duration of 200 as and with varied central energies. The IR field is 800 nm in wavelength, 3 fs in FWHM duration, and 10^{12} W/cm^2 in peak intensity. (Reprinted from S. Nagele, R. Pazourek, J. Feist, and J. Burgdörfer, *Phys. Rev. A*, **85**, 033401 (2012) [53]. Copyrighted by the American Physical Society.)

strongly differs from τ^W. The difference is the CLC delay (red solid line) that coincides with the analytical formula Equation 7.75.

For a general atom, the core potential is comprised of a short-range part and a long-range Coulomb part. The Wigner delay $\tau^W = \frac{d}{dE}\arg d(E)$ depends on both the Coulomb phase shift σ_l and the short-range phase shift δ_l. The single-photon transition dipole within the single-active-electron approximation for initial s and p states are given in Equations 7.48 and 7.49. The total delay can be calculated by

$$\tau^S(E) = \tau^W(E) + \tau^{CLC}(E). \quad (7.76)$$

On the other hand, one can use TDSE to simulate streaking traces for different targets or different ionization channels in one target, and then extract the corresponding streaking time delay τ^S, as shown in Figure 7.20(b), where photoionization from Ne 2s and 2p subshells are considered. The agreement between the simulated and calculated τ^S validates the separation of the streaking time delay by Equation 7.76.

Recall that, in the RABITT situation, there is also a separation of the measured XUV+IR two-photon time delay into a single-photon Wigner-like delay and an IR-induced CC delay (see Equation 7.27). Although the CC delay in the RABITT case is derived from second-order perturbation theory, it is in excellent agreement over a wide range of electron energy with the CLC delay introduced in the case of streaking [52]. One key to understanding this remarkable agreement is the intensity independence of τ^{CLC} (Equation 7.75) indicating that the CLC contribution to the time shift is present in both the single-photon and multiphoton regimes for the IR field. The IR intensities used in the numerical simulations of streaking spectrograms in Pazourek et al. [52] are within $10^{11}{\sim}10^{12}$ W/cm^2 (which is somewhat too low to get a good streaking spectra) in order to achieve good convergence. In this IR intensity range, the second-order perturbation theory is valid as well. However, whether the equality still holds at intensity of the order of 10^{13} W/cm^2 is not clear.

Equation 7.76 can be generalized to many-electron atoms with the inclusion of electron correlation. The CLC delay is the same as in the single-electron case, and one can still relate the Wigner delay to the energy derivative of the phase of the single-photon transition dipole $d(E)$. However, the calculation of $d(E)$ becomes more complicated, especially when interchannel couplings are included. Moreover, if the initial state before photoionization or the final core state after ionization has a permanent dipole moment, the dipole-laser-coupling mechanism will lead to an additional time delay τ^{dLC}, therefore Equation 7.76 should be modified to $\tau^S = \tau^W + \tau^{CLC} + \tau^{dLC}$ [54].

7.6.4 The Controversy of 20 as in the Time-Delay Difference between 2p and 2s Photoionization of Ne

In a pioneering experiment by Schultze et al. [55], the "photoionization time delay" from the $2p$ and $2s$ subshells of Ne was extracted by the attosecond streaking method. The IAP used in the experiment was centered at 106 eV with an FWHM bandwidth of 14 eV, which supports an FWHM duration below 200 as if it is TL. Near-single-cycle IR pulses (750 nm in wavelength, 3.3 fs in FWHM duration) were used, with peak intensities on the order of 10^{13} W/cm^2. The measured trace is shown in Figure 7.21(a), from which one can easily distinguish the photoelectrons that come from $2p$ and $2s$ subshells by electron energy. Unlike calculating the first moment of the trace as in TDSE simulations (see Section 7.6.8), the time delay was retrieved by the FROG–CRAB algorithm since in experiments the XUV and IR fields are not accurately known. Assuming the trace can be modeled by Equation 7.53, where the total wave packet is comprised of two individual wave packets corresponding to the $2p$ and $2s$ electrons, i.e., $\tilde{\chi}(E) = \tilde{\chi}_{2p}(E)+\tilde{\chi}_{2s}(E)$, by applying FROG–CRAB to this trace one can retrieve the two wave packets $\tilde{\chi}_{2p}(E)$ and $\tilde{\chi}_{2s}(E)$ since they are well separated in the energy domain. Then, according to Equation 7.52 one should compare the group delay of the two wave packets at the same photon energy Ω such that the XUV group delay is supposed to cancel. In this way, the Wigner delay difference between $2p$ and $2s$ electrons can be retrieved,

$$\tau_{2p}^{W}(\Omega) - \tau_{2s}^{W}(\Omega) = \frac{d}{d\Omega}\arg\tilde{\chi}_{2p}(\Omega) - \frac{d}{d\Omega}\arg\tilde{\chi}_{2s}(\Omega). \tag{7.77}$$

The right-hand side of Equation 7.77 is the *group delay difference* between $2p$ and $2s$ wave packets retrieved by FROG–CRAB. Figure 7.21(b) shows the trace reconstructed. Detailed comparisons between the measured and reconstructed traces at two particular delays are given in Figure 7.21(d), where good agreement between these two traces is demonstrated. The extracted intensity and group delay of the $2p$ and $2s$ electron wave packets are shown in Figure 7.21(c), from which an average group delay difference of 20 as was derived. This

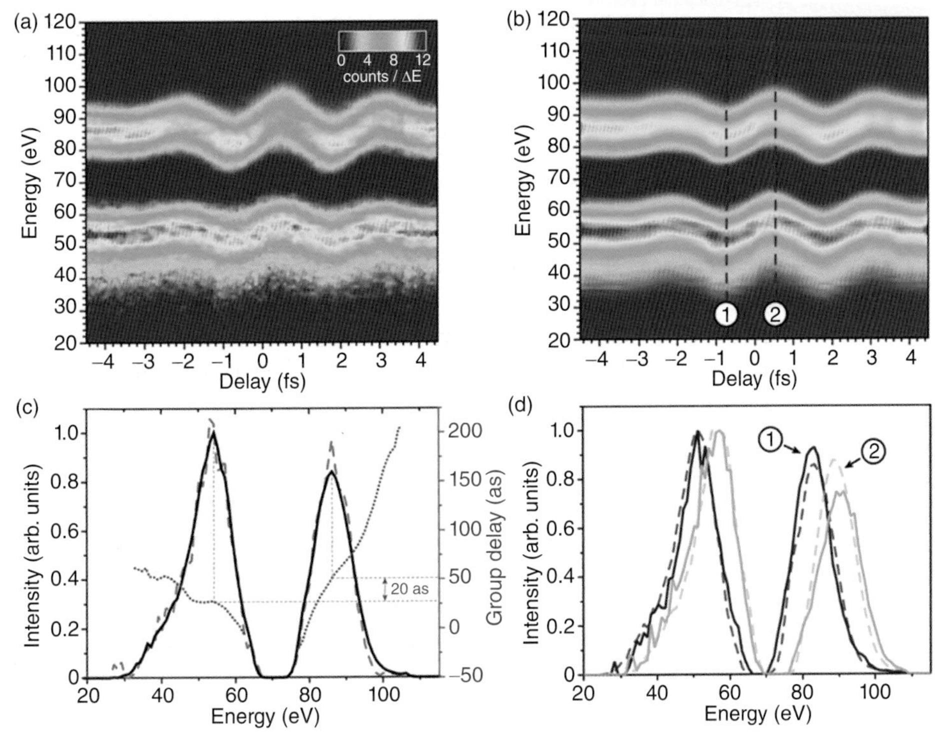

Figure 7.21 (a) Attosecond-streaking spectrogram composed of a series of photoelectron energy spectra recorded by releasing $2s$ and $2p$ electrons from Ne with an attosecond XUV pulse in the presence of a strong near-IR few-cycle laser field. (b) The spectrogram reconstructed by FROG–CRAB. (c) The retrieved intensity (black solid line) and group delay (red dotted line) of the $2s$ and $2p$ wave packets. The retrieved intensity spectra are in excellent agreement with the measured ones (gray dashed line). (d) Comparisons between the reconstructed and measured streaked spectra at two delays (straight dashed lines in b), which exhibit the largest positive and negative shifts of the electron-energy distribution. (From M. Schultze et al., *Science*, **328**, 1658, (2010) [55]. Reprinted with permission from AAAS.)

group-delay difference has been construed to mean that the emission of the $2p$ electron is retarded by 20 as with respect to the $2s$ emission.

Schultze's measurement has triggered many theoretical studies on the origin of the extracted (not "the observed" as stated often in the literature) 20 as time delay. The single-active-electron calculation predicts a Wigner delay difference $\tau_{2p}^{W} - \tau_{2s}^{W}$ of $4 \sim 5$ as at $\Omega = 105$ eV. Compared with the time delay extracted from the experiment, it has the same sign but smaller magnitude. Electron correlation is then considered. Up to now, all the many-electron calculations that approximately account for electron correlation effects agree reasonably well with each other, leading to Wigner-delay differences of less than 10 as. Note that the FROG–CRAB retrieval method is based on SFA Equation 7.46 where the CLC effect is not included. The limitation of the SFA model may lead to the discrepancy between the calculations and the extraction. However, for the high-photon-energy region used in this experiment, the effect of the Coulomb interaction should be small. Based on the discussions in Section 7.6.3, one can estimate the CLC contribution to the delay difference between $2p$ and $2s$ electrons around $\Omega = 105$ eV to be roughly 3 as. Even though we added the CLC part to the Wigner part, the theoretical prediction still cannot reproduce the delay derived from the measurement.

7.6.5 Accuracy of Atomic-Dipole Phase Retrieval using FROG–CRAB

In view of this persistent discrepancy, one has to reexamine how the 20 as time-delay difference was derived from the streaking data. How accurately can the atomic-dipole phase be retrieved from the FROG–CRAB method in view of the approximations discussed in Section 7.5.2? In the derivation of the time-delay difference, according to Equation 7.77, it was assumed that the XUV group delay can be totally canceled out such that the measurement becomes independent of the XUV pulse. This is equivalent to assuming that Equation 7.52 holds accurately without any error. Since the phase of the wave packet retrieved by the FROG–CRAB may not satisfy Equation 7.52 exactly, especially when the XUV is chirped, Equation 7.77 may not be a good approximation. Therefore the group-delay difference derived from the experimental data between the two wave packets deviates from the Wigner-delay difference. One way to test the accuracy of the retrieval of atomic-dipole phase using FROG–CRAB is to simulate streaking spectra generated by a TL XUV pulse against one generated with a chirped XUV pulse. This was carried out in Wei et al. [47] for a model Ne atom.

The results of SFA simulations within the single-active-electron approximation are given in Figure 7.22. Two XUV pulses with the same $\Omega_0 = 105$ eV, $\Delta\Omega = 9$ eV are used to

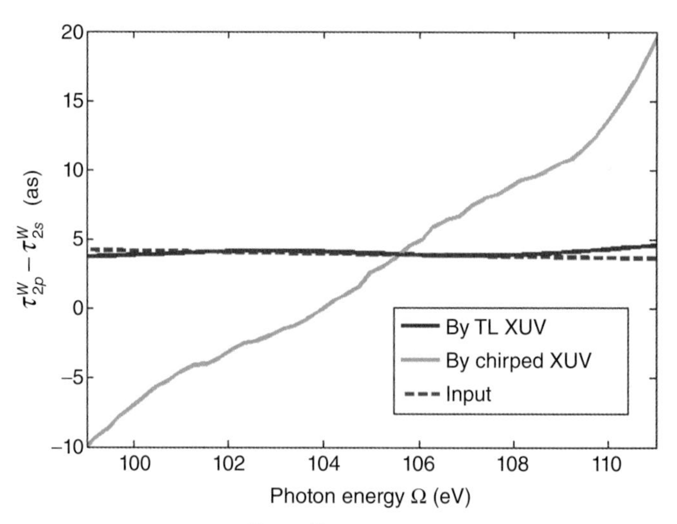

Figure 7.22 FROG–CRAB-retrieved Wigner delay difference $\tau^W_{2p} - \tau^W_{2s}$ in the photoionization of Ne from SFA simulations. The solid blue line is retrieved by using an 190 as TL XUV pulse, and the solid green line by using a 280 as chirped XUV. The red dashed line is the input Wigner-delay difference used in the simulation. In the energy domain both XUV pulses are centered at 105 eV, with an FWHM bandwidth of 9 eV. It is found that accurate atomic-time delay is difficult to obtain if the XUV is chirped. (Reprinted from Hui Wei, Toru Morishita, and C. D. Lin, *Phys. Rev. A*, **93**, 053412 (2016) [47]. Copyrighted by the American Physical Society.)

generate the streaking trace for Ne $2p$ and $2s$ photoionization. The TL pulse has a duration of 190 as while the chirped XUV has a duration of 280 as. The IR field is 800 nm in wavelength, 6.2 fs in FWHM duration, and 10^{12} W/cm^2 in intensity. The LSGPA FROG–CRAB is used to extract the time delay from the simulated trace. From Figure 7.22, when the TL pulse is used, the retrieved Wigner delay difference agrees well with the input data used in the simulation. However, when the chirped pulse is used, the retrieved time delay varies significantly with photon energy, from -8 to 18 as across the XUV bandwidth. This example shows that the retrieved group-delay difference from the experiment may depend on the XUV chirp, due to the limited accuracy of the FROG–CRAB retrieval method. The XUV chirp will lead to errors of several attoseconds in the retrieved time delay, which is detrimental to the accurate time-delay studies. Therefore, a nearly TL XUV pulse is required in time-delay measurements. It is also worth noting that a time-delay error of 10 as amounts to a phase error of 0.07 radians within an energy interval of 5 eV. To obtain a sub-ten-attosecond time delay, the retrieved atomic-dipole phase from experimental data has to be extremely accurate. In view of such complications, an error of about 10 as from the data of Schultze et al. [55] is probably not a cause for alarm. The XUV pulse obtained from the HHG process always contains a certain degree of attochirp. In fact, the streaking spectra shown in Figure 7.21(a) resemble those from a chirped pulse, instead of a transform-limited pulse, as can be seen by comparing Figure 7.12(a) with Figure 7.12(b). At the same time, the streaking spectra are not very sensitive to the atomic-dipole phase, and thus agreement between Figure 7.21(b) and Figure 7.21(a) does not prove that the atomic-dipole phase is accurately retrieved. More examples of such calibration tests can be found in Wei et al. [47].

7.6.6 Photoionization Time Delay between Ar and Ne

The photoionization time delay between Ar and Ne in the lower-energy region has been studied by Sabbar et al. [56]. In their experiment, a gas mixture consisting of Ar and Ne was ionized by an IAP of 12 eV bandwidth centered at a photon energy of 35 eV, in the presence of a moderately strong (about 3×10^{12} W/cm^2) IR field. The few-cycle IR laser pulse has a center wavelength of 735 nm with a pulse duration of 6 fs. Coincidence detection was applied to separate the photoelectrons belonging to the different targets. The measured streaking traces are shown in Figure 7.23(a). The FROG–CRAB was used to extract the wave packets for Ar and Ne photoelectrons, respectively. Since the two photoelectron spectra overlap in energies, they had to patch the Ar and Ne traces together by shifting one trace up along the energy axis, and use the combined trace as the input of the FROG–CRAB retrieval algorithm. In this way, the same gating function $G(t)$ and therefore the same time zeros were used to retrieve the two wave packets.

The reconstructed trace is shown in Figure 7.23(b). Figure 7.23(c) plots the group delay of the retrieved wave packets for Ar (red dotted line) and Ne (blue dot-dashed line) as functions of XUV photon energy. Following the same idea as that in Equation 7.77, the difference between the two measured group delays should be the Wigner-delay difference between Ar and Ne. However, since the electron energy is below 35 eV, the effect of the CLC that is beyond the SFA model should be included. To account for this effect, Sabbar

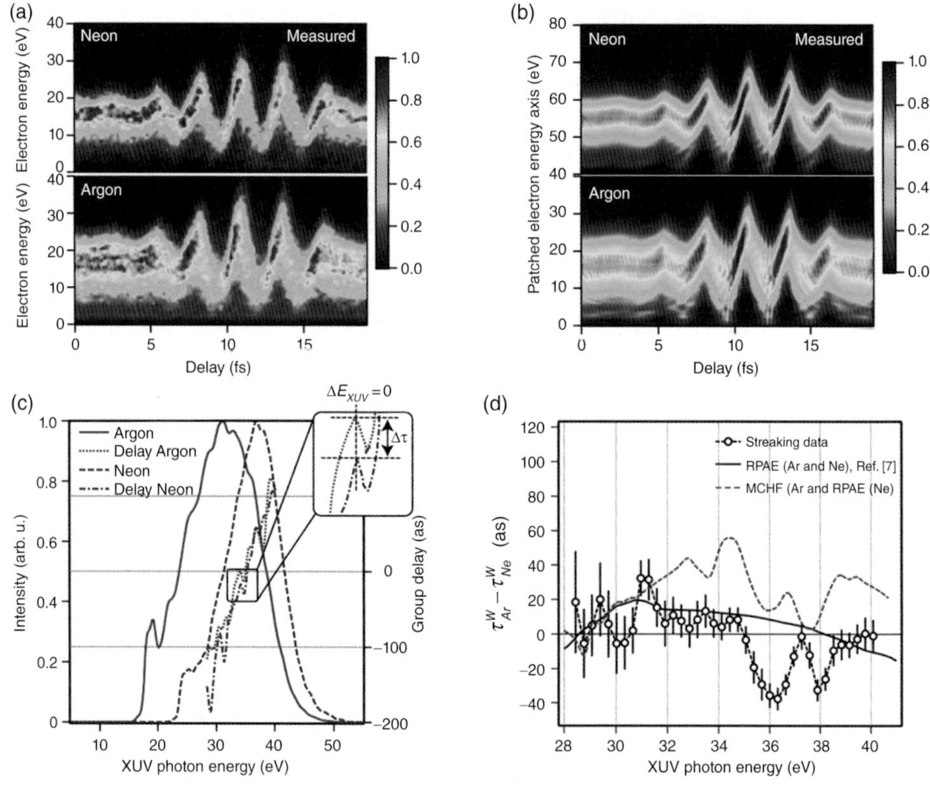

Figure 7.23 (a) Measured spectrograms for photoelectrons from valence orbitals of Ar and Ne, using the coincidence-detection technique. (b) Reconstructed spectrogram by FROG–CRAB. Before applying the FROG–CRAB algorithm the two traces are patched together by shifting one trace up along the energy axis in order to ensure the same time zero used in the retrieval of Ar and Ne wave packets. (c) Intensity and group delay for Ar and Ne wave packets retrieved by FROG–CRAB. The horizontal axis is chosen to be XUV photon energy. The inset shows the group-delay difference corresponding to the same XUV photon energy. (d) The black open circles connected by a dashed line represent the averaged Wigner-delay difference of 33 independent measurements. The CLC part $\Delta\tau^{CLC}$ has been subtracted from the retrieved group-delay difference by using FROG–CRAB. The black solid line is computed using one-photon matrix elements within the RPAE for both Ar and Ne electrons. The magenta dashed line has been obtained using MCHF for Ar and RPAE for Ne, taking resonances into account. (Figure adopted with permission from M. Sabbar et al., *Phys. Rev. Lett.*, **115**, 133001 (2015) [56]. Copyrighted by the American Physical Society.)

et al. used the time delay τ^S in Equation 7.76. They assumed the retrieved group-delay difference between Ar and Ne wave packets is

$$\tau^S_{Ar}(\Omega) - \tau^S_{Ne}(\Omega) = [\tau^W_{Ar}(\Omega) - \tau^W_{Ne}(\Omega)] + \Delta\tau^{CLC}(\Omega), \tag{7.78}$$

where the CLC part is given by

$$\Delta\tau^{CLC}(\Omega) = \tau^{CLC}(E = \Omega - I_{p,Ar}) - \tau^{CLC}(E = \Omega - I_{p,Ne}). \tag{7.79}$$

Therefore they removed the CLC part from the FROG–CRAB-retrieved group-delay difference and treated the new result as the extracted Wigner delay $\tau^W_{Ar} - \tau^W_{Ne}$. This quantity was compared to results obtained from elaborate many-electron theory calculations like the RPAE and the multi-configuration Hartree–Fock (MCHF), shown in Figure 7.23(d). One can only say that the agreement is fair.

The comparison in Figure 7.23(d) is further complicated by the fact that there are many sharp resonances originating from the shake-up thresholds in Ar from 35 to 39 eV. While the MCHF calculation would have included contributions from these resonances, these resonances are not included in the RPAE theory. In the streaking theory employed in the FROG–CRAB-retrieval algorithm, resonances were not included. In addition, the SFA model used in FROG–CRAB is not valid for low-energy photoelectrons (see Figures 7.13(a,b)). In view of such complications, any interpretations of the discrepancy in Figure 7.23(d) will be fortuitous.

7.6.7 Challenge of Accurately Retrieving Wigner Time Delays from Streaked Low-Energy Photoelectrons

In Figure 7.24, the retrieved photoionization delay difference between Ar and Ne from a TDSE simulation is shown. We use a 160 as TL XUV pulse with $\Omega_0 = 40$ eV and $\Delta\Omega = 11.5$ eV. The 800 nm IR field has an intensity of 10^{13} W/cm^2 and a FWHM duration of 4.4 fs. Single-electron model potentials for Ar and Ne given by Tong and Lin [57] are used in the TDSE calculation. The Ar trace is shifted upward by 60 eV and patched together with the Ne trace. FROG–CRAB is then used to extract Ar and Ne photoelectron wave packets from the combined trace. The blue line in Figure 7.24 is the group-delay difference between the two retrieved wave packets. The CLC part $\Delta\tau^{CLC}$ has been removed from the

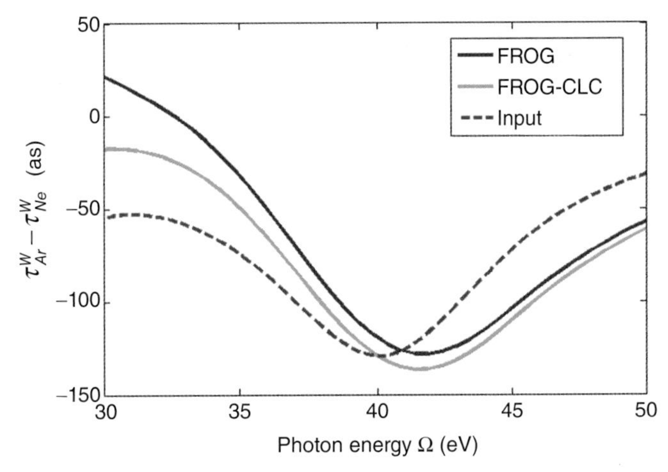

Figure 7.24 The red dashed line is the input Wigner-delay difference $\tau^W_{Ar} - \tau^W_{Ne}$ used in the TDSE simulation. The blue solid line is the group-delay difference between the Ar and Ne wave packets retrieved from the patched trace by using FROG–CRAB. The green solid line plots the result after subtracting the CLC part $\Delta\tau^{CLC}$ from the retrieved group-delay difference. The difference between the red and green lines shows that Wigner time-delay difference for low-energy electrons cannot be accurately retrieved using FROG–CRAB. (Reprinted from Hui Wei, Toru Morishita, and C. D. Lin, *Phys. Rev. A*, **93**, 053412 (2016) [47]. Copyrighted by the American Physical Society.)

blue line and the new result is plotted in green. Even after subtracting the CLC part, around the Cooper minimum of Ar (42 eV according to the model potential used here), the errors between the retrieved time delay and the input Wigner delay cannot be eliminated. This example demonstrates the limitations of the SFA-based, time-delay retrieval method used in FROG–CRAB in the region of low-electron energy.

7.6.8 Validity of Extracting Photoionization Time Delay from the First Moment of the Streaking Trace

In the TDSE simulations mentioned in Section 7.6.3 and in most streaking experiments, the photoionization time delay is extracted from the first moment $\bar{p}(t_d)$ or $\bar{E}(t_d)$, namely the center of momentum (COM) or center of energy (COE) of the spectrogram. This method was first employed by Cavalieri et al. [58], who found that the COE of the streaking trace of the 4f and of the conduction band of the tungsten crystal has a temporal shift of 110 ± 70 as. What is the meaning of this temporal shift? Subsequently, theoretical simulations with simple atomic targets [52] revealed that this shift has two contributions: one from the difference in the Wigner time delay for the photoelectron ejected from the two shells, and the other due to the difference in the CLC term. In other words, this temporal shift is the difference in $\tau^W(\Omega_0) + \tau^{CLC}(\Omega_0)$, where Ω_0 is the central photon energy.

Suppose an IAP with a central photon energy Ω_0 is used. The streaking trace is centered around $p_0 = \sqrt{2(\Omega_0 - I_p)}$. The shift between $\bar{p}(t_d)$ or $\bar{E}(t_d)$ and the IR vector potential $-A(t_d)$ is treated as the time delay at the photon energy $\Omega = \Omega_0$. In most simulations, the IAPs are assumed to be TL. What happens if the IAP has some attochirp, or in the energy domain, a GDD? Can the first moment of the spectrogram still be used to obtain the time delay difference?

To answer these questions, simulations with known IAP and IR for a one-electron model Ne or Ar have been carried out. The dipole amplitude and phase for each model atom can be accurately calculated. For the XUV, two cases were considered: one was a TL pulse with the FWHM of 228 as, the other had chirp with the FWHM of 386 as obtained by assuming that it has a GDD of 2.57×10^4 as^2. The central energy of the XUV was 40 eV with an FWHM of 8 eV. The IR was 800 nm, 4.4 fs in the FWHM. The spectrogram was obtained by solving the TDSE with the known XUV and IR fields. From the calculated spectrograms, the COE at each time delay was calculated. They are shown in Figure 7.25(a). There is a small shift of the peak of the COE between the TL and the chirped XUV for neon because of the GDD of the chirped XUV pulse (the two pulses have been chosen to have zero group delay). There is a large shift between Ne and Ar due to the more rapid phase variation in the transition dipole of Ar than in Ne. Expanding the COEs on the scale of attoseconds, the shift of the peak position of the COE for each case is shown in Figures 7.25(b) and (c), respectively. The results demonstrate that the chirp in the XUV pulse does affect the extracted time delay. In fact, simulations have also been carried out with similar XUV pulses except that the central-photon energies were at 35 eV and 30 eV, respectively. Theoretically, the total delay $\tau^W(\Omega_0) + \tau^{CLC}(\Omega_0)$ for each pulse and target can be calculated. In Figure 7.26(a), the Wigner delay and CLC delay are plotted against the electron energy for the Ne target. Their sum, $\tau^W(\Omega_0) + \tau^{CLC}(\Omega_0)$, is plotted against

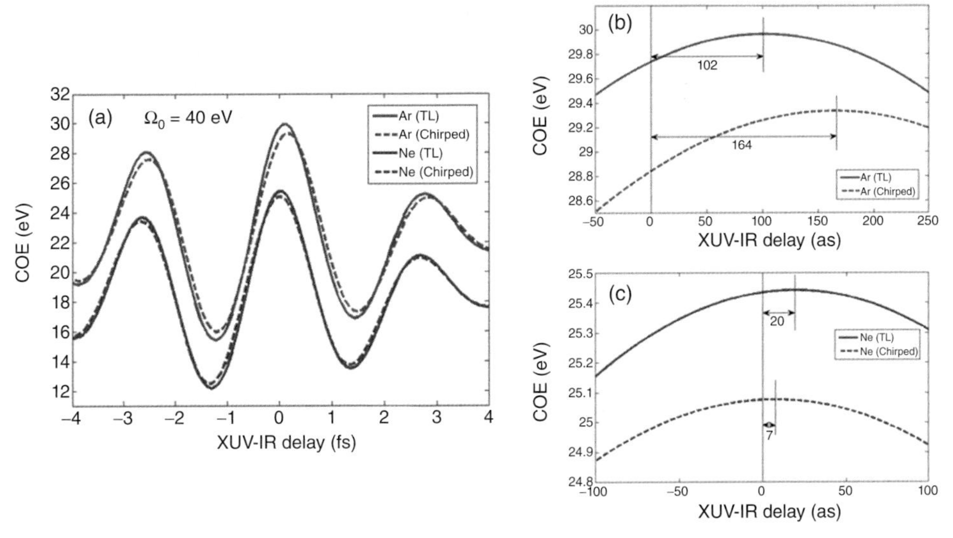

Figure 7.25 (a) COE of TDSE spectrograms for Ar and Ne targets using a TL and a chirped XUV pulse. The central energy of the XUV pulse is $\Omega_0 = 40$ eV. The FWHM bandwidth is $\Delta\Omega = 8$ eV. The FWHM duration is 228 as for the TL pulse and 386 as for the chirped pulse (GDD $= 2.57 \times 10^4$ as^2). Zoom-in plots for (b) Ar and (c) Ne near $t_d = 0$ are also shown. $t_d = 0$ is the peak of the vector potential $-A(t_d)$ used in this simulation, and therefore one can extract the time delay from the peak position of the COE curve.

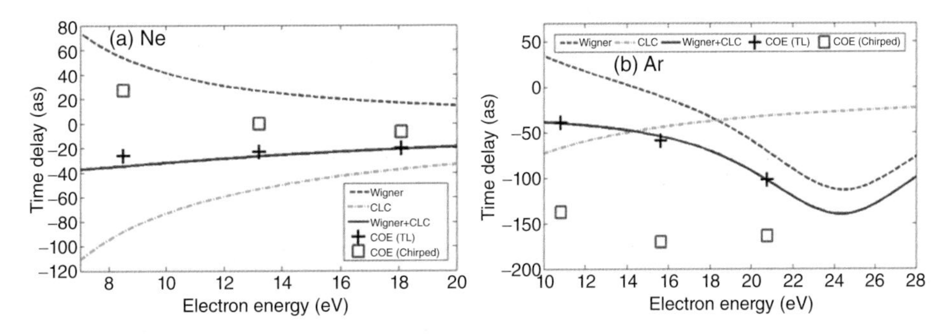

Figure 7.26 Comparison between the input and extracted time delays for (a) Ne and (b) Ar. XUV pulses are centered at $\Omega_0 = 30, 35, 40$ eV, respectively. TL pulses are 228 as in FWHM duration, while chirped ones are 386 as. Black crosses are the time delays extracted from COEs using TL pulses. Magenta squares are extracted using the chirped pulses. The input Wigner delay $\tau^W(E)$ is shown in red dashed lines. Green dot-dashed lines are the CLC term $\tau^{CLC}(E)$ and blue solid lines are the total time delay $\tau^W(E) + \tau^{CLC}(E)$.

the electron energy, as shown on the blue line. For the TL pulse, the retrieved values from the COE curves (crosses) agree quite well with the sum, but not so if the XUV is chirped (magenta squares). Similarly, in Ar, for the TL pulse the extracted time delays from the COE curves agree with the value of $\tau^W(\Omega_0) + \tau^{CLC}(\Omega_0)$, but not so when the XUV has a chirp as shown in Figure 7.26(b). For Ne, the first-moment method can have an error from 30 to 60 as if the XUV is chirped. However, with Ar, that error goes from 60 to 120 as when the photon energy is varied. The numerical values of the present simulations are shown in

Table 7.1 Time delays extracted by comparing the peak positions of the calculated COE of the TDSE spectrograms with the peak position of $-A(t_d)$. TL and chirped XUV pulses centered at photon energies $\Omega = 30, 35, 40$ eV are used. Target is Ne.

XUV central energy (eV)	30	35	40
$\tau^W + \tau^{CLC}$ (as)	−35	−27	−21
$\tau^{COE,TL}$ (as)	−26	−23	−20
$\tau^{COE,Chirped}$ (as)	27	0	−7

Table 7.2 Time delays extracted by comparing the peak positions of the calculated COE of the TDSE spectrograms with the peak position of $-A(t_d)$. TL and chirped XUV pulses centered at photon energies $\Omega = 30, 35, 40$ eV are used. Target is Ar.

XUV central energy (eV)	30	35	40
$\tau^W + \tau^{CLC}$ (as)	−40	−54	−103
$\tau^{COE,TL}$ (as)	−39	−59	−102
$\tau^{COE,Chirped}$ (as)	−137	−170	−164

Tables 7.1 and 7.2. In a streaking experiment, the XUV pulse is expected to have some degrees of attochirp. Clearly, the larger the chirp the larger the error will be. This result shows that the widely used first-moment method for obtaining the "time-delay difference," even after the correction of the CLC delay, cannot be used to extract information about the difference in the Wigner delay of the two photoionization events under the same streaking field. Such a discrepancy would cast doubts on the accuracy of the delay difference reported in the literature. A large volume of such works before 2015 can be found in Pazourek et al. [52]; the newer ones can be identified from the cited articles.

7.6.9 Is Photoionization Time Delay a Misnomer?

So far only the practical issues of extracting photoionization time delays from the streaking experiments have been addressed. Recall that, without the IR field, what one can get from an experiment with XUV alone is the standard photoelectron spectrum given by the modulus square of the photoelectron wave packet, $\tilde{\chi}(E) = \tilde{E}_{XUV}(\Omega)d(E)$ (see Equation 7.52). If the XUV pulse is known, then the amplitude of $d(E)$ is known. Thus, one of the goals of streaking experiments is retrieving the phase of $d(E)$. Once the phase of the transition dipole is known, one can obtain the Wigner time delay from Equation 7.72. In most of the reported attosecond-streaking experiments, instead of the phase of $d(E)$, the "measured" time delay is reported. However, time is not an operator in quantum mechanics so it cannot be measured. Then what is the meaning of a measured time delay? Photoionization time delay has been recognized as a controversial topic, as summarized in a review article [52]. There have been a lot of reported experiments and many hot debates, or rather, confusions. What, precisely, is a photoionization time delay, or equivalently, a Wigner time delay? Is it a "time delay" as we know it classically? Recall that a Wigner time delay is defined through Equation 7.68. That definition actually is meaningless in energy-domain measurements, where the energy is precisely defined and successive measurements are not coherent. Thus it is not possible to obtain the energy derivative of the phase shift in Equation 7.68 unless the ionizing XUV is a coherent pulse. For a wave packet, however, the phase should be specified over the whole spectral bandwidth, not just at a single energy point. In other words, Wigner's definition was meant to give a conceptually classical "meaning" to the abstract quantum-scattering phase shift. It was not meant to be a quantity that can be measured in the laboratory. At this point, readers may recall a similar "debate" discussed in Section 6.1.11, where tomographic imaging was used to retrieve a molecular orbital. Both the Wigner time delay and molecular orbitals are conceptual objects defined through the phase of the dipole. Such derived quantities, if performed without additional approximations, should be as harmless as the original phase shift. However, if the "name" has classical meaning, it entails confusion and gives rise to fruitless debates. From the streaking experiment, in principle, one can retrieve the phase of $d(E)$, or equivalently, $\tau^W(E)$, over the whole bandwidth of the photoelectron since the latter is defined as the derivative of the phase shift with respect to the electron energy. Thus, there is no uncertainty in the derived $\tau^W(E)$. It is confusing only because it was called "photoionization time delay" and time delay means different things to different people.

To avoid confusion as well as unnecessary debates (within the framework of quantum mechanics), it is advisable to replace photoionization time delay by photoionization group delay. This is consistent with the description of the phase of a light pulse (see Equation 7.44), where the group delay is defined. If the photoionization group delay is specified over the whole bandwidth of the photoelectron wave packet, then the phase of the transition dipole is also determined except for an unimportant constant. Given $\tilde{\chi}(E) = \tilde{E}_{XUV}(\Omega)d(E)$, with the spectral amplitude and phase determined, quantum mechanics tells us how to construct this wave packet in the coordinate space as it evolves in time (see Exercise 7.5, Section 8.1.3, and Figure 8.3). In the spirit of quantum mechanics, this is a complete specification of a quantum system after an atom is photoionized by an XUV pulse. What the streaking experiment can measure is the photoionization group delay, not the photoionization time delay.

Finally, does the streaking experiment allow one to retrieve the whole photoelectron wave packet by an XUV pulse for, say, even the simplest one-electron atom? According to the perturbation theory, such an electron wave packet is trivial to calculate, including its phase, if the XUV pulse is known. But in real experiments, the XUV phase is not known. If the RABITT method is used to determine the phase of the XUV, the theory requires that (i) the IR is weak so that the two-photon ionization theory is applicable and (ii) Equation 7.27 is valid. The relation in Equation 7.27 is empirical. It is known to be invalid when there is a nearby resonance, as shown in Section 7.3.3. APT is also less desirable for the purpose of retrieving the phase or the Wigner group delay since the phases are available only at sparsely distributed energy points. If the XUV pulse is an IAP, there are two ways to retrieve the Wigner group delay from the experimental streaking trace. The simplest one is the first-moment method discussed in Section 7.6.8. Unfortunately, so far this method can retrieve only the Wigner group delay at one energy point, which does not allow the reconstruction of the phase of the whole wave packet. The accuracy of this method has also been shown to be questionable for chirped XUV pulses. It also relies on the validity of Equation 7.76, which is empirical. The other method for dipole-phase retrieval from the streaking spectrogram is based on FROG–CRAB. This method is based on the SFA model, which is only valid when the photoelectron energy is high, say, higher than 40 eV. It also has intrinsic errors, as addressed in Section 7.6.5. In addition, at higher energies, the phase shift or the Wigner group delay are small, making the retrieval more challenging. For streaking involving electrons below 40 eV, the SFA theory fails so the FROG–CRAB cannot be used. At present, there is no known retrieval method that can directly and accurately extract the dipole phase from the spectrogram.

The discussion in the previous paragraph says that the streaking experiment docs not allow the retrieval of the phase of the whole electron wave packet. It may seem frustrating that, for such a simple photoionization problem, the phase of the transition dipole still cannot be "observed" experimentally. However, this limitation is the true nature of the quantum world. In the quantum world, each measurement is a projection that only gives partial information about the system, while the Hilbert space where the quantum system resides is infinite. In the end, to understand the dynamics of a system, e.g., to produce a "molecular movie" or an "electron movie," one is not expected to "see" the whole wave packet. Indeed, incomplete information from observations is not foreign even in classical physics. In a real classical movie, what the audience sees is the part that the director wants them to see. To the audience, these projected images tell the whole story. This is also true in the quantum world, where the experimentalists play the role of directors. Not all projections are equal; only those taken from the best "angles" can tell a good story.

Notes and Comments

The generation of APT and IAP was first reported in 2001. Today, attosecond science is still far from being a mature field. There are only a few laboratories in the world that can generate IAPs routinely, and most of the applications use IAPs with photon energies below 150 eV. Although attosecond pulses with higher photon energies in the water-window region have begun to appear, there are still few applications. The generation and characterization of attosecond pulses continue to be active areas of research in many laboratories. Recent laser-technology developments appear to be quite promising in significantly increasing the photon flux and in extending to higher energies. Still, the characterization of attosecond pulses remains quite difficult and in many cases the IAP and APT are not well characterized in the proof-of-principle experiments. Conceptually, time-domain measurements with attosecond pulses also encounter many difficulties. In attosecond physics one is dealing with a pulse. The phase of the pulse is an essential element either as a function of time or a function of frequency. Without the phase there is no attosecond pulse. When an attosecond pulse interacts with matter, it generates an electron wave packet that also has a phase. Unlike strong-field physics, where semiclassical theories may be used successfully at times, in attosecond physics, quantum physics cannot be compromised. While an exact solution of the TDSE may be possible for a simple system, to compare with any experiment, a suitable projection of the wavefunction by a Hermitian operator is needed. From this perspective, time is not to be measured nor is the velocity of an electron. Thus the measurement of photoionization time delay is controversial. In particular, photoionization time delay cannot be determined at the peak of the electron wave packet alone. A wave packet cannot be represented by its parts. A signal is meaningful only if the whole pulse is measured. Similarly, the molecular orbital (or Dyson orbital) of a single electron is not to be measured since all the electrons in a many-electron system are indistinguishable. Attosecond experiments do not measure the time, they only "measure" the change of a system on an attosecond timescale. The real measurement is still the detector in the laboratory that has a picosecond temporal resolution.

Another present challenge in attosecond physics is that almost all the experiments with attosecond pulses are carried out together with a moderately intense IR pulse. To retrieve phase information, the IR pulses have to interact with the medium nonlinearly. Except for methods like RABITT, where a second-order perturbation theory can be carried out accurately, there are few simple models that can predict the outcome of attosecond experiments. Due to the broadband nature, it is probably unlikely that any attosecond experiments can be "completely" understood. In the future, the challenge is to identify experiments where the interpretations can be simple. This is particularly true if attosecond pulses are used to study complex systems. In the real world, a complex system is never fully understood (or observed). Clearly this points out that relevant, simpler models will become essential in attosecond physics. To reach this goal, well-thought-out experiments and modeling will be needed and experiments should be carried out with well-characterized parameters.

Exercises

7.1 Following the procedure from Equations 7.3 to 7.6, derive the oscillation of the sideband signal as expressed in Equation 7.7 with the A and B coefficients given by Equations 7.8 and 7.9, and the atomic phase is given by Equation 7.10, respectively.

7.2 Explain why one cannot retrieve the absolute value of the Wigner time delay from a single XUV+IR photoelectron spectrogram. According to the original definition of Wigner delay from Equations 7.68 or 7.72, it appears that there is no such ambiguity. How can one reconcile this discrepancy?

7.3 List the approximations imposed on the RABITT phase-retrieval method.

7.4 Derive the temporal gate of the PG method, Equation 7.31.

7.5 Consider the wave packet of forward electrons coming from a single-photon-ionization process. The wave packet is approximately described by

$$\psi(x,t) = \int \tilde{E}_{XUV}(\Omega) d(E) e^{-iEt} e^{ipx} dE, \tag{7.80}$$

in which $\Omega = E + I_p$ is the photon energy, and $p = \sqrt{2E}$ is the momentum of the electron. The electric field $\tilde{E}_{XUV}(\Omega) = U(\Omega)e^{i\Phi(\Omega)}$. $U(\Omega)$ is given by Equation 7.36

with $U_0 = 1$, $\Omega_0 = 50$ eV, $\Delta\Omega = 10$ eV. $\Phi(\Omega) = \frac{1}{2}\beta^X(\Omega - \Omega_0)^2$, where β^X is the GDD of the XUV pulse. The transition dipole is modeled by

$$d(E) = e^{i\left[\tau^W(E-E_0)+\frac{1}{2}\beta^W(E-E_0)^2\right]}, \tag{7.81}$$

where τ^W is the Wigner delay at $E = E_0$, and β^W is a GDD related to the dipole phase. Here, $I_p = 20$ eV, $E_0 = 30$ eV, $\tau^W = 50$ as. Generate the electron wave packet for (a) $\beta^X = \beta^W = 0$, (b) $\beta^X = 0$, $\beta^W = -5{,}000\ \text{as}^2$, (c) $\beta^X = 15{,}000\ \text{as}^2$, $\beta^W = -5{,}000\ \text{as}^2$. In each case, plot $|\psi(x,t)|^2$ as a function of x at $t = 0, 300, 600, 900, 1{,}200$ as, respectively. Then, plot the motion of the peak position of the wave packet x_{peak} versus t for cases (a), (b), and (c) in one figure. When $x_{peak} = 0$, what are the corresponding t? Compare them with τ^W. When $t = 200, 400, 600, 800$ as, what are the corresponding x_{peak}?

References

[1] A. McPherson, G. Gibson, H. Jara, et al. Studies of multiphoton production of vacuum-ultraviolet radiation in the rare gases. *J. Opt. Soc. Am. B*, **4**(4):595–601, Apr. 1987.

[2] M. Ferray, A. L'Huillier, X. F. Li, L. A. Lompre, G. Mainfray, and C. Manus. Multiple-harmonic conversion of 1064 nm radiation in rare gases. *J. Phys. B-At. Mol. Opt.*, **21**(3):L31, 1988.

[3] K. C. Kulander and B. W. Shore. Calculations of multiple-harmonic conversion of 1064-nm radiation in Xe. *Phys. Rev. Lett.*, **62**:524–526, Jan. 1989.

[4] D. Strickland and G. Mourou. Compression of amplified chirped optical pulses. *Opt. Commun.*, **56**(3):19–221, 1985.

[5] P. M. Paul, E. S. Toma, P. Breger, et al. Observation of a train of attosecond pulses from high harmonic generation. *Science*, **292**(5522):1689–1692, 2001.

[6] M. Hentschel, R. Kienberger, C. Spielmann, et al. Attosecond metrology. *Nature*, **414**(6863):509–513, Nov. 2001.

[7] H. G. Muller. Reconstruction of attosecond harmonic beating by interference of two-photon transitions. *Appl. Phys. B: Lasers Opt.*, **74**(1):s17–s21, 2002.

[8] V. Véniard, R. Taïeb, and A. Maquet. Phase dependence of (N+1)-color (N>1) ir-uv photoionization of atoms with higher harmonics. *Phys. Rev. A*, **54**:721–728, Jul. 1996.

[9] Y. Mairesse, A. de Bohan, L. J. Frasinski, et al. Attosecond synchronization of high-harmonic soft X-rays. *Science*, **302**(5650):1540–1543, 2003.

[10] R. López-Martens, K. Varjú, P. Johnsson, et al. Amplitude and phase control of attosecond light pulses. *Phys. Rev. Lett.*, **94**:033001, Jan. 2005.

[11] J. M. Dahlström, D. Guénot, K. Klünder, et al. Theory of attosecond delays in laser-assisted photoionization. *Chem. Phys.*, **414**:53–64, 2013.

[12] A. Dalgarno and J. T. Lewis. The exact calculation of long-range forces between atoms by perturbation theory. *Proc. R. Soc. London, Ser. A*, **233**(1192):70–74, 1955.

[13] E. S. Toma and H. G. Muller. Calculation of matrix elements for mixed extreme-ultraviolet-infrared two-photon above-threshold ionization of argon. *J. Phys. B-At. Mol. Opt.*, **35**(16):3435, 2002.

[14] J. M. Dahlström, T. Carette, and E. Lindroth. Diagrammatic approach to attosecond delays in photoionization. *Phys. Rev. A*, **86**:061402, Dec. 2012.

[15] K. Klünder, J. M. Dahlström, M. Gisselbrecht, et al. Probing single-photon ionization on the attosecond time scale. *Phys. Rev. Lett.*, **106**:143002, Apr. 2011.

[16] J. A. R. Samson and J. L. Gardner. Photoionization cross sections of the outer *s*-subshell electrons in the rare gases. *Phys. Rev. Lett.*, **33**:671–673, Sep. 1974.

[17] D. Guénot, K. Klünder, C. L. Arnold, et al. Photoemission-time-delay measurements and calculations close to the 3*s*-ionization-cross-section minimum in Ar. *Phys. Rev. A*, **85**:053424, May 2012.

[18] C. Palatchi, J. M. Dahlström, A. S. Kheifets, et al. Atomic delay in helium, neon, argon and krypton. *J. Phys. B-At. Mol. Opt.*, **47**(24):245003, 2014.

[19] M. Kotur, D. Guénot, A Jiménez-Galán, et al. Spectral phase measurement of a Fano resonance using tunable attosecond pulses. *Nat. Commun.*, **7**:10566, Feb. 2016.

[20] V. Gruson, L. Barreau, Á. Jiménez-Galan, et al. Attosecond dynamics through a Fano resonance: monitoring the birth of a photoelectron. *Science*, **354**(6313):734–738, 2016.

[21] E. Goulielmakis, M. Schultze, M. Hofstetter, et al. Single-cycle nonlinear optics. *Science*, **320**(5883):1614–1617, Jun. 2008.

[22] P. B. Corkum, N. H. Burnett, and M. Y. Ivanov. Subfemtosecond pulses. *Opt. Lett.*, **19**(22):1870–1872, Nov. 1994.

[23] G. Sansone, E. Benedetti, F. Calegari, et al. Isolated single-cycle attosecond pulses. *Science*, **314**(5798):443–446, Oct. 2006.

[24] Z. Chang. Single attosecond pulse and xuv supercontinuum in the high-order harmonic plateau. *Phys. Rev. A*, **70**:043802, Oct. 2004.

[25] H. Mashiko, S. Gilbertson, C. Li, et al. Double optical gating of high-order harmonic generation with carrier-envelope phase stabilized lasers. *Phys. Rev. Lett.*, **100**:103906, Mar. 2008.

[26] X. Feng, S. Gilbertson, H. Mashiko, et al. Generation of isolated attosecond pulses with 20 to 28 femtosecond lasers. *Phys. Rev. Lett.*, **103**:183901, Oct. 2009.

[27] K. Zhao, Q. Zhang, M. Chini, Y. Wu, X. Wang, and Z. Chang. Tailoring a 67 attosecond pulse through advantageous phase-mismatch. *Opt. Lett.*, **37**(18): 3891–3893, Sep. 2012.

[28] J. Li, X. Ren, Y. Yin, et al. Polarization gating of high harmonic generation in the water window. *Appl. Phys. Lett.*, **108**(23):231102, 2016.

[29] H. Vincenti and F. Quéré. Attosecond lighthouses: how to use spatiotemporally coupled light fields to generate isolated attosecond pulses. *Phys. Rev. Lett.*, **108**:113904, Mar. 2012.

[30] K. T. Kim, C. Zhang, T. Ruchon, et al. Photonic streaking of attosecond pulse trains. *Nat. Photon.*, **7**(8):651–656, Aug. 2013.

[31] C. Zhang, G. G. Brown, K. T. Kim, D. M. Villeneuve, and P. B. Corkum. Full characterization of an attosecond pulse generated using an infrared driver. *Sci. Rep.*, **6**:26771, May 2016.

[32] N. Dudovich, O. Smirnova, J. Levesque, et al. Measuring and controlling the birth of attosecond XUV pulses. *Nat. Phys.*, **2**(11):781–786, Nov. 2006.

[33] T. Pfeifer, A. Jullien, M. J. Abel, et al. Generating coherent broadband continuum soft-x-ray radiation by attosecond ionization gating. *Opt. Express*, **15**(25):17120–17128, Dec. 2007.

[34] E. J. Takahashi, P. Lan, O. D. Mücke, Y. Nabekawa, and K. Midorikawa. Attosecond nonlinear optics using gigawatt-scale isolated attosecond pulses. *Nat. Commun.*, **4**:2691, Oct. 2013.

[35] G. Sansone, L. Poletto, and M. Nisoli. High-energy attosecond light sources. *Nat. Photon.*, **5**(11):655–663, Nov. 2011.

[36] P. D. Keathley, S. Bhardwaj, J. Moses, G. Laurent, and F. X. Kärtner. Volkov transform generalized projection algorithm for attosecond pulse characterization. *New J. Phys.*, **18**(7):073009, 2016.

[37] M. T. Hassan, T. T. Luu, A. Moulet, et al. Optical attosecond pulses and tracking the nonlinear response of bound electrons. *Nature*, **530**(7588):66–70, Feb. 2016. Letter.

[38] S. M. Teichmann, F. Silva, S. L. Cousin, M. Hemmer, and J. Biegert. 0.5-keV soft X-ray attosecond continua. *Nat. Commun.*, **7**:11493, May 2016.

[39] J. Rothhardt, S. Hädrich, Y. Shamir, et al. High-repetition-rate and high-photon-flux 70 eV high-harmonic source for coincidence ion imaging of gas-phase molecules. *Opt. Express*, **24**(16):18133–18147, Aug. 2016.

[40] H. Wang, Y. Xu, S. Ulonska, J. S. Robinson, P. Ranitovic, and R. A. Kaindl. Bright high-repetition-rate source of narrowband extreme-ultraviolet harmonics beyond 22eV. *Nat. Commun.*, **6**:7459, Jun. 2015.

[41] J. Itatani, F. Quéré, G. L. Yudin, M. Yu. Ivanov, F. Krausz, and P. B. Corkum. Attosecond streak camera. *Phys. Rev. Lett.*, **88**:173903, Apr. 2002.

[42] M. Kitzler, N. Milosevic, A. Scrinzi, F. Krausz, and T. Brabec. Quantum theory of attosecond XUV Pulse measurement by laser dressed photoionization. *Phys. Rev. Lett.*, **88**:173904, Apr. 2002.

[43] V. S. Yakovlev, J. Gagnon, N. Karpowicz, and F. Krausz. Attosecond streaking enables the measurement of quantum phase. *Phys. Rev. Lett.*, **105**:073001, Aug. 2010.

[44] D. J. Kane. Recent progress toward real-time measurement of ultrashort laser pulses. *IEEE J. Quantum Electron.*, **35**(4):421–431, Apr. 1999.

[45] J. Gagnon, E. Goulielmakis, and V. S. Yakovlev. The accurate FROG characterization of attosecond pulses from streaking measurements. *Appl. Phys. B: Lasers Opt.*, **92**(1):25–32, 2008.

[46] Y. Mairesse and F. Quéré. Frequency-resolved optical gating for complete reconstruction of attosecond bursts. *Phys. Rev. A*, **71**:011401, Jan. 2005.

[47] H. Wei, T. Morishita, and C. D. Lin. Critical evaluation of attosecond time delays retrieved from photoelectron streaking measurements. *Phys. Rev. A*, **93**:053412, May 2016.

[48] M. Chini, S. Gilbertson, S. D. Khan, and Z. Chang. Characterizing ultrabroadband attosecond lasers. *Opt. Express*, **18**(12):13006–13016, Jun. 2010.

[49] X. Zhao, H. Wei, Y. Wu, and C. D. Lin. Phase-retrieval algorithm for the characterization of broadband single attosecond pulses. *Phys. Rev. A*, **95**:043407, Apr. 2017.

[50] H. Wei, A.-T. Le, T. Morishita, C. Yu, and C. D. Lin. Benchmarking accurate spectral phase retrieval of single attosecond pulses. *Phys. Rev. A*, **91**:023407, Feb. 2015.

[51] K. E. McCulloh and George Glockler. The electronic emission spectrum of $C^{13}O^{16}$. *Phys. Rev.*, **89**:145–147, Jan. 1953.

[52] R. Pazourek, S. Nagele, and J. Burgdörfer. Attosecond chronoscopy of photoemission. *Rev. Mod. Phys.*, **87**:765–802, Aug. 2015.

[53] S. Nagele, R. Pazourek, J. Feist, and J. Burgdörfer. Time shifts in photoemission from a fully correlated two-electron model system. *Phys. Rev. A*, **85**:033401, Mar. 2012.

[54] R. Pazourek, J. Feist, S. Nagele, and J. Burgdörfer. Attosecond streaking of correlated two-electron transitions in Helium. *Phys. Rev. Lett.*, **108**:163001, Apr. 2012.

[55] M. Schultze, M. Fiess, N. Karpowicz, et al. Delay in photoemission. *Science*, **328**(5986):1658–1662, Jun. 2010.

[56] M. Sabbar, S. Heuser, R. Boge, et al. Resonance effects in photoemission time delays. *Phys. Rev. Lett.*, **115**:133001, Sep. 2015.

[57] X. M. Tong and C. D. Lin. Empirical formula for static field ionization rates of atoms and molecules by lasers in the barrier-suppression regime. *J. Phys. B-At. Mol. Opt.*, **38**(15):2593, 2005.

[58] A. L. Cavalieri, N. Muller, T. Uphues, et al. Attosecond spectroscopy in condensed matter. *Nature*, **449**(7165):1029–1032, Oct. 2007.

8 Probing Electron Dynamics with Isolated Attosecond Pulses

8.1 Description of Electron Dynamics and Measurements

8.1.1 Electron Wave-Packet Dynamics after Ionization by an Attosecond XUV Pulse

The emergence of attosecond technology has opened numerous opportunities for "probing" electron dynamics at attosecond timescales. Various experiments have been carried out and exceptional time resolution down to a few tens of attoseconds have been claimed using tools such as attosecond photoelectron streaking, high-harmonic spectroscopy, attoclocks, and attosecond transient absorption spectroscopy. In addition, charge migration and hole hopping at attosecond timescales derived from theoretical calculations have drawn a great deal of attention.

Since an electron is not a classical particle, the dynamics of an electron is governed by quantum mechanics. The time-dependent Schrödinger equation (TDSE) describes how the wavefunction (or state vector) evolves in time exactly, but information about the state vector is not directly measurable. In an experiment after the external field is over, the time-dependent wavefunction $\Psi(\mathbf{r},t)$ can be expressed as

$$\Psi(\mathbf{r},t) = \sum_i a_i(0)e^{-iE_it}\phi_i(\mathbf{r}), \tag{8.1}$$

where each $\phi_i(\mathbf{r})$ is the eigenstate of the field-free Hamiltonian with eigenvalue E_i and the summation implies sum over discrete states and integration over continuum states. Each $a_i(0)$ here is the complex scattering amplitude. The complex time-dependent wavefunction Equation 8.1 is normally called an *electron wave packet*. The modulus square of Equation 8.1 describes the time evolution of electron density.

Consider a wave packet consisting of only two eigenstates $|1\rangle$ and $|2\rangle$. Its evolution with time is completely specified if the amplitude and phase of $a_1(0)$ and $a_2(0)$ are known. Since scattering amplitudes are complex quantities and have no classical analog, one may prefer to use charge density to describe the electron dynamics. In the present simple example, the charge density is given by

$$\begin{aligned} N(\mathbf{r},t) &= |a_1(0)\,\phi_1(\mathbf{r})|^2 + |a_2(0)\,\phi_2(\mathbf{r})|^2 \\ &\quad + 2\mathrm{Re}\left[a_1^*(0)a_2(0)\phi_1^*(\mathbf{r})\phi_2(\mathbf{r})\right]\cos\left[(E_1-E_2)\,t\right] \\ &\quad - 2\mathrm{Im}\left[a_1^*(0)a_2(0)\phi_1^*(\mathbf{r})\phi_2(\mathbf{r})\right]\sin\left[(E_1-E_2)\,t\right], \end{aligned} \tag{8.2}$$

which shows that charge density oscillates trivially with a frequency $|E_1 - E_2|$. Knowledge of this oscillation frequency alone gives limited information about the electron wave packet. This simple example demonstrates that determination of the dynamics of an electron wave packet requires the determination of complex scattering amplitudes. Without the phase information, there is no knowledge of dynamics. Thus the description of a time-dependent quantum system in terms of electron density (or its equivalent) is incomplete. For example, knowledge of the evolution of electron density like Equation 8.2 at a prior time does not predict its future behavior.

If the wave packet is composed of continuum states in a bandwidth of ΔE centered around E_0 only, the complete description of the wave packet is the determination of the scattering amplitude and phase within the bandwidth. The density profile of such a wave packet (see the example in Figure 8.3) will move outward as the time is increased. Here, time is a parameter. The density profile will broaden and its shape will change with time. Unlike classical mechanics, knowing the density profile at a given time does not allow the determination of the density profile at a later time. This again points out that the density of an electron wave packet in coordinate space is not a useful quantity for describing a wave packet. Similar to an isolated attosecond pulse, an electron wave packet can be completely represented by its complex scattering amplitude as in Equation 7.33. The determination of the electron wave packet after the pump pulse is not fundamentally different from the determination of phase in an isolated attosecond pulse.

An important tool for studying a dynamic system is transient photoabsorption spectroscopy. Full knowledge of the time-dependent wavefunction is needed to calculate the time-dependent dipole moment $\mathbf{d}(t) = \langle\Psi(\mathbf{r},t)|\mathbf{r}|\Psi(\mathbf{r},t)\rangle$. The Fourier transform of $\mathbf{d}(t)$ is related to the photoabsorption cross-section. In turn, the photoabsorption cross-section is easily determined by passing the light through a gas medium. In fact, attosecond-transient photoabsorption is one of the most powerful experimental tools for attosecond physics to date. This topic will be discussed starting with Section 8.2.

8.1.2 Ultrafast Autoionization Dynamics of Fano Resonances

Fano resonance, as first discussed in Chapter 1, is well known in scattering phenomena in photoionization, electron collisions, and other processes. It was treated in the seminal 1961 paper by Fano [1] in terms of the interaction between a discrete bound state and a degenerate continuum. Photoabsorption populates both the discrete and continuum states. Autoionization of the discrete state to the continuum creates interference to result in asymmetric spectral lineshape. Based on the energy–time uncertainty relation, the lifetime of such a resonance was derived from the measured width. Typical resonance width of 0.1 eV and less would give the lifetime of a few to a few tens of femtoseconds. Such a timescale, in principle, can be readily probed with attosecond pulses. In fact, a number of experiments have reported the extraction of the lifetime of a Fano resonance pumped by an attosecond XUV pulse and probed with an intense infrared (IR) pulse [2, 3]. With the attosecond pulses available one may begin to ask how a Fano resonance is built up as a function of time. This problem was considered in Chu and Lin [4] and others [5] based on the time-independent theory of Fano [1].

According to Equation 1.124, the eigenstate of the Hamiltonian can be expressed as

$$|\psi_E\rangle = a_E\,|\alpha\rangle + \int b_{EE'}\,|\beta_{E'}\rangle\, dE', \tag{8.3}$$

where the coefficients α_E and $b_{EE'}$ are given by Equations 1.125 and 1.126 and $|\alpha\rangle$ and $|\beta_E\rangle$ are the bound state and the continuum states of the two noninteracting subspaces, respectively. Consider the 2s2p(^{1}P) state of helium specifically. It lies at about 60.1 eV from the ground state and has a width of 37 meV and $q = -2.7$. By using a short XUV to excite it and assuming that at $t = 0$ the pulse is turned off, the subsequent time evolution of the wave packet near the resonance for $t > 0$ can be expressed as

$$\big|\Psi(t)\big\rangle = c_g^{(0)} e^{-iE_g t}\,|g\rangle + \int c_E^{(0)} e^{-iEt}\,|\psi_E\rangle\, dE. \tag{8.4}$$

Here, the integral covers the bandwidth of the resonance. Such a resonance can be measured a long time later by projecting the wavefunction to the eigenstate (Equation 8.3 of energy E) with the probability given by $|c_E^{(0)}|^2$. By stitching the measurements at each energy point together, the Fano resonance profile is obtained. Such measurements provide no information about the buildup of the resonance.

In terms of the basis set $|\alpha\rangle$ and $|\beta_E\rangle$, the resonance part in Equation 8.4 can also be expressed as

$$\Psi_{ex}(\mathbf{r},t) = d_\alpha(t)|\alpha\rangle + \int d_E(t)|\beta_E\rangle dE. \tag{8.5}$$

Following the procedure of Chu et al. [4], the coefficients $d_\alpha(t)$ and $d_E(t)$ can be obtained analytically. Define scaled time by $s = \Gamma t$ and scaled energy by $\epsilon = 2(E - E_r)/\Gamma$. The coefficients for the bound and the continuum states are given by

$$d_\alpha(s) = d_\alpha^{(0)}\left(1 - \frac{i}{q}\right) e^{-\frac{1}{2}s}, \tag{8.6}$$

$$d_E(s) = d_E^{(0)} \frac{1}{\epsilon + i}\left[(q+\epsilon)e^{-\frac{i}{2}\epsilon s} - (q-i)e^{-\frac{1}{2}s}\right], \tag{8.7}$$

where the q parameter is defined by

$$q \equiv \frac{d_\alpha^{(0)}}{\pi V d_E^{(0)}}. \tag{8.8}$$

Equation 8.6 shows that the bound state decays exponentially. In Equation 8.7, the continuum amplitude exhibits the interference between the direct ionization part and the decay part. As s goes to infinite at large time, the bound state part vanishes and the modulus square of the continuum part gives the Fano resonance profile. For finite time s, the modulus square of the continuum coefficient, Equation 8.7, gives the resonance profile at the intermediate time. Figure 8.1 shows the evolving resonance profiles calculated using Equation 8.7 for four different values of q's. Each profile begins with a large bandwidth at the initial time in accordance with energy-time uncertainty relation. As time progresses, the lineshape narrows, but it is accompanied by prominent oscillations on both sides of the wings before it settles into the final familiar Fano form.

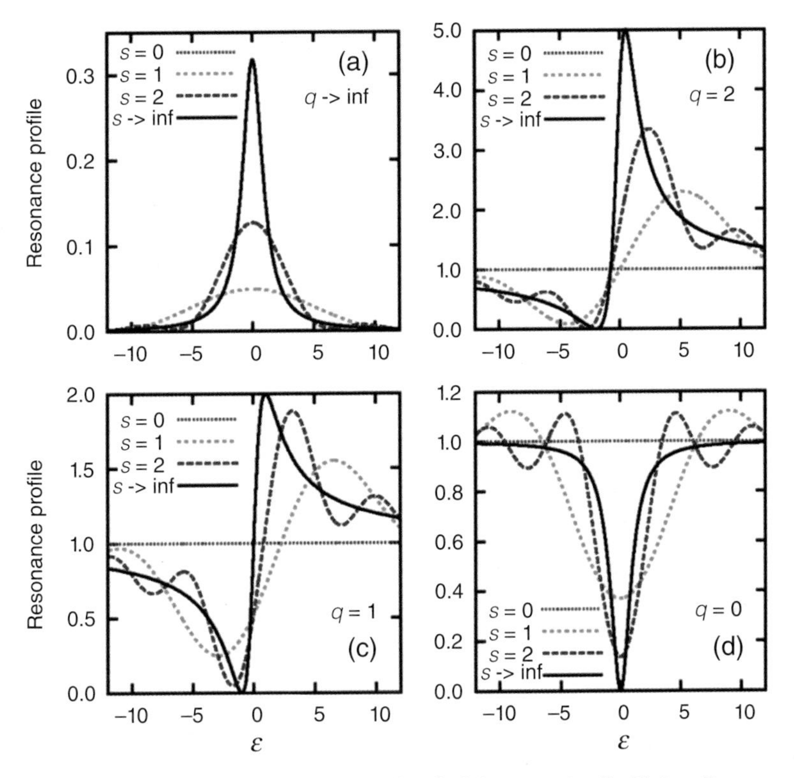

Figure 8.1 Buildup of Fano resonance lineshape from the beginning till its final shape, in units of its lifetime. The q-parameter of the final lineshape is indicated on each frame. The time is scaled as $s = \Gamma t$ and energy is scaled as $\epsilon = 2(E - E_r)/\Gamma$. (Reprinted from W. C. Chu and C. D. Lin, *Phys. Rev. A*, **82**, 053415 (2010) [4]. Copyrighted by the American Physical Society.)

8.1.3 Evolution of a Fano Resonance Wave Packet in the Coordinate Space

The wave packet described in Equation 8.5 can be used to construct the time-dependent electron density distribution of a Fano resonance. Here, an example of the 2p4s resonance of the beryllium atom generated by photoionization from the $2s^2$ ground state is considered. This resonance is known to have $E_r = 2.789$ eV above the ionization threshold, $\Gamma = 0.174$ eV or lifetime of 3.78 fs, and $q = -0.52$. Using a 2 fs (bandwidth 0.912 eV) pump pulse that has the carrier frequency right at the resonance, the time evolution of the Fano profile from $t = 0$ right at the end of the pulse to 40 fs can be calculated later from Equation 8.7. The time evolutions of the resonance profiles are shown in Figure 8.2. Note that the minimum in the Fano profile does not show up till at about 4.5 fs, close to the lifetime of the resonance of 3.78 fs. By 10 fs, the resonance profile becomes quite close to the final shape.

Equation 8.5 allows the construction of the time evolution of the electron density distribution of this resonance. Writing out the wave packet explicitly, the wavefunction to be calculated is

$$\Psi_{ex}(\mathbf{r}_1, \mathbf{r}_2; t) = d_{2p4s}(t)\,\phi_{2p4s}(\mathbf{r}_1, \mathbf{r}_2) + \int d_{E'}(t)\phi_{2sE'p}(\mathbf{r}_1, \mathbf{r}_2)\,dE'. \tag{8.9}$$

To simplify the presentation, one calculates the one-electron density as

$$\rho(r_1;t) = \int \left|\Psi_{ex}(\mathbf{r}_1, \mathbf{r}_2;t)\right|^2 r_1^2 r_2^2 d\Omega_1 d\Omega_2 dr_2 \tag{8.10}$$

where the integration is over all the angles and the radial part of one of the electrons. At large distance, $\rho(r_1;t)$ represents the electron density of the autoionized electron.

Figure 8.3 displays the electron density at selective time steps from $t = 0$ to $t = 160$ fs. At $t = 0$ right after the pulse is over. At the short distance, it is not possible to distinguish which electron has distance r_1 since the two electrons are indistinguishable. This shows that charge oscillation or migration of an electron in a many-electron system has limited meaning in a dynamic system. As time increases, the wave packet moves outward, but the trailing portion continues to show oscillations due to the interference of the direct electron wave and the autoionized electron wave. In the lower frame and at larger times, the outgoing wave packet spreads out but maintains nearly the same shape that reflects the Fano lineshape seen in Figure 8.2(b). The higher energy portion of the wave packet maintains the same profile at the leading edge but for the trailing edge additional oscillations due to interference of the two pathways can be observed. We comment that the results shown in Figure 8.3 cannot be measured directly.

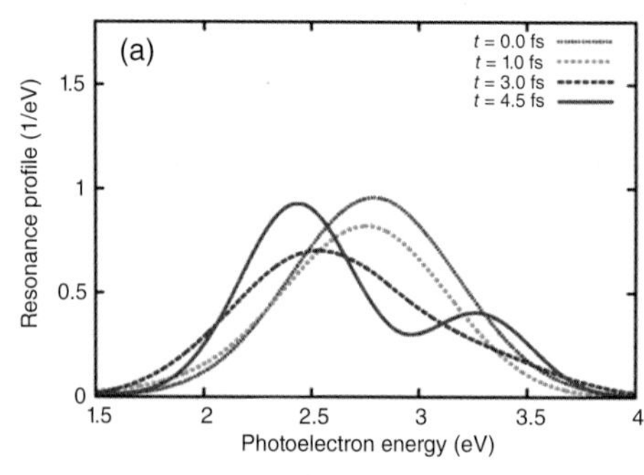

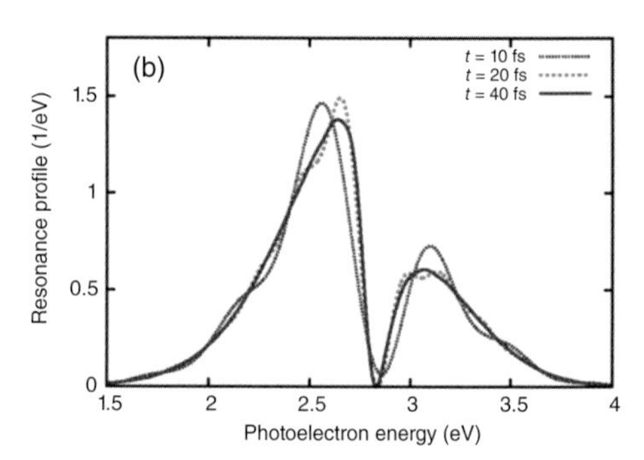

Figure 8.2 Evolution of the profile of the 2p4s (^{1}P) resonance of Be. The lifetime of this resonance is 3.78 fs. (Reprinted from W. C. Chu and C. D. Lin, *Phys. Rev. A*, **82**, 053415 (2010) [4]. Copyrighted by the American Physical Society.)

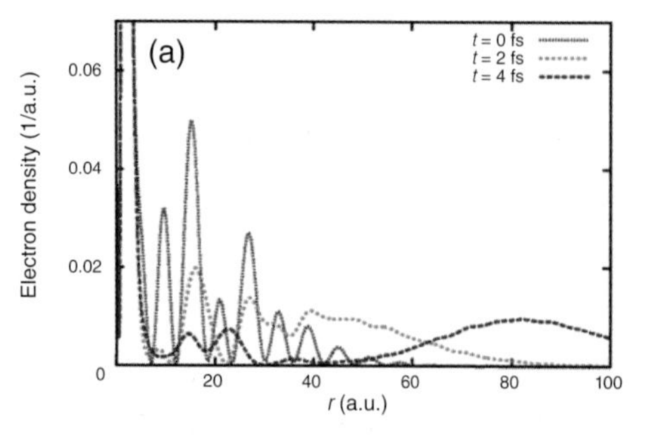

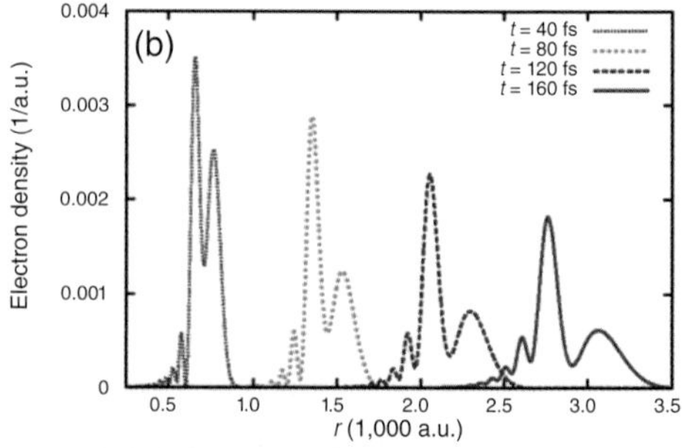

Figure 8.3 The evolution of the radial density of the outgoing autoionized electron versus time for the 2p4s resonance of Be. (Reprinted from W. C. Chu and C. D. Lin, *Phys. Rev. A*, **82**, 053415 (2010) [4]. Copyrighted by the American Physical Society.)

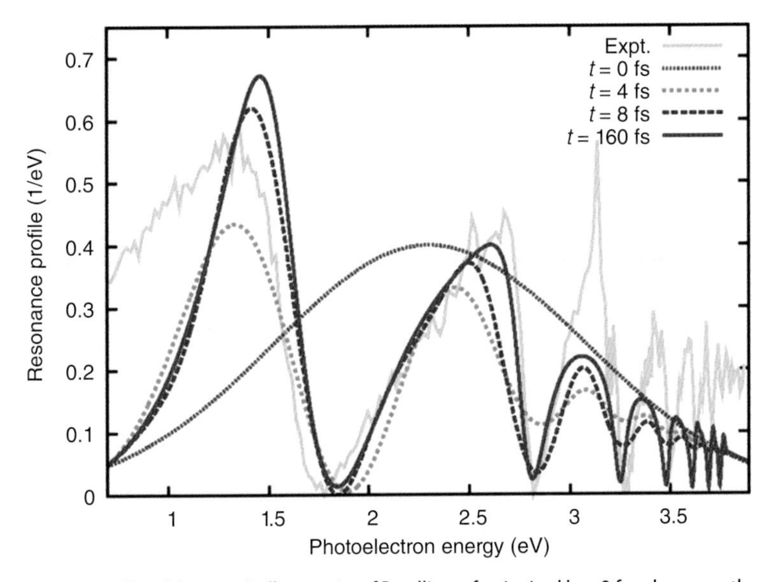

Figure 8.4 Calculated resonance profile of the 2pns Rydberg series of Beryllium after ionized by a 2 fs pulse versus the propagation time after the pulse is over (at $t = 0$). Each higher resonance takes its final lineshape only after it has propagated over a time period longer than its lifetime. (Reprinted from W. C. Chu and C. D. Lin, *Phys. Rev. A*, **82**, 053415 (2010) [4]. Copyrighted by the American Physical Society.)

In atoms or molecules, Fano resonances along the regular Rydberg series generally have the same shape parameter q. Figure 8.4 shows how the absorption spectra along the 2pns series of beryllium atom would look like if it can be measured from the end of the 2 fs excitation pulse. At $t = 0$, the spectrum is given by the Gaussian shape of the ionizing pulse. As time is increased, the low-lying states gradually reach the final Fano shape that would agree well with the experiment. For the higher Rydberg states, resonance features do not appear until the spectra are measured at a time much later than the lifetime of the resonance. Thus, these resonances would look quite different than the ones observed in synchrotron experiments. Note that, for high-n states, the spectral strength at $t = 160$ fs is much weaker than the one from the synchrotron data as discrete components of these resonances have not yet fully decayed.

8.1.4 Measuring the Time-Dependent Buildup of a Fano Resonance

The time-dependent Fano resonance profiles discussed so far are wholly based on Schrödinger quantum mechanics. To measure the Fano profile $|d_E(t_c)|^2$ at time $t = t_c$, the autoionization process at that time must be terminated instantaneously. This can be done by removing the bound state part of the wave packet or promoting the continuum wave packet to new eigenstates that have different energy (and symmetry) such that the bound state cannot decay to it. In either case, this would require a typical attosecond pulse that has a duration much less than the lifetime of the resonance. Take the $2s2p$ resonance of

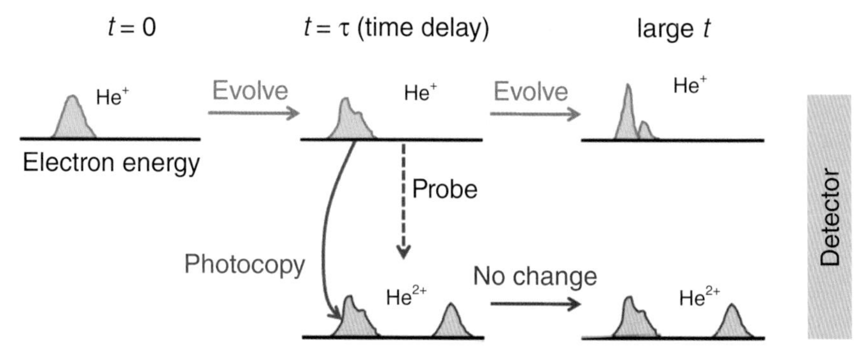

Figure 8.5 A scheme for probing the time evolution of a Fano resonance experimentally using two attosecond pulses. In this scheme, photoelectron spectra are measured. See the text for description of the method.

helium as the example (see Figure 8.5). An attosecond pulse is used to excite the doubly excited state. At $t = 0$, the Fano profile is sketched on the far left of the figure. At a later time, say $t = 4$ fs, the profile has evolved to the middle figure. If the autoionization is "allowed" to continue indefinitely, the usual Fano profile will be observed at the detector as shown. Typical electron detectors take more than hundreds of picoseconds to make a measurement that is far longer than the lifetime of the resonance, and thus only the final profile is registered. However, if a 250 eV 100 as pulse is used to ionize the 1s electron at $t = 4$ fs, then the $1sEp$ continuum is promoted to a $E'pEp$ double continuum wave packet. The profile of the latter is sketched in the bottom half of the figure. Since the bound part of the resonance cannot decay to this channel, the spectral profile of the Fano resonance at $t = 4$ fs is not modified anymore at later times as sketched in the bottom figure. This method requires two attosecond pulses that are not yet available.

8.1.5 First Experimental Observation of the Ultrafast Buildup of a Fano Resonance in the Time Domain

Fano resonances can also be measured by observing the absorption of the XUV pulse through a dilute gas medium. The time-dependent induced dipole by the XUV light pulse is calculated below. Assuming that the resonance is created at $t = 0$, the wave packet for $t > 0$ consisting of the ground state and a single resonance can be written as

$$|\Psi(t)\rangle \approx e^{-iE_g t}|g\rangle + C_\alpha(t)|\alpha\rangle + \int C_E(t)|\beta_E\rangle dE, \tag{8.11}$$

as in Section 8.1.2. The coefficients satisfy the coupled equations

$$\dot{C}_E(t) = -iVC_\alpha(t) - iEC_E(t), \tag{8.12}$$

$$\dot{C}_\alpha(t) = -iE_r C_\alpha(t) - iV\int C_E(t)dE. \tag{8.13}$$

The coefficients are given in Equations 8.6 and 8.7. Assuming that the electric field is polarized along the z-axis, the induced dipole for $t > 0$ is

$$d(t) = \langle\Psi(t)|z|\Psi(t)\rangle = C_\alpha(t)e^{iE_g t}\langle\alpha|z|g\rangle^* + \int C_E(t)e^{iE_g t}\langle\beta_E|z|g\rangle^* dE + c.c.$$

$$= C_\alpha^{(0)}\langle\alpha|z|g\rangle^* e^{-i\Omega_r t}\left\{\left(1-\frac{i}{q}\right)e^{-\frac{\Gamma}{2}t} + \frac{1}{(\pi V q)^2}\int\frac{(q+\varepsilon)e^{-i\frac{\Gamma}{2}\varepsilon t} - (q-i)e^{-\frac{\Gamma}{2}t}}{\varepsilon + i}dE\right\}$$

$$+ c.c., \tag{8.14}$$

where $\Omega_r = E_r - E_g$ is the resonance frequency. The rotating-wave approximation can be applied to drop the complex conjugate part for XUV absorption. With the help of $\int_{-\infty}^{\infty}\frac{1}{\varepsilon+i}d\varepsilon = -i\pi$, Equation 8.14 can be simplified to

$$d(t) \propto i\left[2\delta(t) + \frac{\Gamma}{2}(q-i)^2 e^{-\frac{\Gamma}{2}t}e^{-i\Omega_r t}\right]. \tag{8.15}$$

The photoabsorption cross-section (or optical density (OD)) is proportional to the imaginary part of the Fourier transform of $d(t)$. By focusing on the narrow energy region near the resonance, one obtains

$$\sigma(\Omega) \propto \mathrm{Im}\left[\int_0^\infty d(t)e^{i\Omega t}dt\right] \propto \mathrm{Re}\left[1 + \frac{\Gamma}{2}(q-i)^2\int_0^\infty e^{-\frac{\Gamma}{2}t}e^{i\Delta\Omega t}dt\right] \tag{8.16}$$

$$\propto \mathrm{Re}\left[1 + \frac{(q-i)^2}{1-i\varepsilon}\right] \propto \frac{(q+\varepsilon)^2}{1+\varepsilon^2}, \tag{8.17}$$

where $\Delta\Omega = \Omega - \Omega_r$ and $\varepsilon = \frac{\Delta\Omega}{\Gamma/2}$.

If it is assumed that autoionization terminates at $t = \tau$, the induced dipole in Equation 8.15 becomes

$$d(t) \propto \begin{cases} i\left[2\delta(t) + \frac{\Gamma}{2}(q-i)^2 e^{-\frac{\Gamma}{2}t}e^{-i\Omega_r t}\right] & 0 \le t < \tau \\ 0 & t \ge \tau. \end{cases} \tag{8.18}$$

Then, the upper integration in Equation 8.16 is limited to τ instead of infinity. The absorption cross-section becomes

$$\sigma(\Omega,\tau) \propto \mathrm{Re}\left[1 + \frac{(q-i)^2}{1-i\varepsilon}\left(1 - e^{-\frac{\Gamma}{2}\tau}e^{i\Delta\Omega\tau}\right)\right]$$

$$= \frac{(q+\varepsilon)^2}{1+\varepsilon^2} - e^{-\frac{\Gamma}{2}\tau}\frac{(1+q^2)}{\sqrt{1+\varepsilon^2}}\cos[\Delta\Omega\tau + \varphi(\varepsilon)], \tag{8.19}$$

in which

$$\tan\varphi(\varepsilon) = \frac{q^2\varepsilon - \varepsilon - 2q}{q^2 - 1 + 2q\varepsilon}. \tag{8.20}$$

Equation 8.19 shows that photoabsorption cross-sections can be used to probe the time evolution of the buildup of a Fano resonance if one can use an intense delta pulse to fully remove the decaying part of the resonance. Conditions very similar to this limit have been used in the experiment reported by Kaldun et al. [6].

Figure 8.6 shows the experimental result of the buildup of the $2s2p\ ^1P$ resonance of helium. In the experiment, a 150 as XUV pulse was used to photoionize the helium with a photon energy range from 50 to 72 eV. Then a 7 fs, 740 nm, near-infrared (NIR) laser with intensity of about 10^{13} W/cm^2 was used to rapidly ionize the $2s2p$ part of the resonance. By changing the time delay between the NIR with respect to the XUV, the absorption spectrum from each time delay is put together to show the buildup of the $2s2p$ resonance.

To demonstrate that the transient photoabsorption spectrogram shown in Figure 8.6 can indeed be interpreted as a "movie" of the buildup of the Fano resonance even though the experiment is a measurement of the photoabsorption spectra by a two-color pulse directly, Figure 8.7 displays the lineout of the resonance profiles at a few time delays from the experiment, the *ab initio* TDSE calculation of the absorption spectra, and the analytical theory predicted by Equation 8.19. The good agreement of the analytical theory with the other two results confirms that this interpretation is correct in spite of the expected small discrepancies when the NIR pulse is overlapping with the XUV pulse. Note that the predicted and observed buildup of Fano resonances are generally applicable, i.e., independent of the species. They only depend on the typical Fano resonance parameters: the lifetime, shape parameter q, and resonance energy.

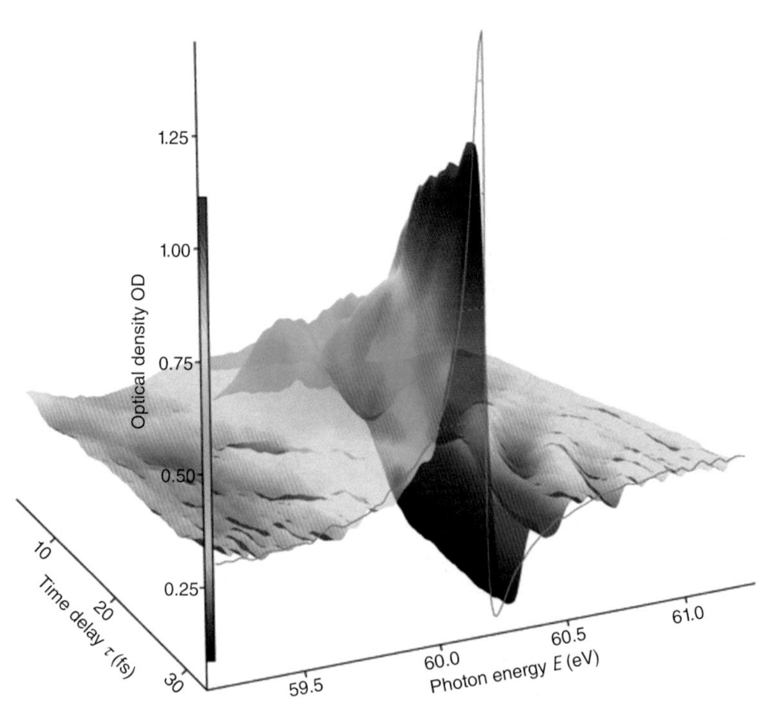

Figure 8.6 Experimental transient absorption spectrogram of the helium $2s2p$ resonance as a function of photon energy E and time delay $\tau > 0$ after the resonance is excited by a 150 as XUV pulse. At $t = \tau$, an intense 7 fs NIR laser pulse was used to completely ionize the bound part of the resonance within the first few femtoseconds of the NIR pulse. (From A. Kaldun et al., *Science*, **354**, 738 (2016) [6]. Reprinted with permission from AAAS.)

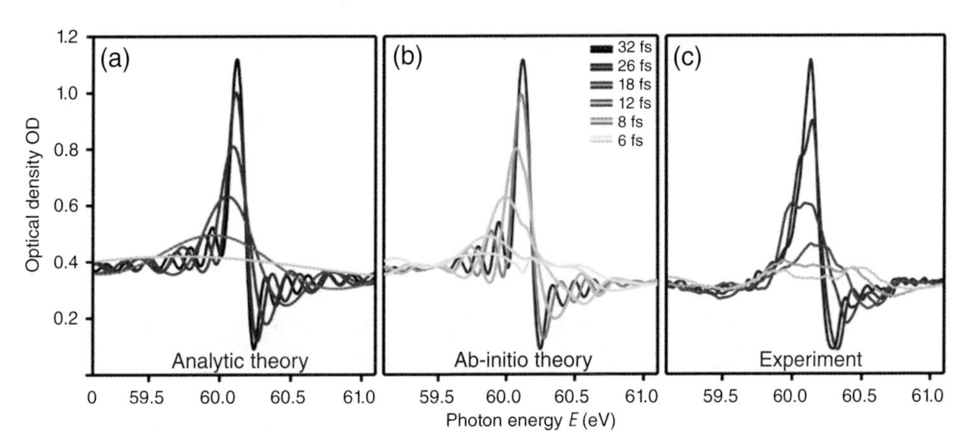

Figure 8.7 Comparison between analytic theory, *ab initio* calculation, and experimental results for the helium 2*s*2*p* Fano line formation. (a) Absorption spectra calculated for a series of time delays between XUV and NIR according to the analytic expression of Equation 8.19. (b) Numerically simulated absorption spectra for a 7 fs FWHM NIR pulse with peak intensity of 20 TW/cm^2. (c) Experimentally recorded spectra. (From A. Kaldun et al., *Science*, **354**, 738 (2016) [6]. Reprinted with permission from AAAS.)

8.2 Attosecond Transient Absorption Spectroscopy

8.2.1 Introduction

In attosecond transient absorption spectroscopy (ATAS), an attosecond XUV pulse typically excites an electron wave packet, which is followed by a moderately intense IR pulse some time later. For photoabsorption, the XUV alone generates a time-dependent dipole moment $d(t)$. The Fourier transform of this $d(t)$ in the frequency domain $d(\omega)$ gives the photoabsorption spectrum. In the presence of the IR field, the electron wave packet is modified and so is the time-dependent dipole $d(t)$. Experimentally, the effect of the IR will be reflected in the photoabsorption spectra. By measuring the change of the photoabsorption spectra versus the time delay between the two pulses, an ATA spectrogram similar to a photoelectron-streaking spectrogram is obtained albeit with the advantage of better spectral resolution.

Today, ATAS experiments are mostly studied with the isolated attosecond pulses (IAPs) obtained from high-order harmonics generated with the IRs. Routine generation of IAPs is still quite challenging. Typical photon energies of the IAP range from 20 to 110 eV and pulse durations from about 100 to 500 as. The IAPs are mostly generated from titanium-sapphire lasers with wavelengths close to 800 nm. The XUV and IR are conveniently phase matched and synchronized, but determination of the absolute time delay is difficult. In ATAS it is desirable that the IR is not intense enough to ionize the target from the ground state. Thus IR peak intensity is typically maintained below 1.0×10^{13} W/cm^2. At intensities within the range of 10^{11} W/cm^2 and 10^{12} W/cm^2, the IR can interact with the excited states nonlinearly and modify the amplitude and phase of the electron wave packet. In a typical

ATA experiment today, the IAP is to excite the atom and create a complex electron wave packet. To extract the phase of the wave packet, a nonlinear IR probe is used. On the other hand, the interval where the two pulses overlap is also of great interest. In this case, the stronger IR is said to dress the atoms and modify the medium's polarizability. The ATAS spectra have been shown to exhibit similar features like those observed in nonlinear optics such as electromagnetically induced transparency (EIT), coherent population transfer, and others [7].

8.2.2 Formulation of ATAS at the Single Atom Level

In ATAS, there is exchange of energy between the light (the two-color field) and the atom. First, a generalized cross-section for the nonlinear interaction between the light field and an atom [8] is formulated. The Hamiltonian of an atom in the two-color field can be written as

$$H = H_A + \varepsilon(t)z. \tag{8.21}$$

Here, $\varepsilon(t)$ is the two-color field consisting of an IAP in the XUV regime and an IR laser with wavelength near 800 nm. Both XUV and IR are linearly polarized along the z-direction. The electron in the atom is described by H_A. The time-dependent Schrödinger equation is

$$i\frac{\partial}{\partial t}|\Psi\rangle = H|\Psi\rangle. \tag{8.22}$$

The rate of change of the energy E of the atom is calculated as

$$\frac{dE}{dt} = \frac{d}{dt}\langle\Psi|H|\Psi\rangle = \langle\Psi|\frac{\partial H}{\partial t}|\Psi\rangle = \langle\Psi|z|\Psi\rangle\frac{\partial\varepsilon}{\partial t} = -\mu(t)\frac{\partial\varepsilon}{\partial t}, \tag{8.23}$$

where the induced dipole in the time domain is expressed by $\mu(t)$. Using the relation $\varepsilon^*(\omega) = \varepsilon(-\omega)$, one can write

$$\Delta E = -\int_{-\infty}^{\infty} \mu(t)\frac{\partial\varepsilon}{\partial t}dt = -\int_{-\infty}^{\infty} i\omega\mu(\omega)\varepsilon^*(\omega)d\omega = \int_{0}^{\infty} \omega S(\omega)d\omega, \tag{8.24}$$

where

$$S(\omega) = -2\mathrm{Im}\left[\mu(\omega)\varepsilon^*(\omega)\right]. \tag{8.25}$$

A generalized absorption cross-section $\bar{\sigma}(\omega)$ can be defined as the ratio of energy absorbed per unit time per unit frequency divided by the incident intensity at a given frequency

$$\bar{\sigma}(\omega) = \frac{\omega S(\omega)}{I_0(\omega)} = \frac{\omega S(\omega)}{c|\varepsilon(\omega)|^2/4\pi} = 4\pi\alpha\frac{\omega S(\omega)}{|\varepsilon(\omega)|^2}, \tag{8.26}$$

where c is given by the inverse of the fine-structure constant α. This generalized absorption cross-section can be used for both weak- and strong-field light pulses.

8.2.3 Propagation of Light in the Transmission Medium

When light interacts linearly (proportional to the intensity) with the medium, the absorption is described by Beer's law

$$N(\omega, x) = N(\omega, 0)e^{-\rho\sigma(\omega)x}, \tag{8.27}$$

where $N(\omega, 0)$ and $N(\omega, x)$ are the intensities at the entrance and at position x of the medium, ρ is the number density, and $\sigma(\omega)$ is the absorption cross-section. This expression requires that the cross-section is only a function of energy and has no spatial and temporal dependence. For the XUV+IR experiments, the response of the medium is more complicated, where the propagation of light is treated similarly to the propagation of harmonics generated in the medium. There is one difference: the IR intensity in ATAS is typically one or two orders weaker such that the IR is not tightly focused. Thus, one can assume that the electric field is independent of the transverse coordinates. By expressing time in the moving frame $t' = t - x/c$, the Maxwell equation of wave propagation is reduced to

$$\frac{\partial E(x, t')}{\partial x} = -\frac{\rho}{c\varepsilon_0}\frac{\partial \mu(x, t')}{\partial t'}. \tag{8.28}$$

In the propagation the XUV and the IR can be calculated separately. The input to Equation 8.28 on the right-hand side is the induced dipole from each atom.

8.2.4 Calculation of Single-Atom-Induced Transition Dipole

To obtain the induced dipole by the XUV+IR pulse at each time delay, one can, in principle, solve the TDSE, evaluate $d(t) = \langle\Psi(t)|z|\Psi(t)\rangle$, and then take the Fourier transform to obtain $d(\omega)$. Such calculations can be carried out accurately if the target is modeled as a one-electron atom. In some cases, it can also be done for the two-electron helium atom [9]. For simple systems, these calculations may be used as "experimental data" where the XUV and IR pulses are "exactly" known. They may serve to identify the origin of some features that appear in the experiments where full knowledge of the XUV and IR pulses are not generally available.

Extending the TDSE method to many-electron atoms and molecules is straightforward. However, numerically, it is much more difficult to achieve convergence. Recall that in ATAS experiments one can achieve meV energy resolution, and thus energy levels and many-electron wavefunctions have to be calculated accurately. ATAS theory for many-electron systems have been formulated using the density matrix theory [10], the multi-configuration time-dependent Hartree–Fock (MCTDHF) theory [11], and the Floquet method [12], but very few calculations have ever been carried out to the extent that can be compared with experiments. In the Floquet method, the XUV can be viewed as a probe of the IR-dressed states. Since the IR intensity in ATAS experiment is on the order of 10^{12} W/cm^2 or less, the interaction of IR with the target is in the multiphoton ionization regime, which is conveniently treated in terms of dressed-atom states or Floquet states.

8.3 General Features of ATA Spectra for Atoms below the First Ionization Threshold

8.3.1 Results from TDSE Calculations

Attosecond transient photoabsorption spectra generated by an attosecond XUV pulse in the presence of a few-femtosecond IR pulse have been widely studied on rare gas atoms in the last few years. Before discussing experimental data, the ATA spectrogram of helium calculated by solving the TDSE in Chen et al. [13] is shown in Figure 8.8. These numerical data are used to identify prominent spectral features where the XUV and IR are clearly specified. The calculation was carried out for a 330 as XUV pulse with central energy at 25 eV dressed by an 800 nm, 11 fs, peak intensity 3×10^{12} W/cm^2 IR laser on a helium atom, which was treated under the single-active-electron approximation. In the figure, the spectrogram is shown for photon energy from 19 eV to 26 eV. On the vertical scale, the positions of the "bright" $1snp$ (or np) states that can be populated by the XUV alone are shown. Also displayed are the positions of the so-called *laser-induced states* (LIS). These are intermediate states that can be reached from "dark states" like $2s$, $3s$, $3d$ by absorbing or emitting one IR photon. Thus the $2s$ final state can be reached from the $2s-$ state by absorbing one IR photon or by emitting an IR photon from the $2s+$ state. Alternatively, they

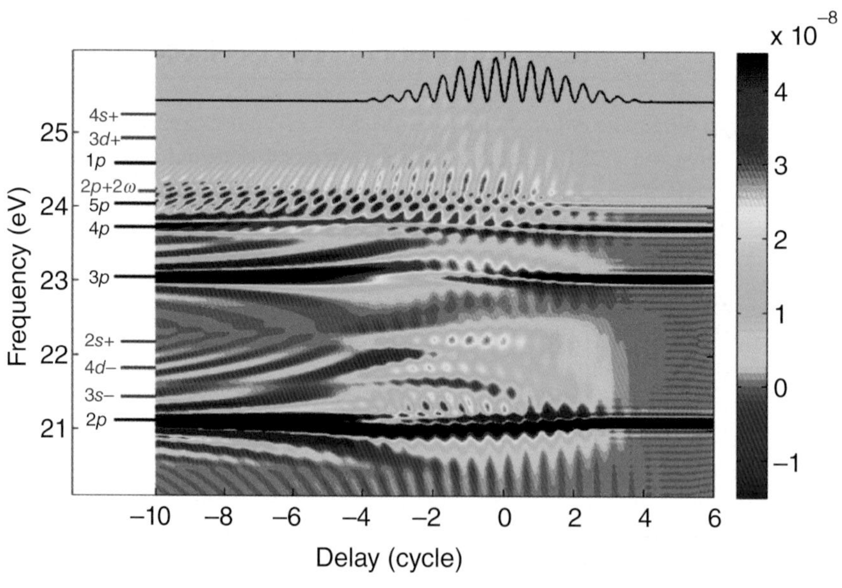

Figure 8.8 A typical ATA spectrogram of He calculated by solving the TDSE. A 330 as IAP with central energy of 25 eV and an 800 nm IR pulse with intensity 3×10^{12} W/cm^2 at varying time delays are used to generate the ATA spectra. The IR laser field is sketched at the top of the figure. Negative time delay means XUV comes in first. This figure is used in the text to help explain important features of the observed ATA spectra. (Reprinted from S. Chen, M. Wu, M. B. Gaarde, and K. J. Schafer, *Phys. Rev. A*, **87**, 033408 (2013) [13]. Copyrighted by the American Physical Society.)

can be understood in the dressed-atom picture, where $2s+$ and $2s-$ would be designated as $(2s, 1)$ and $(2s, -1)$, respectively. The first symbol refers to atomic state and the second refers to the excess or shortage of IR photons in the dressed-atom picture. Thus $2p + 2\omega$ on the vertical axis is $(2p, 2)$ and is reached after absorbing two IR photons from the $2p$. The horizontal axis is the time delay in terms of the optical cycle of the 800 nm IR laser. For negative time delay the XUV comes before the IR. The IR electric field is shown on the very top of the graph.

There are a few prominent features in the computed spectrogram: (i) In the overlapping region, the spectral lines of bright states like $2p$, $3p$, and $4p$ are shifted. They are due to the AC Stark shift. (ii) In the overlapping region, the spectral shapes of the bright states are broadened. (iii) For negative time delay, pronounced hyperbolic curves appear for both the dark and bright states in the nonoverlapping region. (iv) Between $2p$ and $3p$, light-induced states $2s+$, $3d-$, and $3s-$ are clearly visible and similar weaker features are seen for the $2s-$ state. (v) There is clear evidence of half-cycle modulation (or 2ω of the IR) with respect to the time delay in the overlapping region.

8.3.2 AC Stark Shift and Broadening of Bright States: One-Level Model

A bound electron in the presence of an intense laser field with frequency ω_L far from the atomic level $|a\rangle$ will experience an AC Stark shift. For a pulse of the form $\varepsilon_L(t) = \varepsilon_0(t)\cos(\omega_L t)$, the time-dependent wavefunction can be expanded as

$$|\Psi_{\rm NIR}(\mathbf{r}, t)\rangle = \sum_k c_k(t) e^{-iE_k t} |\psi_k(\mathbf{r})\rangle, \tag{8.29}$$

using the complete basis set $|\psi_k\rangle$ of the atomic Hamiltonian. This equation can be rewritten as

$$|\Psi_{\rm NIR}(\mathbf{r}, t)\rangle = \sum_k |c_k(t)| e^{-i\int_{-\infty}^{t}[E_k + \delta E_k(t')]dt'} |\psi_k(\mathbf{r})\rangle. \tag{8.30}$$

In this form, one defines a sub-cycle Stark shift [14]. Under the second-order perturbation approximation, the shift is

$$\begin{aligned}\delta E_a(t) = -i\sum_{k\neq a} d_{ak}\varepsilon_0(t)\cos(\omega_L t)e^{i\omega_{ak}t}\int_{-\infty}^{t} d_{ka}\varepsilon_0(t')\\ \times \cos(\omega_L t')e^{i\omega_{ka}t'}dt'.\end{aligned} \tag{8.31}$$

The integral can be evaluated analytically if the field envelope is given by a double-exponential shape, $\varepsilon_0(t) = \varepsilon_p e^{-|t|/\tau_p}$,

$$\begin{aligned}\delta E_a(t) &= \frac{1}{2}\varepsilon_0(t)^2 \sum_{k\neq a}\left[\frac{\omega_{ka}|d_{ka}|^2}{\omega_{ka}^2 - \omega_L^2}\cos^2(\omega_L t) - i\frac{\omega_L|d_{ka}|^2}{\omega_{ka}^2 - \omega_L^2}\sin(2\omega_L t)\right]\\ &= \frac{1}{2}\varepsilon_0(t)^2[\alpha_a\cos^2(\omega_L t) - i\gamma_a\sin(2\omega_L t)],\end{aligned} \tag{8.32}$$

where

$$\alpha_a = \sum_{k \neq a} \frac{\omega_{ka}|d_{ka}|^2}{\omega_{ka}^2 - \omega_L^2}, \tag{8.33}$$

and

$$\gamma_a = \sum_{k \neq a} \frac{\omega_L|d_{ka}|^2}{\omega_{ka}^2 - \omega_L^2}. \tag{8.34}$$

In this formulation, the real part of $\delta E_a(t)$ is related to the AC Stark shift and the imaginary part is related to the decay amplitude. The results show that an intense laser field can cause level shift and broadening of a bound state. These features should appear when an extreme ultraviolet (EUV) light is used to probe a laser-dressed atom near the absorption line. In Figure 8.8 each bright state in the calculated spectra indeed exhibits such features. The above formulae were derived for weak IR fields such that second-order perturbation theory can be applied. One expects such general features when the IR intensity is increased as well.

8.3.3 Laser-Induced States and Autler–Townes Splitting: Two-Level Model

The model above is not valid if two levels $|1\rangle$ and $|2\rangle$ are separated by a frequency ω_{12} nearly equal to ω_L. In this case, the system is similar to the well-known Rabi two-level problem. The time-dependent wavefunction can be expressed as

$$|\Psi(t)\rangle = c_1(t)e^{-iE_1t}|1\rangle + c_2(t)e^{-iE_2t}|2\rangle + c_g(t)e^{-iE_gt}|g\rangle. \tag{8.35}$$

Once the wavefunction Equation 8.35 is obtained, it is straightforward to obtain $d(t)$ from which $d(\omega)$ and the ATA spectra are calculated. Figure 8.9 shows the typical spectrograms for the case where $\omega_L = 1.55$ eV, for $\omega_{12} = 2.48$ eV (a), and $\omega_{12} = 1.55$ eV (b), respectively.

In Figure 8.9(a), where the resonance condition does not meet, one can observe the broadening and the AC Stark shift of the bright states. A light-induced state is also observed from this model when the coupling laser intensity is strong. In Figure 8.9(b), where the resonance condition is met, the laser coupling splits a single bright state into two. This is called Autler–Townes splitting. In both cases, the two-level model also predicts sub-cycle ($2\omega_L$) oscillation versus the time delay and the existence of higher-order LISs.

8.3.4 Comparison of Theoretical and Experimental ATA Spectrograms

The ATA spectrogram for photon energies between 20 and 24 eV for helium has been reported experimentally and compared to theoretical calculations by solving the TDSE (see Figure 8.10). Theoretical calculations clearly show that there are three LISs between $1s2p$ and $1s3p$ main lines ($3s$-, $3d$-, and $2s$+) from the bottom to the top. However, the spectral shapes observed in the experiment do not compare well with the TDSE simulation. Since the XUV and laser-pulse intensity are usually not well characterized

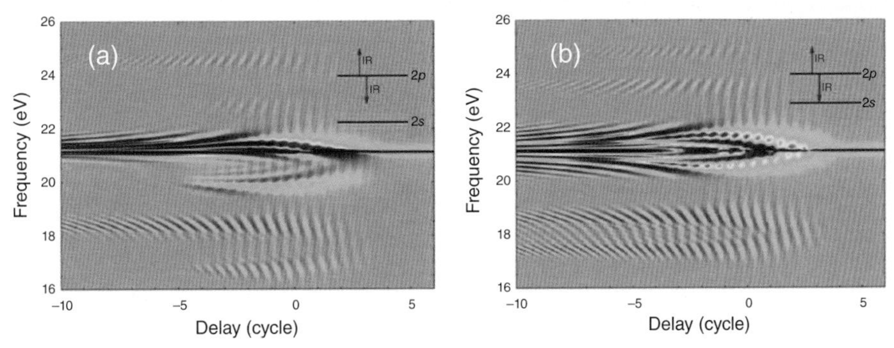

Figure 8.9 ATA spectrogram calculated using an IR-coupled two-level model. Intensity of IR is 1.2×10^{13} W/cm^2, $\omega_L = 1.55$ eV. The $2p$- to $2s$-level energy difference is taken to be 2.48 eV in (a) and 1.55 eV in (b). In (a) clear LIS can be seen and in (b) Autler–Townes doublet splitting emerges. (Reprinted from M. Wu et al., *J. Phys. B-At. Mol. Opt.*, **49**, 062003 (2016) [15]. Copyrighted by IOP publishing.)

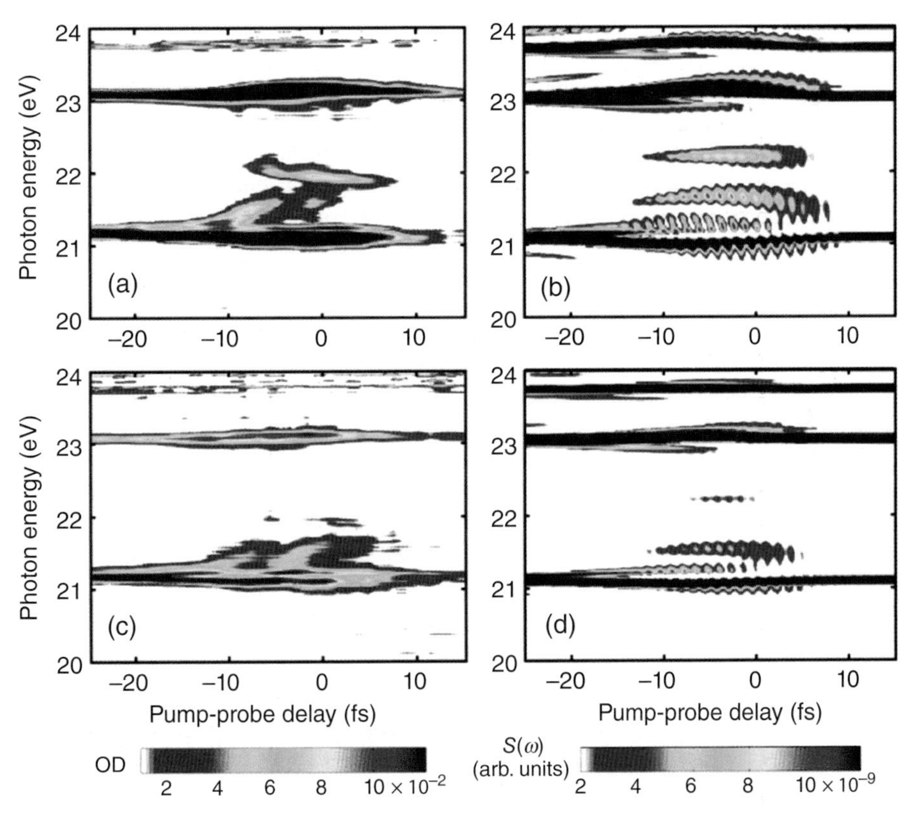

Figure 8.10 Experimental ATA spectra (left) and calculated single-atom frequency dependent response (right) as a function of time delay. Negative time delay means XUV comes in first. XUV is 400 as and IR intensity is 1.6×10^{12} W/cm^2 in (a), (b) and 4.8×10^{11} W/cm^2 in (c), (d), with the pulse duration of 12 fs. The two main horizontal lines are $1s2p$ and $1s3p$ of helium. LISs are clearly seen in between. (Reprinted from S. Chen et al., *Phys. Rev. A*, **86**, 063408 (2012) [16]. Copyrighted by the American Physical Society.)

in the experiment, precise agreement at the quantitative level is difficult. Furthermore, experiments measure light transmission through the gas cell while calculations are for the single-atom response only. In Section 8.2.3, theory of the propagation of the XUV in the medium was discussed. In Section 8.5, effects of propagation on the spectra will be addressed. It is noted that, in the same spectral range of helium, ATA spectrograms have been reported in other experiments [14, 17]. While the general experimental features can be understood by TDSE-level simulations, agreement in the fine details has not been possible.

8.4 ATA Spectrogram for Autoionizing States

8.4.1 Model Three-Level System Coupled by Two Ultrashort Pulses

The ATA spectra are often taken with XUV attosecond pulses with central energy from 20 to 120 eV so far. These energies lie above the first ionization threshold of most atoms and molecules, where many autoionizing states are present and a fair number of ATA experiments have already been carried out for He, Ne, Ar, and Xe atoms. Here, focus is placed on the ATA spectra near the $2s2p$ autoionizing state of helium as these spectra have been studied extensively in many experiments. First consider a three-state model that includes the ground state, one bright autoionizing state, and one dark autoionizing state, respectively.

Consider the autoionizing three-level system coupled by two ultrashort pulses as indicated in Figure 8.11. $|E_1\rangle$ can be considered as the continuum part of the $2s2p\ ^1P$ and $|E_2\rangle$ can be considered as that for the $2s^2\ ^1S$ or $2p^2\ ^1S$, while $|a\rangle$ and $|b\rangle$ are their respective bound parts, and $|g\rangle$ is the ground state. In this truncated space, the time-dependent wavefunction can be expressed as

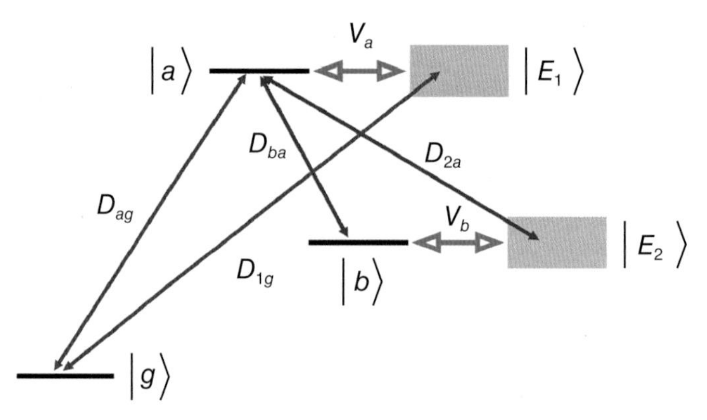

Figure 8.11 Scheme of a model three-level system coupled by a short XUV and an IR pulse. Each autoionizing state is written as a bound part and a continuum part. Couplings between the states included are indicated by arrows, V for autoionization and D for dipole coupling by the light field.

$$|\Psi(t)\rangle = e^{-iE_g t} c_g(t)|g\rangle + e^{-iE_X t}\left[d_a(t)|a\rangle + \int d_{E_1}(t)|E_1\rangle dE_1\right]$$
$$+ e^{-iE_L t}\left[d_b(t)|b\rangle + \int d_{E_2}(t)|E_2\rangle dE_2\right], \qquad (8.36)$$

where E_g is the ground state energy, and $E_X \equiv E_g + \omega_X$ and $E_L \equiv E_g + \omega_X - \omega_L$ represent the central energies pumped by the pulses. The total Hamiltonian is $H(t) = H_A + H_X(t) + H_L(t)$, where H_A is the atomic Hamiltonian and $H_X(t)$ and $H_L(t)$ are the dipole interactions of the atomic system with the XUV and laser, respectively. Both pulses are assumed to be in the form of $E(t) = F(t)e^{i\omega t} + F^*(t)e^{-i\omega t}$, where ω is the carrier frequency and $F(t)$ is a cosine-square type function. Since the fast-oscillating terms are factored out, the $c(t)$ and $d(t)$ coefficients are smooth functions of time. All the V and D are taken as constants because the continuum waves vary slightly across the resonances. By neglecting the coupling of $|E_1\rangle$ to $|b\rangle$ and $|E_2\rangle$, the Schrödinger equation gives the coupled equations for all the coefficients, including the continuum ones, by

$$i\dot{d}_{E_1}(t) = (E_1 - E_X)d_{E_1}(t) - D_{1g}F_X^*(t)c_g(t) + V_a d_a(t), \qquad (8.37)$$
$$i\dot{d}_{E_2}(t) = (E_2 - E_L)d_{E_2}(t) - D_{2a}F_L^*(t)d_a(t) + V_b d_b(t). \qquad (8.38)$$

Using the adiabatic approximation [18, 19], i.e, assuming that the change of the continuum waves is slow by setting $\dot{d}_{E_1}(t) = \dot{d}_{E_2}(t) = 0$, then Equations 8.37 and 8.38 can be used to eliminate the continuum explicitly to result in coupled linear equations

$$i\dot{c}_g(t) = -i\frac{\gamma_g(t)}{2}c_g(t) - \lambda_a F_X(t)d_a(t), \qquad (8.39)$$
$$i\dot{d}_a(t) = -\lambda_a F_X^*(t)c_g(t) - \left[\delta_X + i\frac{\Gamma_a + \gamma_a(t)}{2}\right]d_a(t) - \lambda_b F_L(t)d_b(t), \qquad (8.40)$$
$$i\dot{d}_b(t) = -\lambda_b F_L^*(t)d_a(t) - \left[\delta_X + \delta_L + i\frac{\Gamma_b}{2}\right]d_b(t), \qquad (8.41)$$

for the bound-state coefficients, where $\lambda_a \equiv D_{ag} - i\pi V_a D_{1g}$ and $\lambda_b \equiv D_{ba} - i\pi V_b D_{2a}$ are the complex dipole-matrix elements combining the direct bound–bound transitions and the indirect transition going through the continua; $\delta_X \equiv \omega_X - (E_a - E_g)$ and $\delta_L \equiv \omega_L - (E_b - E_a)$ are the detunings of the fields; $\gamma_g(t) \equiv 2\pi|D_{1g}F_X(t)|^2$ and $\gamma_a(t) \equiv 2\pi|D_{2a}F_L(t)|^2$ are the laser-induced broadenings. Note that, in λ_a and λ_b, the two D values and V uniquely determine the q-parameter, as shown by Equation 8.8. The energy independence of the continua across the resonances removes the AC Stark shifts between $|g\rangle$ and $|E_1\rangle$ and between $|a\rangle$ and $|E_2\rangle$. By solving Equations 8.39 through 8.41, the bound-state part of the wavefunction in Equation 8.36 is obtained. With the bound coefficients $c_g(t)$, $d_a(t)$, and $d_b(t)$, we return to Equations 8.37 and 8.38 for the "second iteration" of the continuum coefficients. Note that Fano parameters and the dipole-matrix elements can be taken from the literature or from typical atomic structure calculations independently. For later analysis, we also introduce a generalized Rabi frequency defined by

$$\Omega(t) \equiv \sqrt{|D_{ba}E_L(t)|^2 + |\delta_L|^2}, \qquad (8.42)$$

which gives an estimate of the coupling strength between the bound parts of the two autoionizing states. The coupling strength over the whole pulse is represented by the pulse area

$$A \equiv \int_{-\infty}^{\infty} \Omega(t)dt. \tag{8.43}$$

Note that the pulse area is only calculable in a good resonance condition where $\Omega(t)$ is finite in time. For a π-pulse, i.e., $A = \pi$, the population in a two-state system is thoroughly transferred from one state to the other.

The three-level equations for autoionizing states presented here can be carried out numerically once the atomic parameters of the states involved are obtained. Once the wavefunctions for the autoionizing states are calculated, it is straightforward to obtain the photoelectron spectra or the transient photoabsorption spectra. For this model to work, the IR intensity should not be too high. Otherwise, ionization of the bound part of the autoionizing states or coupling to other bound or autoionizing states should be included.

In the limit of long pulses, the present three-level system has been widely used to study interesting phenomena such as EIT and coherent population trapping [7] in the optical region. These are effective all-optical control methods for light propagation in a medium. For autoionizing states, these can also be used to control the lineshape of the resonance. The following subsections describe selective examples on how short IR pulses can be used to manipulate photoelectron spectra or ATA spectra.

8.4.2 Time-Delayed Photoelectron and Photoabsorption Spectra

To make sure that the three-level model works well, consider the resonant coupling of the $2s2p(^1P)$ and $2s^2(^1S)$ autoionizing states (AISs) with a 540 nm laser pulse. The binding energies for both AISs are high so ionization by the laser is totally disregarded; thus, the main dynamics are the Rabi oscillation between the two AISs. By adjusting the coupling pulse, the resonance profiles can be manipulated more flexibly and forcefully. Thus, they provide a tool for the coherent control of electrons and photons.

Consider an attosecond pulse with duration 100 as, peak intensity 10^{10} W/cm^2, and central photon energy of 60 eV, used to excite a helium atom in the presence of a visible light pulse at 540 nm, duration 9 fs, and intensity 0.7×10^{12} W/cm^2. The photoelectron spectrogram and the ATA spectra calculated with the three-state model are shown in the top frame of Figure 8.12. For positive time delays, the XUV is ahead of the visible. Clearly, the photoelectron spectra and the XUV absorption spectra have very similar features. At large negative time delay (around -10 fs) where XUV comes after the visible, the standard lineshape of the $2s2p$ resonance can be seen. When the visible and the XUV overlap and for a large positive time delay where the intense visible comes after the XUV, the Fano resonance is severely modified by the Rabi coupling between $2s2p$ and $2s^2$. Rabi coupling depends on the time delay as well as the peak IR intensity. The bottom frame of Figure 8.12 shows how electron profiles as well as absorption profiles change with the intensity of the visible at the fixed time delay of $t_0 = 15$ fs. The laser-free resonance profile

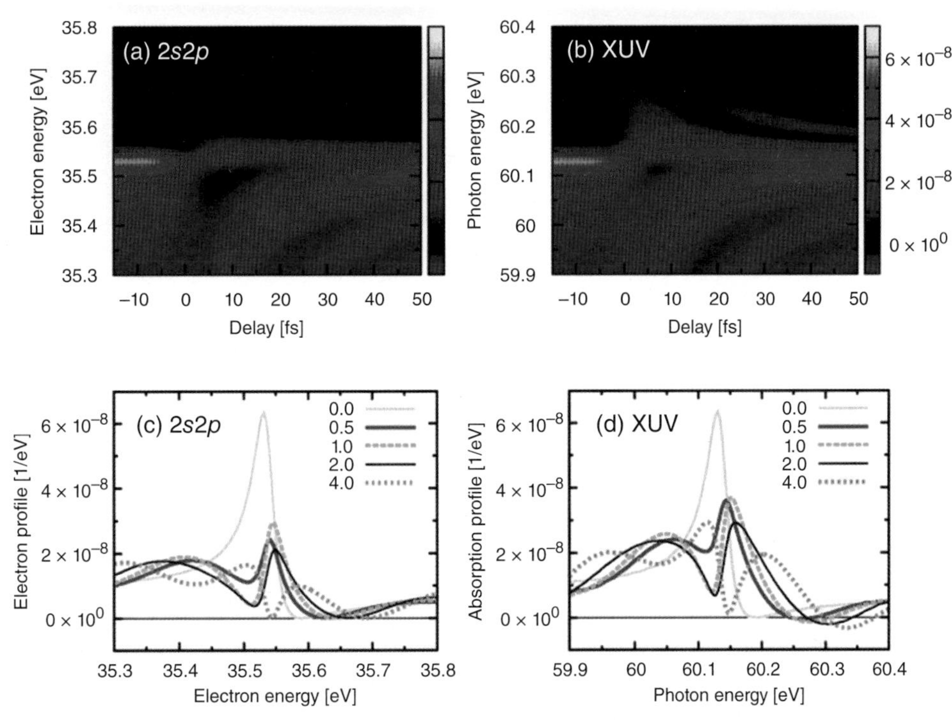

Figure 8.12 Photoelectron and photoabsorption spectra predicted by the three-level model: top frames for the time-delay dependence and bottom frames for the intensity dependence at a fixed time delay. Positive delay means IR comes after the XUV. Intensity of the IR is in units of 10^{12} W/cm^2. (Figure adapted with permission from W. C. Chu and C. D. Lin, *Phys. Rev. A*, **85**, 013409 (2012) [19]. Copyrighted by the American Physical Society.)

(shown in gray lines) are compared to profiles calculated with dressed laser fields. Note that electron and absorption profiles look quite similar. This demonstrates that both are due to the strong modification of the bound part of the wavefunction of the AISs. In other words, atomic lineshape of an AIS can be controlled by a moderate, short IR laser if the near-resonant condition is met. Such control is no different from the familiar EIT where a strong absorption can be changed to strong transmission by detuning the coupling IR laser.

To see the role of Rabi coupling in modifying the electron and absorption spectra, Figure 8.13 shows what happens to the lineshape when the effective pulse area (Equation 8.43) reaches 2π at $t_0 = 3.5$ fs. At this specific time delay, the electron has gone through one full Rabi cycle: going from $2s2p$ to $2s^2$ and then back to $2s2p$. On the left column of Figure 8.13, note that, at this particular time delay, the Fano resonance in the electron spectra changes its lineshape by flipping q to $-q$, (i.e., a left–right flip), while the absorption profile flips from up to down (i.e., absorption to emission).

Another interesting consequence of the three-level model is what happens to Autler–Townes splitting when the IR couples the two levels resonantly but the IR pulse is short. (See Section 8.3.3.)

For the short pulses, the IR has a spectral bandwidth. Although Autler–Townes doublets have been observed with XUV and IR for the $2s2p$ state of helium [21], the durations for

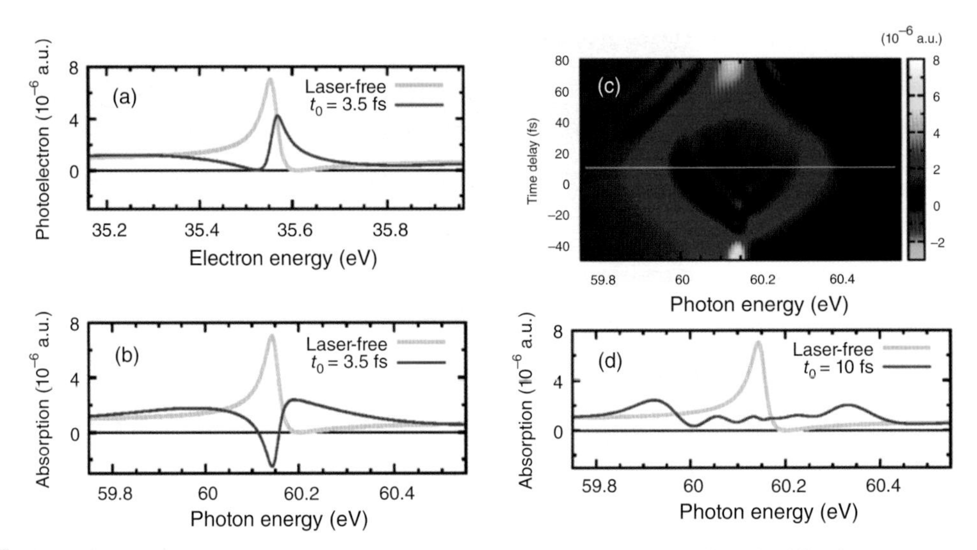

Figure 8.13 A coupling laser can change the lineshape dramatically when the Rabi oscillation goes through a full cycle at $A = 2\pi$ (see Equation 8.43). In the photoelectron spectrum, the resonance flips left and right (a); in the absorption spectrum, the resonance flips up and down (b). (c) An absorption spectrum shows the Autler–Townes doublet splitting for a 50 fs IR dressing pulse. The largest splitting occurs at the time delay when the Rabi frequency (Equation 8.42) is largest (see the lineout for $t = 10$ fs in (d)). (Figure adapted with permission from W. C. Chu and C. D. Lin, *Phys. Rev. A*, **87**, 013415 (2013) [20]. Copyrighted by the American Physical Society.)

both pulses are on the order of 40–80 fs. In the three-level simulation with a 200 as XUV pulse and 50 fs dressing-laser pulse, the signature of the Autler–Townes doublet can be observed where the splitting depends on the time delay, as seen in the right column of Figure 8.13. The splitting is largest when the Rabi frequency (Equation 8.42) is at the peak value that occurs at $t_0 = 10$ fs in the model (see Figure 8.13(d)).

8.4.3 Short-Pulse Approximation

Three-Level Model

The three-level model can be calculated numerically, but it is of interest to look at the short-pulse limit when both the IR and XUV are taken as delta functions. This limit is valid for the attosecond XUV pulse but a bit more stretched for the IR pulse, which normally has a duration of a few to tens of femtoseconds. However, for AISs that have lifetime greater than, say, 20 fs, this would be a reasonable approximation. In the short-pulse approximation, we assume that the amplitude for the ground state is $c_g = 1$. The dipole between $|a\rangle$ and $|E_2\rangle$ is zero (since the excitation energy is very small) such that one can take $q_b \to \infty$. The XUV is turned on at $t = 0$, and the IR is turned on at $t = \tau$ with a duration of δt. The wavefunction between $t = 0$ and τ is straightforward, but the wavefunction for $t > \tau + \delta t$ is needed. By defining

$$\Omega_X(t) \equiv A_X\delta(t), \quad \Omega_L(t) \equiv A_L\delta(t-\tau), \tag{8.44}$$

the bound-state, coupled equations become

$$i\dot{d}_a(t) = -\left(1 - \frac{i}{q_a}\right)\frac{\Omega_X(t)}{2} - i\frac{\Gamma_a}{2}d_a(t) - \frac{\Omega_L(t)}{2}d_b(t), \tag{8.45}$$

$$i\dot{d}_b(t) = -\frac{\Omega_L(t)}{2}d_a(t) - i\frac{\Gamma_b}{2}d_b(t). \tag{8.46}$$

These two equations can be solved for $t > \tau + \delta t$,

$$d_a(t) = \frac{i}{2}\left(1 - \frac{i}{q_a}\right)A_X \cos\left(\frac{A_L}{2}\right)\exp\left(-\frac{\Gamma_a}{2}t\right), \tag{8.47}$$

$$d_b(t) = \frac{i}{2}\left(1 - \frac{i}{q_a}\right)A_X \sin\left(\frac{A_L}{2}\right)\exp\left(\frac{\Gamma_b - \Gamma_a}{2}\tau\right)\exp\left(-\frac{\Gamma_b}{2}t\right). \tag{8.48}$$

By substituting these amplitudes into the continuum equations, an analytic expression for $d_{E_1}(t)$ can be solved. By taking t to infinity, the electron spectrum versus the time delay τ is obtained. Using scaled parameters $A = A_X^2/2\pi q_a^2\Gamma_a$, $T_0 = \Gamma_a\tau/2$, and scaled energy $\varepsilon = 2(E - E_a)/\Gamma_a$, and defining $c = 1 - \cos(A_L/2)$, the electron spectrum is given as

$$P(\varepsilon, A_L, T_0) = \frac{A}{1+\varepsilon^2}\left\{(q_a+\varepsilon)^2 - ce^{-T_0}2(q_a+\varepsilon)\Big[q_a\cos(\varepsilon T_0) + \sin(\varepsilon T_0)\Big] + \left(ce^{-T_0}\right)^2\left(q_a^2+1\right)\right\}. \tag{8.49}$$

This equation is reduced to Fano resonance $P(\varepsilon) = A\frac{(q_a+\varepsilon)^2}{1+\varepsilon^2}$ if $A_L = 0$ (or $c = 0$). For a very small time delay $T_0 \to 0$, $P(\varepsilon) \to A$ if $A_L = \pi$ (or $c = 1$), and $P(\varepsilon) \to A\frac{(-q_a+\varepsilon)^2+4}{1+\varepsilon^2}$ if $A_L = 2\pi$ (or $c = 2$). In the last case, the lineshape is similar to a resonance of $-q_a$.

A similar derivation can be obtained for transient absorption spectra. Expressing it in terms of scaled quantities, the spectral response is given by

$$S(\varepsilon, A_L, T_0) = \frac{A}{1+\varepsilon^2}\left\{(q_a+\varepsilon)^2 - ce^{-T_0}\Big[(q_a^2 + 2q_a\varepsilon - 1)\cos(\varepsilon T_0) - (\varepsilon q_a^2 - 2q_a - \varepsilon)\sin(\varepsilon T_0)\Big]\right\}. \tag{8.50}$$

Similarly, this equation is reduced to Fano resonance $S(\varepsilon) = A\frac{(q_a+\varepsilon)^2}{1+\varepsilon^2}$ if $A_L = 0$ (or $c = 0$). For $T_0 \to 0$, $S(\varepsilon) \to A$ if $A_L = \pi$ (or $c = 1$), and $S(\varepsilon) \to A\left(-\frac{(q_a+\varepsilon)^2}{1+\varepsilon^2} + 2\right)$ if $A_L = 2\pi$ (or $c = 2$). The negative sign in the last case flips the absorption profile from up to down with respect to the horizontal axis. These results agree with the numerical results shown in Figures 8.13(a,b).

The Dipole-Control Model

Section 8.1.5 considers the time-dependent dipole $d(t)$ of a Fano resonance if it is suddenly terminated at $t = \tau$ by the applied intense IR. Here, consider a model where the IR does not fully deplete the bound part of the resonance. Assuming that the IR pulse is given by a delta function instead of Equation 8.15, for $\tau > 0$ (XUV before IR), $d(t)$ can be modeled as

$$d(t) \propto \begin{cases} 2i\delta(t) + i\frac{\Gamma}{2}(q-i)^2 e^{-\frac{\Gamma}{2}t} e^{-i\Omega_r t}, 0 \le t < \tau \\ iA(\tau)\frac{\Gamma}{2}(q-i)^2 e^{-\frac{\Gamma}{2}t} e^{-i\Omega_r t}, t > \tau, \end{cases} \tag{8.51}$$

where the complex gate $A(\tau) = |A(\tau)|\, e^{i\varphi(\tau)}$. The IR can deplete the bound state as well as add a phase $\varphi(\tau)$ to the time-dependent dipole. By defining

$$\Delta\Omega = \Omega - \Omega_r, \quad \varepsilon = 2\Delta\Omega/\Gamma, \quad P(\varepsilon) = q^2 - 1 + 2q\varepsilon, \quad Q(\varepsilon) = 2q + \varepsilon - q^2\varepsilon, \tag{8.52}$$

the photoabsorption cross-section can be obtained as

$$\begin{aligned} \sigma(\Omega,\tau) \propto\ & \frac{(q+\varepsilon)^2}{1+\varepsilon^2} - \frac{e^{-\frac{\Gamma}{2}\tau}}{1+\varepsilon^2} \\ & \times \{[P(\varepsilon)(1 - |A(\tau)|\cos\varphi(\tau)) - Q(\varepsilon)|A(\tau)|\sin\varphi(\tau)]\cos(\Delta\Omega\tau) \\ & + [Q(\varepsilon)(1 - |A(\tau)|\cos\varphi(\tau)) + P(\varepsilon)|A(\tau)|\sin\varphi(\tau)]\sin(\Delta\Omega\tau)\}. \end{aligned} \tag{8.53}$$

One can also consider the situation that if $\tau < 0$ (XUV after IR)

$$d(t) \propto 2i\delta(t) + iA(\tau)\frac{\Gamma}{2}(q-i)^2 e^{-\frac{\Gamma}{2}t} e^{-\Omega_r t}, t \ge 0 \tag{8.54}$$

then the photoabsorption cross-section

$$\sigma(\Omega,\tau) \propto 1 + \frac{|A(\tau)|}{1+\varepsilon^2}\left[P(\varepsilon)\cos\varphi(\tau) + Q(\varepsilon)\sin\varphi(\tau)\right]. \tag{8.55}$$

Comparing Equations 8.50 and 8.53, one finds that the three-level model in the previous subsection is a particular case of the dipole-control model here when $\sin\varphi(\tau) = 0$ and $(1 - |A(\tau)|\cos\varphi(\tau)) = c$. That is, the complex gate simplifies into a real gate $A(\tau) = \cos\frac{A_L}{2}$, where the pulse area A_L is a measurement of the IR-coupling between states $|a\rangle$ and $|b\rangle$. A complex $A(\tau)$ can be applied to model the on or off resonance couplings to many levels from the autoionizing state of interest, and thus the effect of ionization and the AC Stark shift can also be included.

Comparing the Dipole-Control Model to Experiments

Equation 8.53 implies that, for a fixed time delay τ, the absorption profile changes with $|A(\tau)|$ and $\varphi(\tau)$. Let us consider weak coupling to other states such that $|A(\tau)| \approx 1$. Then the effect of IR is to introduce an additional phase factor $e^{i\varphi(\tau)}$ to the dipole after $t = \tau$. In the experiment by Ott et al. [22], absorption profiles of doubly excited states of He from 64.0 to 65.4 eV were measured with and without the IR field. By turning on the 7 fs IR field at a small time delay (about 5 fs) after the XUV pump, they found that the IR-free Fano profiles (Figure 8.14(a)) change into Lorentzian profiles (Figure 8.14(b)). The change of profile can be understood as the change of $\varphi(\tau)$ according to the dipole-control model. Take the sp_{24+} state as an example. From Equation 8.53, the result in Figure 8.14(c) shows that, when $\varphi(\tau) = 0$ (no IR), the profile is the standard Fano resonance with $q = -2.55$. As $\varphi(\tau)$ goes from -180° to 180°, the change of the absorption profile, which is equivalent to a change in q, can be seen clearly. The phase $\varphi(\tau)$ at a fixed time delay depends on the

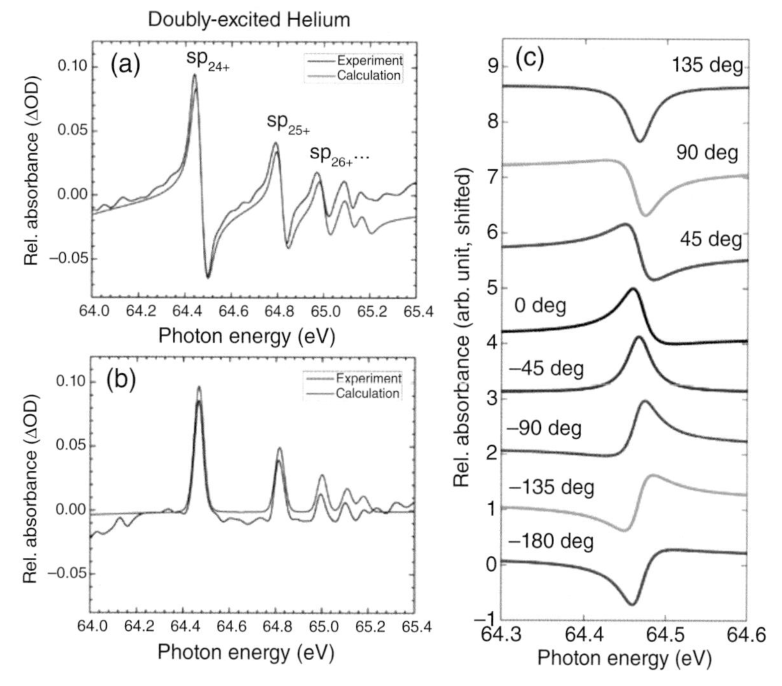

Figure 8.14 (a) Field-free (static) absorption spectrum of doubly excited states of the $N = 2$ series (the sp+ series) in He. The well-known Fano absorption profiles are observed in the transmitted spectrum of a broadband attosecond pulse. (b) When a 7 fs laser pulse immediately follows the attosecond pulse (delta-like) excitation (time delayed by $\sim$ 5 fs) at an intensity of 2.0×10^{12} W/cm^2, the Fano absorption profiles are converted to Lorentzian profiles. From [22]. (c) Absorption profiles in the vicinity of the sp_{24+} resonant state calculated according to Equation 8.53 for $q = -2.55, \tau = 5$ fs, $A(\tau) = 1$. Different lines correspond to different $\varphi(\tau)$. (From C. Ott et al., *Science*, **340**, 716 (2013) [22]. Reprinted with permission from AAAS.)

IR intensity. The experimental condition is close to $\varphi(\tau) = -45^o$ where the asymmetric Fano profile is converted to Lorentzian.

From the absorption spectra at different τ, one can use a fitting method to extract the function $\varphi(\tau)$ based on Equation 8.53 or 8.55. If $\varphi(\tau)$ can be interpreted as the ponderomotive shift

$$\varphi(\tau) = -\int_0^\infty U_p(t-\tau)dt = -\int_{-\tau}^\infty U_p(t)dt, \tag{8.56}$$

where $U_p(t) = \frac{E_{IR}^2(t)}{4\omega_{IR}^2}$, then it is possible to characterize the IR pulse duration, as discussed in Blättermann et al. [23], by taking the derivative of the retrieved $\varphi(\tau)$ to τ.

For relatively large τ, where the overlap between XUV and IR is small, it can be assumed that $|A(\tau)| = |A|, \varphi(\tau) = \varphi$ are two numbers independent of τ. Then Equation 8.53 can be rewritten in the form

$$\sigma(\Omega,\tau) \propto A(\varepsilon) - e^{-\frac{\Gamma}{2}\tau}B(\varepsilon)\cos[\Delta\Omega\tau + \phi(\varepsilon)]. \tag{8.57}$$

The expressions of $A(\varepsilon)$, $B(\varepsilon)$, and $\phi(\varepsilon)$ can be easily obtained from Equation 8.53. In particular, at the resonance center $\Delta\Omega = \varepsilon = 0$ from Equation 8.53 one gets

$$\sigma(\Omega = \Omega_r, \tau) \propto q^2 - e^{-\frac{\Gamma}{2}\tau}\{(q^2 - 1)(1 - |A|\cos\varphi) - 2q|A|\sin\varphi\}. \quad (8.58)$$

Equation 8.58 shows that the width Γ or the lifetime $1/\Gamma$ of the AIS can be retrieved from the delay-dependent ATA signal at the resonance center by fitting it to a curve $A - Be^{-\frac{\Gamma}{2}\tau}$. Figure 8.15(a) shows the experimental ATA spectra around the Xe $5s5p^6 6p$ autoionizing resonance as functions of time delay τ [24]. The XUV pulse spans about 19 to 23 eV. The corresponding pulse duration is about 400 as. The IR pulse is 760 nm in wavelength, 24 fs in duration, and 1.3×10^{12} W/cm^2 in intensity. The green curve in Figure 8.15(a) is for the photon energy at the resonance center (20.956 eV). From the exponential decay at large τ, the lifetime of the $5s5p^6 6p$ state was retrieved as (21.9±1.3) fs. This is in good agreement with previous data (21.1±0.5) fs from linewidth measurements. Figure 8.15(a) also includes four curves for photon energies shifted by ±10 meV and ±20 meV from the resonance center. The curves change with $\Delta\Omega$ as given by Equation 8.57. Particularly, at large τ, one can observe the oscillations as $|\Delta\Omega|$ increases due to the factor $\cos[\Delta\Omega\tau + \phi(\varepsilon)]$. The solid lines in Figure 8.15(b) are theoretical calculations using a three-state model. The dots at the large time delay are fitted to the solid curves using Equation 8.57, where A, B, and ϕ are treated as fitting parameters. Figure 8.15(c) is the direct calculation using Equation 8.53, assuming $A(\tau) = 0$, i.e., the AIS is totally depleted by the IR pulse. We can see good agreements between the experimental results and the theoretical calculations. In the future, it would be of interest to use Equation 8.53 or 8.57 to extract additional parameters like q. Moreover, the τ-dependence of the parameters

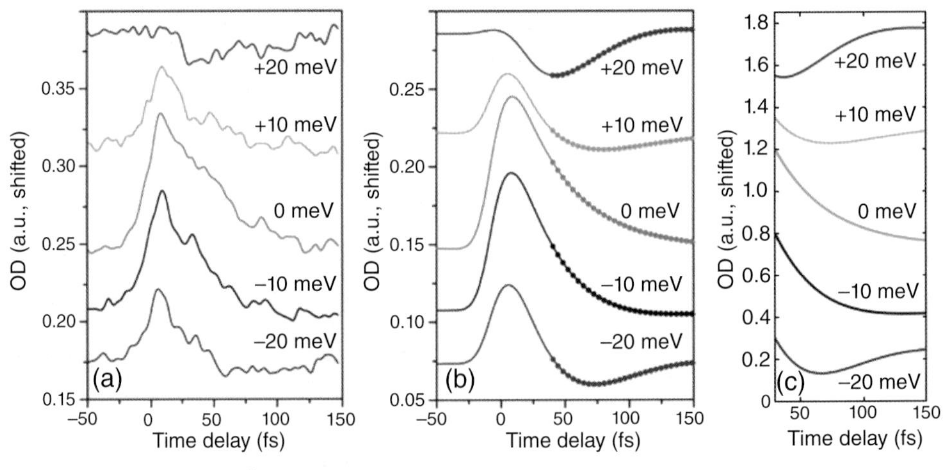

Figure 8.15 ATA signals around the Xe $5s5p^6 6p$ autoionizing resonance as a function of time delay τ. Different curves correspond to different $\Delta\Omega$. $\Delta\Omega = 0$ meV is considered the resonance center. (a) Experimental measurements. (b) Solid lines are theoretical results using a three-state model. Dots are fitted to these solid lines using Equation 8.57. From [24]. (c) Direct calculation according to Equation 8.53, for the delay region $\tau > 25$ fs, with parameters $\Omega_r = 20.956$ eV, $\Gamma = 31.2$ meV, $q = 0.23$, and $|A(\tau)| = 0$. (Reprinted from Birgitta Bernhardt et al., *Phys. Rev. A*, **89**, 023408 (2014) [24]. Copyrighted by the American Physical Society.)

can be included and then it may be possible to retrieve the amplitude and phase of the delay-dependent gate $A(\tau)$.

8.5 Propagation of ATA Spectra in the Gas Medium

In a dilute gas the propagation of the XUV absorption is described by Beer's law (see Equation 8.27). In the linear regime, an intensity-independent absorption cross-section can be defined. For ATA spectroscopy, where the XUV co-propagates with an intense dressing IR field, as shown in Equation 8.26, a generalized absorption cross-section that depends on the laser intensity can be defined. The propagation of the XUV in such an IR-coupled medium is accounted for by solving the Maxwell equation. While most of the theoretical calculations for the ATA spectra are based on the single-atom model, experimentally, the ATA spectra are obtained after the light has gone through the gas medium. While it is generally true that the single-atom model is valid for a dilute gas, when comparing with experimental measurements, propagation in the medium should be taken into account. This section shows how propagation can modify the ATA spectra, which also affects the XUV pulse in the time domain in the medium as well as after the medium.

Figure 8.16 shows the transient absorption spectra on helium near the $1s2p$ excited state after a 500 as XUV pulse with center energy 22.5 eV, peak intensity at 10^9 W/cm^2 in the dressing IR field (5 fs, 780 nm, 2×10^{12} W/cm^2) in a gas pressure of 50 mbar with propagation length (a) $x = 2.5$ μm, which can be regarded as the single-atom response, and (b) $x = 1$ mm. The single-atom ATA spectrogram in (a) is significantly modified after propagation as shown in (b). These results are compared to the experimental data shown in (c). Note that propagation affects the ATA spectrogram non-monotonically versus the time delay. While (b) is in reasonable agreement with the experimental data shown in (c), it is to be noted that the time axes in (b) and (c) are very different. Accurate determination of the gas pressure is also very difficult. It was stated in Pfeiffer et al. [25] that the gas

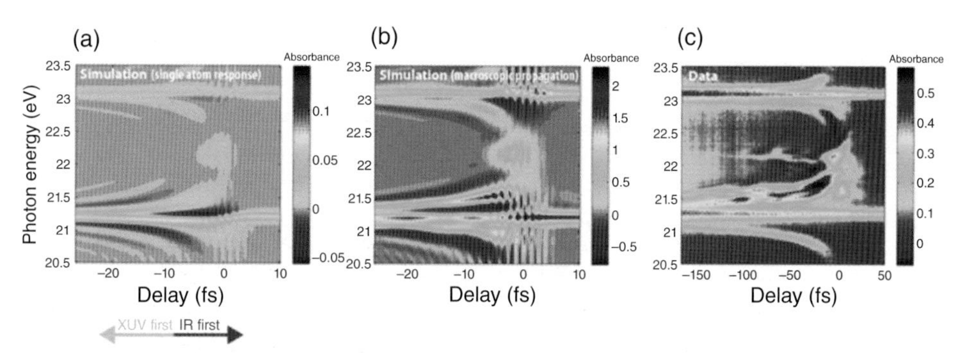

Figure 8.16 Transient absorption spectra (a) calculated within the single-atom model, and (b) after propagation in a medium of gas pressure of 50 mbar. The experimental data is shown in (c). Target is helium and XUV is near the $1s2p$ state of helium. Other parameters are given in the text. (Reprinted from A. N. Pfeiffer et al., *Phys. Rev. A*, **88**, 051402 (2013) [25]. Copyrighted by the American Physical Society.)

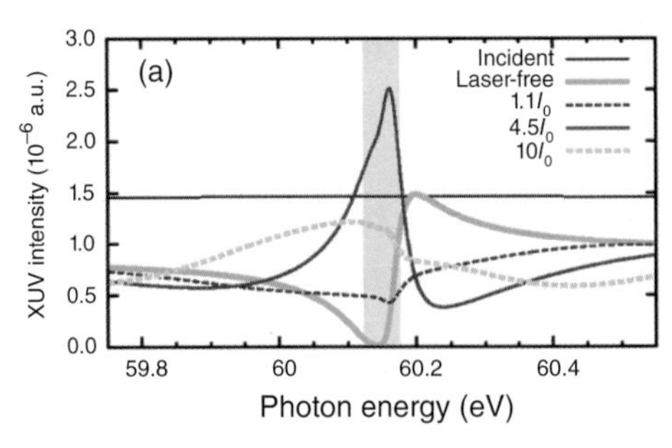

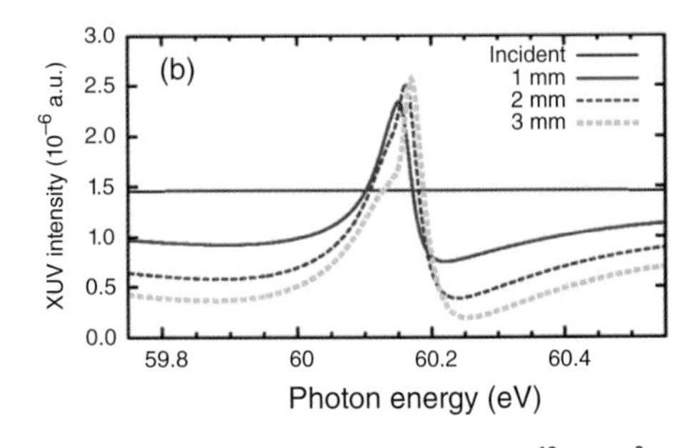

Figure 8.17 (a) XUV spectra near the $2s2p$ resonance of helium with a 9 fs overlapping ($\tau = 0$) 540 nm laser. $I_0 = 10^{12}$ W/cm^2. For IR intensity at 4.5×10^{12} W/cm^2, the Fano line shape is inverted, such that strong absorption near 60.1 eV becomes strong emission. (b) The emission profile versus the propagation length for intensity of 4.5×10^{12} W/cm^2. Only the wings of the lineshape are reduced as the propagation distance is increased. This leads to shaping the XUV pulse to a narrowband light at the expense of longer pulse duration. (Figure adapted with permission from W. C. Chu and C. D. Lin, *Phys. Rev. A*, **87**, 013415 (2013) [20]. Copyrighted by the American Physical Society.)

pressure in the experiment may range from 10 to 240 mbar instead of the 50 mbar used in the simulation.

Section 8.4.2 demonstrated that the $2s2p$ Fano resonance lineshape can be changed drastically if the IR dressing laser is nearly resonant with the $2s^2$ state. In Figure 8.17(a), the theoretical Fano profile is shown near the $2s2p$ resonance when the helium atom is dressed in an IR at intensity of 1.1, 4.5, and 10×10^{12} W/cm^2 for the case that the IR and XUV have zero time delay. At 4.5×10^{12} W/cm^2, the Rabi oscillation has gone through a full cycle, and thus the Fano profile flips from up to down. In this case, instead of a strong absorption, the $2s2p$ state becomes a strong emission. This is similar to the EIT phenomenon in optical transitions. Near the center of the resonance, the transmitted light is stronger than the incident light. The emission feature survives the propagation in the gas medium. Figure 8.17(b) shows that the emission peak does not decrease with the propagation length, while the wings are strongly absorbed as the propagation distance increases. This example shows that the XUV pulse can be shaped in a gas medium with proper dressing IR pulses. The emerged XUV pulse would have bandwidth comparable to the width of the autoionizing Fano resonance. This shaping of the transmitted XUV pulse implies that the time profile of the incident XUV is also significantly altered. (For an example, see figure 4 of [15]).

8.6 ATA Spectroscopy for Small Molecules

8.6.1 Introduction

The ATA spectra have been studied extensively for rare gas atoms, and particularly for helium. The time-delayed IR pulse can be thought of as providing a means to probe the electron wave packet initiated by the XUV pulse if the two pulses do not overlap. If the two pulses overlap in time, the IR plays the role of controlling the XUV photoabsorption of the target. In either case, the nonlinear interaction of IR with the system is necessary if it is to play the role of the "probe" such that the phase of the wave packet generated by the XUV can be extracted. If the two pulses overlap in time, the ATA spectra would induce new spectral features like LIS, lineshape modification, and Autler–Townes splitting. In principle, the ATA spectra can also be generated from molecular targets. Due to the complexity of molecular spectra, only a limited number of ATA spectra have been reported so far. Two examples will be given in Section 8.6.2 and 8.6.3.

8.6.2 ATA Spectra of H_2 Molecules in the 12–17 eV Region: Retrieving Vibrational Wave Packets

The ATA spectra of H_2 in the photon energy region of 12–17 eV have been reported in Cheng et al. [26]. Experimentally, a 750 as vacuum ultraviolet (VUV) pulse is used to excite bound states of H_2 molecules in the presence of a 730 nm, 5 fs, and intensity 5×10^{12} W/cm^2 IR laser in a 20 mbar H_2 gas. Figure 8.18 shows the essential molecular potential curves relevant to this experiment together with experimental and theoretical ATA spectrograms. Within the 11–17 eV range, the electronic curves that can be reached from the ground state of H_2 by the VUV light are indicated. The experimental ATA spectrogram is shown in (c). The vibrational states associated with each electronic state are indicated (color-coded like the potential curve in (a)). The ATA spectrogram obtained by solving the TDSE is shown in (d). To begin with, the two spectrograms from the experiment and from the TDSE calculations do not match very well with the experimental spectral lines that lie about 1 eV higher than the theoretical data. The origin of the discrepancy is not entirely clear. Note that the experimental data is taken at 20 mbar, while the TDSE calculation was carried out only for a single molecule. In Cheng et al. [26], it was speculated that a possible saturation effect may affect the absorption spectra, but neither propagation nor saturation effects are expected to shift the energy levels too significantly. This could be due to the lack of convergence in the TDSE calculation.

Instead of trying to understand the whole spectrogram, here the focus is on what one can learn from such an experiment. Following Cheng et al. [26], it is demonstrated that the ATA spectra can be used to extract the time evolution of a subset of the molecular wave packet on a particular electronic surface. Specifically, consider the ATA spectrum associated with the $B^1\Sigma_u^+$ electronic state. It lies near the wing of the VUV spectrum, and thus is only weakly excited (see Figure 8.18(b)). On the other hand, vibrational states associated with the $B'^1\Sigma_u^+$ curve are favorably populated since they lie near the peak of the VUV spectrum. In the presence of a relatively intense IR, as depicted in Figure 8.18(b), emission of two IR photons may drive the vibrational states from B' to B. Treating this process perturbatively, in the presence of the IR, the time-dependent amplitude of the vibrational state $|\nu_B\rangle$ (of curve B of electronic state B) at time t is

$$c_{\nu_B}(t,\tau) = c_{\nu_B}^{(1)} + c_{\nu_B}^{(3)}(t,\tau). \tag{8.59}$$

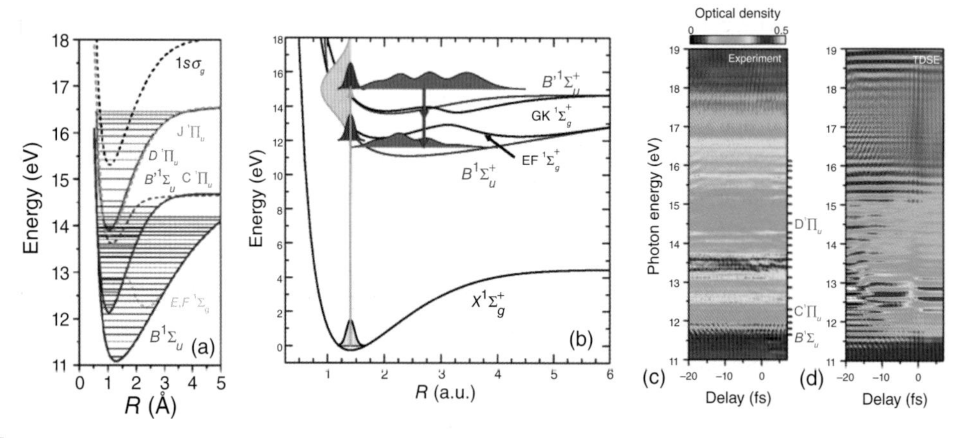

Figure 8.18 ATA spectra of molecular hydrogen from 11 to 19 eV. (a) The hydrogen bound-state manifold consisting of several electronic excited states and vibronic states within each electronic state. (b) The VUV pulse alone excites mostly the vibrational wave packet of the $B'^1\Sigma_u^+$ curve and weakly the vibronic states of the $B^1\Sigma_u^+$ curve. In the presence of the dressing IR pulse, the vibronic states of the $B'^1\Sigma_u^+$ can be coupled to the vibronic states of the $B^1\Sigma_u^+$ via emission of two IR photons through the other two electronic GK and EF states, to create interference with vibronic states directly reached by the VUV alone. From the experimental time-delay interference spectra for each vibrational state in the $B^1\Sigma_u^+$, the amplitude and phase of the vibrational wave packet in the $B'^1\Sigma_u^+$ can be retrieved (see Figure 8.19). (c,d) The measured and TDSE-simulated ATA spectrograms in the 11–19 eV region. There is an 1 eV discrepancy in the highest optical-density band. Note that dense gas target was used in the experiment and the simulation has to make a number of simplifications that may contribute to the discrepancy. (Figure adapted with permission from Yan Cheng et al., *Phys. Rev. A*, **94**, 023403 (2016) [26]. Copyrighted by the American Physical Society.)

Let $|B'\rangle|\chi_{B'}(t)\rangle$ be the wave packet associated with the electronic state B' after the VUV pulse, where

$$|\chi_{B'}(t)\rangle = \sum_{\nu} c^{(1)}_{\nu_{B'}}(t)|\nu_{B'}\rangle. \tag{8.60}$$

The coefficient $c^{(3)}_{\nu_B}(t,\tau)$ for populating vibrational state $|\nu_B\rangle$ from the vibrational states of B' via emission of two IR photons can be written as

$$c^{(3)}_{\nu_B}(t,\tau) \cong e^{-i\omega_{\nu_B,g}(t-\tau)}\Phi_{\mathrm{IR}}(t-\tau)\sum_{\nu_{B'}} A^{(2)}_{\nu_{B'}\to\nu_B}c^{(1)}_{\nu_{B'}}(\tau), \tag{8.61}$$

where $\omega_{\nu_B,g}$ is the excitation energy to the vibrational state $|\nu_B\rangle$ by the VUV, $\Phi_{\mathrm{IR}}(x) = \int_{-\infty}^{x} e^{-t^2/2}dt$ is an envelope function to describe the smoothed sharp activation of the IR pulse at time τ, and $A^{(2)}_{\nu_{B'}\to\nu_B}$ is the two-photon matrix element for making the transition from $|\nu_{B'}\rangle$ to $|\nu_B\rangle$. Since the IR-coupling duration is short, during which time the internuclear distance hardly changes, one can approximate

$$A^{(2)}_{\nu_{B'}\to\nu_B} \simeq A^{(2)}_{B'\to B}\langle\nu_B|\nu_{B'}\rangle, \tag{8.62}$$

where $\langle \nu_B | \nu_{B'} \rangle$ is the Franck–Condon overlap factor. Under these simplifications, one arrives at

$$c_{\nu_B}^{(3)}(t,\tau) \simeq e^{-i\omega_{\nu_B,g}(t-\tau)} \Phi_{\mathrm{IR}}(t-\tau) A_{B' \to B}^{(2)} \langle \nu_B | \chi_{B'}(\tau) \rangle. \tag{8.63}$$

From Equation 8.59, the photoabsorption spectra versus the time delay can then be obtained. The result can be expressed in the general form

$$f_\nu(\tau) = P_\nu(\tau) + Q_\nu(\tau) \cos\left[\omega\tau + \varphi_\nu(\tau)\right], \tag{8.64}$$

where the first term is a smooth background. The amplitude and phase of the oscillatory terms $Q_\nu(\tau)$ and $\varphi_\nu(\tau)$ can be retrieved from the experimental data for each $|\nu_B\rangle$. The vibrational wave packet of the VUV pulse alone on the B' potential surface can then be obtained from

$$|\chi_{B'}(t)\rangle = \sum_\nu |\nu_B\rangle Q_\nu(t) e^{i\varphi_\nu(t)}. \tag{8.65}$$

Based on the TDSE calculation, the time evolution of the vibrational wave packet is shown in the upper frame of Figure 8.19. The lower frame shows the vibrational wave

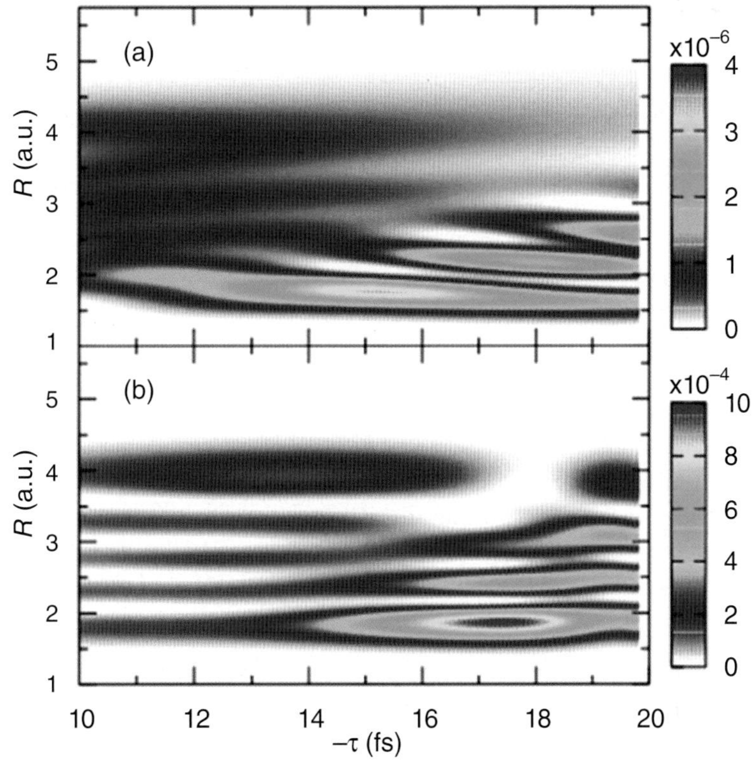

Figure 8.19 (a) Nuclear wave packet in the $B'^1\Sigma_u^+$ potential curve generated from the VUV pulse obtained from the TDSE calculation using experimental parameters. (b) The same nuclear wave packet retrieved from the ATA spectra. (Figure adapted with permission from Yan Cheng et al., *Phys. Rev. A*, **94**, 023403 (2016) [26]. Copyrighted by the American Physical Society.)

packet versus the time delay beginning 10 fs after the VUV pulse, derived using the experimental ATA spectrum using the model described above. There is a reasonable degree of agreement demonstrating that, under favorable circumstances, ATAS is indeed capable of probing the vibrational wave packet initiated by the VUV pulse. On the other hand, to validate the procedure, it is desirable that a different probe pulse with a somewhat different laser intensity or with a different wavelength be used to obtain new ATA spectra. If the same wave packet can be retrieved, then it would establish the validity of the method more generally.

8.6.3 ATA Spectrogram for N_2 Molecules between 14 and 18.5 eV: Modeling the Spectra

High-resolution photoabsorption spectra of N_2 molecules near the first few ionization thresholds have been studied extensively in the literature. Naturally, N_2 is a good candidate for experimentalists to investigate its ATA spectrogram. Since an accurate full *ab initio* calculation for the photoabsorption of such a system is already complicated, a study of the ATA of N_2 will exemplify the challenge facing the understanding of ATA spectra for complex systems, including large polyatomic molecules and solid materials.

In Reduzzi et al. [27], single-attosecond pulses are generated from an Xe target and filtered using either indium plates to obtain photons centered around 15 eV, or tin plates to obtain photons centered around 18 eV. The low-energy attosecond pulse will be called the *low-energy excitation* (LEE) pulse, and the high-energy attosecond pulse will be called the *high-energy excitation* (HEE) pulse. Without the IR, the high-resolution photoabsorption spectrum of N_2 in the 14–19 eV region has been well studied (see Figure 8.20). In Figure 8.20, the LEE region is shown in green dotted lines. Here lies the bound Rydberg states converging to the $X^2\Sigma_g^+$ ground state ($I_p = 15.58$ eV) and the first excited $A^2\Pi_u$ state ($I_p = 16.94$ eV) of the N_2^+ ion. The levels from 17.0 to 18.6 eV, marked in blue lines, are two series of AISs that converge to the second ionic excited state ($B^2\Sigma_u^+$, $I_p = 18.75$ eV).

In the experiment, both the LEE and HEE were used to excite N_2 molecules in the presence of time-delayed IR, and the ATA spectra were analyzed. The attosecond pulses have a duration of about 1 fs, and the IR intensity is estimated to be 10^{12} W/cm^2. Figure 8.21 shows the measured ATA spectrogram (left panel) using the LEE pulse, but the interesting ATA spectra are in the range from 16.8 to 18.6 eV. The AISs in this region are only very weakly populated by the LEE pulse directly, but are significantly populated in the presence of the IR laser pulse by absorbing two IR photons in addition to the LEE photon. In the overlapping region of the two pulses, the two main peaks at about 17.1 eV and 17.9 eV (identified as the $3d\sigma_g$ and $4d\sigma_g$ resonances in Figure 8.20) show clear depletion similar to those observed in atoms. The depletion is accompanied by broadening of the spectral lineshape. Half-optical cycle oscillation with respect to the time delay can also be clearly seen. Similar to atomic resonances, the lifetime of the AIS can be extracted from the ATA spectrum. Based on the experimental data, from the decay curve of the resonance at large negative time delays (IR comes after EUV), the lifetimes of the $3d\sigma_g{}^1\Sigma_u^+$ and $3d\pi_g{}^1\Pi_u$ resonances were extracted yielding 10.2 fs for the former and 13.8 fs for the latter. These

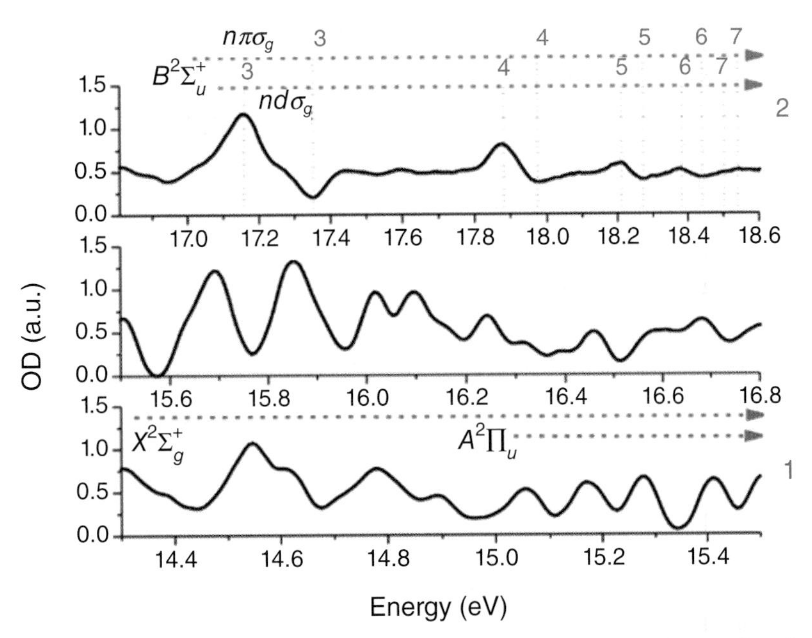

Figure 8.20 Enlarged view of the optical density of a nitrogen molecule from the 14.4 to 18.6 eV region. The green lines (lower panel) indicate excited levels converging to the ground $X^2\Sigma_g^+$ ($I_p = 15.58$ eV) and the first excited $A^2\Pi_u$ ($I_p = 16.94$ eV) state of the N_2^+ ion. The blue lines in the upper panel are two series of AISs that converge to the second ionic excited state ($B^2\Sigma_u^+$, $I_p = 18.75$ eV). Using the LEE pulse alone the AISs are weakly populated. In the presence of a synchronized IR pulse, absorption of two additional IR photons can populate these AISs and the spectra in the 16.8 to 18.6 eV region will vary with the time delay (see Figure 8.21). (Reprinted from M. Reduzzi et al., *J. Phys. B-At. Mol. Opt.*, **49**, 065102 (2016) [27]. Copyrighted by IOP publishing.)

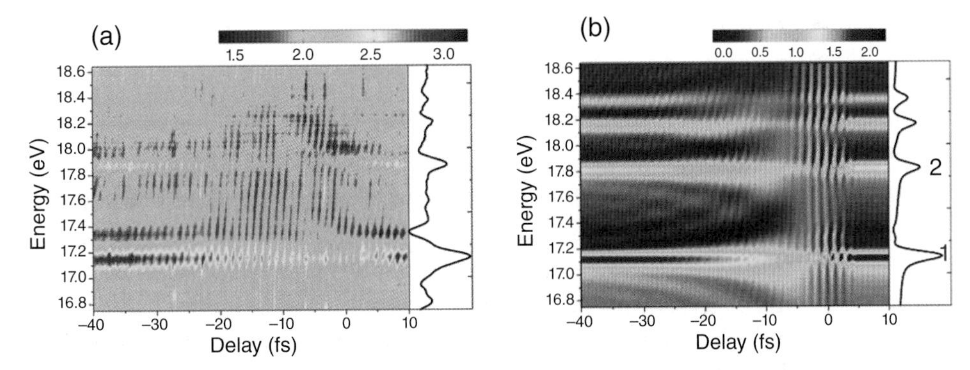

Figure 8.21 ATA spectrogram generated by the LEE pulse in the presence of the IR. The observed spectrogram in the spectral region of the Fano resonances from 16.8 to 18.6 eV (left) is compared to the simulated result (right). The convoluted cross-section is shown on the right-hand side for reference. The lifetimes of the two Fano resonances near mark 1 and 2 have also been extracted from the experimental spectrogram. (Reprinted from M. Reduzzi et al., *J. Phys. B-At. Mol. Opt.*, **49**, 065102 (2016) [27]. Copyrighted by IOP publishing.)

lifetimes are consistent with the resonance widths obtained from the EUV photoionization measurements.

How well can the experimental ATA spectrogram be modeled theoretically? In Reduzzi et al. [27], a model was proposed to carry out simulations without utilizing full quantum chemistry calculations. In this model, the total time-dependent wavefunction is expanded as

$$|\Psi(t)\rangle = e^{-iE_g t} c_g(t)|g\rangle + e^{-i(E_g+\omega_X)t} \times \left[\sum_m c_m(t)|b_m\rangle + \sum_n c_n(t)|f_n\rangle + \int c_E(t)|E\rangle\rangle\right]. \tag{8.66}$$

Such an expansion includes the ground state, bound states, and AISs. For the modeling in Reduzzi et al. [27], all the bound states $|b_m\rangle$ from 13.98 to 15.48 eV and the first seven AISs of the two series shown in Figure 8.20 between 17.0 and 18.4 eV were included. Each AIS is written in terms of the bound component and the continuum part, in the same way used in the atomic target (see Equation 8.66). To carry out the calculations, the energy levels of the bound states are taken from the experiment. For AISs, the widths and q-parameters are taken from the experimental data. The relative strengths of the dipole-matrix elements between the bound excited states and the AISs are assumed to be proportional to $V_n V_m$, where V_j is related to the autoionization width $\Gamma_j = 2\pi V_j^2$. The overall strength between the two groups of bound and AISs is adjusted to maximize the contrast of the interference fringes in the spectra as seen in the experimental data. Using such a simplified model, the simulated results (the right panel of Figure 8.21) can reproduce most of the main features in the experiment. Future improved simulations can incorporate dark states if they are expected to have strong coupling with bright states. Unlike TDSE-type calculations, the parameters entering the expansion in Equation 8.66 can be calculated from quantum chemistry codes. With additional efforts, extension of Equation 8.66 to more complex molecules is doable if the bandwidth of the VUV is limited to a few eVs and the IR intensity is not ionizing the excited molecules significantly.

8.7 Elements of Probing Attosecond Electron Dynamics and Wave-Packet Retrieval

8.7.1 Introduction

Since the first report of the generation of IAPs at the dawn of the twenty-first century, it has been realized that such pulses can be used to study time-resolved electron dynamics in a matter akin to the time-resolved tracking of the atomic motion in a molecule enabled by the advent of femtosecond laser pulses. Similar to how tracing the dynamics of atoms in a molecule is often described as the making of a "molecular movie," tracing the time evolution of an electron wave packet has been considered to be making an "electron movie." Is an "electron movie" the correct way to describe the dynamics of an electron wave packet? In a classical movie, the objects in the movie are localized classical particles.

While nuclei and electrons in a molecule are quantum particles, the vibrational wave packets are mostly rather localized, especially for the heavier atoms, such that a molecular movie still conveys some reality. Can the dynamics of an electron wave packet be perceived as an "electron movie," implying that the electron(s) can go from here to there? Can one talk about charge migration or hole migration only, for example, when one describes an electron wave packet?

Let us start with a simple model. Assume that a wave packet in helium is created at $t = 0$, where the initial wave packet is given by $|\psi(t=0)\rangle = a_1|1s^2\,{}^1S\rangle + a_2|1s2s\,{}^1S\rangle$. Quantum mechanics tells us how this wave packet will evolve in time, $|\psi(t)\rangle = a_1 e^{-iE_1 t}|1s^2\,{}^1S\rangle + a_2 e^{-iE_2 t}|1s2s\,{}^1S\rangle$, where E_1 and E_2 are the energies of the two states, respectively. Roughly speaking, it can be said that one of the electrons is oscillating between the $1s$ and $2s$ states, but how does the two-electron wave packet (instead of the one-electron wave packet) change in time? Figure 8.22(a) shows that the two-electron density oscillates between zero and six atomic units, executing breathing motion with a period of 200 as. This picture is probably as close as one can get to an "electron movie" made on a computer.

Most quantum measurements are conducted in the energy domain. For example, one can use an intense 100 as, central energy 95.2 eV, attosecond pulse to double ionize the wave packet at different delay times. Figure 8.22(b) shows that double-ionization probability will oscillate with time similar to the electron density in the small R region in (a). This is the case because double ionization occurs efficiently only when both electrons are close to the helium nucleus. Measurement is a projection of the wave packet, not the density, onto the eigenstate(s) of the operator. For example, the complex electron wave packet is used to obtain the atomic-dipole moment, not the density. In other words, a description of a dynamic system is a statement on the time-dependent wave packet, not the evolution of the electron density. The determination of the phase of the wave packet is compulsory for the description of an electron wave packet.

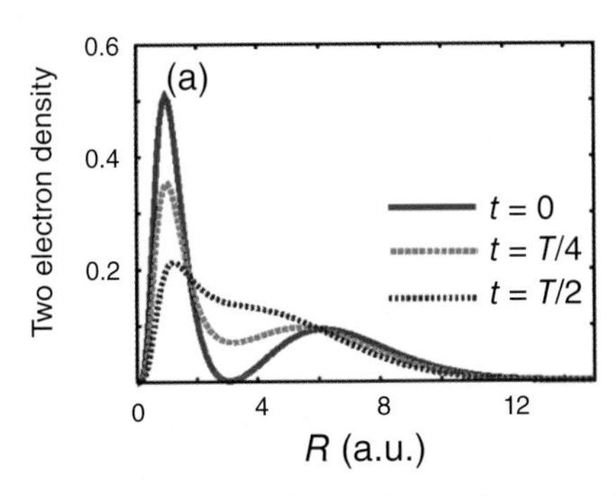

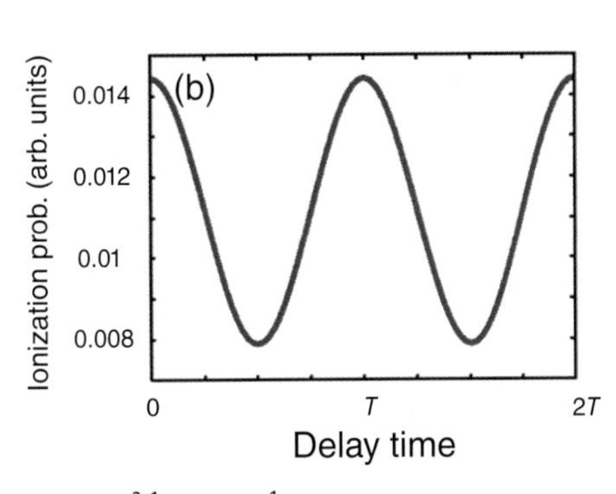

Figure 8.22 (a) Time-dependent two-electron density of a wave packet made of $1s^2\,{}^1S + 1s2s\,{}^1S$ in a helium atom. The electron density oscillates with time and $T = 200$ as is the oscillation period. (b) By double ionizing the helium with a 95.2 eV, 100 as pulse, the total ionization yield shows oscillation similar to the electron density near the small R region. In general, a probe pulse may be sensitive to a portion of the wave packet only. (Figure adapted with permission from T. Morishita, S. Watanabe, and C. D. Lin, *Phys. Rev. Lett.*, **98**, 083003 (2007) [28]. Copyrighted by the American Physical Society.)

8.7.2 Probing Electron Correlation with Attosecond Pulses

Attosecond pulses are powerful tools for probing electron correlation in atoms, molecules, and solids. Recall that, in the shell model (or the mean-field theory), the many-electron effect is approximated by a time-averaged effective screening potential and electron correlation is neglected. In what way can the correlated motion between the electrons be manifested if it is to be probed in the time domain by an attosecond pulse? Here, doubly excited states of helium are used as examples where electron correlation is known to be very important.

In Sections 8.1 and 8.4, it was noted that doubly excited states decay within a few femtoseconds. If one can generate a wave packet consisting of doubly excited states, then the two-electron wave-packet dynamics can be investigated by attosecond pulses.

Take the simplest example. Suppose a two-electron wave packet is made of $2s^2\,{}^1S + 2p^2\,{}^1S$ at $t = 0$. We want to know how this wave packet evolves in time before it fully decays. Since the total angular momentum is zero for this two-electron wave packet, we can describe the wave packet in three spatial coordinates only, r_1, r_2, and θ_{12}. In fact, r_1, r_2 can be replaced with a hyperradius R defined by $R = \sqrt{r_1^2 + r_2^2}$ and a hyperangle α by $\alpha = \tan^{-1}(r_2/r_1)$. The θ_{12} is the angle between the two electrons with respect to the nucleus.

Define the bending vibrational-density distribution of each doubly excited state by

$$\rho_j^{\text{vib}}(\alpha, \theta_{12}) = \int |\varphi_j|^2 d\Omega_r dR, \tag{8.67}$$

where integration over the three Euler angles Ω_r is just multiplication by a constant. For the two doubly excited states, the vibrational-density distributions as a function of α and θ_{12} are shown in Figure 8.23. For $2s^2\,{}^1S$ the density peaks at $\theta_{12} = \pi$, while for $2p^2\,{}^1S$ the density peaks at large and small θ_{12} with a minimum in between. The former is like a linear molecule with the two electrons on opposite sides of the nucleus, while the latter is like a bending molecule undergoing large-amplitude vibration. Clearly, for each individual state, the density distribution is stationary. A wave packet made of the two states will execute bending vibrational motion. From the energy difference between the two states, the oscillation period is 980 as.

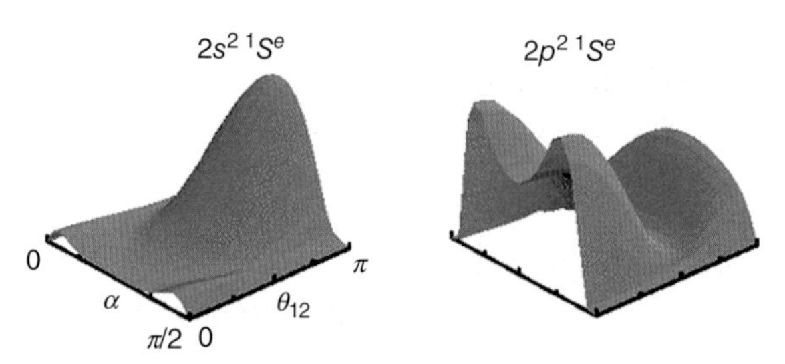

Figure 8.23 Density plots of $2s^2\,{}^1S$ and $2p^2\,{}^1S$ doubly excited states at a fixed hyperradius. (Figure adapted with permission from T. Morishita, S. Watanabe, and C. D. Lin, *Phys. Rev. Lett.*, **98**, 083003 (2007) [28]. Copyrighted by the American Physical Society.)

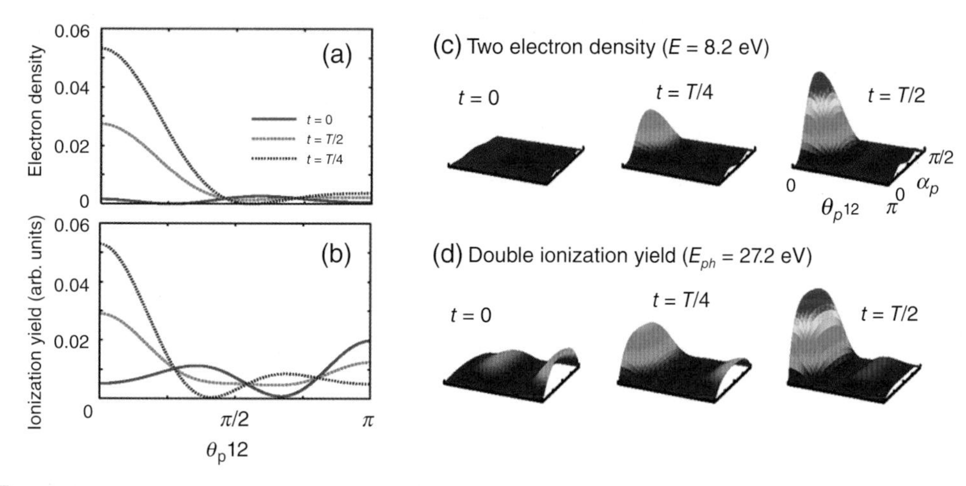

Figure 8.24 Time dependence of the two-electron momentum distribution of the bending vibrational wave packet and similar distributions for the two photoelectrons after the wave packet is double ionized by an attosecond pulse. (a) and (c) for the wave packet in momentum space; (b) and (d) the double ionization yields (see details in the text). (Figure adapted with permission from T. Morishita, S. Watanabe, and C. D. Lin, *Phys. Rev. Lett.*, **98**, 083003 (2007) [28]. Copyrighted by the American Physical Society.)

How can an attosecond pulse be used to probe such bending vibration, i.e., what observable quantities are to be measured? Figure 8.24(a) shows the electron density of the wave packet in momentum space against the angle $\theta_{p_{12}}$ at $t = 0$, $T/4$ and $T/2$ for the total energy of the two electrons at 8.2 eV. The bending vibration is clearly seen as the time evolves. In (b), the double-ionization yields from this wave packet ionized by a 200 as, 27.2 eV attosecond probe pulse, averaged over the total electron energy, are shown against $\theta_{p_{12}}$ for $\theta_{p_1} = -\theta_{p_2}$, where the angles are measured from the polarization direction and $\theta_{p_{12}}$ is the angle between the two momentum vectors of the two outgoing electrons. The evolution of the momentum space-bending vibrational wave packet in (a) is quite similar to the double-ionization yield in (b). This shows that double ionization can be used to retrieve the two-electron vibrational wave packet in the momentum space. The relief plots in (c) and (d) show the two-dimensional momentum space $(\theta_{p_{12}}, \alpha_p)$, where α_p is the relative momentum of the two-electron wave packets. These plots are to be compared to the ones from the momentum distributions of the two-electrons ionized from the wave packet. Clearly, such measurements are not yet possible with the attosecond pulses available today. To probe electron correlation, the momentum distributions of at least two electrons should be measured simultaneously.

8.7.3 Complete Mapping of the Time-Dependent Wave Packet of D_2^+ with Attosecond XUV Pulses

Existing pump-probe experiments involving an attosecond pulse usually employ it as a pump pulse in connection with an IR pulse as a probe. Here, we describe how an attosecond probe pulse allows accurate retrieval of the whole vibrational wave packet for the simplest D_2^+ ion with attosecond temporal resolution and sub-angstrom spatial resolution.

Consider the double ionization of the D_2 molecule in a pump-probe arrangement sketched in Figure 8.25(a). First, a pump pulse ionizes D_2 to D_2^+ ion. On the $1s\sigma$ potential surface of the D_2^+ ion, a vibrational wave packet is created. The retrieval of the evolution of this wave packet is our goal. One possible method is to use a probe pulse to ionize the D_2^+ wave packet again. Information on the internuclear separation R can then be retrieved from the kinetic energy release (KER) of the D^+ ions following the breakup of D_2^{++}. In this example, the KER $E = 1/R$, where R is the internuclear separation at the instant when ionization of D_2^+ occurs. From the distribution of the KER, the nuclear wave packet is directly mapped.

To see how the method works, assume that the pump pulse ionizes the D_2 from the ground state following the Franck–Condon principle. The time evolution of the vibrational wave packet can be readily calculated. Figure 8.25(b) shows this vibrational wave packet (versus R) on the $1s\sigma$ potential surface. While the wave packet oscillates with a period of 21 fs, pronounced interference patterns are observed after one cycle as the traveling waves

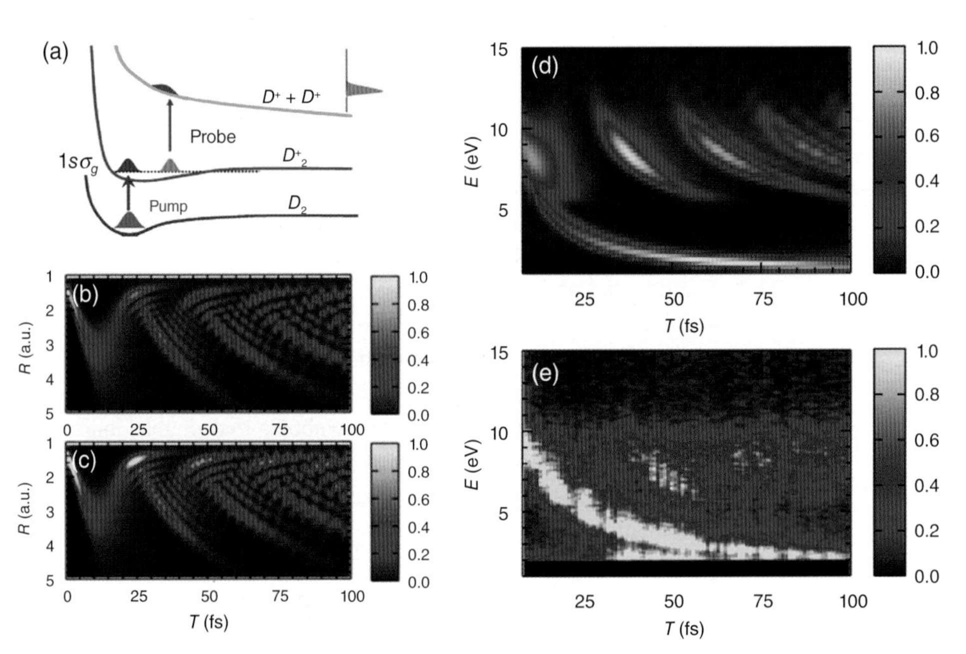

Figure 8.25 (a) Schematic of the generation and the retrieval of D_2^+ vibrational wave packet in a pump-probe experiment. The D_2^+ wave packet is created by a pump pulse by ionizing D_2. The subsequent motion of the wave packet is probed by an attosecond XUV pulse or by a sub-10 femtosecond IR pulse. (b) The evolution of the vibrational wave packet on the $1s\sigma$ potential curve assuming the Franck–Condon principle for ionization by the pump pulse. (c) Reconstructed wave packet retrieved from probing with a 300 as XUV pulse. (d) Reconstructed wave packet if the probe pulse is a 10 fs IR laser. (e) Experimental KER data obtained from experiment with 10 fs IR-probe pulse from [29]. The low-energy curve shown for the 10 fs IR probe is due to ionization from the $2p\sigma$ potential curve. (Figure adapted with permission from X. M. Tong and C. D. Lin, *Phys. Rev. A*, **73**, 042716 (2006) [30]. Copyrighted by the American Physical Society.)

and reflected waves propagate together and interfere. These interference features are readily washed out if a few-femtosecond IR pulse is used as the probe.

Figure 8.25(c) shows that the vibrational wave packet can be retrieved precisely if a 300 as XUV pulse is used to ionize the vibrational wave packet. In contrast, if a 10 fs IR pulse is used as the probe, the theory predicts that the retrieved vibrational wave packet will be similar to the one shown in (d). This figure is compared to the one obtained from the experiment [29] shown in Figure 8.25(e). With the 10 fs probe pulse, the detailed feature of the vibrational wave packet is mostly washed out. In (d) and (e) the evolution of the vibrational wave packet on the $2p\sigma$ curve is also shown.

The advantage of using an attosecond XUV pulse for probing the structure of a dynamic system is that the interaction of the probe with the wave packet may be treated using perturbation theory. Since the probe theory is simple, the wave packet can be easily and accurately retrieved as sketched below.

Assume that the vibrational wave packet is given by $\chi_g(R,\tau)$ for $t = \tau$ and the attosecond pulse is given by

$$I(\omega) \propto e^{-\tau_x^2(\omega-\omega_x)^2/(4\ln 2)}. \tag{8.68}$$

The central energy of the XUV pulse is taken at 50 eV, the pulse duration is 300 as, and the intensity is given by Equation 8.68. At a given R, the ionization probability is

$$P_{ion}(R) \propto \int_{E_b}^{\infty} \sigma(E_b,\omega) I(\omega) d\omega, \tag{8.69}$$

where E_b is the ionization potential of D_2^+ at a given R and σ is the ionization cross-section when the photon energy is ω. For the present purpose, an approximate scaled hydrogenic photoionization cross-section will be used

$$\sigma(E_b,\omega) = \frac{2^7}{3Z_s^2} \frac{1}{(1+\alpha^2)^4} \frac{e^{-4/\alpha}\tan^{-1}\alpha}{1-e^{-2\pi/\alpha}}, \tag{8.70}$$

where $\alpha = \sqrt{\omega/E_b - 1}$ and $Z_s = \sqrt{2E_b}$. To obtain the ionization rate from a wave packet at time τ after the pump pulse, one calculates

$$\frac{dP(R,\tau)}{dR} \propto |\chi_g(R,\tau)|^2 P_{ion}(R). \tag{8.71}$$

With respect to the KER spectra E where $E = 1/R$,

$$\frac{dP(E,\tau)}{dE} = R^2 \frac{dP(R,\tau)}{dR}. \tag{8.72}$$

This shows that the modulus square of the vibrational wave packet $|\chi_g(R,\tau)|^2$ can be directly extracted from Equation 8.71. If the probe pulse is long, then Equation 8.71 should be replaced by an integration over the pulse duration, thus averaging out the detail of the interference feature in the vibrational wave packet.

Note that, in the present method, only the $|\chi_g(R,\tau)|^2$ is retrieved from the probe pulse because the probe interacts with the wave packet linearly. Since the wave packet does not follow the classical equation of motion, knowledge of the density $F(R,t) = |\chi_g(R,t)|^2$ at a given "initial time" t is unable to determine the density $F(R,\tau)$ at later times.

8.8 Probing Attosecond Electron Dynamics of Complex Molecules

8.8.1 Complexity of Attosecond Electron Dynamics in Large Molecules

The previous sections discussed how attosecond pulses, in conjunction with moderately intense IR lasers, were used to study electron dynamics in simple atoms using transient absorption spectroscopy or streaked photoelectron spectra. Can one extend the method to large molecules? What can one learn from such experiments?

It is always of great interest to ask how a molecule relaxes after one of its electrons is removed by an XUV photon or by other means. Clearly, the remaining electrons readjust themselves on a timescale of tens to hundreds of attoseconds and the nuclei rearrange their positions on a timescale of a few to hundreds of femtoseconds, depending on the molecules. In Section 4.4.5, laser-induced electron diffraction was used to demonstrate that the distance between two oxygen nuclei shrinks by 0.1 Å in about 5 fs after an electron is removed by tunnel ionization. In another example, the C–H bond distance changes by 1.2 Å in about 9 fs if an electron is removed from $C_2H_2^+$. While such sub-angstrom change in bond length in a few femtoseconds is just becoming accessible to experimentation with femtosecond lasers, it is tempting to speculate whether one can "observe" electron dynamics on an attosecond timescale.

In quantum mechanics, electron dynamics are described by a complex wave packet. In molecules, this wave packet is expressed as a superposition of eigenstates of molecules under the Born–Oppenheimer (BO) approximation. At the tens-femtosecond timescale, the rotational motion of the molecule can be treated as frozen and the BO states are expressed as product of electronic eigenstates and vibrational eigenstates. At the tens to hundreds of attoseconds timescale, one may further approximate that the nuclear motion is frozen such that the short time behavior of the complex wave packet is represented by a time-dependent electron wave packet. In a naive way, one can then ask how this electron wave packet evolves on the attosecond timescale. Quantum chemistry provides a recipe on how to formulate such a theory even for a large molecule.

Instead of using complex electron wave packets, many theoretical papers have focused on describing electron dynamics in terms of electron density or in terms of hole density in the case of the removal of an electron from the molecule. In these studies, it was assumed that an electron is suddenly removed from the N-electron molecule. The wavefunctions of the remaining $(N-1)$ electrons in the cation are then expressed as the superposition of BO eigenstates of the ion. This sudden approximation simplifies the calculation significantly since the internuclear motion is frozen in this model. One can calculate the change of one-electron density versus time defined by [31]

$$Q(\mathbf{r},t) = \langle\Psi_0|\hat{\rho}(\mathbf{r},t)|\Psi_0\rangle - \langle\Phi_i|\hat{\rho}(\mathbf{r},t)|\Phi_i\rangle, \tag{8.73}$$

where Ψ_0 is the ground-state wavefunction of the neutral molecule, Φ_i is the nonstationary state of the cation, and $\hat{\rho}(\mathbf{r},t)$ is the electron-density operator.

The change of charge density versus time has been called *charge migration* or *hole migration*. Since the nonstationary state of the cation is made of bound eigenstates, one can expect that the charge cloud oscillates with the timescale inversely proportional to the energy difference of each pair of the component bound states. If the largest energy difference of the dominant eigenstates populated is 1 eV, then the oscillation period will be about 4.2 fs. Since electron (or hole) does not move from one point to another classically, *charge oscillation* is probably a better name than charge migration. Many such calculations for various molecules have been carried out in the literature and typical oscillation periods between 2 fs and 6 fs have been reported. The results have been shown to depend on molecular conformation, symmetry, and substitutions.

While charge oscillation reported from these calculations is of interest to some extent, it has not addressed how such oscillation can be determined or extracted from experiments. Position information of the nuclei is not available unless molecules are exposed to scattering with high-energy (tens to hundreds keV) electrons or photons. Fast removal of an outer-shell electron by high-energy photons so that sudden approximation is applicable is not realistic since high-energy photons remove mostly inner-shell electrons only. In our opinion, hole density $Q(\mathbf{r}, t)$ is not a useful quantity for describing the dynamics of an electron wave packet. As discussed in the earlier sections of this chapter, transient absorption spectroscopy is a powerful tool for attosecond physics, but the dipole cannot be calculated with hole density. For electron dynamics, the time dependence relies on the phase of the complex electron wave packet. Without the phase, there is no electron dynamics.

8.8.2 Experiment on Electron Dynamics in Phenylalanine Initiated by an Attosecond Pulse

Stimulated by interest from early experiments on the fragmentation of peptide chains, isolated attosecond pulses were used to study the prompt ionization of the amino acid phenylalanine and its subsequent evolution probed by a 4.5 fs IR pulse. Figure 8.26 summarizes the main ingredients from such an experiment [32]. First, a sketch of the phenylalanine molecule is given in (a) where R is a benzyl group and the immonium is sketched in (b). Upon prompt ionization by a 300 as EUV pulse with photon energy between 15 and 35 eV, an electron is removed. The evolution of the wave packet of the cation of phenylalanine is probed by a time-delayed 4.5 fs IR pulse that ionizes the cation. The resulting dication then dissociates. Figure 8.26(d) shows the yield of the doubly charged immonium ions versus the time delay. The normalized yield exhibits an exponential relaxation time of 25 fs. However, the red fitted curve on the exponential relaxation curve shows a sinusoidal oscillation with a period of 4.3 fs (frequency 0.234 PHz) (see Figure 8.26(d)). This oscillatory curve is shown in (e) after a smooth exponential curve is subtracted. This 4.3 fs oscillation is the main experimental result. What is the origin of such a fast oscillation? In Calegari et al. [32], it was attributed to hole migration.

To justify this interpretation, hole dynamics induced by the EUV attosecond pulse was calculated theoretically using perturbation theory. This is fully justified due to the low intensity of the EUV pulse. Because of the high-photon energy and the large spectral width of the pulse, a large manifold of ionization channels is open, and the cation-electron wave packet is a superposition of many one-hole cationic states. This wave packet is expected to oscillate with many beating frequencies. Assuming that the nuclear geometry does not

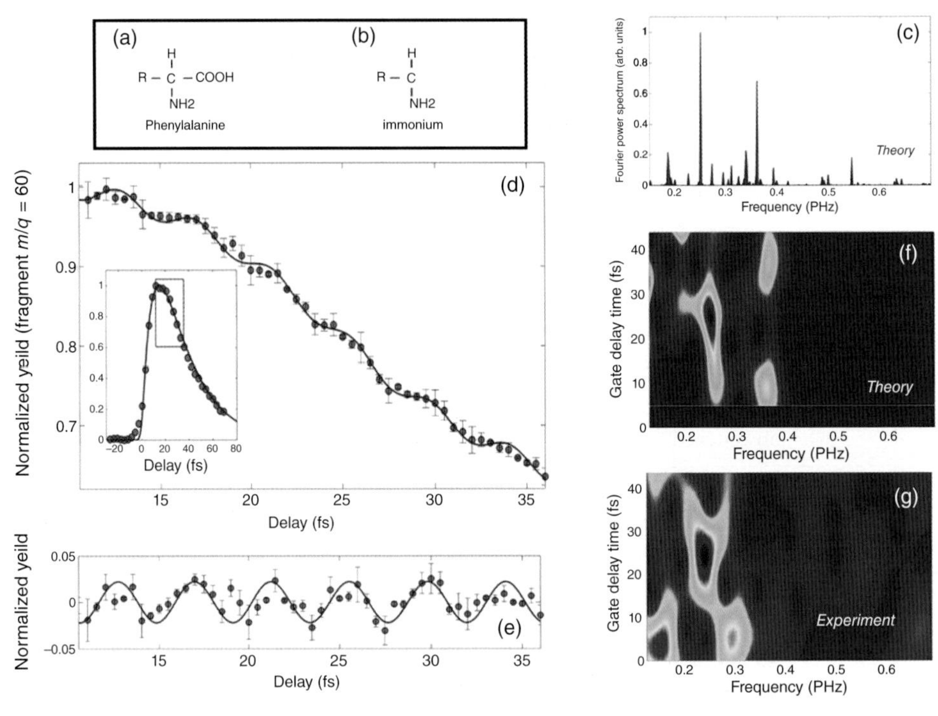

Figure 8.26 (a) Sketch of the phenylalanine molecule where R is a benzyl group. (b) The doubly charged immonium ions after the breakup of the COOH group. In the experiment, phenylalanine is first ionized by the EUV attosecond pulse. The cation is then further ionized by an IR probe pulse at different time delays. The ionization signal of doubly charged immonium ions versus time delay is shown in (d). The inset in (d) shows the signal from the rise to the decay; (e) shows the oscillation of the experimental data and the fitted curve after a smooth exponential decay curve is subtracted. (c) Fourier power spectrum of the calculated hole density from theory. (f) The spectrogram obtained from the calculated hole density. (g) The spectrogram obtained from the fitted oscillatory curve in (e). The model attempts to draw a correlation between (f) and (g), i.e., the oscillation of the decay curve is related to the oscillation of the hole density. (From F. Calegari et al., *Science*, **346**, 336 (2014) [32]. Reprinted with permission from AAAS.)

change, these frequencies can be obtained from molecular structure calculations. In (c) of Figure 8.26 the Fourier power spectrum of the calculated hole density is displayed. A dominant peak occurs at 0.25 PHz and another at about 0.36 PHz.

A more detailed Fourier analysis of the oscillation on the experimental exponential curve in (d) using a Gaussian window function with width of 10 fs is shown in (g), which is compared to the one from the theoretical simulations in (f). The agreement between the simulation and the experimental result is only fair. The beat frequency at short time delay from the theory (0.36 PHz) is higher than from the data (0.30 PHz). For delay time around 20–30 fs, the beat frequency from the experiment is about 0.25 PHz. This agrees well with the simulation. Simulation of the hole density near the amine (-NH_2) group indicates that the beat frequency is consistent with the frequency obtained in the dication data. But in what way the hole density oscillation near the amine affects the breakup of COOH from the doubly charged immonium ions (if any) is left unexplained.

Experiments with biological molecules pose many technical challenges. Simulations with *ab initio* time-dependent theory for large molecules are extremely complicated with unknown accuracy. In particular, the simulation is unable to treat the ionization and the fragmentation of the cation wave packet by the 4.5 fs IR pulse at the long time delays covered in the experiment. Including nuclear degrees of freedom in theoretical calculations in a nonlinear probe field is already challenging for a small molecule. To understand attosecond electron dynamics of a large molecule, various probe tools should be used. With the limited tools available today, it is still a long way before attosecond electron dynamics in complex molecules will be understood in spite of the tremendous progress in the past decade.

Notes and Comments

In this final chapter, the electron wave packet created by an IAP was investigated by a photoabsorption spectroscopy method. To study the dynamics of the electron wave packet, the IAP-induced dipole is "perturbed" in the presence of a time-delayed, moderate IR pulse. In ATAS, the imaginary part of $d(\omega, \tau)$ is determined over the spectral bandwidth of the IAP for a range of time delays τ. With ATAS, by using high-resolution spectrometers, high-temporal as well as high-spectral resolution can be simultaneously achieved. Thus ATAS is a very powerful tool for studying the dynamics of sharp spectral features originating from the bound states or the unstable AISs. Since the IR coupling to the electron wave packet has to be high enough to induce nonlinear coupling in order to probe the phase of the electron wave packet, ATAS often exhibits complicated structures. Such complications render it nearly impossible to extract the electron wave packet from the ATA spectra. Instead, the effort seen in this chapter has been centered on identifying new spectral features generated through the strong coupling with the IR pulse. So far, understanding of ATA spectra has largely relied on full *ab initio* calculations. As illustrated in this chapter, only under special circumstances (such as with helium and hydrogen molecules) can the whole ATA spectra be studied theoretically. On the other hand, certain aspects of ATAS can be understood with simple models even for complex systems. Future studies of ATAS for complex systems would benefit from asking pertinent questions and identifying goal-specific probe pulses.

Exercises

8.1 Following [4], derive Equations 8.6 and 8.7.
8.2 Reproduce Figure 8.1 for $q = -2$ and compare it for $q = 2$.
8.3 Derive Equation 8.15 from Equation 8.14.
8.4 Derive Equations 8.19 and 8.20.
8.5 Apply Equation 8.19 to the $2s2p\ {}^1P^o$ state of helium to generate the transient absorption spectra. Compare this with Figure 8.6.
8.6 The dipole-control model, Equation 8.51, provides a simple general prediction on the effect of a short IR pulse on the profile of a Fano resonance. Reproduce Figures 8.14(c) and 8.15(c).

References

[1] U. Fano. Effects of configuration interaction on intensities and phase shifts. *Phys. Rev.*, **124**:1866–1878, Dec. 1961.

[2] M. Drescher, M. Hentschel, R. Kienberger, et al. Time-resolved atomic inner-shell spectroscopy. *Nature*, **419**(6909):803–807, Oct. 2002.

[3] S. Gilbertson, M. Chini, X. Feng, S. Khan, Y. Wu, and Z. Chang. Monitoring and controlling the electron dynamics in helium with isolated attosecond pulses. *Phys. Rev. Lett.*, **105**:263003, Dec. 2010.

[4] W.-C. Chu and C. D. Lin. Theory of ultrafast autoionization dynamics of Fano resonances. *Phys. Rev. A*, **82**:053415, Nov. 2010.

[5] T. Mercouris, Y. Komninos, and C. A. Nicolaides. Time-dependent formation of the profile of the He $2s2p\ ^1P^o$ state excited by a short laser pulse. *Phys. Rev. A*, **75**:013407, Jan. 2007.

[6] A. Kaldun, A. Blättermann, V. Stooß, et al. Observing the ultrafast buildup of a Fano resonance in the time domain. *Science*, **354**(6313):738–741, 2016.

[7] M. Fleischhauer, A. Imamoglu, and J. P. Marangos. Electromagnetically induced transparency: optics in coherent media. *Rev. Mod. Phys.*, **77**:633–673, Jul. 2005.

[8] M. B. Gaarde, C. Buth, J. L. Tate, and K. J. Schafer. Transient absorption and reshaping of ultrafast XUV light by laser-dressed helium. *Phys. Rev. A*, **83**:013419, Jan. 2011.

[9] L. Argenti, Á. Jiménez-Galán, C. Marante, C. Ott, T. Pfeifer, and F. Martín. Dressing effects in the attosecond transient absorption spectra of doubly excited states in helium. *Phys. Rev. A*, **91**:061403, Jun. 2015.

[10] S. Pabst, A. Sytcheva, A. Moulet, A. Wirth, E. Goulielmakis, and R. Santra. Theory of attosecond transient-absorption spectroscopy of krypton for overlapping pump and probe pulses. *Phys. Rev. A*, **86**:063411, Dec. 2012.

[11] X. Li, D. J. Haxton, M. B. Gaarde, K. J. Schafer, and C. W. McCurdy. Direct extraction of intense-field-induced polarization in the continuum on the attosecond time scale from transient absorption. *Phys. Rev. A*, **93**:023401, Feb. 2016.

[12] X. M. Tong and N. Toshima. Controlling atomic structures and photoabsorption processes by an infrared laser. *Phys. Rev. A*, **81**:063403, Jun. 2010.

[13] S. Chen, M. Wu, M. B. Gaarde, and K. J. Schafer. Quantum interference in attosecond transient absorption of laser-dressed helium atoms. *Phys. Rev. A*, **87**:033408, Mar. 2013.

[14] M. Chini, B. Zhao, H. Wang, Y. Cheng, S. X. Hu, and Z. Chang. Subcycle ac stark shift of helium excited states probed with isolated attosecond pulses. *Phys. Rev. Lett.*, **109**:073601, Aug. 2012.

[15] M. Wu, S. Chen, S. Camp, K. J Schafer, and M. B Gaarde. Theory of strong-field attosecond transient absorption. *J. Phys. B-At. Mol. Opt.*, **49**(6):062003, 2016.

[16] S. Chen, M. J. Bell, A. R. Beck, et al. Light-induced states in attosecond transient absorption spectra of laser-dressed helium. *Phys. Rev. A*, **86**:063408, Dec. 2012.

[17] M. Chini, X. Wang, Y. Cheng, et al. Sub-cycle oscillations in virtual states brought to light. *Sci. Rep.*, **3**:1105, Jan. 2013.
[18] L. B. Madsen, P. Schlagheck, and P. Lambropoulos. Laser-induced transitions between triply excited hollow states. *Phys. Rev. Lett.*, **85**:42–45, Jul. 2000.
[19] W.-C. Chu and C. D. Lin. Photoabsorption of attosecond XUV light pulses by two strongly laser-coupled autoionizing states. *Phys. Rev. A*, **85**:013409, Jan. 2012.
[20] W.-C. Chu and C. D. Lin. Absorption and emission of single attosecond light pulses in an autoionizing gaseous medium dressed by a time-delayed control field. *Phys. Rev. A*, **87**:013415, Jan. 2013.
[21] Z.-H. Loh, C. H. Greene, and S. R. Leone. Femtosecond induced transparency and absorption in the extreme ultraviolet by coherent coupling of the He 2s2p (1Po) and 2p2 (1Se) double excitation states with 800 nm light. *Chem. Phys.*, **350**(13):7, 2008.
[22] C. Ott, A. Kaldun, P. Raith, et al. Lorentz meets Fano in spectral line shapes: a universal phase and its laser control. *Science*, **340**(6133):716–720, 2013.
[23] A. Blättermann, C. Ott, A. Kaldun, et al. In situ characterization of few-cycle laser pulses in transient absorption spectroscopy. *Opt. Lett.*, **40**(15):3464–3467, Aug. 2015.
[24] B. Bernhardt, A. R. Beck, X. Li, et al. High-spectral-resolution attosecond absorption spectroscopy of autoionization in xenon. *Phys. Rev. A*, **89**:023408, Feb. 2014.
[25] A. N. Pfeiffer, M. J. Bell, A. R. Beck, H. Mashiko, D. M. Neumark, and S. R. Leone. Alternating absorption features during attosecond-pulse propagation in a laser-controlled gaseous medium. *Phys. Rev. A*, **88**:051402, Nov. 2013.
[26] Y. Cheng, M. Chini, X. Wang, et al. Reconstruction of an excited-state molecular wave packet with attosecond transient absorption spectroscopy. *Phys. Rev. A*, **94**:023403, Aug. 2016.
[27] M. Reduzzi, W.-C. Chu, C. Feng, et al. Observation of autoionization dynamics and sub-cycle quantum beating in electronic molecular wave packets. *J. Phys. B-At. Mol. Opt.*, **49**(6):065102, 2016.
[28] T. Morishita, S. Watanabe, and C. D. Lin. Attosecond light pulses for probing two-electron dynamics of helium in the time domain. *Phys. Rev. Lett.*, **98**:083003, Feb. 2007.
[29] A. S. Alnaser, B. Ulrich, X. M. Tong, et al. Simultaneous real-time tracking of wave packets evolving on two different potential curves in H_2^+ and D_2^+. *Phys. Rev. A*, **72**:030702, Sep. 2005.
[30] X. M. Tong and C. D. Lin. Attosecond xuv pulses for complete mapping of the time-dependent wave packets of D_2^+. *Phys. Rev. A*, **73**:042716, Apr. 2006.
[31] J. Breidbach and L. S. Cederbaum. Universal attosecond response to the removal of an electron. *Phys. Rev. Lett.*, **94**:033901, Jan. 2005.
[32] F. Calegari, D. Ayuso, A. Trabattoni, et al. Ultrafast electron dynamics in phenylalanine initiated by attosecond pulses. *Science*, **346**(6207):336–339, 2014.

Appendix Constants, Conversion Factors, Atomic Units and Useful Formulae

A.1 Useful Constants

Electron charge $= 4.8 \times 10^{-10}$ esu $= 1.6 \times 10^{-19}$ C
Electron mass $= 9.1 \times 10^{-28}$ gm
$\hbar = 1.05 \times 10^{-27}$ ergs $\cdot$ s $= 6.58 \times 10^{-16}$ eV·s
Avogadro number $= 6.022 \times 10^{23}$ /mol
Boltzmann's constant, k $= 1.380658 \times 10^{-23}$ J/K $= 8.617 \times 10^{-5}$ eV/K

A.2 Energy Conversion Factors

1 eV = 8065.54 cm^{-1}
1 au = 27.21 eV = 2 Ry
1 degree kelvin = 0.0862 meV = = 0.695 cm^{-1}
1 Kcal/mol = 43.4 meV
1 THz photon $\rightarrow$ 4.13 meV

A.3 Momentum, Wavelength, and Energy

Photon momentum: k (au) $= 2.7 \times 10^{-4}$ E (eV); k = 1 au $\Rightarrow$ 3.7 keV
de Broglie wavelength for electron: 1 Å $\rightarrow$ 150 eV; k = 1 au $\Rightarrow$ 13.6 eV

A.4 Atomic Units (au): $e = \hbar = m = 1$

Unit of length = a_0 = Bohr radius = 0.529 Å
Unit of velocity = v_0 = electron velocity in the first Bohr orbital = αc
Unit of time = $a_0/\alpha c$ = 24.2 asec (1 fs = 41 au)
Unit of frequency $= v_0/a_0 = 4.13 \times 10^{16}\ \text{sec}^{-1}$
Unit of electric field $= 5.14 \times 10^{9}$ V/cm $= e/a_0^2$
Unit of magnetic field $= 2.35 \times 10^{5}$ Tesla

Unit of magnetic-dipole moment = $e\hbar/2m$ = 5.79 × 10^{-5} eV/Tesla
Unit of electric-dipole moment = ea_0 = 2.542 Debye

A.5 Shorthand Notations

10^9 = giga; 10^{12} = tera; 10^{15} = peta ; 10^{18} =exa; 10^{21} = zetta; 10^{24} = yocto
10^{-9} = nano; 10^{-12} = pico; 10^{-15} = femto; 10^{-18} = atto; 10^{-21} = zepto;
10^{-24} = yotta

A.6 Lasers

For 800 nm, → ω = 0.057 au → 1.551 eV
Intensity for peak E-field at 1 au (linear polarized) → 3.54 × 10^{16} W/cm^2
Peak electric field at intensity of 10^{12} W/cm^2 is 0.27 V/Å
$U_P = E^2/4\,\omega^2 = 9.33\, I(10^{14}\,\text{W/cm}^2)\, \lambda^2$ (in μm)
(U_P = 6 eV for 800 nm at 10^{14} W/cm^2)
Keldysh parameter: $\gamma = \sqrt{I/(2U_P)}$
Time-frequency width relation:

$$\Delta\omega = \frac{4\ln 2}{\tau}\sqrt{1+\xi^2} \tag{A.1}$$

where ξ is the linear chirp.
Gaussian pulse:

$$I(\omega) = e^{-\tau^2(\omega-\omega_0)^2/(4\ln 2)},\ E(t) = e^{-2\ln 2(t^2/\tau^2)}, \tag{A.2}$$

where τ is FWHM in time domain.
Width in eV for a Gaussian pulse with τ = 1 fs is 1.83 eV.

A.7 Oscillator Strength and Spontaneous Emission Rates

$A = 2\cdot(E^2/c^3)\, f\,(4.13\times10^{16})$ 1/sec, where E is transition energy in au,
c = 137.036, and f is the oscillator strength.

A.8 Pressure

1 atm = 1.01 × 10^5 Pa = 760 Torr = 1.013 bar
1 mbar = 0.75 Torr
1 psi = 51.7 Torr = 0.068 atm

Further Reading

Atkins, Friedman P. W. Atkins and R. S. Friedman, *Molecular quantum mechanics*, 4th ed. Oxford University Press, New York, 2005.

Bransden, Joachain B. H. Bransden and C. J. Joachain. *Physics of atoms and molecules*, 2nd ed. Prentice Hall, Upper Saddle River, NJ, 2003.

Boyd R. J. Boyd. *Nonlinear optics*. Academic Press, 2nd edition, 2003.

Chang Z. Chang. *Fundamentals of attosecond optics*. CRC Press, Boca Raton, FL, 2011.

Demtröder W. Demtröder. *Atoms, molecules and photons: an introduction to atomic-, molecular- and quantum physics*. Springer, Berlin, 2010.

Hargittai, Hargittai I. Hargittai and M. Hargittai. *Stereochemical applications of gas-phase electron diffraction*. VCH, New York, 1988.

Harris, Bertolucci D. C. Harris and M. D. Bertolucci. *Symmetry and spectroscopy: an introduction to vibrational and electronic spectroscopy*. Oxford University Press, New York, 1978.

Herzberg, 1945 G. Herzberg. *Molecular spectra and molecular structure: II. infrared and Raman spectra of polyatomic molecules*. D. Van Nostrand Company. Inc, Princeton, 1945.

Herzberg, 1950 G. Herzberg. *Molecular spectra and molecular structure: I. spectra of diatomic molecules*, 2nd ed. D. Van Nostrand Company. Inc, 1950.

Joachain, Kylstra, Potvliege C. J. Joachain, N. J. Kylstra, and R. M. Potvliege. *Atoms in intense laser fields*. Cambridge University Press, Cambridge, 2014.

Plaja, Torres, Zar L. Plaja, R. Torres, and A. Zaïr, eds. *Attosecond physics. Springer series in optical sciences.* Springer, Heidelberg, 2013.

Quack, Merkt Quack, M., and Merkt, F., eds. *Handbook of high-resolution spectroscopy*. John Wiley & Sons, New York, 2011.

Siegman A. E. Siegman. *Lasers*. University Science, Mill Valley, CA, 1986.

Saleh, Teich B. E. A. Saleh and M. C. Teich. *Fundamentals of photonics*. John Wiley & Sons, Inc., New York, 1991.

Schultz, Vrakking T. Schultz and M. Vrakking, eds. *Attosecond and XUV physics: ultrafast dynamics and spectroscopy*, Wiley-VCH Verlag, Berlin, 2014.

Solutions for Selected Problems

1.2.

State of Li	Binding Energy Calculated (eV)	Binding Energy from NIST (eV)
2s	5.37	5.39
2p	3.54	3.54
3s	2.02	2.02
4f	0.85	0.85

1.4.

$$V(r) = -\frac{Z_c + a_1 e^{-a_2 r} + a_3 r e^{-a_4 r} + a_5 e^{-a_6 r}}{r}, \tag{A1}$$

with parameters $a_1 = 16.039$, $a_2 = 2.007$, $a_3 = -25.543$, $a_4 = 4.525$, $a_5 = 0.961$, $a_6 = 0.443$, and $Z_c = 1$. The calculated DCS for e-Ar^+ collision is plotted in Figure A1.

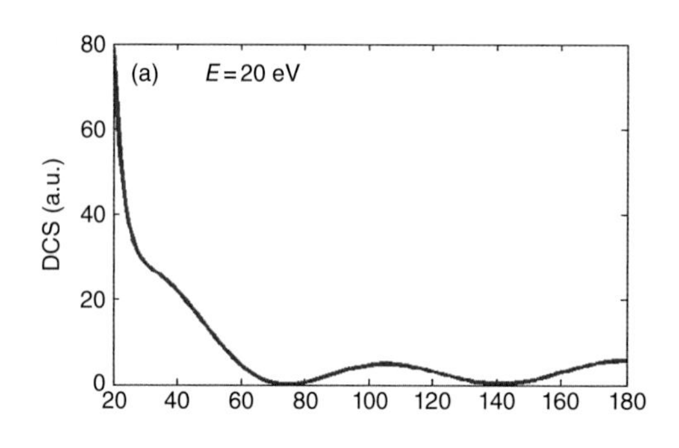

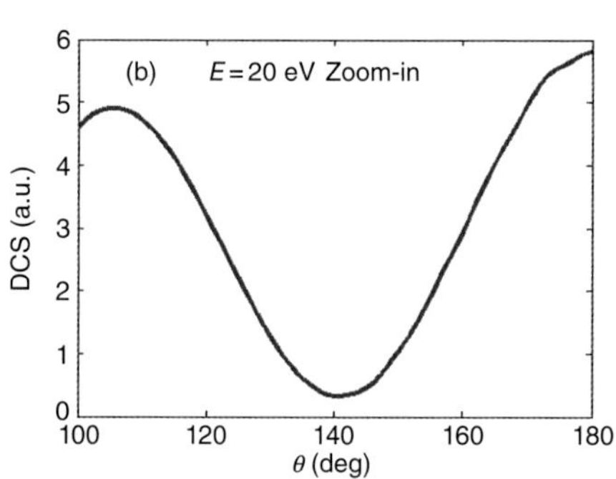

Figure A1 DCS for e-Ar^+ collision at the incident energy of 20 eV (a) over the angular range of 20°–180° and (b) zoom in for the angular range of 100°–180°. Figure corresponds to Problem 1.4.

1.7. $\tau_{3p} = \left(\bar{W}^s_{1s,3p} + \bar{W}^s_{2s,3p}\right)^{-1}$, $\tau_{3d} = \left(\bar{W}^s_{2p,3d}\right)^{-1}$.
The lifetimes scale with Z^{-4}.

1.8. For $\omega = 40$ eV, $S_0 = 0.2840$, $S_2 = 0.0379$, $\eta_{E0} = 0.2675$, $\eta_{E2} = 1.3207$, $\beta = -0.2206$;
For $\omega = 50$ eV, $S_0 = 0.2068$, $S_2 = 0.1443$, $\eta_{E0} = 0.1016$, $\eta_{E2} = 1.4624$, $\beta = 0.1986$.

Lifetime	H	He^+	Ne^{9+}
$3p$ state	5.26 ns	329 ps	526 fs
$3d$ state	15.4 ns	965 ps	1.54 ps

1.10.

(a) The $2s2p\ ^1P$ state can autoionize into $1sEp\ ^1P$ state, which consists of an ion $He^+(1s)$ and a continuum electron with the angular-momentum quantum number $l = 1$. $E_r = 60.15$ eV, $\Gamma = 37.4$ meV, $q = -2.75$. The autoionization life time is 17.6 fs.
(b) The $2s2p\ ^1P$ state can decay radiatively to $1sns\ ^1S$ and $1snd\ ^1D$ states. The most likely state it will decay to is the $1s2s\ ^1S$ state. The radiative decay lifetime is roughly 100 ps.
(c) The $2p^2\ ^3P$ state cannot decay by autoionization. However, this state can decay radiatively into $1snp\ ^3P$ state.

1.12. By analyzing the MO diagram, we see that the HOMO of the N_2 is a bonding orbital, whereas it is an antibonding orbital in O_2 and F_2. Therefore by removing an electron from the HOMO, N_2 becomes less bonded, and the bond length increases. The situation is opposite for O_2 and F_2. The LUMO for these molecules are antibonding, so adding an electron into this orbital will increase the bond lengths.

1.13. This molecule belongs to C_{3v} symmetry. It has $5 \times 3 = 15$ degrees of freedom, and $15 - 6 = 9$ vibrational degrees of freedom, among which there are three A_1 modes and three E (doubly degenerate) modes. By checking the character table we find that all these modes are both infrared and Raman active.

1.14. The ground-state term is ^1A_g. The excited-state term can be obtained by the direct product of b_{1u} and b_{2g}. This gives B_{3u}. The spin state can be singlet or triplet: $^1B_{3u}$ and $^3B_{3u}$. The transition to the triplet state is weak since it is spin forbidden, whereas the transition to the singlet state is strong since it is spin allowed.

1.16. Hint: Using the same notation as in section 20.1 in [Siegman], a ray crossing the transverse plane at z is completely characterized by the coordinate of its crossing point x and the angle x'. To derive equations 4 and 5 in section 20.1, (x_2, x_2') and (x_1, x_1') can be related by the $ABCD$ matrix in the paraxial approximation as

$$\begin{pmatrix} x_2 \\ x_2' \end{pmatrix} = \begin{pmatrix} A & B \\ C & D \end{pmatrix} \begin{pmatrix} x_1 \\ x_1' \end{pmatrix}, \tag{A2}$$

where $AD - BC = 1$ for a lossless system.

To derive equation 6 in section 20.1, according to the Snell's law

$$n_1\theta = n_1 \frac{x_1}{R_1} = x_1'. \tag{A3}$$

Equation 7 in section 20.1 can be derived in the same manner.

Comparing the optical path in the free space in Equation 1.279 of the current book with the optical length derived in equation 12 in section 20.1 of [Siegman], the Huygens–Fresnel kernel in Equation 1.282 can be written in the same general form as that for the free-space situation.

1.17. Hint: The following *ABCD* matrices are needed. (1) Free-space propagation:

$$\mathbf{M} = \begin{pmatrix} 1 & d \\ 0 & 1 \end{pmatrix}, \tag{A4}$$

where d needs to change according to the experimental setups. (2) Transmission through a thin lens:

$$\mathbf{M} = \begin{pmatrix} 1 & 0 \\ -1/f & 1 \end{pmatrix}. \tag{A5}$$

2.4.

$$|\varepsilon(\omega)|^2 = \frac{E_0^2\tau^2}{8\pi \ln 2\sqrt{1+\xi^2}} e^{-\frac{4\ln 2(\omega-\omega_0)^2}{(\Delta\omega)^2}}, \tag{A6}$$

where $\xi = \frac{\alpha\tau^2}{4\ln 2}$, the FWHM spectral bandwidth $\Delta\omega = \frac{4\ln 2}{\tau}\sqrt{1+\xi^2}$.

2.6. For Ar, when FWHM = 25 fs, $I_S = 3.16 \times 10^{14}$ W/cm^2; when FWHM = 5 fs, $I_S = 4.72 \times 10^{14}$ W/cm^2.
For Xe, when FWHM = 25 fs, $I_S = 1.12 \times 10^{14}$ W/cm^2; when FWHM = 5 fs, $I_S = 1.61 \times 10^{14}$ W/cm^2.

2.7. For Ar, when FWHM = 25 fs, $I_S = 4.26 \times 10^{14}$ W/cm^2; when FWHM = 5 fs, $I_S = 7.73 \times 10^{14}$ W/cm^2.

3.7. Using the MO-ADK theory Equation 3.31, one can calculate the ionization rate from HOMO of N_2 and O_2 as functions of the alignment angle θ between the laser polarization direction and the molecular axis.
For N_2 HOMO($3\sigma_g$) $I_p = 15.6$ eV, $C_{00} = 2.68$, $C_{20} = 1.10$, $C_{40} = 0.06$.
For O_2 HOMO($1\pi_g$) $I_p = 12.3$ eV, $C_{21} = 0.52$, $C_{41} = 0.03$.
Use an 800 nm laser with 10^{13} W/cm^2 intensity, we can obtain the ionization rate from MO-ADK as shown in Figure A2.

4.4. The MCF for isotropically distributed N_2 molecules is $\gamma = \frac{\sin(qR)}{qR}$. The N–N bond length at the equilibrium position is $R_0 = 1.10$ Å. Here we plot the MCF for $R = R_0, 2R_0, 5R_0$ using 150 eV electrons in Figure A3. It is possible to retrieve the bond length during the dissociation process by fitting the measured MCF, or taking the inverse Fourier transform.

5.3. Hint: The derivation can be found in the tutorial by Le et al., *J. Phys. B-At. Mol. Opt.*, **49**, 053001 (2016), equations 57 through 61

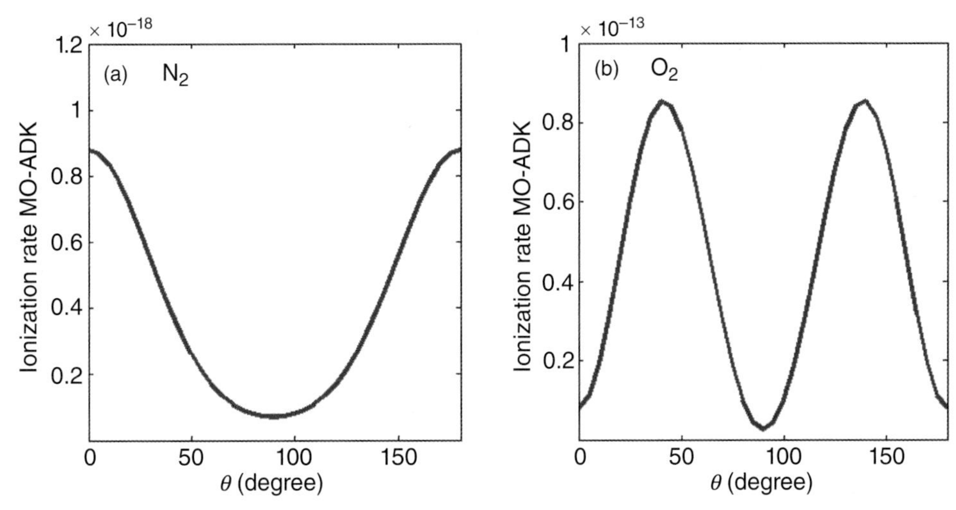

Figure A2 Ionization rate from (a) N_2 and (b) O_2 as functions of the alignment angle θ, calculated by MO-ADK theory. Figure corresponds to Problem 3.7.

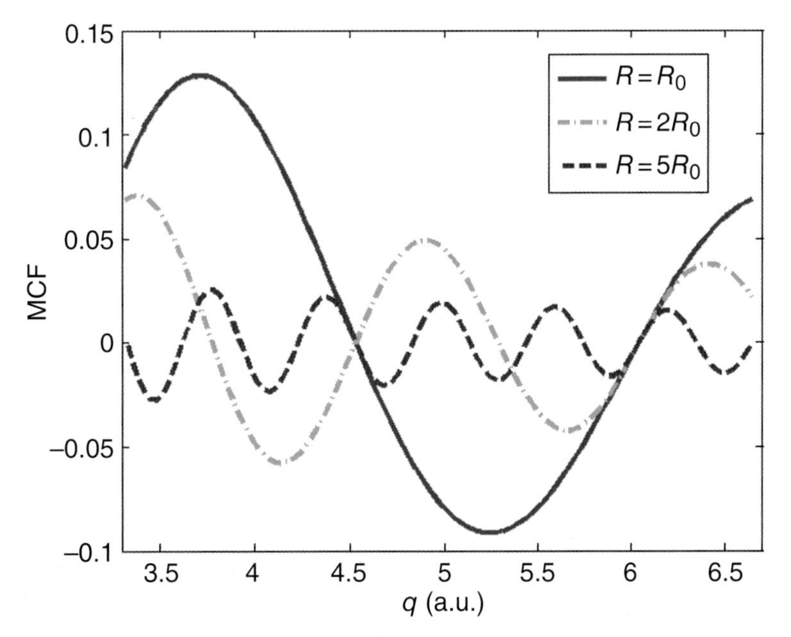

Figure A3 MCF for isotropically distributed N_2 molecules with N–N bond length $R = R_0$, $2R_0$, and $5R_0$ using 150 eV electrons. Figure corresponds to Problem 4.4.

5.5. Hint: In the integration, treat A_q, ρ, and Δk as constants, respectively. To derive equation 1 in Constant et al., *Phys. Rev. Lett.*, 82, 1668 (1999), replace Δk by L_{coh}, and replace κ_q by L_{abs}.

5.6. Hint: The "critical" ionization fraction is 4% (1.5%) for Ar, 1% (0.4%) for Ne, and 0.5% (0.2%) for He at the fundamental wavelength of 0.8 μm (1.3 μm), respectively.

5.7. Hint: When calculating the phase mismatch due to the induced-dipole phase, for harmonic order $q = 21$, the coefficients are $\alpha^q_{i=S} \approx 1\times10^{-14}$ rad cm^2/W ("short" trajectory) and $\alpha^q_{i=L} \approx 24\times10^{-14}$ rad cm^2/W ("long" trajectory), and for harmonic order $q = 63$, this coefficient is $\alpha^q_{i=S,L} \approx 13.7\times10^{-14}$ rad cm^2/W.

7.2. The Wigner delay defined in Equation 7.68 or 7.72 is an additional time taken by the photoelectron to hit the detector compared to the case when there is no core potential. This quantity is absolute in concept but can never be measured in real experiments. On the other hand, in a single XUV+IR experiment, the XUV pulse is unknown, which means we do not know the exact experimental situation in which the peak of XUV and IR overlap. The XUV–IR delay t_d is used for identifying photoelectron spectra but does not have absolute meaning. At $t_d = 0$ the peak of XUV and IR can be separate; in other words, if we let $E_{IR}(t)$ peak at $t = 0$, the XUV pulse $E_{XUV}(t)$ does not necessarily peak at $t = 0$. There is a group delay corresponding to the XUV pulse that is unknown. From a single streaking spectrogram it is possible to extract the group delay of the photoelectron wave packet, which consists of the XUV group delay and the Wigner delay. However, since the XUV group delay is unknown, the Wigner delay cannot be retrieved.

7.4. The electric field of the left and right circular, polarized pulses can be expressed as

$$\mathbf{E}_l(t) = E_0F(t)(\hat{x} + i\hat{y})e^{-i\omega t}, \tag{A7}$$

$$\mathbf{E}_r(t) = E_0F(t)(\hat{x} - i\hat{y})e^{-i\omega t}. \tag{A8}$$

If the two pulses are combined with a time delay T_d (assume T_d is an integer number n of the optical period $2\pi/\omega$), the combined field is

$$\begin{aligned}\mathbf{E}(t) &= \mathbf{E}_l(t + T_d/2) + \mathbf{E}_r(t - T_d/2)\\ &= (-1)^n e^{-i\omega t}E_0[F(t + T_d/2)(\hat{x} + i\hat{y}) + F(t - T_d/2)(\hat{x} - i\hat{y})]\\ &= (-1)^n e^{-i\omega t}E_0\{\hat{x}[F(t + T_d/2) + F(t - T_d/2)]\\ &\quad + i\hat{y}[F(t + T_d/2) - F(t - T_d/2)]\}.\end{aligned} \tag{A9}$$

The time-dependent ellipticity is given by

$$\varepsilon(t) = \frac{|F(t + T_d/2) - F(t - T_d/2)|}{|F(t + T_d/2) + F(t - T_d/2)|}. \tag{A10}$$

For Gaussian envelope $F(t) = e^{-2\ln 2\frac{t^2}{\tau^2}}$ we have

$$\begin{aligned}\varepsilon(t) &= \frac{|e^{-2\ln 2\frac{(t+T_d/2)^2}{\tau^2}} - e^{-2\ln 2\frac{(t-T_d/2)^2}{\tau^2}}|}{|e^{-2\ln 2\frac{(t+T_d/2)^2}{\tau^2}} + e^{-2\ln 2\frac{(t-T_d/2)^2}{\tau^2}}|}\\ &= \frac{|e^{-2\ln 2\frac{T_d}{\tau^2}t} - e^{2\ln 2\frac{T_d}{\tau^2}t}|}{|e^{-2\ln 2\frac{T_d}{\tau^2}t} + e^{2\ln 2\frac{T_d}{\tau^2}t}|} = \frac{|1 - e^{-4\ln 2\frac{T_d}{\tau^2}t}|}{|1 + e^{-4\ln 2\frac{T_d}{\tau^2}t}|}.\end{aligned} \tag{A11}$$

For small $|t|$ such that $(4\ln 2)\frac{T_d}{\tau^2}|t| \ll 1$,

$$\varepsilon(t) \approx 2\ln 2\frac{T_d}{\tau^2}|t|. \tag{A12}$$

Therefore, the time interval δt_{PG} wherein $\varepsilon(t) < \varepsilon_{th}$ can be estimated is

$$\delta t_{PG} = \frac{\varepsilon_{th}}{\ln 2}\frac{\tau^2}{T_d}. \tag{A13}$$

7.5. In Figure A4 we plot the modulus square of the wave packet $|\psi(x,t)|^2$ as functions of x at $t = 0, 300, 600, 900, 1{,}200$ as, for case (a), (b), and (c), respectively.

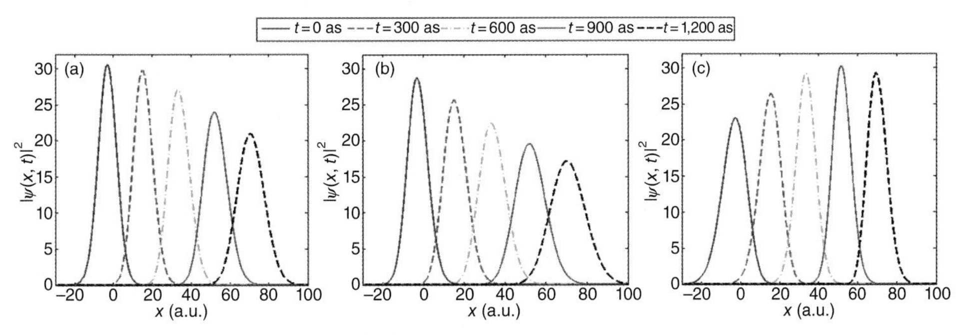

Figure A4 Modulus square of the wave packet $|\psi(x,t)|^2$ as functions of x at $t = 0, 300, 600, 900, 1{,}200$ as, for case (a) $\beta^X = \beta^W = 0$, (b) $\beta^X = 0$, $\beta^W = -5{,}000\ \text{as}^2$ and (c) $\beta^X = 15{,}000\ \text{as}^2$, $\beta^W = -5{,}000\ \text{as}^2$. Figure corresponds to Problem 7.5.

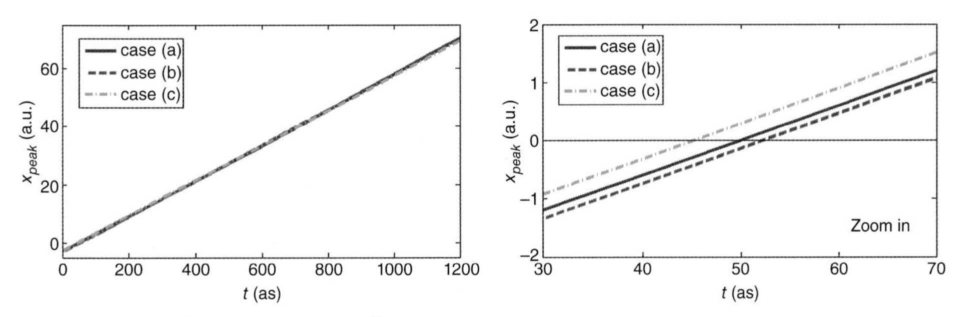

Figure A5 x_{peak} against t for cases (a), (b), and (c), respectively. The right panel shows the zoom in of the left panel near $x_{peak} = 0$. Figure corresponds to Problem 7.5.

Then we plot x_{peak} versus t for cases (a), (b), and (c), respectively in Figure A5. From the zoom-in plot we can see when $x_{peak} = 0$, (a) $t = 50.0$ as, (b) $t = 52.3$ as, and (c) $t = 45.4$ as. Only in case (a) the time corresponding to $x_{peak} = 0$ equals τ^W.
When $t = 200$ as, (a) $x_{peak} = 9.01$, (b) $x_{peak} = 8.94$, (c) $x_{peak} = 9.41$;
When $t = 400$ as, (a) $x_{peak} = 21.11$, (b) $x_{peak} = 21.15$, (c) $x_{peak} = 21.45$;
When $t = 600$ as, (a) $x_{peak} = 33.30$, (b) $x_{peak} = 33.43$, (c) $x_{peak} = 33.38$;
When $t = 800$ as, (a) $x_{peak} = 45.58$, (b) $x_{peak} = 45.76$, (c) $x_{peak} = 45.35$.

8.3. We first work out the energy integral in Equation 8.14:

$$\begin{aligned}
&\int \frac{(q+\varepsilon)e^{-i\frac{\Gamma}{2}\varepsilon t}-(q-i)e^{-\frac{\Gamma}{2}t}}{\varepsilon+i}dE \\
&=\frac{\Gamma}{2}\int \frac{(q-i)e^{-i\frac{\Gamma}{2}\varepsilon t}+(\varepsilon+i)e^{-i\frac{\Gamma}{2}\varepsilon t}-(q-i)e^{-\frac{\Gamma}{2}t}}{\varepsilon+i}d\varepsilon \\
&=\frac{\Gamma}{2}\left\{\int e^{-i\frac{\Gamma}{2}\varepsilon t}d\varepsilon+(q-i)\int \frac{e^{-i\frac{\Gamma}{2}\varepsilon t}}{\varepsilon+i}d\varepsilon-(q-i)e^{-\frac{\Gamma}{2}t}\int \frac{1}{\varepsilon+i}d\varepsilon\right\} \\
&=2\pi\delta(t)-(q-i)\frac{\Gamma}{2}(2\pi i)e^{-\frac{\Gamma}{2}t}+(q-i)\frac{\Gamma}{2}(i\pi)e^{-\frac{\Gamma}{2}t} \\
&=2\pi\delta(t)-i\pi(q-i)\frac{\Gamma}{2}e^{-\frac{\Gamma}{2}t}
\end{aligned} \tag{A14}$$

Insert Equation A14 into Equation 8.14 and drop the complex conjugate part:

$$\begin{aligned}
d(t) &= C_\alpha^{(0)}\langle\alpha|z|g\rangle^* e^{-i\Omega_r t}\left\{\left(1-\frac{i}{q}\right)e^{-\frac{\Gamma}{2}t}+\frac{1}{(\pi Vq)^2}\left[2\pi\delta(t)-i\pi(q-i)\frac{\Gamma}{2}e^{-\frac{\Gamma}{2}t}\right]\right\} \\
&= C_\alpha^{(0)}\langle\alpha|z|g\rangle^* e^{-i\Omega_r t}\frac{1}{\pi V^2q^2}\left\{2\delta(t)+\frac{\Gamma}{2}(q-i)^2e^{-\frac{\Gamma}{2}t}\right\} \\
&= iE_{XUV}(\Omega_r)|\langle\alpha|z|g\rangle|^2\frac{1}{\pi V^2q^2}\left\{2\delta(t)+\frac{\Gamma}{2}(q-i)^2e^{-\frac{\Gamma}{2}t}e^{-i\Omega_r t}\right\} \\
&\propto i\left\{2\delta(t)+\frac{\Gamma}{2}(q-i)^2e^{-\frac{\Gamma}{2}t}e^{-i\Omega_r t}\right\}.
\end{aligned} \tag{A15}$$

8.4.

$$\begin{aligned}
\sigma(\Omega,\tau) &\propto \mathrm{Re}\left[1+\frac{(q-i)^2}{1-i\varepsilon}\left(1-e^{-\frac{\Gamma}{2}\tau}e^{i\Delta\Omega\tau}\right)\right] \\
&=\frac{1}{1+\varepsilon^2}\mathrm{Re}\left[1+\varepsilon^2+(1+i\varepsilon)(q^2-1-2iq)(1-e^{-\frac{\Gamma}{2}\tau}\cos(\Delta\Omega\tau)-ie^{-\frac{\Gamma}{2}\tau}\sin(\Delta\Omega\tau))\right] \\
&=\frac{1}{1+\varepsilon^2}\mathrm{Re}\left[1+\varepsilon^2+(P(\varepsilon)+iQ(\varepsilon))(1-e^{-\frac{\Gamma}{2}\tau}\cos(\Delta\Omega\tau)-ie^{-\frac{\Gamma}{2}\tau}\sin(\Delta\Omega\tau))\right],
\end{aligned} \tag{A16}$$

where $P(\varepsilon)=q^2+2q\varepsilon-1$, $Q(\varepsilon)=q^2\varepsilon-2q-\varepsilon$. Equation A16 can be further simplified as

$$\begin{aligned}
\sigma(\Omega,\tau) &\propto \frac{1}{1+\varepsilon^2}\left[1+\varepsilon^2+P(\varepsilon)-e^{-\frac{\Gamma}{2}\tau}(P(\varepsilon)\cos(\Delta\Omega\tau)-Q(\varepsilon)\sin(\Delta\Omega\tau))\right] \\
&=\frac{1}{1+\varepsilon^2}\left[(q+\varepsilon)^2-\sqrt{P^2(\varepsilon)+Q^2(\varepsilon)}e^{-\frac{\Gamma}{2}\tau}\cos(\Delta\Omega\tau+\varphi(\varepsilon))\right] \\
&=\frac{(q+\varepsilon)^2}{1+\varepsilon^2}-\frac{\sqrt{(1+\varepsilon^2)(q^2+1)^2}}{1+\varepsilon^2}e^{-\frac{\Gamma}{2}\tau}\cos(\Delta\Omega\tau+\varphi(\varepsilon)) \\
&=\frac{(q+\varepsilon)^2}{1+\varepsilon^2}-\frac{(q^2+1)}{\sqrt{1+\varepsilon^2}}e^{-\frac{\Gamma}{2}\tau}\cos(\Delta\Omega\tau+\varphi(\varepsilon)),
\end{aligned} \tag{A17}$$

where $\tan\varphi(\varepsilon)=\frac{Q(\varepsilon)}{P(\varepsilon)}=\frac{q^2\varepsilon-2q-\varepsilon}{q^2+2q\varepsilon-1}$.

Index